Sustainable Forests: Ecosystems and Management

Sustainable Forests: Ecosystems and Management

Edited by Aduardo Hapke

SYRAWOOD
PUBLISHING HOUSE
New York

Published by Syrawood Publishing House,
750 Third Avenue, 9th Floor,
New York, NY 10017, USA
www.syrawoodpublishinghouse.com

Sustainable Forests: Ecosystems and Management
Edited by Aduardo Hapke

International Standard Book Number: 978-1-68286-416-6 (Hardback)

Cataloging-in-publication Data

Sustainable forests : ecosystems and management / edited by Aduardo Hapke.
p. cm.
Includes bibliographical references and index.
ISBN 978-1-68286-416-6
1. Sustainable forestry. 2. Ecosystem management. 3. Forest management. 4. Forest ecology. I. Hapke, Aduardo.
SD387.S87 S87 2017
634.92--dc23

Printed in China.

TABLE OF CONTENTS

Preface VII

Chapter 1 **Effects of Litter Manipulation on Litter Decomposition in a Successional Gradients of Tropical Forests in Southern China** 1
Hao Chen, Geshere A. Gurmesa, Lei Liu, Tao Zhang, Shenglei Fu, Zhanfeng Liu, Shaofeng Dong, Chuan Ma, Jiangming Mo

Chapter 2 **Variable Strength of Forest Stand Attributes and Weather Conditions on the Questing Activity of *Ixodes ricinus* Ticks over Years in Managed Forests** 10
Ralf Lauterbach, Konstans Wells, Robert B. O'Hara, Elisabeth K. V. Kalko, Swen C. Renner

Chapter 3 **Succession Influences Wild Bees in a Temperate Forest Landscape: The Value of Early Successional Stages in Naturally Regenerated and Planted Forests** 17
Hisatomo Taki, Isamu Okochi, Kimiko Okabe, Takenari Inoue, Hideaki Goto, Takeshi Matsumura, Shun'ichi Makino

Chapter 4 **Gap Dynamics and Structure of Two Old-Growth Beech Forest Remnants in Slovenia** 25
Tihomir Rugani, Jurij Diaci, David Hladnik

Chapter 5 **Response of Soil Respiration to Acid Rain in Forests of Different Maturity in Southern China** 38
Guohua Liang, Xingzhao Liu, Xiaomei Chen, Qingyan Qiu, Deqiang Zhang, Guowei Chu, Juxiu Liu, Shizhong Liu, Guoyi Zhou

Chapter 6 **Patterns of Detection and Capture are Associated with Cohabiting Predators and Prey** 48
Billie T. Lazenby, Christopher R. Dickman

Chapter 7 **The Distribution and Abundance of an Island Population of Koalas (*Phascolarctos cinereus*) in the Far North of their Geographic Range** 64
Denise C. McGregor, Sarah E. Kerr, Andrew K. Krockenberger

Chapter 8 **Mapping the Global Emergence of *Batrachochytrium dendrobatidis*, the Amphibian Chytrid Fungus** 73
Deanna H. Olson, David M. Aanensen, Kathryn L. Ronnenberg, Christopher I. Powell, Susan F. Walker, Jon Bielby, Trenton W. J. Garner, George Weaver, The *Bd* Mapping Group, Matthew C. Fisher

Chapter 9 **Soil Temperature Triggers the Onset of Photosynthesis in Korean Pine** 86
Jiabing Wu, Dexin Guan, Fenhui Yuan, Anzhi Wang, Changjie Jin

Chapter 10 **Interactive Effects of Nitrogen and Phosphorus on Soil Microbial Communities in a Tropical Forest** 93
Lei Liu, Tao Zhang, Frank S. Gilliam, Per Gundersen, Wei Zhang, Hao Chen, Jiangming Mo

Chapter 11 **Separating the Effects of Environment and Space on Tree Species Distribution: From Population to Community** ... 103
Guojun Lin, Diana Stralberg, Guiquan Gong, Zhongliang Huang, Wanhui Ye, Linfang Wu

Chapter 12 **Biomass Allocation Patterns across China's Terrestrial Biomes** ... 113
Limei Wang, Longhui Li, Xi Chen, Xin Tian, Xiaoke Wang, Geping Luo

Chapter 13 **Habitat Selection of a Large Carnivore along Human-Wildlife Boundaries in a Highly Modified Landscape** ... 122
Chihiro Takahata, Scott Eric Nielsen, Akiko Takii, Shigeyuki Izumiyama

Chapter 14 **Spatial Variation in Soil Properties among North American Ecosystems and Guidelines for Sampling Designs** ... 135
Henry Loescher, Edward Ayres, Paul Duffy, Hongyan Luo, Max Brunke

Chapter 15 **Interannual Variation of the Bowen Ratio in a Subtropical Coniferous Plantation in Southeast China, 2003-2012** ... 166
Yakun Tang, Xuefa Wen, Xiaomin Sun, Huimin Wang

Chapter 16 **Rates of Litter Decomposition and Soil Respiration in Relation to Soil Temperature and Water in Different- Aged *Pinus massoniana* Forests in the Three Gorges Reservoir Area, China** ... 176
Wenfa Xiao, Xiaogai Ge, Lixiong Zeng, Zhilin Huang, Jingpin Lei, Benzhi Zhou, Maihe Li

Chapter 17 **Richness of Lichen Species, Especially of Threatened Ones, is Promoted by Management Methods Furthering Stand Continuity** ... 187
Steffen Boch, Daniel Prati, Dominik Hessenmöller, Ernst-Detlef Schulze, Markus Fischer

Chapter 18 **Dissecting Variation in Biomass Conversion Factors across China's Forests: Implications for Biomass and Carbon Accounting** ... 196
Yunjian Luo, Xiaoquan Zhang, Xiaoke Wang, Yin Ren

Chapter 19 **Scaling Relationships between Leaf Mass and Total Plant Mass across Chinese Forests** ... 206
Shanshan Xu, Yan Li, Genxuan Wang

Permissions

List of Contributors

Index

PREFACE

There has been tremendous rise in environmental concerns over the past decade. Sustainable forestry aims to overlook the management of forests in a manner which is least harmful for the environment and fruitful to mankind at the same time. This book contains some path-breaking studies in the field of sustainable forestry and management and highlights some of its key concepts and applications. While understanding the long-term perspectives of the topics, the book makes an effort in highlighting their impact as a modern tool for the growth of the discipline. It will help the readers in keeping pace with the rapid changes in this field. This book will serve as a valuable source of reference to a broad spectrum of readers including environmentalists, ecologists, researchers, conservationists, professionals and students associated with the field of forestry at various levels.

This book has been the outcome of endless efforts put in by authors and researchers on various issues and topics within the field. The book is a comprehensive collection of significant researches that are addressed in a variety of chapters. It will surely enhance the knowledge of the field among readers across the globe.

It gives us an immense pleasure to thank our researchers and authors for their efforts to submit their piece of writing before the deadlines. Finally in the end, I would like to thank my family and colleagues who have been a great source of inspiration and support.

Editor

Effects of Litter Manipulation on Litter Decomposition in a Successional Gradients of Tropical Forests in Southern China

Hao Chen[1,5], Geshere A. Gurmesa[2,5,6], Lei Liu[1,3], Tao Zhang[1,4], Shenglei Fu[1], Zhanfeng Liu[1], Shaofeng Dong[1], Chuan Ma[1], Jiangming Mo[1]*

1 Key Laboratory of Vegetation Restoration and Management of Degraded Ecosystems, South China Botanical Garden, Chinese Academy of Sciences, Guangzhou, China, **2** Department of Geosciences and Natural Resource Management, University of Copenhagen, Copenhagen, Denmark, **3** State Key Laboratory of Urban and Regional Ecology, Research Center for Eco-Environmental Sciences, Chinese Academy of Sciences, Beijing, China, **4** Institute of Tropical Pratacultural Science, Zhanjiang Normal University, Zhanjiang, China, **5** University of Chinese Academy of Sciences, Beijing, China, **6** Sino-Danish Center for Education and Research, Aarhus, Denmark

Abstract

Global changes such as increasing CO_2, rising temperature, and land-use change are likely to drive shifts in litter inputs to forest floors, but the effects of such changes on litter decomposition remain largely unknown. We initiated a litter manipulation experiment to test the response of litter decomposition to litter removal/addition in three successional forests in southern China, namely masson pine forest (MPF), mixed coniferous and broadleaved forest (MF) and monsoon evergreen broadleaved forest (MEBF). Results showed that litter removal decreased litter decomposition rates by 27%, 10% and 8% and litter addition increased litter decomposition rates by 55%, 36% and 14% in MEBF, MF and MPF, respectively. The magnitudes of changes in litter decomposition were more significant in MEBF forest and less significant in MF, but not significant in MPF. Our results suggest that change in litter quantity can affect litter decomposition, and this impact may become stronger with forest succession in tropical forest ecosystem.

Editor: Ben Bond-Lamberty, DOE Pacific Northwest National Laboratory, United States of America

Funding: This study was founded by National Key Basic Research 973 Programme (2011CB403204) and National Natural Science Foundation of China (No: 41273143). The funders had no role in study design, data collection and analysis, decision to publish, or preparation of the manuscript.

Competing Interests: The authors have declared that no competing interests exist.

* E-mail: mojm@scib.ac.cn

Introduction

Litter decomposition is a key process that regulates nutrient recycling in ecosystems, influences net ecosystem carbon (C) storage, and is the first step in the formation of soil humus [1]. Litter decomposition can be affected by many environmental factors including the physical environment (e.g. temperature, moisture, and soil pH), nutrients availability and activities of decomposers in the soil [2]. On the other hand, litter quantity can alter microclimate, number and dynamics of decomposer organisms and nutrient availability in forest floor and mineral soil [3]. Thus, the quantity of the litter itself has an impact on litter decomposition. It has been reported that global climate changes due to rising atmospheric CO_2 concentration and temperature can increase net primary production (NPP) and consequently litter production in forest ecosystems [4–6]. In contrast, litter inputs are also likely decreased due to extensive deforestation and cultivation [7]. Therefore, evaluating the effects of these changes in litter inputs on litter decomposition is crucial for our understanding of ecosystem nutrient supplies and future global C cycle.

Despite the substantial number of studies on decomposition in a wide range of ecosystems, the influence of a sustained change in litter quantity on litter decomposition has not been well addressed [3]. Existing evidence on the effects of litter quantity on the leaf litter decomposition is indirect, and the results are inconsistent and incomparable. For example, comparison of litter decomposition before and after hurricanes showed that the large inputs of litter following hurricanes would cause accelerated or slower decomposition rate [8,9]. Similarly, results from the labile-C (glucose, cellulose, or root exudates) addition experiments have shown that labile-C addition could have both positive and negative effect on leaf litter decomposition [10–13]. Moreover, potentially decreased litter quantity after understory removal also showed both decreasing [14,15] and increasing [16] effects on litter decomposition.

Compared to the methods mentioned above, litter manipulation (removal or addition) experiment is considered to be a direct way for studying effects of litter quantity on ecosystem processes [17]. Using this method, recent studies have reported that increased litter input would accelerate soil C release and decreased soil C content in tropical forests due to the "priming effect" [18–21]. However, most of these studies assessed the effect of litter quantity on the decomposition of soil organic matter and only few studies addressed the effect on decomposition of fresh leaf litter. In addition, more studies focused on the effect of litter removal [22,23] and to our knowledge only one studied effects of litter addition on litter decomposition in forest ecosystem [3], highlighting the need for more complete comparative study on effects of both litter addition and removal on leaf litter decomposition.

While estimates of decomposition after litter manipulation are commonly reported in single forest type, forests at different successional stages are likely to present distinct responses. The Dinghushan Biosphere Reserve (DHSBR) consists of three typical forest types in southern China at different forest successional stages in terms of age and exposure to human disturbances; hence it provides an excellent opportunity to study the response of litter decomposition to litter manipulation along a forest successional gradient. The three forest types are a pioneer masson pine forest (MPF), a transitional coniferous and broad-leave mixed forest (MF) and a climax monsoon evergreen broad-leaved forest (MFBF) [24]. Previous studies showed that these forests have large variation in litter production, nutrient status in plant and soil, and some other environment factors [24–26]. This study presents the results of litter decomposition experiment where we quantitatively measured the effects of litter removal and addition on litter decomposition in three forests in DHSBR. Our objectives were to examine the effects of litter removal and addition on litter decomposition in tropical forests and to compare these effects among forests at different successional stages. We hypothesized that (1) litter removal would decrease litter decomposition but litter addition would increase litter decomposition in each forest types due to the priming effect; and (2) response of litter decomposition to litter removal/addition would vary among the three forests.

Materials and Methods

Ethics Statement

No specific permits were required for the described field studies. This research station (DHSBR) belongs to South China Botanical Garden, Chinese Academy of Sciences, which also supported the study. We confirmed that the location is not privately owned. We also confirmed that the field studies did not involve endangered or protected species. Data will be made available upon request.

Study Site

The study was conducted in Dinghushan Biosphere Reserve (DHSBR), which is in the middle part of Guangdong province, southern China (112°33' E and 23°10' N). Average annual precipitation is 1927 mm, with 75% occurring from March to August and only 6% from December to February [27]. Mean annual relative humidity is 80%. Mean annual temperature is 21.0°C with the lowest and highest average temperature being in January (12.6°C) and July (28.0°C) [27]. Soil type is lateritic red earth (Ultisols in the USDA soil taxonomy or Acrisols in the FAO soil classification) formed from sandstone [28].

There are three main forest types in the reserve; a pioneer masson pine forest (hereafter named as MPF), a coniferous and broad-leave mixed forest (hereafter named as MF) and a monsoon evergreen broad-leaved forest (hereafter named as MEBF) [24]. The three forest types form distinct successional gradients [24,29]. The MPF belongs to the first stage of the successional processes and occurs in the transition zone (periphery) of the reserve at an elevation of about 200m. *Pinus massoniana* Lamb is the dominant species in MPF (Table S1), which was planted in the 1930s. Top soil (0–30 cm) texture in MPF is medium gravel-medium loam (Ultisols in the USDA soil taxonomy), and the capacity of field moisture and wilting coefficient in the soil expressed as gravimetric water content are 26.0% and 10.9%, respectively [30]. The MF is distributed on areas next to the MPF and towards the core areas of the reserve at an elevation of about 200–300 m. It was developed by a gradual invasion of the originally planted MPF by some pioneer broadleaf species through natural succession [31]. The plant composition in MF has greatly been changed. Dominant tree species in the mixed forest are *Pinus massoniana* Lamb, *Schima superba* Chardn. & Champ., *Cryptocarya chinensis* Hance, *Craibiodendron kwangtungense* S. Y. Hu, *Lindera metcalfiana* Allen, and *Cryptocarya concinna* Hance (Table S1). The texture of top soil in MF is medium gravel-heavy loam with field capacity and wilting point at 25.3% and 8.2% gravimetric water content, respectively [30]. The MEBF is distributed in the core area of the reserve at an elevation varying from 250 to 300 m. It has been protected from direct human impact for more than 400 years [28]. Major species in MEBF are *Castanopsis chinensis* Hance, *Schima superba* Chardn. & Champ., *Cryptocarya chinensis* (Hance) Hemsl., *Machilus chinensis* (Champ. ex Benth.) Hemsl., *Syzygium rehderianum* Merr. & Perry in the canopy layer (Table S1) and *Calamus rhabdicladus* Burret, *Ardisia quinquegona* Bl. and *Hemigramma decurrens* (Hook.) Copel in the understory layer. The texture of top soil in MEBF is light gravel-heavy loam. The water retention capacity of top soil is the highest with field moisture capacity and wilting coefficient of 34.6% and 11.4% gravimetric water content, respectively [30]. Litter layers are different among three forests. Litter layer is thin in the MEBF due to the faster decomposition rate, but thick in the MF and MPF. The three forests also vary in litter production, nutrient status in soil and plant, and some other environmental factors. See Table 1 for information on selected site characteristics of the three forest types.

Experimental Design

In each forest, we used a randomized complete block design with five blocks (i.e., n = 5). In each block, we set up three 1 m×1 m plots to be used as a control (CT), a litter removal (L−), and a litter addition (L+) plots making up a total of 15 plots in each forest. There was at least 3 m buffer zone between two adjacent plots to avoid overlapping effects of different treatments. The litter decomposition experiment started in February, 2007, by placing litterbags on the surface of litter layer in the plots. We removed all litter every month from the L− plots and added it to the corresponding L+ plots to cover the litterbags. Litter removal from the L− plots caused low average litter input but did not cause entire absence of litter input to the plots because the plots was receiving litter in the following days of each mouth. The controls received the normal input of aboveground litter.

Litter Collection and Initial Nutrients Analysis

Leaf litter was collected using litter traps and nylon mesh placed on the forest floor under the trees in the study sites in May and June 2006, during which litterfall is peak [32]. Leaf litter of *Schima superba* Chardn. & Champ. (SS) and *Castanopsis chinensis* Hanc (CC) was collected in MEBF and MF. Leaf litter of SS, CC and *Pinus. massoniana* Lamb (PM) was collected in MPF. These species were chosen in this study because they are the dominant tree species and contribute 40–91% the total leaf litter fall in these forests [33]. All litter was air-dried to a constant weight and six sub-samples from each kind of litter were analyzed for initial nitrogen (N) and phosphorus (P) concentrations (Table S2). N concentration was determined by the semimicro-Kjeldahl digestion method followed by the detection of ammonium with a Wescan ammonia analyzer, while total P concentration was analyzed colorimetrically after acidified ammonium persulfate digestion [34].

Litter Decomposition Experiment

Litter decomposition was determined by placing fresh litter in mesh bags in the plots. A total of 2625 litter bags were prepared from 25×25 cm polyvinyl screen with 0.5×0.5 mm mesh in the bottom and 2×2 mm in the top. Leaf litter of each species was mixed before filling in the mesh bags. Each bag was filled with

Table 1. Comparisons of litter production, leaf litter N and P concentration and selected soil properties among MEBF, MF, and MPF.

Forest types	MEBF	MF	MPF
Litter production (Mg ha^{-1} yr^{-1})‡	8.3 (0.64)^a	8.5 (0.62)^a	3.3 (0.57)^b
Leaf litter N (mg g^{-1})†	17.5 (0.12)^a	15.0 (0.10)^b	9.6 (0.07)^c
Leaf litter total P (mg g^{-1})†	0.53 (0.01)^a	0.32 (0.01)^b	0.36 (0.01)^b
Soil N (mg g^{-1})	1.99 (0.18)^a	0.93 (0.08)^b	1.15 (0.16)^b
Soil organic matter (%)	7.3 (0.8)^a	3.7 (0.2)^b	5.2 (0.2)^b
Soil C/N ratio	21.0 (0.6)^a	23.8 (2.2)^a	28.1 (3.9)^b
Soil total P (mg g^{-1})	0.49 (0.03)^a	0.38 (0.01)b	0.44 (0.01)^{ab}
Soil available P (mg kg^{-1})	2.2 (0.5)^{ab}	1.5 (0.5)^a	2.9 (0.2)^b
Soil moisture (%)	22.6 (1.1)^a	16.4 (1.9)^b	15.3 (1.1)^b
Soil temperature (°C)	21.8 (0.36)^a	22.6 (0.37)^b	23.41 (0.39)^c

‡from Zhou et al. (2011); †leaf litter chemical characteristics were average values of main litter species in the each forest, which were measured in 2012 in the control plots. Other values were measured in August 2007 in the control plots. Values are means, standard error in parentheses, n = 5, means not sharing the same superscript letter were statically different at *P*-value of 0.05.

10 g air-dried mass of litter. Only one litter type was put in each bag and the litter bags were evenly distributed among each plot. Litter bags were retrieved at about 3, 6, 9, 12, and 18 months (hereafter named as first, second, third, fourth and last sampling date, respectively) after the start of the study. Five litter bags of each species (a total of 525 bags on each collection) were randomly selected and collected from each plot at each sample date. The average of the 5 litterbags of each species per plot at each sampling date was used in the statistical calculations. After removing roots, soil and other extraneous materials, the leaf residue in each litter bag was oven dried at 45°C for 48 h and weighed. Litter from the last sampling date was measured for N and P concentrations using the method as described above.

Soil and Microbial Sampling

Soil sampling was conducted in August 2008 (1.5 years after litter manipulation). From each plot, 5 soil cores with 2.5 cm inner diameter were collected at random points from a 0–10cm soil depth and combined to create one composite sample per plot. Changes in soil microbial biomass was studied only in MEBF using phospholipid fatty acid (PLFA) analysis as described in Bossio and Scow [35], because samples from the other two forests were polluted during transportation of the samples from fields.

Calculations and Data Analysis

To determine litter decomposition rates we used the following decomposition model [36]: $y = e^{(-kt)}$ (exponential model), where y is the fraction of mass remaining at a specific time t (years), "e" the base of natural logarithm, k the decomposition coefficient ($year^{-1}$). To determine changes in decomposition due to treatment effect, k value changes were given by $(k_{trt}-k_{con})/k_{con}$, where k_{trt} is the decomposition coefficient in treatments (L− or L+), k_{con} the decomposition coefficient in control. Nutrient content was calculated by multiplying the nutrient concentration by the mass remaining. Nutrient content was then reported as a proportion of the initial leaf content [37].

One-way Analysis of Variance (ANOVA) was used to test the difference in decomposition rate (k) among forests or litter species in controls, and to test the difference in PLFAs among treatments in MEBF. Two-way ANOVA was used to determine the interaction effect of litter manipulation treatments and litter types on litter decomposition rate and nutrient remaining in each forest. Repeated measure ANOVA was used to determine the difference in litter mass remaining among forests, litter manipulation treatments and litter species. In all ANOVAs, the block was included as random factors. In addition, data was log transformed to fulfill the requirements of normality and homogeneity of variance. All analyses were conducted using SAS software (SAS Institute Inc., Cary NC, USA). Statistically significant differences were set with P values <0.05 unless otherwise stated. Mean values ± standard errors were reported in the text.

Results

Litter Decomposition Rate in Control Plots

The leaf litter decomposition rates in control plots varied among forest types and tree species (Fig. 1, Table 2). Leaf litter decomposition rate (i.e., k value) was significantly higher in MEBF than in MF and MPF for all possible species comparison. However, no significant difference in litter decomposition rate was found between MF and MPF for all species. Average litter decomposition rates (k) were 1.97, 0.92, and 0.73 for MEBF, MF, and MPF, respectively. When compared within forest type, litter decomposition rate was significantly lower in CC leaf litter than in SS in the MEBF and MF (Table 2, $P = 0.037$ and 0.021, respectively). However, there was no significant difference among the selected three species in the MPF (Table 2, $P = 0.261$).

Effects of Litter Manipulation on Litter Decomposition

Responses of litter decomposition rates to litter removal/addition varied with forest types (Table 2). In MEBF, litter removal significantly decreased litter decomposition rates for both SS and CC ($P = 0.045$ and 0.030, respectively). Repeated measure ANOVA with Turkey' HSD test showed that mass remaining was significantly higher in L− than in control plots in the second, third, fourth and fifth sampling dates for SS leaf litter. For CC leaf litter, significant difference in litter decomposition between L− and control plots was observed only after the last sampling date (Fig. 1). In contrast, litter addition significantly increased litter decomposition rate for both SS and CC ($P = 0.018$ and 0.005, respectively). However, such a response of leaf litter decomposition rates to litter addition showed temporal variation between the two

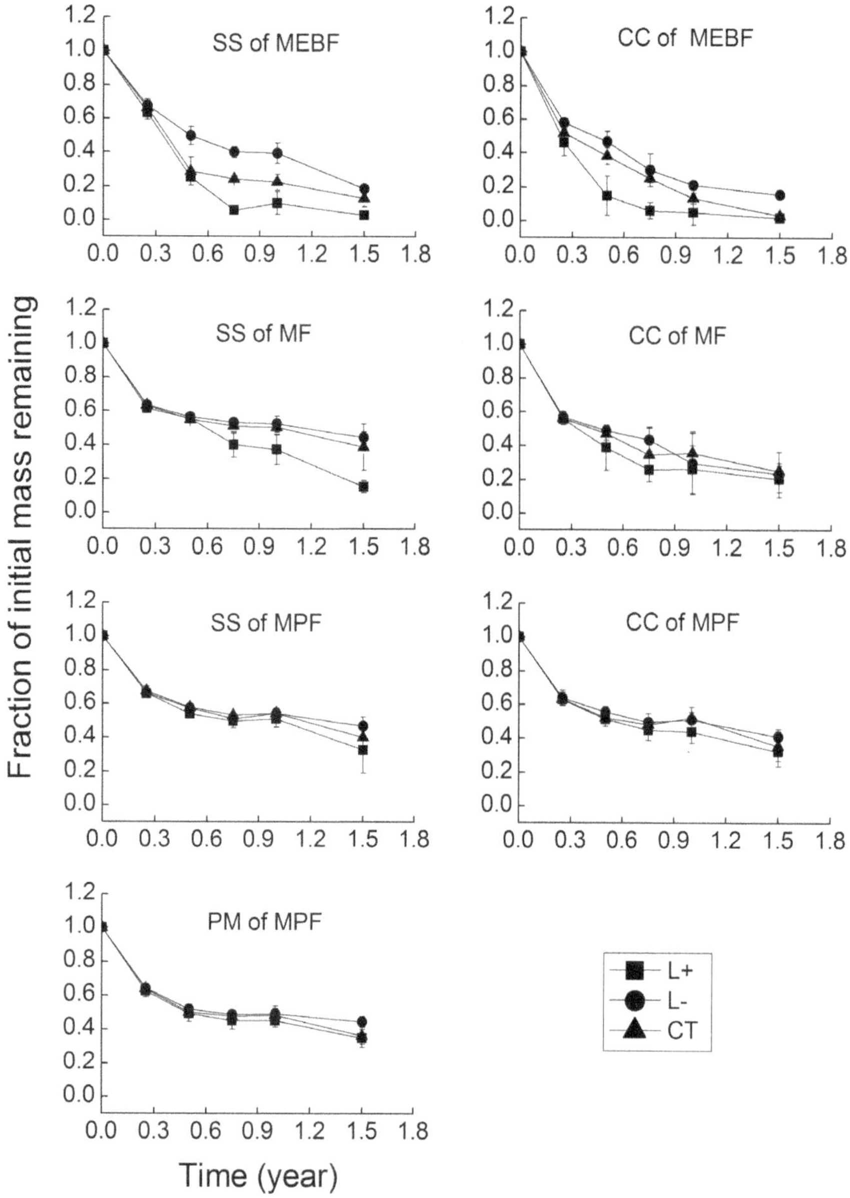

Figure 1. Mass loss of decomposing leaf litter of three dominant tree species in various litter manipulation treatments in the MEBF, MF, and MPF.

species. For SS, significant difference between control and L+ plots was found in the second, third, and fourth sampling dates, whereas the reported response for CC was observed in the third and fourth sampling dates (Fig. 1).

In MF, SS and CC leaf litter showed similar response to litter addition. Litter decomposition rate increased significantly after litter addition for both species (Table 2, $P=0.019$ and 0.012 for SS and CC, respectively). However, the effect of litter removal on litter decomposition varied between the two species (Table 2). Litter removal significantly decreased litter decomposition for SS ($P=0.033$), but not for CC ($P=0.842$). Repeated measure ANOVA with Turkey' HSD test indicated that litter addition significantly increased decomposition of SS leaf litter in the third, fourth and fifth sampling dates, and significantly increased decomposition of CC leaf litter in the second and third sampling dates (Fig. 1). Litter removal significantly decreased SS leaf litter decomposition in the last sampling date, but no significant change in leaf litter decomposition was observed for CC leaf litter decomposition at all sampling dates (Fig. 1).

In MPF, decomposition rate increased slightly with litter addition and decreased slightly with litter removal for all species (Table 2), but one-way ANOVA showed that this increase/ decrease was not significant among treatments for all species. Repeated measure ANOVA also showed no significant difference in mass remaining among treatments over the entire study period for all species. There are no significant interaction effects of litter species and treatments on litter decomposition rate in all the three forests ($P>0.05$ for all).

The relative change in k values in response to the treatments decreased in the order: MEBF>MF>MPF (Fig. 2). Average decrease in litter decomposition rate as a result of litter removal was 27% in the MEBF, which was significantly higher than the

Table 2. Comparisons of decomposition rates (*k* values) between treatments and litter species in each forest.

Species	Treatments	MEBF		MF		MPF	
		K	R^2	*K*	R^2	*k*	R^2
SS	CT	1.74 (0.36)^a	0.84 (0.04)	0.80 (0.05)^a	0.60 (0.09)	0.68 (0.04)^a	0.71(0.06)
	L−	1.23 (0.19)^b	0.86 (0.04)	0.66 (0.02)^b	0.57 (0.07)	0.64 (0.02)^a	0.54 (0.09)
	L+	2.80 (0.24)^c	0.86 (0.06)	1.13 (0.05)^c	0.81(0.04)	0.84 (0.09)^a	0.69 (0.07)
CC	CT	2.19 (0.06)^c	0.75 (0.06)	1.04 (0.06)^{cd}	0.68 (0.10)	0.78 (0.05)^a	0.62 (0.09)
	L−	1.63 (0.29)^a	0.80 (0.04)	1.01 (0.05)^{cd}	0.77 (0.07)	0.72 (0.04)^a	0.60 (0.06)
	L+	3.30 (0.25)^d	0.84 (0.04)	1.35 (0.12)^d	0.85 (0.09)	0.89 (0.08)^a	0.75 (0.07)
PM	CT	–		–		0.80 (0.03)^a	0.64 (0.04)
	L−	–		–		0.71 (0.02)^a	0.51 (0.06)
	L+	–		–		0.84 (0.05)^a	0.64 (0.09)

Notes: *k* values and coefficients of determination (R^2) are based on a single negative exponential model. Two-way ANOVA with SNK test was used in each forest respectively to determine the effect of litter treatments and litter types on *k* values. Values are means, standard error in parentheses, n = 5. Under each forest type and for each variable column, means not sharing the same superscript letter were statically different at *P*-value of 0.05.

respective values in MF (10%) and MPF (8%) ($P<0.001$). On the other hand, litter addition increased leaf litter decomposition by 55% in the MEBF, which is significantly larger than the average percent increase in leaf litter decomposition rates in MF (36%) and MPF (14%) ($P<0.001$).

Litter Nutrient Remaining

Nutrient (N and P) remaining in all leaf litter types was measured after the last sampling date and it varied depending on forests and species (Fig. 3). In MEBF, litter removal significantly increased N and P remaining in decomposing leaf litter for both SS and CC. In contrast, litter addition decreased N and P remaining in both species leaf litter, but the decrease in N and P remaining was significant only for SS leaf litter. In MF, litter removal significantly increased N and P remaining and litter addition significantly decreased N and P remaining in SS leaf litter. However, none of the treatments caused significant changes in N and P remaining in decomposing CC leaf litter when compared to the control plots, but both litter removal and litter addition tended to decrease N and P remaining in the CC leaf litter (all $P>0.05$). In MPF, litter removal tended to increase N and P remaining and litter addition tended to decrease N and P remaining in all species, but the magnitude of these changes varied among the treatments and species. For example, litter removal significantly increased N remaining in PM ($P=0.005$), P remaining in SS and CC ($P=0.024$ and $P=0.033$, respectively).

Soil Microbial Biomass

Litter manipulation had no significant effect on soil microbial biomass in the MEBF (Fig. 4, $P=0.074$). The average total PLFAs were 55.40 (6.91), 57.94 (7.9), and 45.34 (6.1) nmol g^{-1} in control, L−, and L+ plots, respectively. Similarly, there was no significant difference among litter manipulation treatments for the Bac PLFAs, Fun PLFAs, and F:B (Fig. 4).

Discussion

In the present study, decomposition rates (*k* values) of leaf litter observed in the controls (ranging from 0.68 to 2.19) were similar to those found in subtropical forests of south China [33,38–39] and other subtropical/tropical forests [40–42], but were slightly lower than that of some other tropical forests with high rainfall [43]. However, the *k* values in our study were higher than those reported from temperate forests [44,45], indicating higher decomposition rate in these humid sub-tropical forests of southern china. In addition, decomposition rate increased with forest succession: MEBF>MF>MPF, which was in agreement with results reported by previous studies in adjacent forests [33,38].

Compared to control plots which received normal litter input, litter removal significantly decreased litter decomposition in the MEBF. Similar effect of litter removal on litter decomposition was reported by a study in an old-growth forest in Panama [3]. Based on available evidence from literature, there are several possible mechanisms that could explain our observation. Litter removal changes the microclimate (e.g. moisture and temperature) in the forest floor [46,47], causes direct nutrient losses and changes in soil physical and chemical characteristics [48,49], and decreases decomposer biomass and activity [3,50]. Even though we did not measure plot microclimate, previous studies have reported that litter removal usually affects soil microclimate, mainly decreasing moisture and greater fluctuations in temperature in forest floor [46,47]. On the other hand, a previous study observed in adjacent forests showed that litter removal had no significant effect on the soil temperature in all the three forests, and significantly decreased

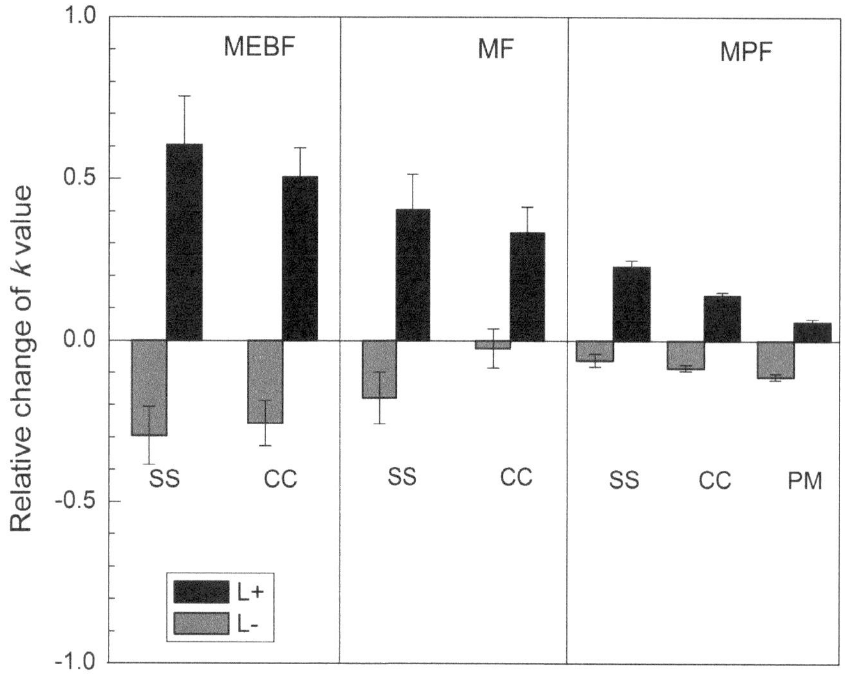

Figure 2. Comparisons of relative change of *k* value between different litter manipulation treatments, litter species and forest types.

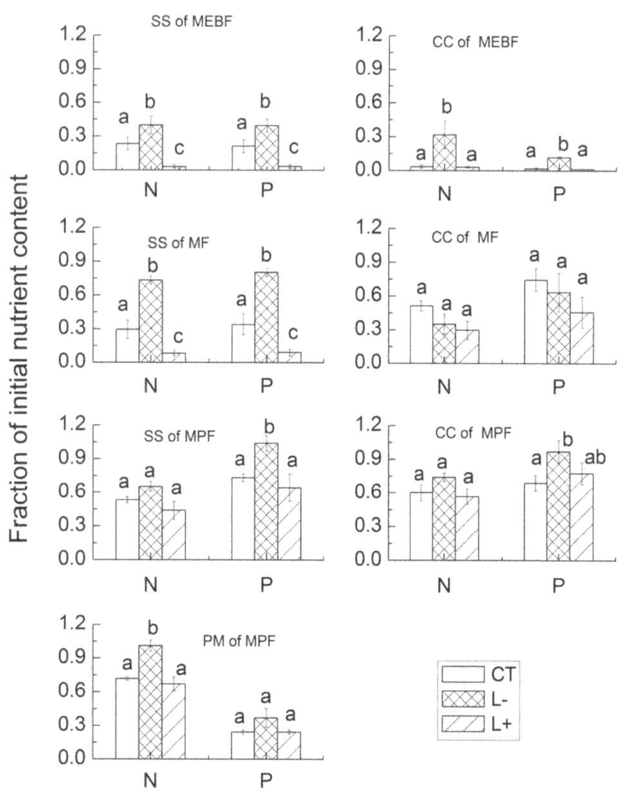

Figure 3. Nutrients remaining in the last stage of the decomposition. Values are means, standard error in parentheses, n = 5, means not sharing the same superscript letter were statically different at *P*-value of 0.05.

soil moisture only in MF [26]. However, the soil temperature and moisture were measured in mineral soil at 5 cm depth [26]. We expect more changes in forest floor microclimate following the litter manipulation, but such data are missing in our study.

Decreased nutrient supply for decomposer microbial communities after litter removal is another likely explanation for the slower decomposition in the L− plots. In this study, litter was removed once a month in L− plots, which could remove 44–73% of litter nutrient input [49]. This decreased nutrient supply might decrease microbial activities in the L− plots. We measured the litter nutrients remaining in the last sampling date and found that the content of N and P increased in L− plots compared to those in controls for all litter. This might indicate that more N and P were immobilized in L− plots, possibly because the microbes need more N and P in L− treatment due to reduced litter input of these nutrients making the microbes nutrient limited as suggested by [36].

However, we did not observe significant and clear changes in soil microbial biomass and microbial community as indicated by PLFA result (Fig. 4). Similar results were also reported from temperate forests, where change in the soil microbial biomass or activity in litter removal treatment were not significant [51]. It has been suggested that absence of expected increase in soil microbial biomass could be due to soil pools buffering the effect of litter manipulation from aboveground [51,52]. However, our data on microbial biomass included only mineral soil because clear forest floor layer in MEBF is usually absent due to fast turn-over of organic materials. Even though data on the forest floor is not available in this study, previous study reported that litter removal decreased decomposers biomass more in forest floor than in mineral soil [3]. The study also showed that litter removal reduced the abundance of meso-arthropods in forest floor because meso- and micro-arthropods, which play significant role in decomposing organic materials, mostly inhabit top forest floor layer. Though our data only for mineral soil might not be very conclusive

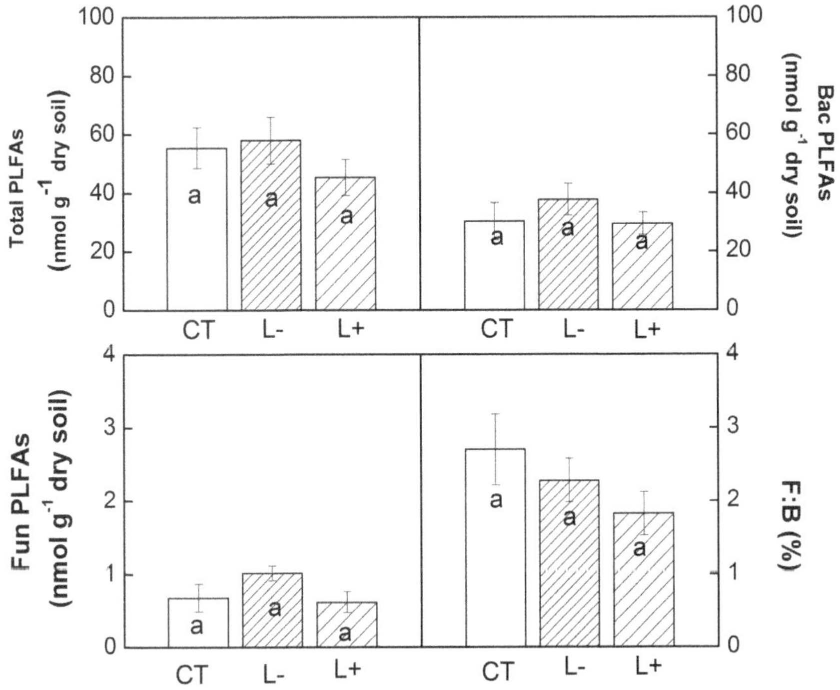

Figure 4. Comparisons of soil microbial PLFAs between treatments in the MEBF. Data from August 2008. Bac PLFAs: Bacterial PLFAs; Fun PLFAs: Fungal PLFAs; F:B (%): the percentage of fungal to bacterial PLFAs. Values are means, standard error in parentheses, n = 5, means sharing the same superscript letter were not statically different (*P*-value ≥0.05).

especially when data is absent for the more active and dynamic litter layer, it showed that microbial activities might not be affected by short term litter manipulation. However, more comprehensive studies including response of microbial dynamics to litter manipulation both in mineral soil and forest floor are needed for further understanding of the subject.

By contrast, litter addition significantly accelerated the decomposition of leaf litter in the MEBF. Our result is inconsistent with results reported by Sayer *et al.* [3], which showed that litter addition did not affect leaf litter decomposition but significantly increased wood litter decomposition. However, our result partially supports the "priming effect" hypothesis which suggests that the addition of fresh organic matter (leaf litter) can stimulate decomposition of the organic matter [53,54]. The reason for this response is not clear in this study. We observed decreased N and P reaming in L+ plots compared to control plots indicating that increased nutrient availability from the additional litter could stimulate the microbe to release litter nutrients, which partially contribute to increased decomposition of leaf litter. In addition, we did not find significant changes in soil microbial biomass and microbial community after the litter addition (Fig. 4), but we still cannot rule out the possibility of the change of microbial biomass and microbial communities in litter layer. For example, Sayer *et al.* [3] found increased abundance of meso-arthropode in forest floor after doubling litter input resulting in increased mass loss of leaf litter. However, we could not attribute our observation to an increase in meso-arthropodes because the size of the mesh we used (0.5mm×0.5 mm) might have reduced the contribution macro-decomposer communities despite possible increases in their abundance. Comparative studies of effects of litter manipulation on litter decomposition with different mesh sizes might provide better understanding on possible confounding effects of mesh size in litter decomposition studies.

The effect of litter manipulation experiment on leaf-litter decomposition in our study varied significantly among the three forest types and exhibited clear pattern with successional gradients. The effect was highest in the climax MEBF and lowest in the MPF whereas MF showed intermediate response (Fig. 2). The reasons for this response pattern are currently not clear. However, several explanations could be suggested. Litter production was found to differ among these three forests, amounting 83, 85 and 33 Mg ha^{-1} yr^{-1} in MEBF, MF, and MPF, respectively (Table 1). Similarly, nutrient concentration and quality of leaf (C/N ratio, lignin content etc.) differs among the three forests. Foliar N and P concentrations in MF and MPF were significantly lower than in MEBF (Table 1). This suggests that more C and other nutrients in MEBF would be removed or input after litter manipulation compared to those in MF and MPF, and thus resulting in more significant response in MEBF. Another possible explanation could be different responses in soil microclimate after litter treatments among the forests, especially for soil moisture. Soil moisture is a more important factor than soil temperature for the fastest decomposition rate in the MEBF because Table 1 showed that the MEBF had the highest moisture and the lowest temperature in soil than other two forests (Table 1). Because of this higher initial moisture, we believe that the thorough litter removal/addition may change the soil moisture of the MEBF in a larger extent compared to other forests with lower initial moisture. In addition, for the MF and MPF, the long-term history of relatively more exposure to human disturbance, lower litter production and less canopy cover also might have reduced the responses to our short-term litter manipulation experiment.

Conclusion

Our results showed that litter removal decreased litter decomposition and litter addition increased litter decomposition in three tropical forests at different successional stages. However, the effect was highest in the climax MEBF and lowest in the MPF whereas MF showed intermediate response, suggesting that the change in litter quantity can affect litter decomposition in tropical forests and this impact may become stronger with forest succession in the studied tropical forests. Our results can provide relevant information on how future climate changes modify accumulation of organic matter (C) in tropical forest and consequently affect nutrient cycle, and for any sudden events in the forest such as hurricane and tree diseases which also changes quantity of litter input to the forest floor. However, recent mechanisms are still far from clear, thus we also suggest similar studies in temperate and other tropical forest ecosystems to further strengthen the findings.

Supporting Information

Table S1 Indices of the tree structure in the three tropical forest types. The survey was conducted in February 2007.

Table S2 Initial chemical characteristics of three studied leaf litter.

Acknowledgments

We thank Xiaoming Fang and Wei Zhang for their skilful assistance in field work, Guohua Liang for his assistance in laboratory work, and Drs. Per Gundersen and Weixing Zhu for their invaluable suggestions in experiment and earlier version of the manuscript.

Author Contributions

Conceived and designed the experiments: JMM HC SLF ZFL. Performed the experiments: CM SFD LL TZ HC. Analyzed the data: HC GAG. Contributed reagents/materials/analysis tools: SFD CM HC. Wrote the paper: HC. English editing: GAG.

References

1. Hobbie SE, Vitousek PM (2000) Nutrient limitation of decomposition in Hawaiian forests. Ecology 81: 1867–1877.
2. Chapin FS, Matson PA, Mooney HA (2002) Principles of terrestrial ecosystem ecology. New York, USA: Springer.
3. Sayer EJ, Tanner EVJ, Lacey AL (2006) Effects of litter manipulation on early-stage decomposition and meso-arthropod abundance in a tropical moist forest. Forest Ecology and Management 229: 285–293.
4. Hickler T, Smith B, Prentice IC, MjÖFors K, Miller P, et al. (2008) CO_2 fertilization in temperate FACE experiments not representative of boreal and tropical forests. Global Change Biology 14: 1531–1542.
5. Norby RJ, DeLucia EH, Gielen B, Calfapietra C, Giardina CP, et al. (2005) Forest response to elevated CO_2 is conserved across a broad range of productivity. Proceedings of the National Academy of Sciences of the United States of America 102: 18052–18056.
6. Raich JW, Russell AE, Kitayama K, Parton WJ, Vitousek PM (2006) Temperature influences carbon accumulation in moist tropical forests. Ecology 87: 76–87.
7. Holmes KW, Chadwick OA, Kyriakidis PC, de Filho EPS, Soares JV, et al. (2006) Large-area spatially explicit estimates of tropical soil carbon stocks and response to land-cover change. Global Biogeochemical Cycles 20: GB3004.
8. Ostertag R, Scatena FN, Silver WL (2003) Forest floor decomposition following hurricane litter inputs in several Puerto Rican forests. Ecosystems 6: 261–273.
9. Sullivan NH, Bowden WB, McDowell WH (1999) Short-Term Disappearance of Foliar Litter in Three Species Before and After a Hurricane. Biotropica 31: 382–393.
10. Barantal S, Schimann H, Fromin N, Hattenschwiler S (2012) Nutrient and Carbon Limitation on Decomposition in an Amazonian Moist Forest. Ecosystems 15: 1039–1052.
11. Boberg J, Finlay RD, Stenlid J, Nasholm T, Lindahl BD (2008) Glucose and ammonium additions affect needle decomposition and carbon allocation by the litter degrading fungus *Mycena epipterygia*. Soil Biology & Biochemistry 40: 995–999.
12. Chigineva NI, Aleksandrova AV, Tiunov AV (2009) The addition of labile carbon alters litter fungal communities and decreases litter decomposition rates. Applied Soil Ecology 42: 264–270.
13. Kuzyakov Y, Hill PW, Jones DL (2007) Root exudate components change litter decomposition in a simulated rhizosphere depending on temperature. Plant and Soil 290: 293–305.
14. Liu ZF, Wu JP, Zhou LX, Lin YB, Fu SL (2012) Effect of understory fern (*Dicranopteris dichotoma*) removal on substrate utilization patterns of culturable soil bacterial communities in subtropical Eucalyptus plantations. Pedobiologia 55: 7–13.
15. Wu JP, Liu ZF, Wang XL, Sun YX, Zhou LX, et al. (2011) Effects of understory removal and tree girdling on soil microbial community composition and litter decomposition in two Eucalyptus plantations in South China. Functional Ecology 25: 921–931.
16. Matsushima M, Chang SX (2007) Effects of understory removal, N fertilization, and litter layer removal on soil N cycling in a 13-year-old white spruce plantation infested with Canada bluejoint grass. Plant and Soil 292: 243–258.
17. Sayer EJ (2006) Using experimental manipulation to assess the roles of leaf litter in the functioning of forest ecosystems. Biological Reviews 81: 1–31.
18. Crow SE, Lajtha K, Bowden RD, Yano Y, Brant JB, et al. (2009) Increased coniferous needle inputs accelerate decomposition of soil carbon in an old-growth forest. Forest Ecology and Management 258: 2224–2232.
19. Fontaine S, Bardoux G, Abbadie L, Mariotti A (2004) Carbon input to soil may decrease soil carbon content. Ecology Letters 7: 314–320.
20. Leff JW, Wieder WR, Taylor PG, Townsend AR, Nemergut DR, et al. (2012) Experimental litterfall manipulation drives large and rapid changes in soil carbon cycling in a wet tropical forest. Global Change Biology 18: 2969–2979.
21. Sayer EJ, Heard MS, Grant HK, Marthews TR, Tanner EVJ (2011) Soil carbon release enhanced by increased tropical forest litterfall. Nature Climate Change 1: 304–307.
22. Kranabetter JM, Chapman BK (1999) Effects of forest soil compaction and organic matter removal on leaf litter decomposition in central British Columbia. Canadian Journal of Soil Science 79: 543–550.
23. Ruan HH, Li YQ, Zou XM (2005) Soil communities and plant litter decomposition as influenced by forest debris: Variation across tropical riparian and upland sites. Pedobiologia 49: 529–538.
24. Zhou GY, Guan LL, Wei XH, Zhang DQ, Zhang QM, et al. (2007) Litterfall production along successional and altitudinal gradients of subtropical monsoon evergreen broadleaved forests in Guangdong, China. Plant Ecology 188: 77–89.
25. Huang WJ, Zhou GY, Liu JX (2012) Nitrogen and phosphorus status and their influence on aboveground production under increasing nitrogen deposition in three successional forests. Acta Oecologica-International Journal of Ecology 44: 20–27.
26. Tang X, Liu S, Zhou G, Zhang D, Zhou C (2006) Soil-atmospheric exchange of CO_2, CH_4, and N_2O in three subtropical forest ecosystems in southern China. Global Change Biology 12: 546–560.
27. Huang ZF, Fan ZG (1982) The climate of Dinghushan. Tropical and subtropical Forest Ecosystem. Tropical and Subtropical Forest Ecosystem 1: 11–23.
28. Mo HM, Brown S, Peng SL, Kong G (2003) Nitrogen availability in disturbed, rehabilitated and mature forests of tropical China. Forest Ecology and Management 175: 573–583.
29. Peng SL, Wang BS (1993) Forest succession at Dinghushan, Guangdong, China, Botanical Journal of South China 7: 34–42.
30. Zhang BG, Zhuo MN (1985) The physical properties of soil under different forest types in Dinghushan Biosphere Reserve. Tropical and Subtropical Forest Ecosystem Research 3: 1–10 (in Chinese with English abstract).
31. Fang YT, Zhu WX, Mo JM, Zhou GY, Gundersen P (2006) Dynamics of soil inorganic nitrogen and their responses to nitrogen additions in three subtropical forests, south China. Journal of Environmental Sciences 18: 752–759
32. Zhang DQ, Ye WH, Yu QF, Kong GH, Zhang YC (2000) The litter-fall of representative forests of successional series in Dinghushan. Acta Ecologica Sinica 20: 938–944 (in Chinese with English abstract).
33. Mo JM, Brown S, Xue J, Fang Y, Li Z (2006) Response of litter decomposition to simulated N deposition in disturbed, rehabilitated and mature forests in subtropical China. Plant and Soil 282: 135–151.
34. Anderson JM, Ingram JSI (1989) Tropical Soil Biology and Fertility: A Handbook of Methods. Oxford, England: CAB International Wallingford.
35. Bossio DA, Scow KM (1998) Impacts of Carbon and Flooding on Soil Microbial Communities: Phospholipid Fatty Acid Profiles and Substrate Utilization Patterns. Microbial Ecology 35: 265–278.
36. Olson JS (1963) Energy Storage and the Balance of Producers and Decomposers in Ecological Systems. Ecology 44: 322–331.
37. McGroddy ME, Silver WL, de Oliveira RC (2004) The effect of phosphorus availability on decomposition dynamics in a seasonal lowland Amazonian forest. Ecosystems 7: 172–179.
38. Fang H, Mo JM, Peng SL, Li ZA, Wang H (2007) Cumulative effects of nitrogen additions on litter decomposition in three tropical forests in southern China. Plant and Soil 297: 233–242.
39. Chen H, Dong S, Liu L, Ma C, Zhang T, et al. (2013) Effects of experimental nitrogen and phosphorus addition on litter decomposition in an old-growth tropical forest. PLOS ONE (in press).

40. Alhamd L, Arakaki S, Hagihara A (2004) Decomposition of leaf litter of four tree species in a subtropical evergreen broad-leaved forest, Okinawa Island, Japan. Forest Ecology and Management 202: 1–11.
41. Cusack DF, Chou WW, Yang WH, Harmon ME, Silver WL, et al. (2009) Controls on long-term root and leaf litter decomposition in neotropical forests. Global Change Biology 15: 1339–1355.
42. Loranger G, Ponge JF, Imbert D, Lavelle P (2002) Leaf decomposition in two semi-evergreen tropical forests: influence of litter quality. Biology and Fertility of Soils 35: 247–252.
43. Cleveland CC, Reed SC, Townsend AR (2006) Nutrient regulation of organic matter decomposition in a tropical rain forest. Ecology 87: 492–503.
44. Gholz HL, Wedin DA, Smitherman SM, Harmon ME, Parton WJ (2000) Long-term dynamics of pine and hardwood litter in contrasting environments: toward a global model of decomposition. Global Change Biology 6: 751–765.
45. Melillo JM, Aber JD, Muratore JF (1982) Nitrogen and Lignin Control of Hardwood Leaf Litter Decomposition Dynamics. Ecology 63: 621–626.
46. Ponge JF, Arpin P, Vannier G (1993) Collembolan response to experimental perturbations of litter supply in a temperate forest ecosystem. European Journal of Soil Science 29: 141–153.
47. Villalobos-Vega R, Goldstein G, Haridasan M, Franco A, Miralles-Wilhelm F, et al. (2011) Leaf litter manipulations alter soil physicochemical properties and tree growth in a Neotropical savanna. Plant and Soil 346: 385–397.
48. Dzwonko Z, Gawronski S (2002) Effect of litter removal on species richness and acidification of a mixed oak-pine woodland. Biological Conservation 106: 389–398.
49. Mo JM, Brown S, Lenart M, Kong GH (1995) Nutrient Dynamics of a Human-Impacted Pine Forest in a Mab Reserve of Subtropical China. Biotropica 27: 290–304.
50. Li YQ, Xu M, Sun OJ, Cui WC (2004) Effects of root and litter exclusion on soil CO_2 efflux and microbial biomass in wet tropical forests. Soil Biology & Biochemistry 36: 2111–2114.
51. Fahey TJ, Battles JJ, Wilson GF (1998) Responses of early successional northern hardwood forests to changes in nutrient availability. Ecological Monographs 68: 183–212.
52. Fisk MC, Fahey TJ (2001) Microbial biomass and nitrogen cycling responses to fertilization and litter removal in young northern hardwood forests. Biogeochemistry 53: 201–223.
53. Bingeman CW, Varner JE, Martin WP (1953) The Effect of the Addition of Organic Materials on the Decomposition of an Organic Soil. Soil Science Society of America Proceedings 17: 34–38.
54. Kuzyakov Y, Friedel JK, Stahr K (2000) Review of mechanisms and quantification of primingeffects. Soil Biology & Biochemistry 32: 1485–1498.

2

Variable Strength of Forest Stand Attributes and Weather Conditions on the Questing Activity of *Ixodes ricinus* Ticks over Years in Managed Forests

Ralf Lauterbach[1], Konstans Wells[1,2], Robert B. O'Hara[2], Elisabeth K. V. Kalko[1†], Swen C. Renner[1,3*]

1 Institute of Experimental Ecology, University of Ulm, Ulm, Germany, **2** Biodiversity and Climate Research Centre, Frankfurt, Germany, **3** Smithsonian Conservation Biology Institute, National Zoological Park, Front Royal, United States of America

Abstract

Given the ever-increasing human impact through land use and climate change on the environment, we crucially need to achieve a better understanding of those factors that influence the questing activity of ixodid ticks, a major disease-transmitting vector in temperate forests. We investigated variation in the relative questing nymph densities of *Ixodes ricinus* in differently managed forest types for three years (2008–2010) in SW Germany by drag sampling. We used a hierarchical Bayesian modeling approach to examine the relative effects of habitat and weather and to consider possible nested structures of habitat and climate forces. The questing activity of nymphs was considerably larger in young forest successional stages of thicket compared with pole wood and timber stages. Questing nymph density increased markedly with milder winter temperatures. Generally, the relative strength of the various environmental forces on questing nymph density differed across years. In particular, winter temperature had a negative effect on tick activity across sites in 2008 in contrast to the overall effect of temperature across years. Our results suggest that forest management practices have important impacts on questing nymph density. Variable weather conditions, however, might override the effects of forest management practices on the fluctuations and dynamics of tick populations and activity over years, in particular, the preceding winter temperatures. Therefore, robust predictions and the detection of possible interactions and nested structures of habitat and climate forces can only be quantified through the collection of long-term data. Such data are particularly important with regard to future scenarios of forest management and climate warming.

Editor: Robert B. Srygley, USDA-Agricultural Research Service, United States of America

Funding: The work has been funded by the DFG Priority Program 1374 "Infrastructure-Biodiversity-Exploratories" (KA 1241/15-1). The funders had no role in study design, data collection and analysis, decision to publish, or preparation of the manuscript.

Competing Interests: The authors have declared that no competing interests exist.

* E-mail: swen.renner@uni-ulm.de

† Deceased.

Introduction

If we are to understand variation in occurrence patterns and population dynamics, and how these are affected by global change we need to quantify the important environmental forces so that we can improve our predictions and management decisions. The integration of organisms in their environment involves complex interactions with abiotic and biotic factors. For instance, within a network of host-parasite species, the various taxa not only interact with each other, but are also concomitantly affected by the underlying habitat and climate conditions [1,2]. Moreover, variation in environmental conditions has been shown to be a major driver of population fluctuations [3–6]. Although time series analyses of long-term data are a valuable tool for disentangling possible drivers of population fluctuations and spatiotemporal heterogeneity, many studies of particular organisms lack data collected over several years. Hence, studies of the way that focal organism respond to environmental gradients frequently rely on single years or on data sets pooled over a limited number of consecutive sampling years. Generalizations about the response of species to various habitat conditions are also challenged by the inconsistent relative impact of environmental forces over years [7,8]. It is therefore critical to ask how those forces may change over years, and what kind of conclusions can be confidentially drawn from short-term studies.

Ixodid ticks are a major pathogen-transmitting vector in temperate forests. Ticks are ectotherms that spend the largest proportion of their life off-host. They ambush their vertebrate host in vegetation and can survive adverse winter conditions in leaf litter [9]. Abiotic factors, such as temperature, humidity, and elevation, are key factors that determine the development and survival of ticks [10–13]. In particular, microclimatic conditions play an important role in the questing and feeding behavior in ticks, such that questing activity in unfed nymphs decreases in upper vegetation layers with increasingly dry conditions [14].

Another major factor affecting tick occurrence and density is their access to vertebrate host species [15]. Both abiotic conditions and host access, as major sources of variation in tick occurrence and density, interplay in a complex manner with local habitat and vegetation conditions. At the local scale of diverse forest types, vegetation structure has profound effects on the microclimate relevant to ticks [16,17]. Vegetation structures, such as the near-ground vegetation, litter, and the upper soil, are, in turn, largely determined nowadays by forest management practices [18]. In managed and modified temperate production forests, forest stand characteristics such as tree composition, stratification, and the intensity of silvicultural practices are key drivers of species occurrences [19].

Together with habitat conditions, climate (in particular temperature and precipitation) are expected to change considerably during the next few years and decades as part of ongoing global change [20,21]. At a regional scale, climate conditions such as warmer winters and drier spring periods are likely to alter tick survival and distribution [22,23]. On the local scale, tick survival and density can be expected to be mainly affected by the overall regional climate, together with local microclimatic conditions, which are, in turn, driven by local habitat features and topography [24]. Ticks have been found to be less sensitive to drought in deciduous than in coniferous temperate forests [17,25]. Given that managed forest continue to change at the stand level through silvicultural practices [26], and that the management practices induce compositional changes in the vegetation toward drought-tolerant species [27], an understanding of those environmental factors that influence tick occurrence and associated disease risk is important.

In this study, our aim has been to analyze observations from a monitoring program to increase our understanding of the way that climate and habitat conditions have impacted tick density in differently managed forest stands in SW Germany since 2008. One of our main interests has been to investigate the aspect of forest management that most importantly influences questing nymph densities: (1) tree composition, (2) stand type, or (3) forest successional stage. The last-mentioned has rarely been considered to date but comprises a forest classification that is commonly used by forest management authorities. Forest successional stage classification might therefore aid in regional tick and disease risk mapping. We have further assessed whether the relative impact of environmental forces on tick density differs over years and discuss the shortcoming of studies with limited spatiotemporal replicates.

Materials and Methods

Study site and data collection

As part of the interdisciplinary long-term project 'Biodiversity Exploratories' (www.biodiversity-exploratories.de), 50 permanently marked experimental plots in different forest stands in the Schwabian Alb, SW Germany (48°25′ N, 9°23′ E) have been established. All plots comprise relatively homogeneous forest cover over 100×100 m [28]. This region comprises a submontane Karst area (elevation 500–900 m a.s.l.) with annual average temperatures between 6 and 7°C and a vegetation period of about 200 days [28]. The land cover was originally dominated by the common beech (*Fagus sylvatica*) but has substantially changed during the long history of human land use involving the clearance of parts of the forests and various farming practices resulting in a mosaic landscape covered with about 41% forest, 24% grassland, and 22% fields (unpublished data). Forest stands range from the remains of old-growth beech forest, through continuous stands with low-intensive logging, to actively managed deciduous, mixed deciduous-coniferous, and coniferous age-class forests. In addition to beech, common ash (*Fraxinus excelsior*) and sycamore (*Acer pseudoplatanus*) are the most common deciduous tree species in this area. Norway spruce (*Picea abies*) monocultures have been largely planted and comprise about 20% of forest stands.

Questing ticks were sampled in the summer (June to August) between 2008 and 2010 by dragging a white woolen cotton cloth of 80×80 cm in size. The cloth was sewn at one end to a thin metal pole for support and spread over the forest floor and lower ground vegetation. We sampled ticks in a sampling area of 12×12 m in 49 out of 50 of the experimental plots. The same areas were sampled in consecutive years. The site of the sampling areas was chosen next to permanently installed game-exclusion sites (fenced 12×12 m). We systematically avoided areas with dense vegetation growth or large numbers of saplings or piles of dead wood, as the efficiency of drag sampling is largely biased by vegetation height [29]. For sampling, the cloth was systematically dragged over the sampling area, being held as close to the ground as possible with the aid of a telescope rod. We sampled all plots in the same fashion regardless of vegetation type and ground cover. In order to ensure equivalent sampling of all plots, dragging within sampling areas was performed in six intervals of ca. 50 meters (covering different parts of the sampling area) after which we examined the cloth and used fine forceps to remove all ticks. These intervals are relatively large and might not be appropriate to capture as many ticks as possible compared with intervals of only 5 meters [29]. While we acknowledge this shortcoming, we expect the same bias to occur on all plots and thus consider that our sampling provides a meaningful measure of relative tick questing activity. Ticks were sampled during the daytime (08:00 to 17:00) on days without rain. Because of the large-scale design of the research platform and workload, we were not able to sample all plots simultaneously. Different plots were sampled during a period of 53 days in the initial year and of 15 days in the second and third years.

Tick activity changes rapidly throughout the season, and sampling over a longer time period may thus bias density estimates. We found, however, in a preliminary analysis of correcting for sampling time as a random effect in the analysis (outlined below) only a negligible effect of sampling time on the number of ticks collected under various environmental conditions; we thus present our results without this random effect.

Ticks were preserved in 80% ethanol for later identification and counting. For habitat classification, we categorized forest stand composition (based on beech and spruce as the dominant overstory tree species, with a threshold level of 70% of dominant trees) into three categories: beech, beech-mixed, and spruce. We further distinguished successional stages (thicket, pole wood, young timber, old timber; see [28]). We estimated herb coverage of the sampling area as the percentage of area covered with any herbaceous plant. We estimated shrub coverage as the coverage of sampling area with all shrubs and tree saplings between 30 and 200 cm in height. Soils were classified as cambisols or leptosols, based on extensive soil profile examination and were included as a covariate. We expected soil attributes to contribute to microclimatic conditions (I. Schöning, Uni Jena; see [28]).

We used gridded weather data from the German Weather Service (*Deutscher Wetterdienst*, [30]), which interpolates daily measured data from weather stations to a spatial grid with 1 km^2 resolution as monthly means. We calculated averaged temperature and averaged precipitation in the winter (December to January) and spring (March to May) from these monthly data preceding our sampling efforts. Here, we assumed precipitation as an indicator of relative humidity or water saturation deficit, which is known to be an important driver of questing behavior in *I. ricinus* [31].

As the precipitation and temperature in the spring was highly positively correlated, we used only winter temperature and winter and spring precipitation in our analysis. We calculated yearly averaged weather conditions as the yearly means of z-scores of weather data.

For quantitative analysis of tick questing activity in relation to environmental covariates, we considered only nymphs, as larvae are commonly spatially clustered.

All data are deposited in the Biodiversity Exploratories Information System (BExIS; http://www.biodiversity-exploratories.de/) and are available upon request from the authors according to project policy.

Data analysis

Hierarchical and multilevel modeling approaches allow inference to be made with regard to group-specific parameters, whilst also estimating the variation between the groups. This allows decomposition of the variation among potential covariates and indicates those that are important for explaining the overall variation [32,33]. With such models, one can explicitly focus on the variance in both the spatial and temporal dimension. In addition, uncertainty in predicting effects of covariates that might themselves vary in relative effects among spatiotemporal dimensions are adequately considered in such models.

We employed a generalized linear mixed model based on Bayesian Markov-Chain Monte Carlo methods to estimate the relative effects of environmental covariates on the spatiotemporal variation in questing nymph density. We used the software package MCMCglmm [34] in R [35]. Within this framework, some parameters describing the corresponding effects of explanatory variables were treated as random variables with assumed hyper-prior distribution, and so posterior distribution of parameters were estimable together with the possibility of specifying the (co)variance of the random effects.

We assumed a generalized linear regression model for investigating the link between the variability in questing nymph density and the underlying environmental covariates.

In a first step, the density of questing nymphs at various sites in the different years can be described as:

Number of nymphs ~ (1|Sites) + Forest Type + Successional Stage + Herb Coverage + Shrub Coverage + Soil + Elevation + Winter Temperature, Winter Precipitation + Spring Precipitation.

where (1|Sites) corresponds to a variable intercept ('mean tick number') which is allowed to vary over sites fitted as random effects. In addition, constant coefficients are estimated for environmental variables. Random noise is added from an assumed Poisson distribution (note that the MCMC approach accounts for the overdispersion in data).

The hierarchical model structure can be most easily explained with the variable intercept α_{Sites} assumption: the level of allowing the mean to vary across sites from a common mean μ_α is modeled from a normal distribution with the respective variance σ_α^2 by $\alpha_{Sites} \sim N(\mu_\alpha, \sigma_\alpha^2)$.

The size of σ_α^2 allows the identification of the divergence of site-specific means from the common mean, i.e., $\sigma_\alpha^2 = 0$ leads to complete pooling across sites, while $\sigma_\alpha^2 >> 0$ suggests that mean tick densities vary greatly across sites, resulting in different posterior distributions for site-specific estimates [33].

As a preliminary characterization of whether overall yearly averaged winter temperature (in a colder winter, for example, temperatures generally dropped at all sites, and our data included more variation among years than within years in weather covariates) have an impact on the relative effects of forest type and stage, we fitted additional random effects by including yearly averaged winter temperature into the intercept and allowing them to vary by forest type and stage:

Number of nymphs ~ (1|Sites) + Forest Type + Successional Stage + Herb Coverage + Shrub Coverage + Soil + Elevation + Winter Temperature + Winter Precipitation + Spring Precipitation + (Yearly Averaged Winter Temperature| Forest Type + Successional Stage).

In principal, in such a random regression framework, an increase in the variance for the added random effects means that the identity of the different habitat types might explain the larger amounts of variation depending on yearly weather conditions.

We also fitted models for single year datasets with removed random effects.

We standardized all numerical explanatory variables so that they each had a mean of zero and standard deviation of one. Posteriors were gathered by running 350,000 iterations, discarding 100,000 iterations, and thinning by 25, resulting in 10,000 draws for posterior estimates. Convergence and mixing was checked visually by running two parallel chains. Results are given as the posterior mode and 95% highest posterior density as the credible interval (CI).

Results

During the three years of sampling, we collected a total of 3,355 larvae and 855 nymphs of the species *Ixodes ricinus*.

In the full models for nymphs (analysis of larvae is not considered because of their clustered occurrence), the posterior densities revealed significant effects of habitat features with fewer nymphs being sampled in older forests stands of timber and pole wood compared with the earlier stage of thicket (Figure 1), which accounted for 17% of the variation in abundances (nymphs). Questing nymph density significantly decreased with increasing herb coverage (explaining 25% of variance).

In the full models for nymphs, winter temperature was associated with an increase in density, with a change of 1°C increasing relative questing nymph density by 82% (Figure 2) and explaining 12% of the variance.

The random effect of yearly averaged weather conditions revealed no considerable effects on habitat identity effects, and we

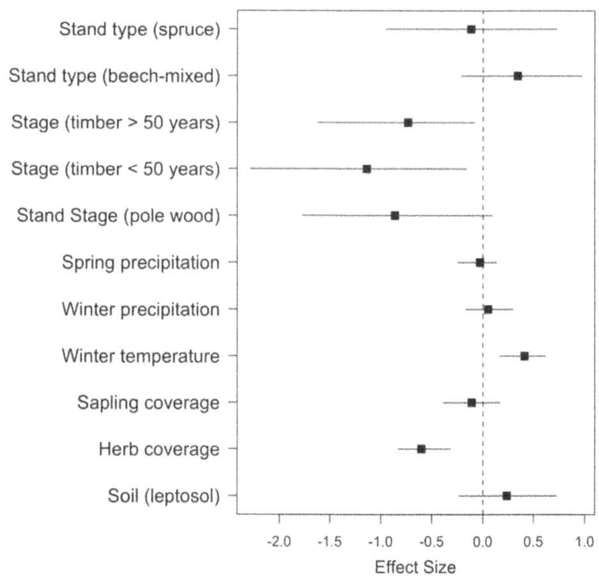

Figure 1. Effect sizes of environmental covariates on questing nymph densities from the full Bayesian MCMC generalized linear model comprising three years of sampling (2008–2010). Effect sizes are given as posterior modes and 95% highest posterior density estimates. Note: the baseline for stand type is 'beech' and for stand stage 'thicket'.

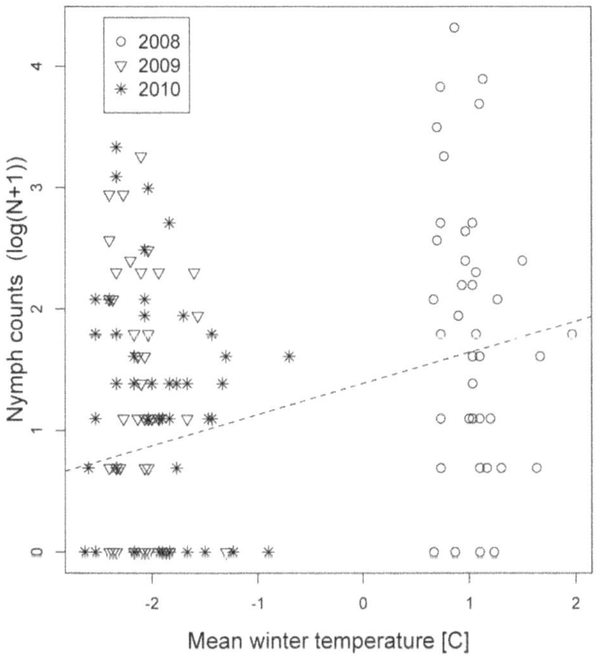

Figure 2. Relationships between the number of counted tick nymphs and mean winter temperatures from 2008–2010. The Y-axis comprises log-transformed (log(counts+1)) count data. The winter in 2008 was considerably warmer than the two following years. Hence, the respective values cluster on the right site of plots. Simple regression lines from the linear model relationships show the trend of increasing nymph densities with milder winter temperatures.

were hence not able to show possible interaction effects between habitat and weather. Since the posterior distributions largely resembled prior distributions, this was most likely attributable to the small sample size of three years.

Single-year analysis of questing nymph density revealed the largest effect size for winter temperature in 2008 with fewer ticks at increasing temperature, explaining 88% of the variance in nymph activity. Notably, this effect contrasts with the winter temperature effect on nymphs from the full model. In the latter, the temperature effect is mainly driven by the pronounced overall yearly differences (Figure 3). In contrast, the effect size was largest for the spring precipitation in 2010, with more nymphs being sampled with increasing spring precipitation, accounting for 57% of the variation in nymph activity during 2010; in 2010, spring precipitation was considerably lower than in the two preceding years.

Discussion

Our study has identified forest successional stage as an important predictor of nymph questing density of *Ixodes ricinus* in production forests in SW Germany. Nymph questing density was strongly linked to the preceding winter temperature, especially to overall yearly changes in weather conditions. The variable prediction strength of environmental covariates across years, the contrasting effect of winter temperature on nymphs from a single year, and the pronounced effects of weather across years suggest that more robust prediction can be achieved through long study periods only. We highlight that this study only examined the relative tick questing density in different forest stands, whereas the strong dependence of tick survival and reproduction on host accessibility makes it clear that tick density is not independent of host abundance and diversity from local to regional scale. Increasing deer abundances in Europe may have caused increases in tick populations recently, and both abiotic and biotic environmental factors are likely to have caused recent increases in tick populations [36].

Increased nymph densities in younger forest stands and the weak effect of stand type is surprising, as tick abundance has previously been suggested to be favored by canopy closure, which is higher in older forest stands. In contrast, a pronounced shrub and herbaceous layer, which was not present at our research sites in younger forest stands, and particular litter types have been suggested to favor *I. ricinus* abundance and survival elsewhere in Europe [11,37,38] and *I. scapularis* in North America [17,39]. Whereas forest classification into stand composition and successional stages comprises rather coarse measures, this classification can be applied at a larger spatial scale through digital forest inventories and management maps. These are readily available for most managed forest stands in Central Europe and elsewhere (see also [11,40]). Managed forests mostly consist of patches with homogeneous tree composition, age structure, and vertical organization. Simplification of the key characteristics might therefore help to predict the consequences for the occurrence and distribution of species on a larger scale and under future scenarios of forest development and global change. Other factors, such as herb and shrub coverage, might influence tick occurrence and survival by being associated with more suitable microhabitat [11,38] but are more difficult to extrapolate to a larger spatial scale. Herb and shrub coverage are largely variable within and across different forest stands.

In contrast to our expectation that tick densities rise with increasing herb and shrub cover, we found the opposite trend for herb coverage. Fewer nymphs were associated with increasing herb cover. We have no explanation for this finding. Perhaps other factors linked with herb coverage, such as specific habitat preferences of rodent host species where herb coverage dominates, might have led to these results. Larger tick populations in young forest stands, for example, might be linked to an increased abundance of rodent hosts such as *Myodes glareolus* [41] or *Apodemus flavicollis* [42] in these forest. Wood mice *Apodemus sylvaticus* were found to aggregate under *Rhododendron* shrubs in southern England, but questing tick density was lower under these shrubs compared to surrounding forest habitat [43]. Variation in host abundance blur the possible impact of habitat features such as herb coverage or other environmental variables. Notably, in addition to the possible role of host densities on tick abundance and questing behavior, the tick burden of rodent hosts was also found to vary with relative humidity and vegetation cover at other study sites in Germany [44]. Moreover, sampling bias such as the less effective sampling of areas with high vegetation might well influence drag sampling results [29].

Our study results are preliminary in that our coarse sampling procedure provides only a first estimate of relative tick questing densities. The number of drag-sampled tick individuals cannot be used to distinguish local tick density and their questing behavior, as both factors determine the number of ticks that are counted. Further, we have no information on important factors such as the abundance of potential host species. The sampling period of three years does indeed highlight that the strength of environmental factors on tick density varies over years, but possible interactions and the covariation of habitat, weather, and host species are hardly detectable from such short-term studies.

Nevertheless, we expect our finding of increased questing nymph densities in young forest stands to be important as an

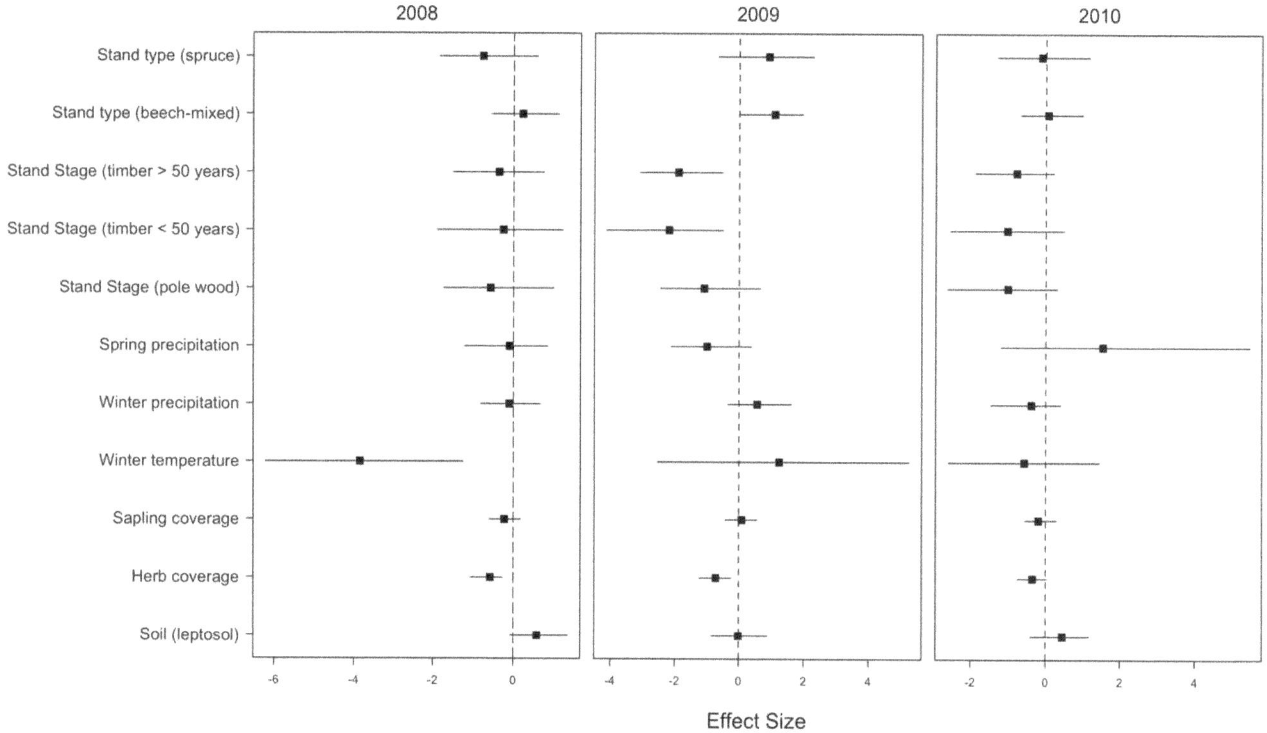

Figure 3. Effect sizes of environmental covariates on nymph densities in each of the three sampling years 2008–2010 as estimated by single-year models. Effect sizes are given as posterior modes and 95% highest posterior density estimates. The baseline for stand type is 'beech', for stand stage 'thicket' and for soil 'cambisols'.

indication that forest management practices have an impact on tick occurrence and therefore also possibly on disease transmission risk on the stand level. In practice, the large number of abiotic and biotic factors that could influence tick questing density suggests that systematic sampling with an accurate sampling procedure and replicates over a time-scale of years is needed. Such a sampling method will directly link local tick abundance to various potential environmental forces. Integrative monitoring efforts ideally would also include surveys that allow accurate abundance estimates of various vertebrate host species, such as rodents and deer [43,45].

This is particularly important, since to date, the environmental factors suggested to influence tick population dynamics are ranked inconsistently in their importance [46,47]. The combination of diverse aspects in a single analysis is also a prerequisite for distinguishing various kinds of direct and indirect relationships typically shaping interactions in disparate environments [48,49]. Only when the different sources of variation are determined, can we understand the genuine functional relationships between interacting organisms and various environmental attributes, which in turn ultimately determine questing tick density in time and space. Variability in the relative importance of environmental forces during the course of our three-year sample study is one major obstacle to disentangling the influence of even a limited number of environmental forces on tick density. We have observed, for example, a significant increase in questing nymph density after the mild winter in 2008. Milder winter temperatures have also been identified as important aspects for the northward expansion of *I. ricinus* [50]. Our short sampling period of three years however urges caution in concluding that a direct relationship exists between winter temperature and tick density. Notably, the increasing winter temperature across sites in 2008 resulted in fewer nymphs in contrast to the increased tick densities associated with the milder yearly winter temperature. This demonstrates that conclusions from single-year studies might differ from long-term studies [9], and moreover, that the effects of environmental forces might differ in spatiotemporal dimensions.

This finding and, in particular, our conclusion that milder weather conditions favor higher tick densities need further research. Such aspects in general have rarely been adequately examined with sufficiently large samples [39]. Notably, the gridded weather data extrapolated from regional weather stations comprise so far only a coarse and preliminary measure. These coarse-grained data might therefore not fully account for actual gradients in microclimatic conditions. Moreover, microclimates in forests are well known to be influenced by local stand attributes [51,52], and the true gradient in microclimatic conditions might therefore be even more pronounced. Tick questing behavior and activity is strongly linked to temperature and humidity [53]. Pronounced yearly weather changes might therefore also alter the relative effect of habitat features over periods of years. During heat waves in the summer, for example, beech forests have been found to have a more moderating effect on maximum daytime temperature than spruce forests [54]. Spruce forests are relatively warmer during minimum temperatures as a consequence of the different amounts of radiation reaching the ground. In young forest stands of thicket and pole wood, which lack pronounced canopy layers, the level of radiation reaching the ground is high and fosters a rise in soil and air temperatures and a decrease in ground humidity during hot weather.

We suggest that the possible variation in the effects of environmental factors on tick density over years needs to be addressed in future research, and that possible covariances in the effects of environmental factors should be considered. If the effective weather and microclimatic conditions differ across habitat types as discussed above, a likely (but to our knowledge poorly investigated) consequence is that the relative effects of habitat features such as tree composition and forest stage will differ with the varying weather conditions. Given the technical limits of accurately fitting the interactions and dynamics of population fluctuations from two to three years of data, as is common for many tick studies, the lack of meaningful random effects in our study is not a surprise. In addition to the need for the monitoring with sufficiently replicated sampling in time and space, field studies of tick abundance and distribution would benefit from research frameworks that assess a suite of relevant and standardized abiotic and biotic covariates to improve the disentanglement of the relative strength and possible relationships between the various factors recorded as influencing tick occurrence to date.

Acknowledgments

The late Elisabeth Kalko unexpectedly past away to our great disbelief and sadness during the preparation of this manuscript.

We thank the managers of the Biodiversity Exploratories, S Gockel, A Hemp, M Gorke, and S Pfeiffer for their work in maintaining the plot and project infrastructure, and F Buscot, M Fischer, D Hessenmöller, KE Linsenmair, J Nieschulze, D Prati, I Schöning, E-D Schulze, M Ayasse, and WW Weisser for their role in setting up the Biodiversity Exploratories project. J Dolezil, J Hailer, M Fellendorf, and H Stegmaier are thanked for technical support. We are grateful to the Deutsche Wetterdienst and J Schröder from the Biodiversity and Climate Research Centre (BiK-F) for providing and preparing weather data. Field-work permits were issued by the responsible state environmental offices of Baden-Württemberg according to §72 BbgNatSchG. We thank the editor and unknown reviewers for their work and improvement of previous versions of the manuscript.

Author Contributions

Conceived and designed the experiments: EKVK SCR KW. Performed the experiments: RL. Analyzed the data: KW RBO. Contributed reagents/materials/analysis tools: EKVK SCR. Wrote the paper: RL KW RBO EKVK SCR.

References

1. Gilbert L (2010) Altitudinal patterns of tick and host abundance: a potential role for climate change in regulating tick-borne diseases? Oecologia 162: 217–225.
2. Ostfeld RS (2009) Climate change and the distribution and intensity of infectious diseases. Ecology 90: 903–905.
3. Sæther BE (1997) Environmental stochasticity and population dynamics of large herbivores: A search for mechanisms. Trends in Ecology and Evolution 12: 143–147.
4. Bjørnstad ON, Grenfell BT (2001) Noisy clockwork: Time series analysis of population fluctuations in animals. Science 293: 638–643.
5. Boyce MS, Haridas CV, Lee CT, Boggs CL, Bruna EM, et al. (2006) Demography in an increasingly variable world. Trends in Ecology and Evolution 21: 141–148.
6. Mutshinda CM, O'Hara RB, Woiwod IP (2009) What drives community dynamics? Proceedings of the Royal Society B: Biological Sciences 276: 2923–2929.
7. Lõhmus A (2003) Are certain habitats better every year? A review and a case study on birds of prey. Ecography 26: 545–552.
8. Wells K, Böhm SM, Boch S, Fischer M, Kalko EKV (2011) Local and landscape-scale forest attributes differ in their impact on bird assemblages across years in forest production landscapes. Basic and Applied Ecology 12: 97–106.
9. Randolph SE (2004) Tick ecology: processes and patterns behind the epidemiological risk posed by ixodid ticks as vectors. Parasitology 129: S37–S65.
10. Perret J-L, Guigoz E, Rais O, Gern L (2000) Influence of saturation deficit and temperature on *Ixodes ricinus* tick questing activity in a Lyme borreliosis-endemic area (Switzerland). Parasitology Research 86: 554–557.
11. Estrada-Peña A (2001) Distribution, abundance, and habitat preferences of *Ixodes ricinus* (Acari: Ixodidae) in northern Spain. Journal of Medical Entomology 38: 361–370.
12. Eisen RJ, Eisen L, Castro MB, Lane RS (2003) Environmentally related variability in risk of exposure to Lyme disease spirochetes in Northern California: Effect of climatic conditions and habitat type. Environmental Entomology 32: 1010–1018.
13. Ogden NH, Maarouf A, Barker IK, Bigras-Poulin M, Lindsay LR, et al. (2006) Climate change and the potential for range expansion of the Lyme disease vector *Ixodes scapularis* in Canada. International Journal for Parasitology 36: 63–70.
14. Randolph SE, Storey K (1999) Impact of microclimate on immature tick-rodent host interactions (Acari: Ixodidae): implications for parasite transmission. Journal of Medical Entomology 36: 741–748.
15. LoGiudice K, Duerr STK, Newhouse MJ, Schmidt KA, Killilea ME, et al. (2008) Impact of host community composition on Lyme disease risk. Ecology 89: 2841–2849.
16. Daniel M, Dusbábek F (1994) Micrometeorological and microhabitat factors affecting maintenance and dissemination of tick-borne diseases in the environment. In: Sonenshine DE, Mather TN, editors. Ecological Dynamics of Tick-Borne Zoonoses: Oxford University Press. 91–138.
17. Schulze TL, Jordan RA (2005) Influence of meso- and microscale habitat structure on focal distribution of sympatric *Ixodes scapularis* and *Amblyomma americanum* (Acari: Ixodidae). Journal of Medical Entomology 42: 285–294.
18. Lindenmayer D, Franklin JF (2002) Conserving forest biodiversity: a comprehensive multiscaled approach. Washington DC: Island Press. 368 p.
19. Paillet Y, Bergès L, Hjältén J, Ódor P, Avon C, et al. (2010) Biodiversity differences between managed and unmanaged forests: Meta-analysis of species richness in Europe. Conservation Biology 24: 101–112.
20. Weltzin JF, Loik ME, Schwinning S, Williams DG, Fay PA, et al. (2003) Assessing the response of terrestrial ecosystems to potential changes in precipitation. BioScience 53: 941–952.
21. Solomon S, Plattner GK, Knutti R, Friedlingstein P (2009) Irreversible climate change due to carbon dioxide emissions. Proceedings of the National Academy of Sciences of the United States of America 106: 1704–1709.
22. Brownstein JS, Holford TR, Fish D (2005) Effect of climate change on Lyme disease risk in North America. EcoHealth 2: 38–46.
23. Ogden NH, St-Onge L, Barker IK, Brazeau S, Bigras-Poulin M, et al. (2008) Risk maps for range expansion of the Lyme disease vector, *Ixodes scapularis*, in Canada now and with climate change. International Journal of Health Geographics 7: 24.
24. Suggitt AJ, Gillingham PK, Hill JK, Huntley B, Kunin WE, et al. (2011) Habitat microclimates drive fine-scale variation in extreme temperatures. Oikos 120: 1–8.
25. Taylor AR, Wolters V (2005) Responses of oribatid mite communities to summer drought: The influence of litter type and quality. Soil Biology & Biochemistry 37: 2117–2130.
26. Bauhus J, Puettmann K, Messier C (2009) Silviculture for old-growth attributes. Forest Ecology and Management 258: 525–537.
27. Lasch P, Lindner M, Erhard M, Suckow F, Wenzel A (2002) Regional impact assessment on forest structure and functions under climate change – The Brandenburg case study. Forest Ecology and Management 162: 73–86.
28. Fischer M, Bossdorf O, Hänsel F, Hemp A, Hessenmöller D, et al. (2010) Implementing large-scale and long-term functional biodiversity research: The Biodiversity Exploratories. Basic and Applied Ecology 11: 473–485.
29. Dobson ADM, Taylor JL, Randolph SE (2011) Tick (*Ixodes ricinus*) abundance and seasonality at recreational sites in the UK: hazards in relation to fine-scale habitat types revealed by complementary sampling methods. Ticks and Tick-borne Diseases 2: 67–74.
30. Müller-Westermeier G (1995) Numerisches Verfahren zu Erstellung klimatologischer Karten. Offenbach, Germany: Deutscher Wetterdienst.
31. Perret J-L, Guerin PM, Diehl PA, Vlimant MI, Gern L (2003) Darkness induces mobility, and saturation deficit limits questing duration, in the tick *Ixodes ricinus*. Journal of Experimental Biology 206: 1809–1815.
32. Wikle CK (2003) Hierarchical Bayesian models for predicting the spread of ecological processes. Ecology 84: 1382–1394.
33. Gelman A, Hill J (2007) Data analysis using regression and multilevel/hierarchical models. New York, USA. Cambridge University Press. 625 p.
34. Hadfield JD (2010) MCMC methods for multi-response generalized linear mixed models: The MCMCglmm R package. Journal of Statistical Software 33: 1–22.
35. R Development Core Team (2010) R: A language and environment for statistical computing. Vienna, Austria: R Foundation for Statistical Computing. http://cran.r-project.org/.
36. Dobson ADM, Randolph SE (2011) Modelling the effects of recent changes in climate, host density and acaricide treatments on population dynamics of *Ixodes ricinus* in the UK. Journal of Applied Ecology 48: 1029–1037.
37. Lindström A, Jaenson TGT (2003) Distribution of the common tick, *Ixodes ricinus* (Acari: Ixodidae), in different vegetation types in Southern Sweden. Journal of Medical Entomology 40: 375–378.
38. Hubálek Z, Halouzka J, Juricová Z, Sikutova S, Rudolf I (2006) Effect of forest clearing on the abundance of *Ixodes ricinus* ticks and the prevalence of *Borrelia burgdorferi* s. l. Medical & Veterinary Entomology 20: 166–172.

39. Lubelczyk CB, Elias SP, Rand PW, Holman MS, Lacombe EH, et al. (2004) Habitat associations of *Ixodes scapularis* (Acari: Ixodidae) in Maine. Environmental Entomology 33: 900–906.
40. Daniel M, Kolar J, Zeman P, Pavelka K, Sadlo J (1998) Predictive map of *Ixodes ricinus* high-incidence habitats and tick-borne encephalitis risk assessment using satellite data. Experimental & Applied Acarology 22: 417–433.
41. Ecke F, Löfgren O, Sörlin D (2002) Population dynamics of small mammals in relation to forest age and structural habitat factors in northern Sweden. Journal of Applied Ecology 39: 781–792.
42. Rosa R, Pugliese A, Ghosh M, Perkins SE, Rizzoli A (2007) Temporal variation of *Ixodes ricinus* intensity on the rodent host *Apodemus flavicollis* in relation to local climate and host dynamics. Vector-Borne & Zoonotic Diseases 7: 285–295.
43. Malo AF, Godsall B, Prebble C, Grange Z, Mccandless S, et al. (2012) Positive effects of an invasive shrub on aggregation and abundance of a native small rodent. Behavioral Ecology doi: 10.1093/beheco/ars202.
44. Kiffner C, Vor T, Hagedorn P, Niedrig M, Ruhe F (2011) Factors affecting patterns of tick parasitism on forest rodents in tick-borne encephalitis risk areas, Germany. Parasitology Research 108:323–335.
45. Gilbert L, Maffey GL, Ramsay SL, Hester AJ (2012) The effect of deer Management on the abundance of *Ixodes ricinus* in Scotland. Ecological Applications 22: 658–667.
46. Ostfeld RS, Canham CD, Oggenfuss K, Winchcombe RJ, Keesing F (2006) Climate, deer, rodents, and acorns as determinants of variation in Lyme-disease risk. PLoS Biology 4: 1058–1068.
47. Ostfeld RS (2011) Lyme disease: the ecology of a complex system. New York: Oxford University Press.
48. Goudard A, Loreau M (2008) Nontrophic interactions, biodiversity, and ecosystem functioning: an interaction web model. The American Naturalist 171: 91–106.
49. Tylianakis JM, Tscharntke T, Lewis OT (2007) Habitat modification alters the structure of tropical host-parasitoid food webs. Nature 445: 202–205.
50. Lindgren E, Talleklint L, Polfeldt T (2000) Impact of climatic change on the northern latitude limit and population density of the disease-transmitting European tick *Ixodes ricinus*. Environmental Health Perspectives 108: 119–123.
51. Chen JQ, Saunders SC, Crow TR, Naiman RJ, Brosofske KD, et al. (1999) Microclimate in forest ecosystem and landscape ecology – Variations in local climate can be used to monitor and compare the effects of different management regimes. BioScience 49: 288–297.
52. Grimmond CSB, Robeson SM, Schoof JT (2000) Spatial variability of micro-climatic conditions within a mid-latitude deciduous forest. Climate Research 15: 137–149.
53. Hubálek Z, Halouzka J, Juoicová Z (2003) Host-seeking activity of ixodid ticks in relation to weather variables. Journal of Vector Ecology 28: 159–165.
54. Renaud V, Rebetez M (2009) Comparison between open-site and below-canopy climatic conditions in Switzerland during the exceptionally hot summer of 2003. Agricultural and Forest Meteorology 149: 873–880.

3

Succession Influences Wild Bees in a Temperate Forest Landscape: The Value of Early Successional Stages in Naturally Regenerated and Planted Forests

Hisatomo Taki[1]*, Isamu Okochi[2], Kimiko Okabe[1], Takenari Inoue[3], Hideaki Goto[4], Takeshi Matsumura[5], Shun'ichi Makino[2]

1 Department of Forest Entomology, Forestry and Forest Products Research Institute, Tsukuba, Ibaraki, Japan, **2** Forestry and Forest Products Research Institute, Tsukuba, Ibaraki, Japan, **3** Tama Forest Science Garden, Forestry and Forest Products Research Institute, Hachioji, Tokyo, Japan, **4** Kyushu Research Centre, Forestry and Forest Products Research Institute, Kumamoto, Kumamoto, Japan, **5** Nasushiobara, Tochigi, Japan

Abstract

In many temperate terrestrial forest ecosystems, both natural human disturbances drive the reestablishment of forests. Succession in plant communities, in addition to reforestation following the creation of open sites through harvesting or natural disturbances, can affect forest faunal assemblages. Wild bees perform an important ecosystem function in human-altered and natural or seminatural ecosystems, as they are essential pollinators for both crops and wild flowering plants. To maintain high abundance and species richness for pollination services, it is important to conserve and create seminatural and natural land cover with optimal successional stages for wild bees. We examined the effects of forest succession on wild bees. In particular, we evaluated the importance of early successional stages for bees, which has been suspected but not previously demonstrated. A range of successional stages, between 1 and 178 years old, were examined in naturally regenerated and planted forests. In total 4465 wild bee individuals, representing 113 species, were captured. Results for total bees, solitary bees, and cleptoparasitic bees in both naturally regenerated and planted conifer forests indicated a higher abundance and species richness in the early successional stages. However, higher abundance and species richness of social bees in naturally regenerated forest were observed as the successional stages progressed, whereas the abundance of social bees in conifer planted forest showed a concave-shaped relationship when plotted. The results suggest that early successional stages of both naturally regenerated and conifer planted forest maintain a high abundance and species richness of solitary bees and their cleptoparasitic bees, although social bees respond differently in the early successional stages. This may imply that, in some cases, active forest stand management policies, such as the clear-cutting of planted forests for timber production, would create early successional habitats, leading to significant positive effects for bees in general.

Editor: Jeff Ollerton, University of Northampton, United Kingdom

Funding: This study was funded as a project of the Research Institute for Humanity and Nature and was also supported in part by the Global Environment Research Fund (S-9) of the Ministry of the Environment, Japan. The funders had no role in study design, data collection and analysis, decision to publish, or preparation of the manuscript.

Competing Interests: The authors have declared that no competing interests exist.

* E-mail: htaki@affrc.go.jp

Introduction

In many temperate terrestrial forest ecosystems, natural disturbances such as avalanches, windstorms, fires, and floods, and human disturbance including tree harvesting, drive the reestablishment of forests [1]. Succession in plant communities, in addition to reforestation following the creation of open sites through harvesting or natural disturbances, can affect forest faunal assemblages. For example, the intermediate disturbance hypothesis suggests that shifts in species richness are likely and the midway stage of succession would contain the highest species richness [2,3]. Previous studies have consistently demonstrated that the species richness of various taxa peak at different times as succession proceeds, indicating that the diverse taxa within a community include species that prefer specific successional stages [4,5]. Among the wide range of forest successional stages, the conservational and ecological importance and the value of the early stages of succession may require more attention [6]. The early successional stage can be characterized by high productivity of plants, compound floral and faunal food webs, and large nutrient changes [6]. Previous studies have shown that this stage in a temperate forest region can have a great diversity in abundance and species richness of flora and fauna [7,8].

Wild bees perform an important ecosystem function in human-altered and natural or seminatural ecosystems as bee species are important pollinators for both crops and wild flowering plants [9]. Pollinators including wild bees can increase the production of approximately 75% of the 115 most essential crops worldwide [10], and the proportion of flowering plants pollinated by animals is estimated to be 87.5% of all flowering plants [11]. The conservation and creation of seminatural and natural land cover at optimal successional stages for wild bees are important to maintain a high abundance and species richness for pollination services [12,13]. Examples of this include human-dominated landscapes

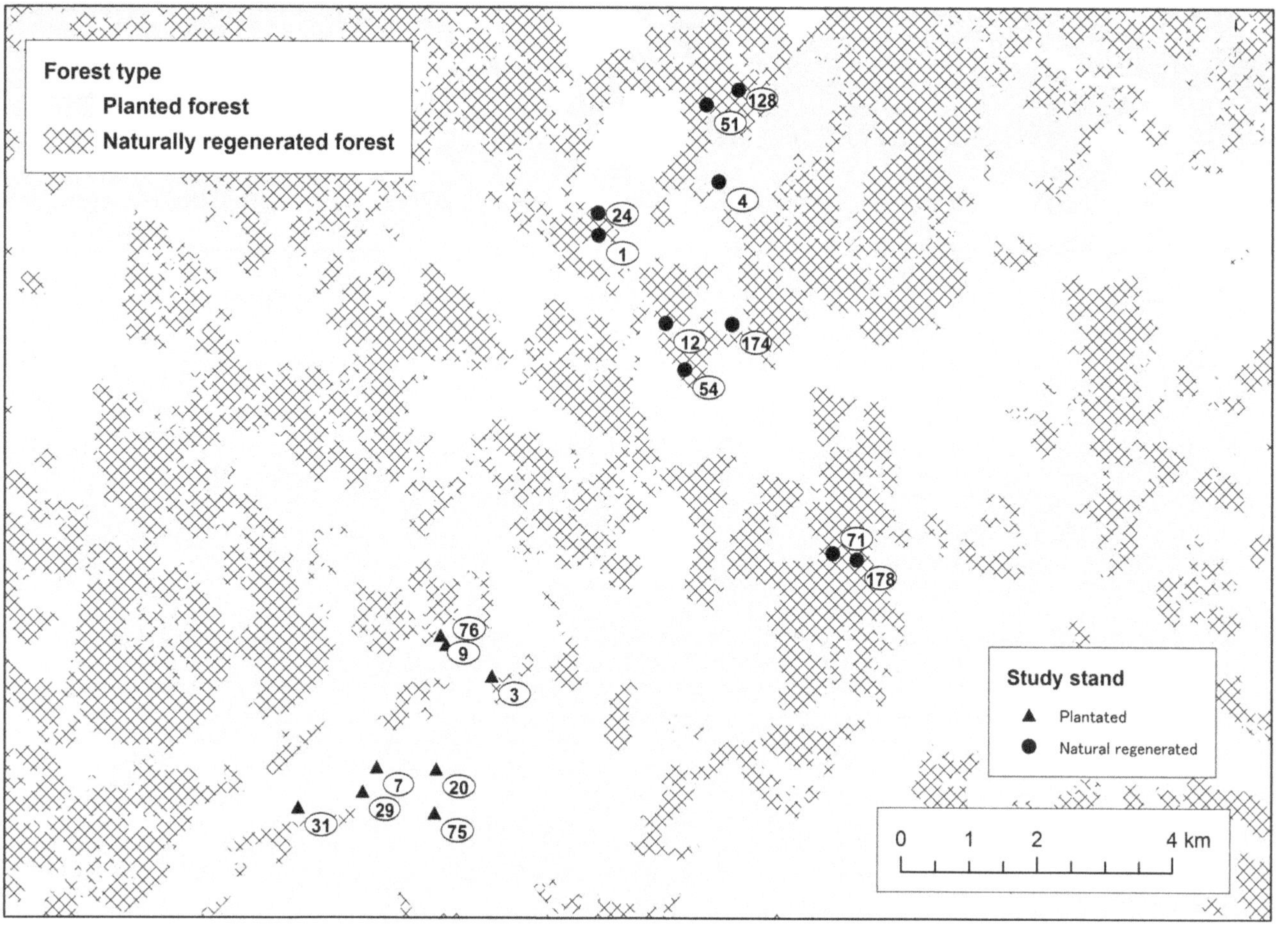

Figure 1. The study region in the northern part of Ibaraki Prefecture in central Japan (approximately 36°50′–7′N, 140°32′–8′E; 500–50 m above sea level) and locations of the study stands within areas of planted conifers and naturally regenerated forests. Numbers indicate reforestation age.

Table 1. Study sites, 10 naturally regenerated and 8 conifer (*Cryptomeria japonica*) planted forest stands, in the northern part of Ibaraki Prefecture in central Japan (approximately 36°50′–57′N, 140°32′–38′E; 500–850 m above sea level).

ID	Reforestation type	Age	Area (ha)
N1	Natural	1	2.53
N4	Natural	4	4.91
N12	Natural	12	4.13
N24	Natural	24	23.56
N51	Natural	51	10.37
N54	Natural	54	14.72
N71	Natural	71	19.11
N128	Natural	128	32.49
N174	Natural	174	11.75
N178	Natural	178	9.29
C3	Conifer planted	3	4.28
C7	Conifer planted	7	6.04
C9	Conifer planted	9	4.9
C20	Conifer planted	20	4.88
C29	Conifer planted	29	14.29
C31	Conifer planted	31	12.07
C75	Conifer planted	75	3.13
C76	Conifer planted	76	2.55
	Mean	52.61	10.28

Stand ages were based on tentative estimations recorded in the National Forest Inventory of Japan.

with a long history of human land use, among which grasslands are a notable seminatural habitat. One study on wild bees, covering a period of 1 to 5 years of secondary succession of European fallow land, found the highest species richness after 2 years [14]. However, another study of wild bees in European limestone quarries found little difference in abundance and species richness among habitats ranging from 1 to 121 years old [15]. This

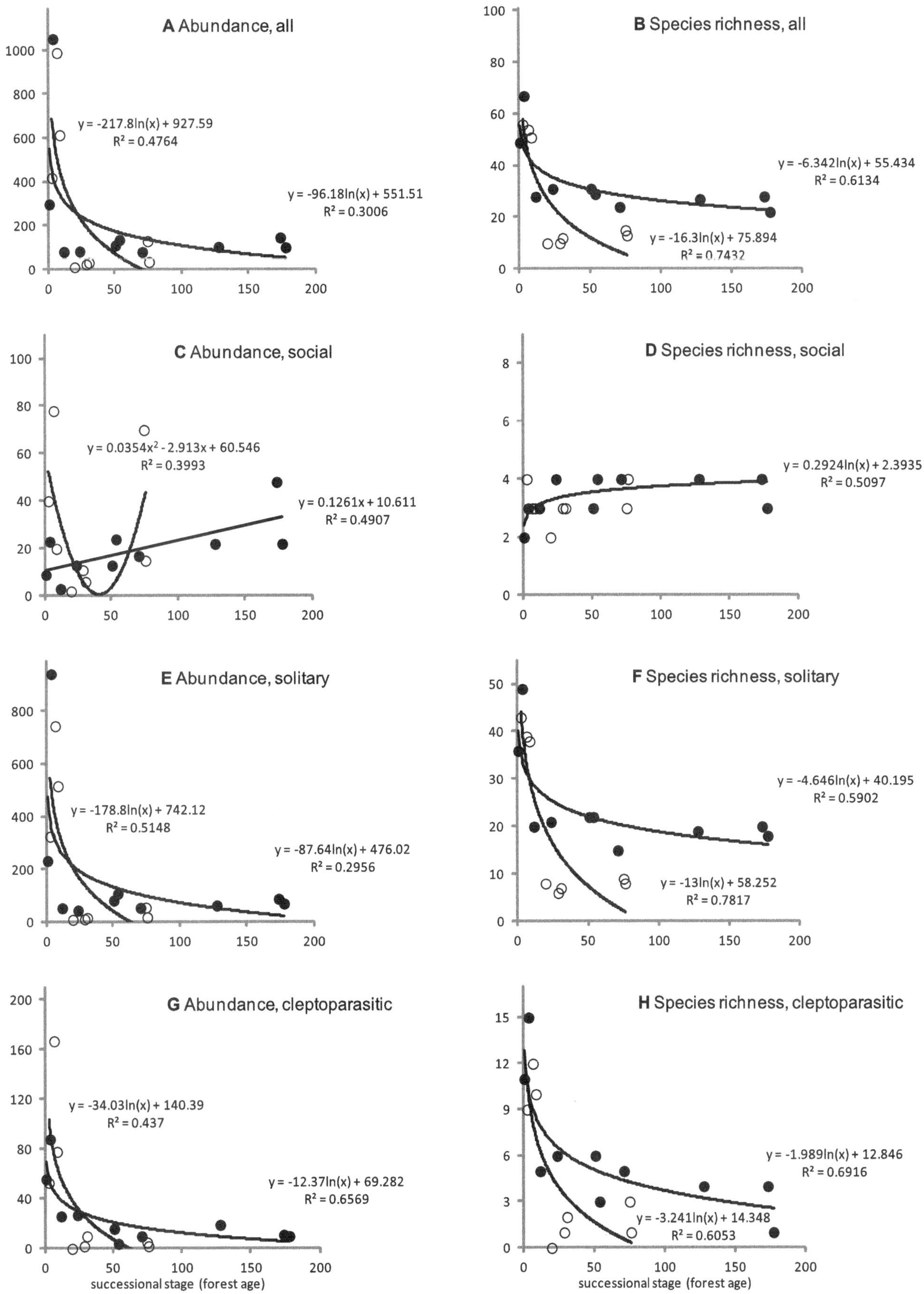
A Abundance, all
y = -217.8ln(x) + 927.59
R² = 0.4764
y = -96.18ln(x) + 551.51
R² = 0.3006
B Species richness, all
y = -6.342ln(x) + 55.434
R² = 0.6134
y = -16.3ln(x) + 75.894
R² = 0.7432
C Abundance, social
y = 0.0354x² - 2.913x + 60.546
R² = 0.3993
y = 0.1261x + 10.611
R² = 0.4907
D Species richness, social
y = 0.2924ln(x) + 2.3935
R² = 0.5097
E Abundance, solitary
y = -178.8ln(x) + 742.12
R² = 0.5148
y = -87.64ln(x) + 476.02
R² = 0.2956
F Species richness, solitary
y = -4.646ln(x) + 40.195
R² = 0.5902
y = -13ln(x) + 58.252
R² = 0.7817
G Abundance, cleptoparasitic
y = -34.03ln(x) + 140.39
R² = 0.437
y = -12.37ln(x) + 69.282
R² = 0.6569
successional stage (forest age)
H Species richness, cleptoparasitic
y = -1.989ln(x) + 12.846
R² = 0.6916
y = -3.241ln(x) + 14.348
R² = 0.6053
successional stage (forest age)

Figure 2. Effects of different successional stages on the abundance and species richness of bees, showing total (A and B), social (C and D), solitary (E and E), and cleptoparasitic (G and H) assemblages, in naturally regenerated and conifer (*Cryptomeria japonica*) planted forests. Y axes indicate the numbers of collected individuals for abundance and the number of collected species for species richness, respectively. Solid circles represent naturally regenerated forest, and open circles represent planted conifer forest.

indicates that even newly created and permanent grassland-like habitats provide important habitats for wild bees

In regions dominated by temperate forests, bee abundance and species richness within the forest area are known to increase with decreasing forest cover in the surrounding landscape [16]. This finding suggests that even if an optimal habitat for bees were created, a rapid decrease in abundance and species richness would occur with habitat succession in regions where forest regeneration to canopy closure occurs rapidly from grassland-like habitats. Rapid forest regeneration to canopy closure of matured and old-growth forest provides a habitat for closed-forest wildlife but presents a trade-off with early successional species [6,8]. Therefore, optimal management of the successional stages of seminatural habitat would help to maintain high abundance and species richness of bees at regional and landscape scales, providing effective pollination services.

We aimed to determine the optimal successional stages, and to propose better management practices of seminatural habitat, for wild bees in a temperate forest region, in order to ensure pollination services in the region. We examined the effects of reforestation stages, from recently harvested to old growth forests between 1 and 178 years old. Bee assemblages were examined for the two different reforestation types of planted conifer and naturally regenerated forests. Plant species composition regularly varies between different types and ages of reforestation [17]. Here, we tested the hypothesis that the different successional stages of seminatural habitats would affect the abundance and species richness of wild bees, but the responses of wild bees might differ depending on the characteristics of the bees (i.e., social or parasitic status). We particularly assessed the importance of the early successional stages for bees, which has been suspected but not previously proven [18]. We conducted this study in Japan, where the most intensive anthropogenic reforestation efforts involve the conversion of natural or seminatural broad-leaved forests to monocultures of conifer trees. To accommodate the high demand for timber after the Second World War, broad-leaved trees were harvested and coniferous species, primarily *Cryptomeria japonica*, *Chamaecyparis obtusa*, and *Larix leptolepis*, were planted until the 1970s. Such conifer plantations now account for approximately 40% of the total forest cover in Japan [19].

Materials and Methods

Study Region and Sites

The study region was the northern part of Ibaraki Prefecture in central Japan (approximately 36°50′–7′N, 140°32′–8′E; 500–50 m above sea level). The annual mean temperature of the study region is 10.7°C, and the mean annual precipitation is 1910 mm [20]. The landscapes have undergone dramatic anthropogenic changes since the Second World War. Intensive planted reforestation conducted in this area has converted the landscape from natural or seminatural forests of broad-leaved trees, dominated by *Quercus serrata*, *Quercus mongolica*, and *Fagus crenata* to monoculture conifer plantations of *Cryptomeria japonica* and *Chamaecyparis obtusa* [21]. For this study, 10 naturally regenerated deciduous forest stands (2.53–32.49 ha) and eight planted conifer (*C. japonica*) forest stands (2.55–14.29 ha) were selected (Table 1; Figure 1). The ages of the naturally regenerated forest stands varied between 1 and 178 (1, 4, 12, 24, 51, 54, 71, 128, 174, and 178) years old after harvest. The ages of planted forest stands varied between 3 and 76 (3, 7, 9, 20, 29, 31, 75, and 76) years old after harvest. The conifer plantation trees in the study region are usually harvested at around 80 to 90 years of age. The ages were based on a tentative estimation of the National Forest of Japan.

Bee Collection

We collected wild bees using standard Townes-type Malaise traps (Golden Owl Publishers; 180 cm long, 120 cm wide, and 200 cm high) [22] in 2002 for naturally regenerated forests and 2003 for planted conifer forests. In each of the selected forest stands, two traps were placed approximately in the center of the forest stand to avoid forest edge effects. A mixture of ethanol and propylene glycol was used to preserve bees captured in the Malaise traps. Trapped wild bees were collected every 2 weeks from late April to early November in both forest types. We brought all collected bees into the laboratory at the Forestry and Forest Products Research Institute, Tsukuba, where voucher specimens were kept, and identified at the species level. We then classified cleptoparasitic *Coelioxys*, *Epeolus*, *Nomada*, and *Sphecodes* species as "cleptoparasitic," social *Apis* and *Bombus* species as "social," and the remaining species as "solitary." Although there may have been some species of primitively eusocial bees in the "solitary" group, such as species in the family Halictidae [23], we tentatively classified them as solitary because of the limited available information on their biology.

Data Analysis

The effects of the successional stage of reforestation for both forest types on the abundance and species richness of bee groups (all, social, solitary, and cleptoparasitic bee assemblages) were determined using ordinary least-squares regression models. We conducted simple and quadratic functions because scatterplots used previously indicated that some results might have nonlinear effects from the successional stages. We disqualified results from the quadratic functions when the vertexes placed within the forest ages (between 3 and 76 years old for planted forest and between 1 and 178 years old for naturally regenerated forest) indicated values below 0 on the y axis, which were biologically illogical situations because bee abundance and species richness became minus values. As well as the simple linear and quadratic regression models, we examined models in which the successional stages of reforestation were square-root and log transformed. We then compared the models using Akaike's information criterion (AIC), the ΔAIC values obtained by the difference from AIC values of the null models. All of the analyses were performed using the R software (version 2.15.0) [24]. Prior to these regression analyses, we used Mantel tests to check for significant spatial autocorrelation among the study sites for the abundance and species richness of bees in naturally regenerated and planted conifer forest stands; no autocorrelation was found (all $p>0.400$). The ade4 package of R, with the number of permutations set to 1000, was used for the Mantel tests [25].

Table 2. Model comparisons, for the abundance and species richness of bee groups (all, social bees, solitary bees, and cleptoparasitic assemblages) in natural regenerated forest.

	All bees				Social				Solitary				Cleptoparasitic			
	abundance		species richness		abundance		species richness		abundance		species richness		abundance		species richness	
	AIC	ΔAIC	AIC	ΔAIC	AIC	ΔAIC	AIC	ΔAIC	AIC	ΔAIC	AIC	ΔAIC	AIC	ΔAIC	AIC	ΔAIC
null	145.369		83.856		81.285		24.169		143.676		78.018		96.530		59.459	
y=x	145.765	0.396	81.645	−2.211	76.538	−4.747	24.399	0.230	143.990	0.314	76.544	−1.474	94.218	−2.312	55.145	−4.314
$y=x+x^2$	145.451	0.082	79.816	−4.040	78.338	−2.947	20.803	−3.366	143.874	0.198	74.005	−4.013	90.397	−6.133	54.247	−5.212
y=log(x)	143.794	−1.575	76.352	−7.504	79.870	−1.415	19.041	−5.128	142.171	−1.504	71.097	−6.921	87.832	−8.698	49.696	−9.763
y=sqrt(x)	144.667	−0.702	79.100	−4.756	77.671	−3.614	22.573	−1.596	142.928	−0.748	74.106	−3.912	91.175	−5.355	52.173	−7.286

The table shows values of AIC and of ΔAIC calculated as the difference from AIC values of the null models.

Results

We captured and identified a total of 4465 wild bee individuals from 113 species, representing 4 social (*Apis cerana*, *Bombus ardens*, *Bombus diversus*, and *Bombus hypocrita*), 88 solitary (32 *Andrena*, 6 *Ceratina*, 4 *Colletes*, 1 *Halictus*, 8 *Hylaeus*, 28 *Lasioglossum*, 4 *Megachile*, 1 *Macropis*, 2 *Osmia*, 1 *Tetralonia*, and 1 *Xylocopa*), and 21 cleptoparasitic (2 *Coelioxys*, 1 *Epeolus*, 13 *Nomada*, and 5 *Sphecodes*) species. Model comparisons for the naturally regenerated forest indicated that the simple linear regression model be selected for the abundance of social bees with the log-transformed regression models selected for the other categories (Table 2). Model comparisons for the planted conifer forest indicated that the lowest AIC value was obtained from the null model of species richness of social bees, and the quadratic regression model was selected for the abundance of social bees, with the log-transformed regression models selected for the other categories (Table 3). The effects of succession on all bees, solitary bees, and cleptoparasitic bees for both forest types exhibited similar trends. Higher abundances and species richness were observed in the early successional stages than in mature successional stages, and higher species richness was observed in naturally regenerated forest than in planted conifer forest as the successional stages progressed (Figure 2). However, the trends of successional effects on social bees were different. Higher abundance and species richness of social bees in naturally regenerated forest were observed as the successional stages progressed, whereas the abundance of social bees in planted conifer forest produced a concave-shaped relationship when plotted (Figure 1).

Discussion

The response of all bees to the different successional stages of naturally regenerated and planted conifer forests indicated that the early successional stages provide important habitats for bees. The majority of collected bees were solitary species. Cleptoparasitic species do not collect pollen and nectar for their offspring and were parasites of the solitary bees collected in the study. Therefore, the trends for all bees may be based on the trends observed for the solitary and cleptoparasitic bees. Although there are several bee species that might depend on habitats provided during the late successional stages of temperate forests [26], it is suspected that most bees prefer open and disturbed habitats to late successional stages [18,23]. Solitary bees have specific habitat requirements and are considered to be more restricted to grassland-like habitats than social bees [27,28]. Forestry practices such as clear-cutting and thinning offer a wide variety of habitat transitions but frequently create more open and early successional habitats, which generate more floral and habitat resources for bees [8]. This implies that many bee species would respond positively to forest management practices in a region where forest regeneration to canopy closure occurs quickly, such as our selected study region [22]. Our results also demonstrated that abundance and species richness were as high in the early successional stage of planted conifer forest as in naturally regenerated forest. If such young planted forests can provide suitable habitats for bees, even in a forest managed for timber production, active clear-cutting of planted forests that generates early succession will have substantial positive effects on early successional species [8].

For example, in Japan, the area of grassland and young forests has declined rapidly [29] to the extent that 68% of the national land area is covered by forest, of which 42% is plantation [30]. Forestry activities in Japan declined after the 1960s and 1970s with the increase in imported timber following international free trade agreements and the introduction of a floating exchange rate

Table 3. Model comparisons for the abundance and species richness of bee groups (all, social bees, solitary bees, and cleptoparasitic assemblages) in conifer (*Cryptomeria japonica*) planted forest.

	All bees				**Social**				**Solitary**				**Cleptoparasitic**			
	abundance		**species richness**		**abundance**		**species richness**		**abundance**		**species richness**		**abundance**		**species richness**	
	AIC	**ΔAIC**	**AIC**	**ΔAIC**	**AIC**	**ΔAIC**	**AIC**	**ΔAIC**	**AIC**	**ΔAIC**	**AIC**	**ΔAIC**	**AIC**	**ΔAIC**	**AIC**	**ΔAIC**
null	119.883		74.847		79.729		18.516		116.102		70.825		90.871		50.644	
$y = x$	118.666	−1.216	72.215	−2.631	81.719	1.989	20.146	1.630	114.341	−1.761	67.709	−3.115	89.577	−1.294	48.900	−1.743
$y = x+x^2$	115.345*	−4.538	58.538*	−16.309	79.653	−0.077	18.751	0.235	110.824*	−5.278	51.292*	−19.533	88.165*	−2.706	42.407*	−8.236
$y = \log(x)$	116.706	−3.176	65.972	−8.875	81.418	1.689	20.431	1.915	112.316	−3.786	60.649	−10.176	88.276	−2.595	45.207	−5.437
$y = \mathrm{sqrt}(x)$	117.410	−2.472	69.494	−5.353	81.683	1.953	20.463	1.947	112.973	−3.129	64.698	−6.126	88.590	−2.281	47.018	−3.626

The table shows values of AIC and of ΔAIC calculated as the difference from AIC values of the null models. Asterisks (*) indicate that the vertexes placed within the forest ages (between 3 and 76 years old for planted conifer forest) were below 0 on the y axis for the best models (quadratic functions), and thus the second-best models were selected.

system [31]. Many of the trees in planted forests are now reaching harvest age, although the wood supply from the forests has slightly increased [29]. As the area of grassland and early successional forests in Japan has declined, a national decline in bird and butterflies species, which are dependent on early successional stages, is also thought to have occurred [32,33]. Trade-offs must be carefully considered when planning the extent and location of forest cutting. These trade-offs can be between bee diversity and other taxonomic diversity, or between pollination and other ecosystem services [34], such as carbon storage, flood control, water provision, and timber production, which are provided by mature naturally regenerated and planted forests. Some active clear-cutting of planted forests, leading to early succession, might become an acceptable policy in Japan, although conservation of primary growth forests and careful rotation planning of where to cut, at both a landscape and national scale, are required.

Wild bees pollinate various crops [35–40]. The area surrounding bee-pollinated crop fields would ideally be a grassland-like habitat, and hence the management and quality maintenance of seminatural and natural ecosystems around agricultural fields are important [41,42]. For the example of almond production in California, it was found that practices within farms, such as organic farming alone, would not sustain pollination services by wild bees, and the presence of high-quality habitats along the edges of the orchards are important [43]. In many temperate terrestrial forest ecosystems, forest regeneration to canopy closure occurs quickly; hence the creation and maintenance of early successional stages of both naturally regenerated and planted conifer forests, surrounding bee-pollinated cropland could help to enhance pollination services. For example, near the present study region, in the northern part of Ibaraki Prefecture in central Japan, common buckwheat, which is a heterostylous crop that depends highly on insect pollination, is a distinctive crop grown by local landholders in mountainous regions [39,44]. Such a crop might benefit from the habitats created for bees.

Although the species richness of the social bees collected was relatively low (four species), our results indicated that the responses of social bee species to the different forest successional stages were different from those of solitary and cleptoparasitic bees. As the successional stages progressed, positive relationships were observed for the abundance and species richness of social bees in naturally regenerated forest. However, as successional stages progressed, the convex relationship for abundance and no relationship for the species richness of social bees were observed in planted conifer forest. In naturally regenerated forest, nesting resources might be more widely available in mature successional stages than in early successional stages. This could include cavities in large trees for *Apis cerana* [45,46] and underground cavities for the *Bombus* species [47]. In planted conifer forests, the early succession stages may provide floral resources for the *Apis* and *Bombus* bees; on the other hand, late successional stages (e.g., the 75- and 76-year-old stands) may have lower floral resources but more underground cavities for nesting sites for *Bombus* species. Planted forests are usually disturbed by active human management, such as pre-commercial tree thinning and understory weed clearing [48]; thus the 20- and 31-year-old forest stands might have contained fewer nesting sites for the *Bombus* species as well as poorer floral resources. It is important to interpret our results considering the effects of wider spatial scales on *Apis* and *Bombus* species [49]. The foraging ranges of these social bees are likely wider than those of the majority of the solitary bees [50–52]. Previous studies have indicated that *Apis cerana* forages over 1 km [53] and the spatial ranges of landscape effects, which influenced its foraging, were from 1 to 3 km [39,41]. Although the foraging ranges of our captured *Bombus* species were unknown, other *Bombus* species have been known to forage for at least 250 m to 1.5 km [54–57]. Therefore, influences at the landscape level rather than at the forest stand level might have influenced the responses of these social bees.

Wild bees were sampled during a single year in each reforestation type using Malaise traps in the present study. There was a risk that this particular sampling method may be biased for or against particular groups of bees, and may not represent sampling completeness of the bee fauna in the study region. However, we sampled 4465 wild bee individuals from 113 species using the Malaise traps. Based on the sampled bees, we can conclude that the early successional stages of both naturally regenerated and planted conifer forests maintain high abundances and species richness of solitary bees and their cleptoparasitic bees, although social bees respond differently in early successional stages. These results may have implications for active forest stand management practices, such as clear-cutting of planted forests for timber production, which would create early successional habitats, leading to significant positive effects on bees in general. However, in addition to the conservation of primary, old growth natural forests, trade-offs between bee diversity and other taxonomic diversity in relation to pollination and other ecosystem services should be considered.

Acknowledgments

We thank T. Kitajima and K. Meshiai for the care of bee samples in the laboratory.

Author Contributions

Conceived and designed the experiments: HT IO KO TI SM. Performed the experiments: HT KO TI HG TM SM. Analyzed the data: HT SM. Wrote the paper: HT SM.

References

1. Oliver CD, Larson BC (1996) Forest Stand Dynamics, Updated Edition. New York: Wiley. 520 p.
2. Grime JP (1973) Competitive exclusion in herbaceous vegetation. Nature 242: 344–347.
3. Connell JH (1978) Diversity in tropical rain forests and coral reefs - High diversity of trees and corals is maintained only in a nonequilibrium state. Science 199: 1302–1310.
4. Pickett STA, Thompson JN (1978) Patch dynamics and the design of nature reserves. Biological Conservation 13: 27–37.
5. Spies TA, Turner MG (1999) Dynamic forest mosaics. In: Hunter MLJ, editor. Maintaing biodiversity in forest ecosystems. Cambridge: Cambridge University Press. pp. 95–160.
6. Swanson ME, Franklin JF, Beschta RL, Crisafulli CM, DellaSala DA, et al. (2011) The forgotten stage of forest succession: early-successional ecosystems on forest sites. Frontiers in Ecology and the Environment 9: 117–125.
7. Fontaine JB, Donato DC, Robinson WD, Law BE, Kauffman JB (2009) Bird communities following high-severity fire: Response to single and repeat fires in a mixed-evergreen forest, Oregon, USA. Forest Ecology and Management 257: 1496–1504.
8. Yamaura Y, Royle JA, Shimada N, Asanuma S, Sato T, et al. (2012) Biodiversity of man-made open habitats in an underused country: a class of multispecies abundance models for count data. Biodiversity and Conservation 21: 1365–1380.
9. Corbet SA, Williams IH, Osborne JL (1991) Bees and the pollination of crops and wild flowers in the European community. Bee World 72: 47–59.
10. Klein AM, Vaissiere BE, Cane JH, Steffan-Dewenter I, Cunningham SA, et al. (2007) Importance of pollinators in changing landscapes for world crops. Proceedings of the Royal Society of London B 274: 303–313.
11. Ollerton J, Winfree R, Tarrant S (2011) How many flowering plants are pollinated by animals? Oikos 120: 321–326.
12. Kremen C, Williams NM, Bugg RL, Fay JP, Thorp RW (2004) The area requirements of an ecosystem service: crop pollination by native bee communities in California. Ecology Letters 7: 1109–1119.
13. Winfree R (2010) The conservation and restoration of wild bees. In: Ostfeld RS, Schlesinger WH, editors. Year in Ecology and Conservation Biology 2010. Malden: Wiley-Blackwell. pp. 169–197.
14. Steffan-Dewenter I, Tscharntke T (2001) Succession of bee communities on fallows. Ecography 24: 83–93.
15. Krauss J, Alfert T, Steffan-Dewenter I (2009) Habitat area but not habitat age determines wild bee richness in limestone quarries. Journal of Applied Ecology 46: 194–202.
16. Winfree R, Griswold T, Kremen C (2007) Effect of human disturbance on bee communities in a forested ecosystem. Conservation Biology 21: 213–223.
17. Fujimori T (2001) Ecological and silvicultural strategies for sustainable forest management. Amsterdam: Elsevier.
18. Winfree R, Bartomeus I, Cariveau DP (2011) Native pollinators in anthropogenic habitats. Annual Review of Ecology, Evolution, and Systematics 42: 1–22.
19. Statistics-Department (2008) The 82nd Statistical Yearbook of Ministry of Agriculture Forestry and Fisheries, Japan 2006~2007. Tokyo: Statistics Department, Ministry of Agriculture Forestry and Fisheries, Japan.
20. Mizoguchi Y, Morisawa T, Ohtani Y (2002) Climate of Ogawa forest reserve. In: Nakashizuka T, Matsumoto Y, editors. Diversity and interaction in a temperate forest community: Ogawa forest reserve of Japan. Tokyo: Springer. pp. 11–18.
21. Makino S, Goto H, Hasegawa M, Okabe K, Tanaka H, et al. (2007) Degradation of longicorn beetle (Coleoptera, Cerambycidae, Disteniidae) fauna caused by conversion from broad-leaved to man-made conifer stands of *Cryptomeria japonica* (Taxodiaceae) in central Japan. Ecological Research 22: 372–381.
22. Makino S, Goto H, Inoue T, Sueyoshi M, Okabe K, et al. (2006) The monitoring of insects to maintain biodiversity in Ogawa forest reserve. Environmental Monitoring and Assessment 120: 477–485.
23. Michener CD (2007) The bees of the world. Baltimore: The Johns Hopkins University Press.
24. R Development Core Team (2012) R, a language and environment for statistical computing. Vienna: R Foundation for statistical computing.
25. Dray S, Dufour AB (2007) The ade4 package: implementing the duality diagram for ecologists. Journal of Statistical Software 22: 1–20.
26. Taki H, Kevan PG, Ascher JS (2007) Landscape effects of forest loss in a pollination system. Landscape Ecology 22: 1575–1587.
27. Banaszak J (1992) Strategy for conservation of wild bees in an agricultural landscape. Agriculture Ecosystems & Environment 40: 179–192.
28. Osborne JL, Williams IH, Corbet SA (1991) Bees, pollination and habitat change in the European Community. Bee World 72: 99–116.
29. Yamaura Y, Oka H, Taki H, Ozaki K, Tanaka H (2013) Sustainable management of planted landscapes: lessons from Japan. Biodiversity and Conservation 21: 3107–3129.
30. FAO (2007) State of the world's forests 2007. Rome: Food and Agriculture Organization of the United Nations.
31. Nagy A (1988) International trade in forest products; Nagy A, editor. Bicester: AB Academic Publishers
32. Yamaura Y, Ikeno S, Sano M, Okabe K, Ozaki K (2009) Bird responses to broad-leaved forest patch area in a plantation landscape across seasons. Biological Conservation 142: 2155–2165.
33. Inoue T (2005) Causes of butterfly decline in Japan. Japanese Journal of Entomology (New Series) 8: 43–64.
34. Chan KMA, Shaw MR, Cameron DR, Underwood EC, Daily GC (2006) Conservation Planning for Ecosystem Services. PLoS Biol 4: e379.
35. Morandin LA, Winston ML (2005) Wild bee abundance and seed production in conventional, organic, and genetically modified canola. Ecological Applications 15: 871–881.
36. Winfree R, Williams NM, Dushoff J, Kremen C (2007) Native bees provide insurance against ongoing honey bee losses. Ecology Letters 10: 1105–1113.
37. Slaa EJ, Sanchez Chaves LA, Malagodi-Braga KS, Hofstede FE (2006) Stingless bees in applied pollination: practice and perspectives. Apidologie 37: 293–315.
38. Sheffield CS, Kevan PG, Westby SM, Smith RF (2008) Diversity of cavity-nesting bees (Hymenoptera: Apoidea) within apple orchards and wild habitats in the Annapolis Valley, Nova Scotia, Canada. Canadian Entomologist 140: 235–249.
39. Taki H, Okabe K, Yamaura Y, Matsuura T, Sueyoshi M, et al. (2010) Effects of landscape metrics on *Apis* and non-*Apis* pollinators and seed set in common buckwheat. Basic and Applied Ecology 11: 594–602.
40. Klein AM, Steffan-Dewenter I, Tscharntke T (2003) Bee pollination and fruit set of Coffea arabica and C-canephora (Rubiaceae). American Journal of Botany 90: 153–157.
41. Taki H, Yamaura Y, Okabe K, Maeto K (2011) Plantation vs. natural forest: Matrix quality determines pollinator abundance in crop fields. Scientific Reports 1: 132.
42. Garibaldi LA, Steffan-Dewenter I, Kremen C, Morales JM, Bommarco R, et al. (2011) Stability of pollination services decreases with isolation from natural areas despite honey bee visits. Ecology Letters 14: 1062–1072.
43. Klein AM, Brittain C, Hendrix SD, Thorp R, Williams N, et al. (2012) Wild pollination services to California almond rely on semi-natural habitat. Journal of Applied Ecology 49: 723–732.
44. Taki H, Okabe K, Makino S, Yamaura Y, Sueyoshi M (2009) Contribution of small insects to pollination of common buckwheat, a distylous crop. Annals of Applied Biology 155: 121–129.
45. Sasaki M (1999) Wonders of the Japanese honeybee - Biology of Northernmost *Apis cerana* -. Tokyo: Kaiyusha.
46. Yoshida T (2000) Rearing method and ecology of Japanese honeybee. Tokyo: Tamagawa University Press.
47. Sakagami S, Katayama E (1977) Nests of some Japanese bumblebees (Hymenoptera, Apidae). Journal of the Faculty of Science, Hokkaido University Series VI Zoology 21: 92–153.
48. Hirata A, Sakai T, Takahashi K, Sato T, Tanouchi H, et al. (2011) Effects of management, environment and landscape conditions on establishment of hardwood seedlings and saplings in central Japanese coniferous plantations. Forest Ecology and Management 262: 1280–1288.
49. Steffan-Dewenter I, Munzenberg U, Burger C, Thies C, Tscharntke T (2002) Scale-dependent effects of landscape context on three pollinator guilds. Ecology 83: 1421–1432.
50. Gathmann A, Tscharntke T (2002) Foraging ranges of solitary bees. Journal of Animal Ecology 71: 757–764.
51. Zurbuchen A, Landert L, Klaiber J, Muller A, Hein S, et al. (2010) Maximum foraging ranges in solitary bees: only few individuals have the capability to cover long foraging distances. Biological Conservation 143: 669–676.

52. Greenleaf SS, Williams NM, Winfree R, Kremen C (2007) Bee foraging ranges and their relationship to body size. Oecologia 153: 589–596.
53. Dyer FC, Seeley TD (1991) Dance dialects and foraging range in three Asian honey bee species. Behavioral Ecology and Sociobiology 28: 227–233.
54. Knight ME, Martin AP, Bishop S, Osborne JL, Hale RJ, et al. (2005) An interspecific comparison of foraging range and nest density of four bumblebee (Bombus) species. Molecular Ecology 14: 1811–1820.
55. Osborne JL, Clark SJ, Morris RJ, Williams IH, Riley JR, et al. (1999) A landscape-scale study of bumble bee foraging range and constancy, using harmonic radar. Journal of Applied Ecology 36: 519–533.
56. Osborne JL, Martin AP, Carreck NL, Swain JL, Knight ME, et al. (2008) Bumblebee flight distances in relation to the forage landscape. Journal of Animal Ecology 77: 406–415.
57. Wolf S, Moritz RFA (2008) Foraging distance in Bombus terrestris L. (Hymenoptera: Apidae). Apidologie 39: 419–427.

Gap Dynamics and Structure of Two Old-Growth Beech Forest Remnants in Slovenia

Tihomir Rugani*, Jurij Diaci, David Hladnik

Biotechnical Faculty, Department of Forestry and Renewable Forest Resources, University of Ljubljana, Ljubljana, Slovenia

Abstract

Context: Due to a long history of intensive forest exploitation, few European beech (*Fagus sylvatica* L.) old-growth forests have been preserved in Europe.

Material and Methods: We studied two beech forest reserves in southern Slovenia. We examined the structural characteristics of the two forest reserves based on data from sample plots and complete inventory obtained from four previous forest management plans. To gain a better understanding of disturbance dynamics, we used aerial imagery to study the characteristics of canopy gaps over an 11-year period in the Kopa forest reserve and a 20-year period in the Gorjanci forest reserve.

Results: The results suggest that these forests are structurally heterogeneous over small spatial scales. Gap size analysis showed that gaps smaller than 500 m^2 are the dominant driving force of stand development. The percentage of forest area in canopy gaps ranged from 3.2 to 4.5% in the Kopa forest reserve and from 9.1 to 10.6% in the Gorjanci forest reserve. These forests exhibit relatively high annual rates of coverage by newly established (0.15 and 0.25%) and closed (0.08 and 0.16%) canopy gaps. New gap formation is dependant on senescent trees located throughout the reserve.

Conclusion: We conclude that these stands are not even-sized, but rather unevenly structured. This is due to the fact that the disturbance regime is characterized by low intensity, small-scale disturbances.

Editor: Ben Bond-Lamberty, DOE Pacific Northwest National Laboratory, United States of America

Funding: Field research and manuscript preparation was partly funded by the Slovenian Ministry of Agriculture, Forestry, and Food through project grant (project V4-0346) and by the Programme research group Forest, Forestry, and Renewable Forest Resources. The funders had no role in study design, data collection and analysis, decision to publish, or preparation of the manuscript.

Competing Interests: The authors have declared that no competing interests exist.

* E-mail: tihomir.rugani@bf.uni-lj.si

Introduction

Virgin and old-growth forests are important ecosystem cornerstones. Although they are well represented in the world (35.7% of forest area), only 2.8% of Europe's (Russian Federation excluded) forest area is classified as primary forest [1]. Human land use has a long history in Europe, where almost no areas have been left to retain their natural, unaltered character [2,3]. The importance of understanding such ecosystems has been emphasized by several researchers [4,5,6].

European beech (*Fagus sylvatica* L.) is one of the most common of the many autochthonic tree species in Europe. Its range extends from Western Europe to the Carpathians in the east, and from Scandinavia in the north to Greece in the south, and is able to thrive in a broad range of soil and climate conditions from the lowlands to the timberline. European beech is a strong competitor and is thus present in a variety of mixed species forests as well as in pure stands [7]. These characteristics make it a dominant tree species in Central Europe. Because of the growing trend towards ecologically-based forest management, an understanding of the natural patterns and processes in beech stands is essential to help inform management practices. A number of studies have been carried out in the few remaining beech old-growth stands in Northern [8,9], Western [10], Central [11,12,13,14,15,16], Eastern [17], and Southern [18,11,19] Europe. Despite this body of work, there is still debate concerning the natural dynamics of beech stands. A common idea is that beech forms homogeneous stands dominated by large canopy trees with a sparse understory and middle layer, often referred to as 'cathedral-like' stands [2]. Two conceptual models have been proposed to explain this structure. One model suggests that after stand replacing events a new wave of even-aged regeneration occupies the canopy [2]. A second model suggests that a 'cathedral-like' stand can develop because beech grows rapidly through the middle height layer when light is available (i.e. in gaps), thereby resulting in an uneven-aged stand that has a uniform height structure and a diverse diameter structure (i.e. a wide range of diameters exist in the canopy layer) [13]. In contrast, a third model may suggest that beech stands have a heterogeneous age and size structure at small to intermediate spatial scales depending on the particular disturbance history of the stand, whereby recruitment to the upper layers of the stand is driven by spatiotemporal variation in light availability.

Many studies carried out in old-growth beech stands in Europe have used traditional approaches that focus on structural characteristics, such as descriptions of size distributions and mapping of development phases [6]. Until recently, few studies in the region have used methods that examine the processes behind the observed patterns. In temperate forests worldwide, numerous studies have shown the influence of gap disturbance processes on forest dynamics [20,21,22,23,24,25]. Similarly, recent work in old-growth beech forests in Europe has provided valuable insight into natural disturbance processes for this forest type [11,14,16,26]. In this study, we use time series data on forest structure and canopy gaps to investigate how two old-growth beech forest remnants in Slovenia fit the three previously mentioned a conceptual model, which has not explicitly been addressed in past work.

Materials and Methods

1. Ethics statement

All necessary permits were obtained for the described field studies. The permits were issued by the Forestry service of Slovenia (Central unit). Gorjanci forest reserve is privately owned and the permit was issued by the owner's legal representative (Roman Zaletelj).

2. Study area

We studied two beech-dominated old-growth forest reserves: the Gorjanci and Kopa reserves. Both reserves are located in the northern part of the Dinaric mountain range, which lies in Slovenia.

The Gorjanci forest reserve is situated on the ridge of the Gorjenci massif, on the northern slope (45°45′N, 15°19′E). The ridge top (Trdinov vrh) is 1178 m a.s.l. and lies on the border between Slovenia and Croatia. The reserve encompasses 22.98 ha and ranges from 990 to 1150 m a.s.l. According to the nearest (1.6 km) meteorological station (Sv. Miklavz, 969 m a.s.l.), it has a continental climate with a mean annual precipitation of 1290 mm and a mean temperature of 12°C from March to September. The topography is steep (8–30°) with localized flat areas. The geology is comprised predominately of dolomite bedrock, but limestone is also locally present [27]. The vegetation is dominated by European beech (*Fagus sylvatica* L.), while other species, such as sycamore maple (*Acer pseudoplatanus* L.) and silver fir (*Abies Alba* Mill.), are less frequent. Large beech trees are present throughout the entire reserve but do not exceed 115 cm in diameter at breast height (d.b.h.) and 45 m in height. The reserve was not managed after World War Two and is considered not to have had any human interventions for at least a century. The forest reserve had not been accessible until a forest road was built to Trdinov vrh after World War Two. During our survey we did not find any remains of cut stumps.

The Kopa forest reserve is situated under the ridge top (Kopa, 1077 m a.s.l.) of the Kocevski Rog massif and has a northeast aspect (45°37′N, 15°02′E). Its size is 14.05 ha, and it ranges from 980 to 1080 m a.s.l. The climate is continental with a mean annual precipitation of 1240 mm and a mean temperature of 12.1°C from March to September, according to the nearest (5.1 km) meteorological station (Planina – Mirna gora, 740 m a.s.l.). The topography is moderately sloped and is characterized by Karst phenomena (e.g. sinkholes). The geology is comprised of limestone with substantial surface rockiness [27]. The vegetation is dominated by beech, while other species, such as sycamore maple, European ash (*Fraxinus excelsior* L.), and silver fir, are less frequent (<10% in the growing stock). The reserve has not been managed for more than 50 years. No cut stumps or other indications of human influence have been found.

3. Field sampling and data analysis

We examined the structural characteristics of the two forest reserves based on data from 24 (Gorjanci) and 21 (Kopa) circular sample plots with a radius of 15 m (i.e., 700 m^2 in size) established on a systematic grid (100×100 m) and measured in 2004 and 2006, respectively. We measured the d.b.h. (1.30 m) of all (live and dead) trees (d.b.h.≥10 cm in Gorjanci and d.b.h.≥5 cm in Kopa), the position of all trees (live and dead), and the height of the three closest trees to the centre of the sample plots. We sampled lying (mean diameter was measured) and standing (d.b.h. was measured) dead trees. In Slovenia the monitoring of forest reserves is conducted in the framework of forest inventories. A drawback is that the circular sample plots used in the forest inventory are too small to draw conclusions on large-scale variation based on the pair correlation function for exploratory data analysis and the commonly used Ripley's K-function in spatial statistics [28,29]. The spatial characteristics of forest stands are often estimated on the basis of evaluating the neighbourhood of a given number of reference trees [30]. Our aim was to assess the spatial distribution of trees at a scale comparable with that of small canopy tree fall. As the usefulness of the nearest neighbour distances for spatial pattern analysis is very limited [31], we analysed the spatial distribution of trees in the sample plots by calculating the distances and variation of these distances from the centre of the plot to all trees in the plot, a method based on the work of [32] and [33] and also presented and tested in [34]. This analysis follows the [35] index as a measure of the departure from randomness in a population of plants [36]. presented the mathematical derivations for a random point process (i.e. homogeneous Poisson process) as an extension of the index to the next nearest neighbours to detect larger-scale heterogeneity by only using the distance to the nearest neighbour.

We calculated average distances Rk to the k-th nearest trees from plot centers and compared them with theoretical distances according to [36]:

$$R_{kT} = \frac{a_k}{\sqrt{\mu}} \tag{1}$$

where a_k is a constant, calculated for particular tree values up to $k = 15$ or the minimum number of trees on the sampling plots:

$$a_k = \frac{k \times (2k)!}{(2^k \times k!)^2} \tag{2}$$

and μ is density, calculated from tree number ($N_{ha}/10\ 000$). Based on the second moment of the mean distance to the nearest trees presented by Thompson, the theoretical variation of distances was calculated:

$$\sigma^2_{Rk} = \frac{k}{\mu \times \left[\frac{1}{\pi} - \frac{k \times ((2k)!)^2}{(2k \times k!)^4}\right]} \tag{3}$$

It was shown by [37] that the actual coefficient of variation depends on the average values of distances between trees and the variance of these differences, whereas the theoretical coefficient of variation, which is calculated for a random distribution of trees, is dependent only on the *k* value, which represents the sequence of distances:

Table 1. Summary of the characteristics of the image data used.

Study area	Acquisition date	Mapping scale	Focal length	Spectral resolution	Sensor
Gorjanci	17.7.2006	1:20 000*	120 mm	CIR	Z/I DMC
	2/9/1986	1:5 000	305 mm	CIR	Zeiss LMK 1000
Kopa	21.7.2006	1:44 000*	120 mm	CIR	Z/I DMC
	18.7.1998	1:17 500	210 mm	PAN	Zeiss LMK 1000

(*) Mapping scale estimate based on the digital mapping camera (DMC) from Z/I Imaging with 12 μm pixel size and 25 cm to 50 cm ground sample distance (GSD). For 2009 the UltraCamXp Vexcel Imaging camera was used (orthoimages with GSD of 25 cm).

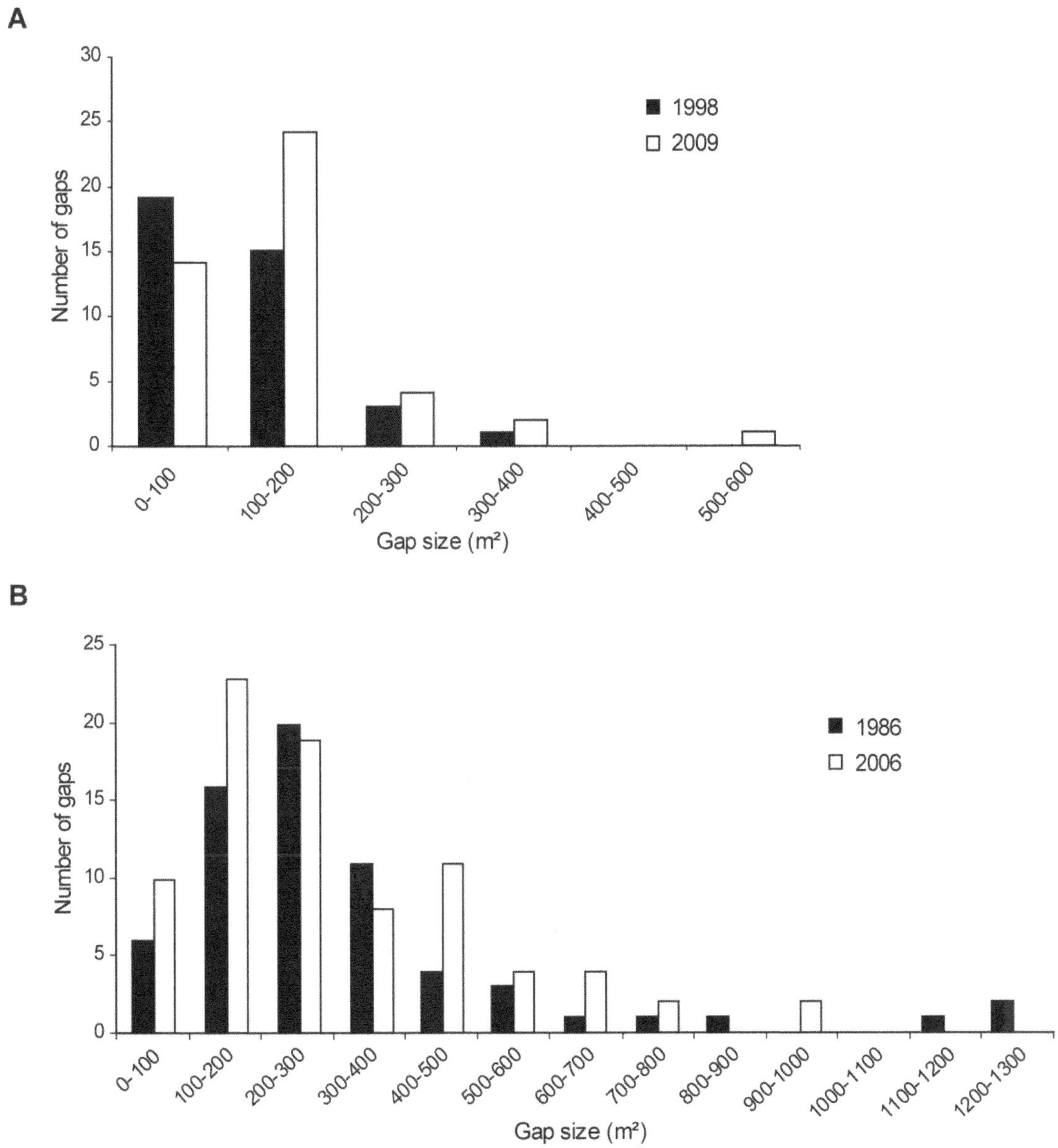

Figure 1. Gap size distribution in the Kopa (A) and Gorjanci (B) forest reserves.

Table 2. Gap characteristics and gap dynamics processes in the Kopa forest reserve.

Year	1998	1998–2009				2009
		Gap dynamics processes				
Gap area (m^2)		New gap formation	Gap expansion	Gap reduction	Gap closure	
Total	4483(319)	2080(148)	1498(107)	−579(41)	−1154(82)	6327(450)
No. of gaps	38(2.7)	22(1.6)	15(1.1)	11(0.8)	12(0.9)	45(3.2)
Gap size ≤200 m^2	3464(247)	2080(148)	1220(86)	−361(26)	−1154(82)	4093(291)
No. of gaps	34(2.4)	22(1.5)	14(1.0)	8(0.6)	12(0.85)	38(2.7)
Gap size >200 m^2	1019(73)	0(0)	278(20)	−218(15)	0(0)	2235(159)
No. of gaps	4(0.3)	0(0)	1(0.1)	3(0.2)	0(0)	7(0.5)

The numbers in parentheses are relative values calculated per forest reserve area.

$$KV_{kt} = \frac{1}{a_k \times \sqrt{\frac{k}{\pi} - a_k^2}} \tag{4}$$

On this basis we may conclude that the coefficient of variation of the distances between trees can be used as an indicator of the spatial distribution of trees. As shown in [34], the distances and variation of these distances on sampling plots can be compared with theoretical values, derived from [36].

We calculated the coefficients of variation of the distances from the centers of the plots to all trees (d.b.h.≥5 cm in Kopa and d.b.h.≥10 cm in Gorjanci), and of dominant trees (trees in the upper canopy layer), and compared them with a theoretical coefficient for random spatial distribution. If the variation of distances from the plot center to the k-tree is positioned above the theoretical line, the distribution of trees shows a tendency towards clustered distribution. If the variation of distances from the plot center to the k-tree is positioned below the theoretical line, the distribution of trees shows a tendency towards uniform distribution. Dominant trees in sample plots were estimated as a sample of the 100 largest (based on diameter) trees per hectare; in 700 m^2 circular sample plots this would require the seven largest trees per plot. The average height of the 100 largest trees per hectare (i.e., quadratic mean diameter (d_{q100}), defined as the diameter corresponding to the mean basal area of the 100 largest trees per hectare) is often used in Europe as an estimate of stand top height (h_{dom}) [38,39]. Based on the measurements in the sampling plots, we calculated the quadratic mean diameter and the regression height of the largest tree sample in both forest reserves (Gorjanci (d_{q100} = 63.8 cm; h_{dom} = 31.9 m) and Kopa (d_{q100} = 64.0 cm; h_{dom} = 34.0 m)). The canopy layer was estimated to exceed 2/3 of the dominant height (22 m) and the middle layer was estimated to exceed 1/3 of the dominant height (11 m). Lower layer trees were all trees under 11 meters in height.

To better understand the structural characteristics and their dynamics, we analyzed the complete inventory data (full calliper-ing of all trees with d.b.h≥10 cm) obtained from the four previous forest management plans for the two management units [40,41]. All live trees (d.b.h.≥10 cm) were measured at approximate 10-year intervals (4 measurements over a period of 31 years in Kopa and 35 years in Gorjanci) in both forest reserves.

To gain a better understanding of the disturbance dynamics, we studied the characteristics of the canopy gap disturbance regime. The predominant disturbance agent in the observed forest reserves was wind, which resulted in stem breakage or uprooting (personal observation). A canopy gap was delineated if regeneration in the gap was shorter than half (20 m) of the upper canopy height. Gaps were delineated from aerial photographs using a digital stereo plotter (Summit Evolution, DAT/EM Systems) for precise delineation of canopy gaps. Digital aerial images from the cyclic aerial survey of Slovenia were used for the delineation of gaps in

Table 3. Gap characteristics and gap dynamics processes in the Gorjanci forest reserve.

Year	1986	1986–2006				2006
		Gap dynamics processes				
Gap area (m^2)		New gap formation	Gap expansion	Gap reduction	Gap closure	
Total	20 903(910)	11 406(496)	4034(176)	−4812(209)	−7157(311)	24 374(1061)
No. of gaps	66(2.9)	482(2.1)	18(0.8)	17(0.7)	31(1.4)	83(3.6)
Gap size ≤200 m^2	3052(133)	3475(151)	1837(80)	−1152(50)	−2067(90)	11310(492)
No. of gaps	23(1.0)	27(1.2)	6(0.3)	2(0.1)	17(0.7)	39(1.7)
Gap size > 200 m^2	17850(777)	7931(345)	2197(96)	−4660(203)	−5090(222)	13064(569)
No. of gaps	43(1.9)	21(0.9)	12(0.5)	15(0.6)	14(0.6)	44(1.9)

The numbers in parentheses are relative values calculated per forest reserve area.

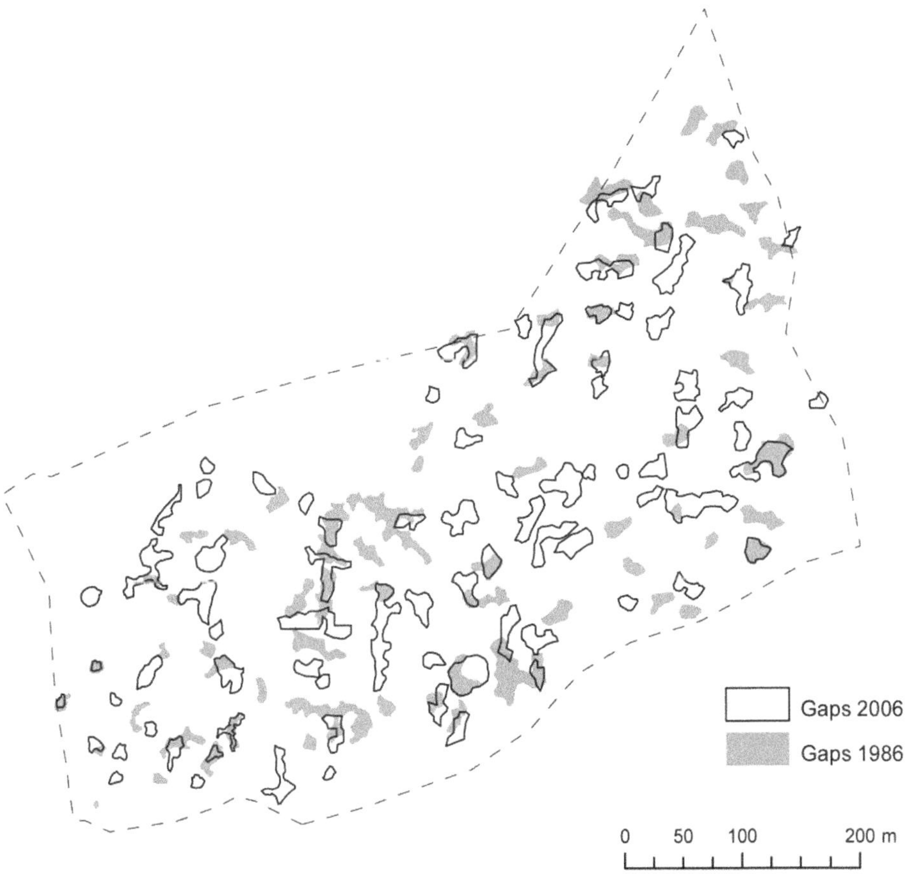

Figure 2. Gap dynamics in the Gorjanci forest reserve in the studied period (1986–2006).

2006. The older panchromatic black-and-white (PAN) and Kodak 2443 color infrared (CIR) aerial film photographs were scanned using 21 μm resolution to achieve comparable resolution (Table 1). During the analysis the stereoscopic pairs of aerial photographs were interpreted and compared with terrain work based on the absolute orientation of stereoscopic photographs which established the reference to the terrain coordinate system. The absolute orientation of the stereopairs was based on at least 10 to 15 ground control points. The residuals of the X and Y coordinates for each point were allowed to be less than 1 m. The root mean square error (RMSE*xyz*) for the stereo model of the Gorjanci forest reserve was 0.36 m and 1.14 m for the Kopa reserve. The delineated gaps were compared with the corresponding ground truth, based on GPS positions of individual gaps. For the 22 gaps in the Gorjanci forest reserve we determined the position of the gaps with a sub-meter GPS (Trimble GeoXT) with differential correction. GPS use within the forest reserve was hindered by canopy cover and terrain, which blocked satellite signals from reaching the GPS receiver, so it was not possible to record the positions of single trees accurately. Capturing data in forest gaps that were clear of direct over-head cover improved the results of post-processed, differentially corrected data (average errors were between 2 and 3 m). In the field we measured the distances from the GPS position within forest gaps to the crown outline of the closest canopy tree. These distances were compared in the GIS environment with the corresponding distances of delineated tree crowns in the stereo model. For 22 ground GPS points the average difference to the crown positions on the delineated gaps was 2.47 m with standard error of 0.42 m.

We delineated gaps larger than 20 m^2 (five meters in diameter), which could have been the result of a small canopy tree fall. This was a methodological constraint since it is difficult to delineate small forest gaps from aerial photographs of shaded hillsides, as it is not possible to clearly distinguish between regeneration in the gap and the trees of the upper canopy height. Trees in small gaps may be shaded or partially hidden by others, and the illumination conditions in the smaller gaps confounded efforts to delineate them successfully using visual stereo-image interpretation. Similar problems of misclassification of the non-dominant tree species in image segmentation and semi-automatic tree species classification approaches have been reported elsewhere [42]. We analyzed a 20-year time series of aerial photographs for the Gorjanci reserve (1986 and 2006) and an 11-year series (1998 and 2009) for the Kopa reserve. This data were used to quantify gap size distribution, gap fraction, and gap dynamics processes. We quantified new gap formation (newly formed gaps in the observed period), gap expansion (gaps that were expanded in the observed period), gap reduction (gaps that decreased in size in the observed period), and gap closure (gaps that closed in the observed period) in the GIS environment (ArcGIS). For each of these four gap dynamic processes, we calculated the number and area of gaps. The annual proportion of the total area covered by new gaps was calculated:

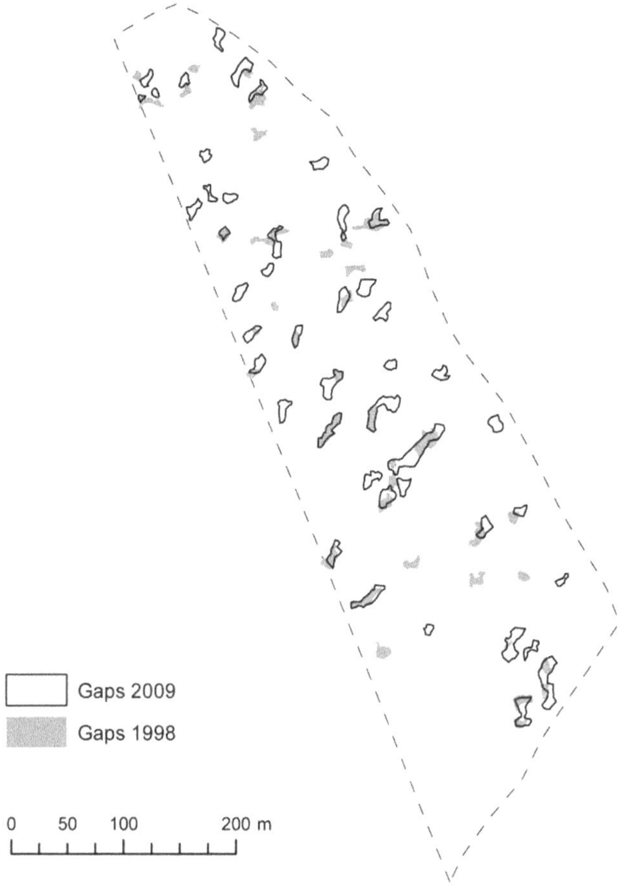

Figure 3. Gap dynamics in the Kopa forest reserve in the studied period (1998–2009).

$$\text{Percent of total area covered by new gaps (annually) (\%)} = \frac{\left(\dfrac{\text{New gap formation area}}{\text{Forest reserve area}} \times 100\right)}{\text{Time between measurements(years)}} \quad (5)$$

The annual proportion of the total area covered by closed gaps was calculated:

$$\text{Percent of total area covered by closed gaps (annually) (\%)} = \frac{\left(\dfrac{\text{Gap closure area}}{\text{Forest reserve area}} \times 100\right)}{\text{Time between measurements(years)}} \quad (6)$$

We also carried out a more detailed analysis of new gap formation in which we studied the relationship between newly formed gaps and the presence of gaps in the first measurement. Newly established gaps were classified into several groups based on their proximity to previously existing gaps. We divided the newly established gaps based on whether they were formed more or less than five meters (minimal diameter of a gap) from a previously existing gap. Another group was composed of gaps formed from previously existing gaps (overlap) that had decreased in size (more or less than 50% in area). Furthermore, gaps formed in the five meter buffer zone of previously existing gaps were classified in a separate group and divided based on whether previously existing gaps decreased in area more or less than 50%. The last group was made up of gaps that coalesced from two previous gaps.

Results

1. Canopy gaps

Delineation of canopy gaps with a digital stereoplotter showed an increase in the gap fraction (proportion of the surface covered by gaps) in both reserves. In the Kopa reserve, the gap fraction increased from 3.2 to 4.5% (41%) during the eleven-year period. Average gap size increased from 118 to 141 m^2 (Figure 1A).

Small gaps (≤200 m^2) were most frequent while medium-sized gaps (200–600 m^2) made up only a small number of gaps (Table 2; Figure 1A). Larger gaps (>600 m^2) were not delineated in this period.

The total change in gap area (169 m^2 per year) was characterized by a large proportion of newly established and expanded gaps. These processes also involved the largest number of gaps. Gap closure and gap reduction represented a lower proportion (in area) but included 61% of all gaps captured in 1998. All gap dynamic processes were predominantly supported by small gaps. The annual proportion of the total area covered by new and closed gaps amounted to 0.15 and 0.08% of the whole forest area, respectively.

In the Gorjanci forest reserve, the gap fraction increased from 9.1 to 10.6% (17%) during the 20-year period. Average gap size decreased from 317 to 294 m^2. The gap size distribution showed a predominance of small gaps (Figure 1A).

Gaps ranging from 100 to 300 m^2 were most frequent, while gaps smaller than 100 m^2 and gaps ranging from 300 to 500 m^2 were less frequent. Larger gaps (>600 m^2) were rare and reached up to 1200 m^2 in size. Even though medium-sized gaps (>200 m^2) were less frequent (Table 3) than small gaps (≤200 m^2) in both measurements (47% in 1986 and 48% in 2006), they represented a dominant proportion in area (77% in 1986 and 74% in 2006).

The total change in gap area (174 m^2 per year) was characterized by a large proportion (in area and in number) of newly established and closed gaps (Table 3, Figure 2). Gap expansion and gap reduction were similar (in area and in number). New gap establishment, gap expansion, and gap closure were dominated by small gaps (in number) and by medium-sized gaps (in area). In contrast, gap reduction was observed mostly in medium-sized gaps (in area and in number). The annual proportion of the total area covered by new and closed gaps amounted to 0.25 and 0.16%, respectively. The gap dynamics were comparable in the Kopa forest reserve (Table 2, Figure 3).

A more detailed analysis of new gap formation (Figure 4) in the Kopa (Table 4) and Gorjanci (Table 5) forest reserves showed that 38% and 41% of gaps formed in the previously undisturbed stand (more than 5 m from the already existing gap) in Kopa and Gorjanci, respectively. In the five meter buffer zone of the previously existing and subsequently closed gap, one gap formed in Kopa and seven gaps in Gorjanci. Most of the gaps were reformed from previously existing gaps, thus partially overlapping the previous gap (47% in Kopa and 41% in Gorjanci). Only a few gaps formed in the five meter buffer zone of the older gaps (four in Gorjanci and one in Kopa). A few gaps (one in Gorjanci and two in Kopa) also coalesced from two smaller gaps. Only one gap in each reserve was enlarged with no reformation of the previous gap edge.

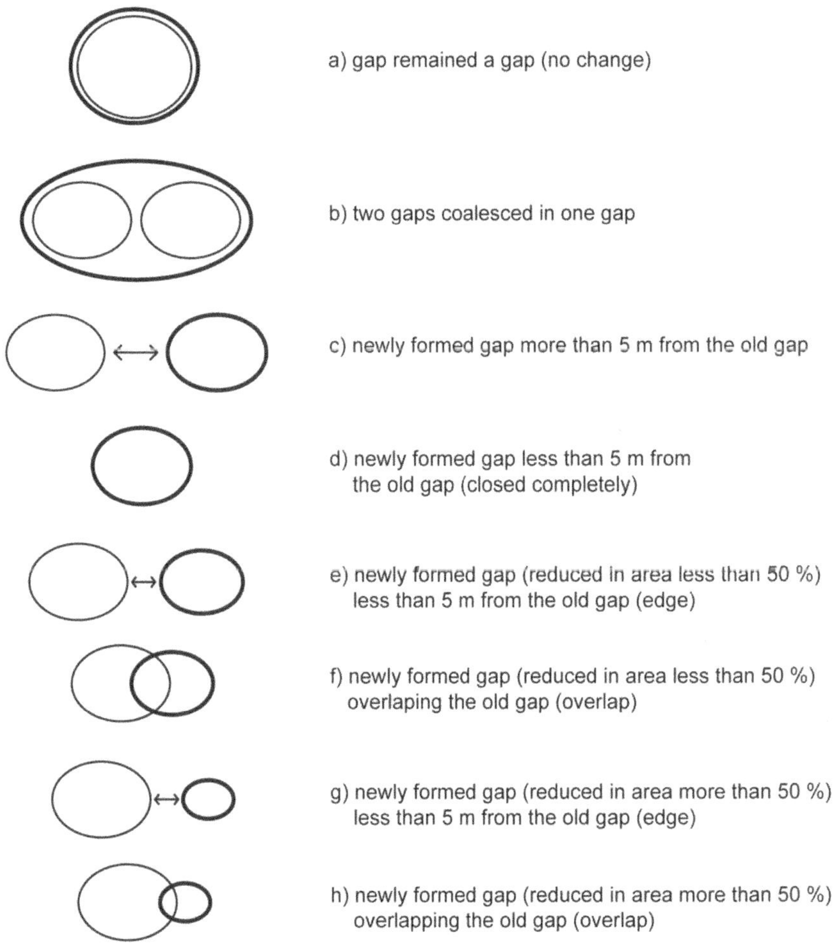

Figure 4. Gap formation processes (basic line represents gaps in the first measurement period, bold line represents gaps in the second measurement period).

2. Forest structure

A complete inventory (2009) of the Gorjanci forest reserve showed that beech was the predominant tree species (97.8% of the total growing stock). Sycamore maple was present in minor proportion (2.2%) and other species, such as silver fir and European rowan (*Sorbus aucuparia* L.), were rare. In the Kopa forest reserve (complete inventory in 2006) beech, sycamore maple, and silver fir accounted for 91.7%, 5.8%, and 2.4% of the total growing stock, respectively, while other tree species such as European ash, small-leaved lime (*Tilia cordata* Mill.), wild cherry (*Prunus avium* L.), goat willow (*Salix caprea* L.), and European rowan, amounted to only 0.1% of the total growing stock.

The analysis of the sample plot data showed a relatively low live tree density in both reserves (Table 6). The variation between the plots was high (coefficient of variation – CV~50%), indicating spatial variability of stem density. The basal area in the Gorjanci and Kopa forest reserves amounted to 41 and 39 $m^2\ ha^{-1}$, respectively. The spatial variability of basal area was observed, with a coefficient of variation of 31% in the Kopa reserve and 27% in the Gorjanci reserve. Similarly, the amount of growing stock was relatively high, amounting to 595 $m^3\ ha^{-1}$ in the Kopa reserve and 650 $m^3\ ha^{-1}$ in the Gorjanci reserve. The analysis of the coarse woody debris (CWD) showed a large dead wood pool in these forest reserves. It comprised 19 and 21% of the total aboveground volume in the Kopa and Gorjanci reserves, respectively. It was spatially unevenly distributed, particularly in the Kopa reserve, where the coefficient of variation amounted to 108%.

A strong tendency towards clustered distribution was observed for all trees with d.b.h.≥5 cm in the Kopa forest reserve. The tendency towards clustered distribution for all trees with d.b.h.≥10 cm was less pronounced in both reserves. The distribution of trees in the upper canopy layer (dominant trees) showed a distinct tendency towards regular distribution in both reserves (Figure 5).

Figure 6 presents a height curve based on 113 measured beech tree heights in both reserves (72 in the Gorjanci reserve and 41 in the Kopa reserve). Tree heights reached up to 41 m. The upper canopy layer (25–40 m height) was constituted by trees of a wide d.b.h. range (from 20 to 110 cm). Trees belonging to the middle layer (10–25 m) ranged from 15 to 30 cm d.b.h.

A complete inventory of all trees (≥10 cm d.b.h.) in the Gorjanci forest reserve showed a decreasing distribution of trees per d.b.h. class (Figure 7). When trees up to 35 cm d.b.h. were examined, a decrease in number was observed over time. A substantial number of trees up to 25 cm d.b.h. were measured in the first measurement (1974). The number of trees with 30 cm≤d.b.h.<75 cm remained relatively constant over the entire measurement period (1974–2009). The number of trees above 75 cm d.b.h. increased over time. This trend was partially

Table 4. Classification of newly established gaps in the Kopa forest reserve.

	1998 gap									
2009 gaps	**Gap (1998) remained a gap (2009)**	**2 gaps (1998) coalesced in 1 gap (2009)**	**Newly formed gap (2009) >5 m from closed gap (1998)**	**Closed completely**	**Partial closure (<50% of area)**		**Partial closure (>50% of area)**		**No change**	
	1 (2.2)	2 (4.4)	17 (38.0)							
	Formed on edge and overlap of 1998 gap			1 (2.2)	10 (22.2)		12 (26.7)		1 (2.2)	
	Edge	Overlap		1 (2.2)	0	10 (22.2)	1 (2.2)	11 (24.5)	0	1 (2.2)

The numbers in parentheses are percentages.

Table 5. Classification of newly established gaps in the Gorjanci forest reserve.

	1986 gap									
2006 gaps	**Gap (1986) remained a gap (2006)**	**2 gaps (1986) coalesced in 1 gap (2006)**	**Newly formed gap (2006) >5 m from closed gap (1986)**	**Closed completely**	**Partial closure (<50% of area)**		**Partial closure (>50% of area)**		**No change**	
	1 (1.2)	1 (1.2)	33 (40.7)							
	Formed on edge and overlap of 1986 gap			7 (8.6)	19 (23.5)		18 (22.2)		1 (1.2)	
	Edge	Overlap		7 (8.6)	2 (2.5)	17 (21)	2 (2.5)	16 (19.7)	0	1 (1.2)

The numbers in parentheses are percentages.

Table 6. Main structural parameters of the Kopa and Gorjanci forest reserves. All parameters measured in the plots were scaled to one hectare.

	Kopa	Gorjanci
Number of sample plots	21	24
Living trees		
Stem density (N ha^{-1}) Average	338 (581)	261
SD	171 (312)	131
CV (%)	51(54)	50
Basal area (m^2 ha^{-1})		
Average	39 (40)	41
SD	12 (12)	11
CV (%)	31 (30)	27
Growing stock (m^3 ha^{-1})		
Average	595 (597)	650
SD	194 (194)	181
CV (%)	33 (32)	28
CWD (m^3 ha^{-1})		
Average	144 (144)	171
SD	155 (155)	70
CV (%)	108 (108)	41
CWD (%)	19 (19)	21

SD...standard deviation, CV...coefficient of variation, CWD...coarse woody debris.
The calculations were made for all trees with d.b.h.≥10 cm. The numbers in parentheses are calculations for all trees with d.b.h.≥5 cm.

masked by the fact that all trees above 75 cm d.b.h. were classified in the same size class (d.b.h.≥75 cm). Growing stock development in the Gorjanci forest reserve showed an increase in the measurement period (Figure 8). This increase was solely due to an increase within the largest-diameter class (d.b.h.≥50 cm).

Similarly, a continuous increase in total growing stock was observed in the Kopa forest reserve from 1985 to 2006 (Figure 9). The increase in growing stock was most pronounced in the largest-diameter class (d.b.h.≥50 cm). The volume of growing stock in the small-diameter class (10 cm≤d.b.h.<30 cm) increased in the last decade measured, while in the previous decades the volume of growing stock was similar. The volume of growing stock in the medium-diameter class (30 cm≤d.b.h.<50 cm) decreased in the first decade and remained at the same level in the last three measurements (21-year period).

Discussion

Gap size analysis in the Kopa reserve showed that small gaps (≤200 m^2) are the dominant driving force of forest dynamics. Medium-sized gaps were rare, and large gaps were not present. The gap fraction increased by 41%, but still remained under 5% of the total area studied. The gap fraction is as low as in some other Southern European beech reserves [11,19]. In the Kopa reserve, medium-sized gaps (>200 m^2) comprised 23% (1998) and 35% (2009) of the total gap area, which is less than what was observed in a similar beech dominated old-growth forests in Slovenia [19]. New gap formation and expansion of small gaps seem to be the main driving forces behind stand dynamics and are mitigated by gap closure. Small gaps are most likely closed by the lateral growth of tree crowns [2,43]. Average gap size in the Kopa reserve was 141 m^2 in 2009 and was less than half the average gap size in the Gorjanci reserve in 2006 (297 m^2). In the Gorjanci reserve, the gap fraction was higher (10.6% in 2006), which is comparable with other beech forest reserves [10,26]. Medium-sized gaps were more frequent here (more than 50% in number) and they constituted three quarters of the total gap area. Large gaps (>1000 m^2) were very rare, even though there have been some reports of large disturbances in beech forest reserves in Slovenia [44], France [10], and Romania [17]. Most of the changes in gap area resulting from gap formation and gap closure occurred in medium-sized gaps, which did not exceed 1200 m^2. Annual rates of gap closure and new gap formation are not as high as those reported in some other forest types [45], but are substantially higher than those found in some beech reserves in Europe [26]. Kenderes et al. [26] presented annual rates of the total area covered by new gaps between 0.05 and 0.09% and annual rates of the total area covered by closed gaps between 0.01 and 0.03%. Our results suggest that gap disturbance regime governed by small gaps can exhibit relatively large changes in gap area over short periods of time. The process of new gap formation is related to the presence of older gaps since a high proportion of the gaps (46–51%) were formed in the vicinity (less than five meters) of the older gaps. This proportion is larger to that of the gaps formed in the previously undisturbed canopy (38–41%). It seems that the process of new gap formation is more frequent in the vicinity of the previously existing gap since the gap area is 10 to 20 times smaller than the area of the undisturbed canopy.

A similar conclusion can be drawn from the variability of stand characteristics estimated in the sampling plots. The variation in basal area and volume per hectare in our study, measured by the coefficient of variation, was between the 20% found in virgin beech forest in Ukraine [13] and the 38% found in beech dominated natural forests in the Czech Republic [46], whereas the variation in tree number was 10 to 13% greater in our study. The high standard deviation (131 to 171) and coefficient of variation (~50%) suggest that these forests are structurally (d.b.h. structure) heterogeneous over small spatial scales. Such high variability has only been detected in Italian [47] and Czech [46] non-managed beech stands, but not in Slovakian ones [48]. Mean tree densities in the Kopa (338 per hectare) and Gorjanci (261 per hectare) forest reserves are comparable with other beech dominated old-growth forests in Europe [6,11,13,14,15,19,16]. Basal area comprises around 40 m^2 ha^{-1} and varies up to 30%, indicating a relatively stable spatial distribution. Higher values (up to 52 m^2 ha^{-1}) have been observed in some comparable forests although lower values have been determined in stands with a shorter history of non-management [49].

The average growing stock of live trees was 595–650 m^3 ha^{-1} and had similar variability as did basal area (28–33%). Much higher values (up to 1268 m^3 ha^{-1}) have been detected in some reserves in Slovakia, Ukraine, and Albania. We observed a high proportion of CWD (19–21% of total growing stock) in the Kopa and Gorjanci reserves. The amount of CWD in both reserves is high compared with similar European beech old-growth forests [50]. CWD was distributed very unevenly in the forest (CV up to 108%), indicating very localized, small disturbance events in the otherwise undisturbed background canopy layer.

Since beech is a shade-tolerant tree species, groups of seedlings can establish under small canopy openings and persist for longer periods of time, thus sustaining themselves only on scarce, moving sun flecks [2], even after canopy gaps are closed. When the spatial distribution of the dominant (upper canopy layer) trees was

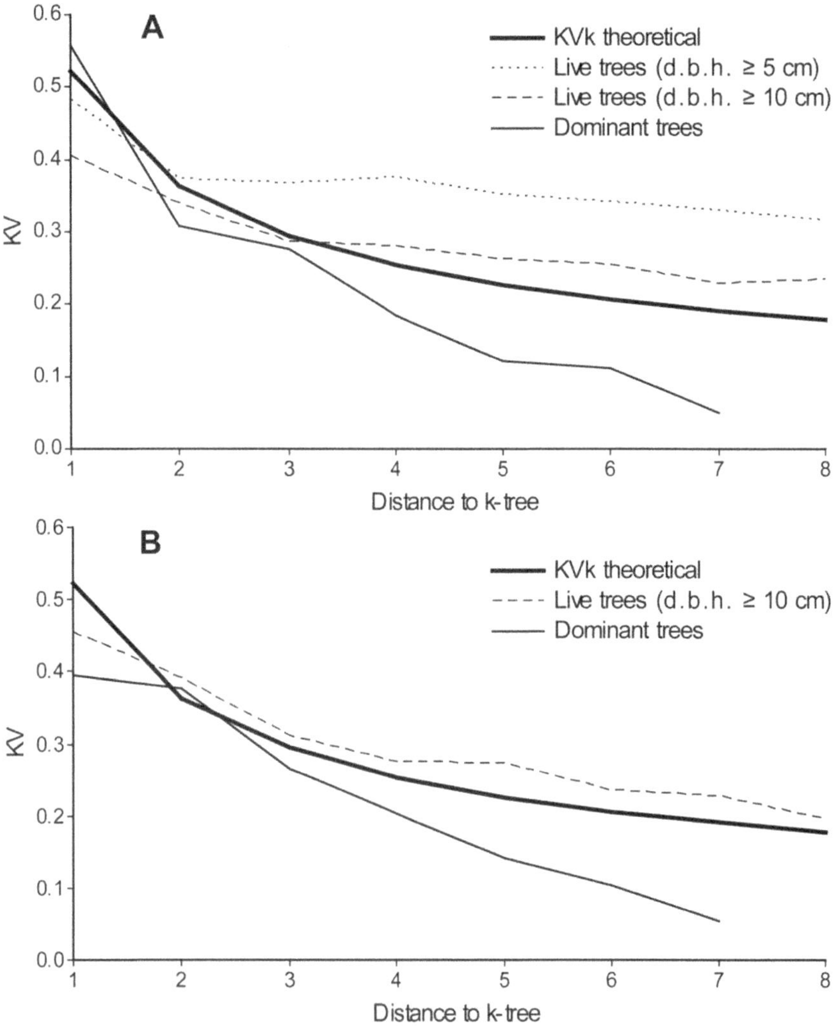

Figure 5. Spatial distribution of live trees in the Kopa (A) and Gorjanci (B) forest reserves. The bold line represents the theoretical (random) distribution of trees.

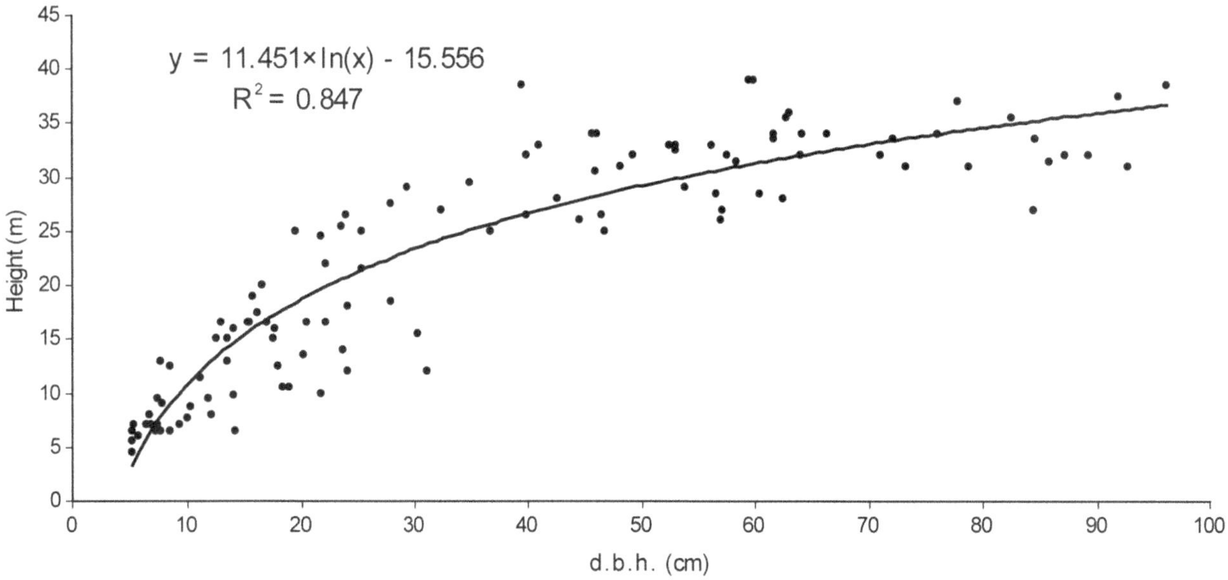

Figure 6. Height curve of beech in the Kopa and Gorjanci forest reserves. "d.b.h." is diameter at breast height.

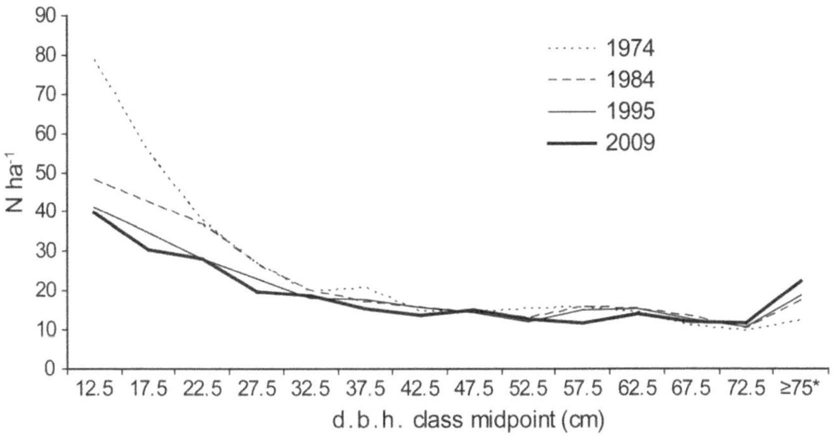

Figure 7. Diameter distribution (frequency of trees per 5-cm diameter classes) of all trees in the Gorjanci forest reserve. Note that all trees with d.b.h.≥75 cm were classified in one 5-cm d.b.h. class (*).

observed, a uniform distribution was evident. This pattern was also observed in some other European old-growth forests [51]. This may be one of the reasons for the 'cathedral like' appearance of pure beech forests.

Tree height analysis showed that the middle layer (11–22 m) was very poorly represented. This could be explained by the fact that beech grows through the pole-sized and intermediate-diameter classes very quickly [6,13]. This may also explain the 'cathedral-like' appearance of beech stands [18]. Successful trees reach the canopy layer (22 m), thus forming a wide diameter distribution of upper storey trees (20–110 cm d.b.h.), similar to other old-growth European beech reserves [13].

The time series of tree diameter distribution in the Gorjanci forest reserve showed a constant decrease in the number of trees smaller than 25 cm in d.b.h. in the measurement period. It is likely that in the preceding period there had been a higher intensity disturbance event, which resulted in favorable light conditions for successful seedling recruitment. Such events have also been detected in beech dominated old-growth forests in Slovakia [6]. It could also be the result of unrecorded felling after World War Two, when a forest road was constructed in the vicinity of the reserve. Because of the probable constant closure of the canopy, we did not find any increase in tree frequency in the larger d.b.h. classes in later periods. This can be attributed to the high mortality among beech saplings under closed canopy. The development of growing stock in the Kopa reserve also indicates a relatively stable rate of increase over the last three measurement periods (1985–

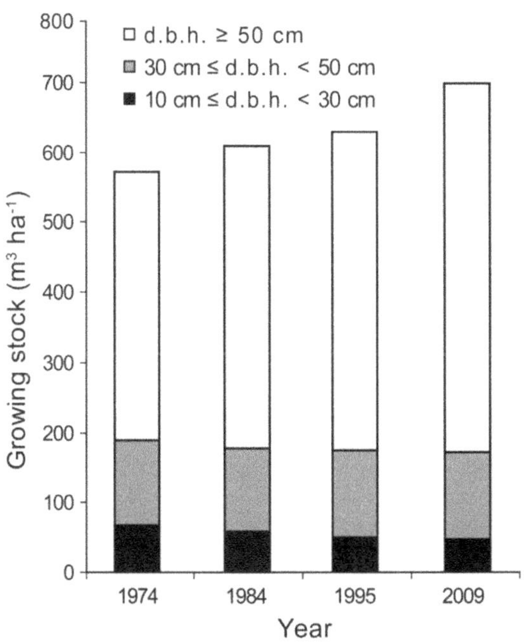

Figure 8. Distribution of growing stock in three d.b.h. classes in the Gorjanci forest reserve. "d.b.h." is diameter at breast height.

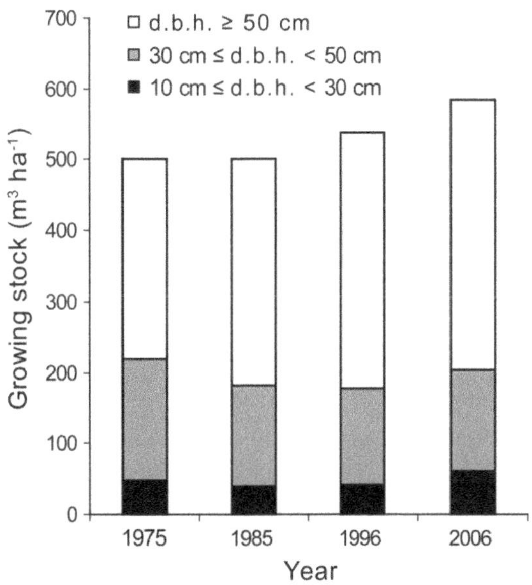

Figure 9. Distribution of growing stock in three d.b.h. classes in the Kopa forest reserve.

2006). The proportion of trees in the growing stock with d.b.h. larger than 50 cm increased, indicating ageing of this forest reserve. An increase in growing stock was also observed in the Gorjanci forest reserve.

Structural characteristics of the forests in our study indicate diverse diameter structure and a less diverse height structure. Consequently, we conclude that these stands are not even-sized (uniform) as proposed by some authors [2], but rather unevenly structured. This is due to the fact that the disturbance regime is characterized by small-scale, low-intensity disturbances that are variable in space and time. These findings largely support the third conceptual model presented in the introduction, although the lack of trees in the middle layer recorded in this study may also lend support to the second model. Because large-scale, high-intensity disturbances are rare, which limits the possibility of synchronous recruitment over larger areas, the first conceptual model is not supported. Yet, it cannot be excluded due to the small size of both reserves and short time interval of measurements. For a better understanding of the disturbance regime in beech dominated old-growth forests, we propose more meta-analyses of similar studies in Europe, as well as analysis of larger areas and the application of dendroecological methods. This would provide an important insight into the disturbance history and age structure of these stands as well as a better understanding of past human land-use practices.

Acknowledgments

We would like to thank the Novo mesto Regional Unit of the Slovenia Forestry Service for all the data and support they provided.

Author Contributions

Conceived and designed the experiments: DH TR JD. Performed the experiments: TR DH. Analyzed the data: TR DH. Contributed reagents/materials/analysis tools: TR DH. Wrote the paper: TR DH JD.

References

1. FAO (2010) Global Forest Resources Assessment 2010. FAO: Rome. 343p.
2. Ellenberg H (1996) Vegetation Mitteleuropas mit den Alpen: in ökologischer, dynamischer und historischer Sicht. Stuttgart: Ulmer. 1095 p.
3. Ellenberg H, Leuschner C (2010) Vegetation Mitteleuropas mit den Alpen. Stuttgart : E. Ulmer. 1333 p.
4. Leibundgut H (1959) Über Zweck und Methodik der Struktur< und Zuwachsanalyse von Urwäldern. Schweiz Z Forstwes 110:111–124.
5. Pruša E (1985) Die böhmischen und mährischen Urwälder - ihre Struktur und Ökologie. Praga: Verlag der Tschechoslowakischen Akademie der Wisenschaften. 573 p.
6. Korpel Š (1995) Die Urwälder der Westkarpaten, Gustav Fischer Verlag: Stuttgart, Jena, New York. 310 p.
7. Peters R (1997) Beech Forests. Dordrecht ; Boston ; London : Kluwer Academic Publishers. 169 p.
8. Emborg J (1998) Understorey light conditions and regeneration with respect to the structural dynamics of a near-natural temperate deciduous forest in Denmark. For Ecol Manage 106: 83–95.
9. Christensen M, Emborg J, Busse Nielsen A (2006) The forest cycle of Suserup Skov- revisited and revised. Eco Bull 52:1–10.
10. Pontailler JY, Faille A, Lemee G (1997) Storms drive successional dynamics in natural forests: a case study in Fontainebleau forest (France). For Ecol Manage 98:1–15.
11. Tabaku V, Meyer P (1999) Gap patterns of Albanian and Central European beech forests. Forstarchiv 70:87–97.
12. Manning DB (2003) Gap Studies in the Weber- stedter Holz, a Near Natural Beech Forest in Central Germany. Forest &Landscape, Denmark. NAT- MAN project. pp. 1–23.
13. Commarmot B, Bachofen H, Bundziak Y, Bürgi A, Ramp B, et al. (2005) Structures of virgin and managed beech forests in Uholka (Ukraine) and Sihlwald (Switzerland): a comparative study. For Snow Landsc Res 79:45–56.
14. Drössler L, von Lüpke B (2005) Canopy gaps in two virgin beech forest reserves in Slovakia. J For Sci 51:446–457.
15. von Oheimb G, Westphal C, Tempel H, Hardtle W (2005) Structural pattern of a near-natural beech forest (*Fagus sylvatica*) (Serrahn, North-east Germany). For Ecol Manage 212:253–263.
16. Drössler L, von Lüpke B (2007) Stand structure, regeneration and site conditions in two virgin beech forest reserves in Slovakia. Allg Forst Jagdztg 178:121–135.
17. Aszalos R, Standovar T (2003) A study of Gap Characteristics in the Nera Forest Reserve using Satellite Image Analysis. For Land, Denmark, pp. 1–14
18. Leibundgut H (1982) Europäische Urwälder der Bergstufe, Bern: P.Haupt. 308 p.
19. Zeibig A, Diaci J, Wagner S (2005) Gap disturbance patterns of a *Fagus sylvatica* virgin forest remnant in the mountain vegetation belt of Slovenia. For Snow Landsc Res 79:69–80.
20. Runkle JR (1981) Patterns of disturbance in some old-growth mesic forests of eastern North America. Ecology 63:1533–1546.
21. Lertzman KP (1992) Patterns of gap-phase replacement in a subalpine, old-growth forest. Ecology 73:657–669.
22. Yamamoto S, Nishimura N (1999) Canopy gap formation and replacement pattern of major tree species among developmental stages of beech (*Fagus crenata*) stands, Japan. Plant Eco 140:167–176.
23. Battles JJ, Fahey TJ (2000) Gap Dynamics Following Forest Decline: A Case Study of Red Spruce Forests. Eco Appl 10:760–774.
24. Nagel TA, Svoboda M (2008) Gap disturbance regime in an old-growth *Fagus-Abies* forest in the Dinaric Mountains, Bosnia-Herzegovina. Can J For Res 38:2728–2737.
25. Kucbel S, Jaloviar P, Saniga M, Vencurik J, Klimas V (2010) Canopy gaps in an old-growth fir-beech forest remnant of Western Carpathians. Eur J For Res 129:249–259.
26. Kenderes K, Kral K, Vrska T, Standovar T (2009) Natural gap dynamics in a Central European mixed beech-spruce-fir old-growth forest. Eco Sci 16:39–47.
27. Mlinšek D, Accetto M, Anko B, Piskernik M, Robič D, et al. (1980) Gozdni rezervati v Sloveniji. Institut za gozdno in lesno gospodarstvo pri Biotehniski fakulteti. Ljubljana. 414 p.
28. Penttinen A, Stoyan D, Henttonen HM (1992) Marked Point Processes in Forest Statistics. For Sci 38:806–824.
29. Fortin M-J, Dale MRT (2005) Spatial analysis: a guide for ecologists, Cambridge University Press. 365 p.
30. Pommerening A (2002) Approaches to quantifying forest structures. Forestry 75:305–324.
31. Dale MRT (1999) Spatial pattern analysis in plant ecology. Cambridge University Press. 326 p.
32. Kotar M (1993) Dolocanje nacina razmestitve dreves v optimalni razvojni fazi gozda. Zbornik gozdarstva in lesarstva 42:121–153.
33. Puhek V (1993) Preizkus veljavnosti metod razmika za ocenjevanje gostote in temeljnice sestojev. Zbornik gozdarstva in lesarstva 42:155–198.
34. Fueldner K (1995) Zur Strukturbeschreibung in Mischbestaenden. Forstarchiv 66:235–240.
35. Clark PJ, Evans FC (1954) Distance to Nearest Neighbor as a Measure of Spatial Relationships in Populations. Ecology 35:445–453.
36. Thompson HR (1956) Distribution of Distance to Nth Neighbour in a Population of Randomly Distributed Individuals. Ecology 37:391–394.
37. Puhek V (1998) Procjena strukturnih elemenata sastojine na osnovu prostornog razporeda stabala. Zagreb: Sveuciliste u Zagrebu, Sumarski fakultet. 194 p.
38. Zingg A (1994) Top heights in mixed stands: their definition and calculation. In: da Costa P, Emilia M, Preushler T (Ed.) Mixed Stands, Research Plots, Measurements and Results, Models. Proceedings from the Symposium of the IUFRO Working Groups S4.01-03/S4.01-04, Lousa Portugal, April 25–29, Universidad de tecnica de Lisboa, Lisboa. pp. 69–79.
39. Sharma M, Amateis RL, Burkhart HE (2002) Top height definition and its effect on site index determination in thinned and unthinned loblolly pine plantations. For Ecol Manage 168:163–175.
40. Anonymous (2007) Forest management plan for FMU Crmosnjice 1977–2016.
41. Anonymous (2010) Forest management plan for FMU Gorjanci 1975–2014.
42. Waser LT, Ginzler C, Kuechler M, Baltsavias E, Hurni L (2011) Semi-automatic classification of tree species in different forest ecosystems by spectral and geometric variables derived from Airborne Digital Sensor (ADS40) and RC30 data. Rem Sens Environ 115:76–85.
43. Lemoine D, Jacquemin S, Granier A (2002) Beech (*Fagus sylvatica* L.) branches show acclimation of xylem anatomy and hydraulic properties to increased light after thinning. Ann For Sci 59:761–766.
44. Marinšek A, Diaci J (2004) Razvoj inicialne faze na vetrolomni povrsini v pragozdnem ostanku Ravna gora (Development of the initial phase after windthrow in the virgin forest remnant Ravna gora). Zbornik gozdarstva in lesarstva 73:31–50.
45. Lorimer CG, Frelich LE (1989) A methodology for estimating canopy disturbance frequency and intensity in dense temperate forests. Can J For Res 19:651–663.
46. Kral K, Janik D, Vrska T, Adam D, Hort L, et al. (2010) Local variability of stand structural features in beech dominated natural forests of Central Europe: Implications for sampling. For Ecol Manage 260:2196–2203.

47. Piovesan G, Di Filippo A, Alessandrini A, Biondi F, Schirone B (2005) Structure, dynamics and dendroecology of an old-growth *Fagus* forest in the Apennines. J Veg Sci 16:13–28.
48. Schütz JP, Saniga M (2011) Modelling the risks of natural stand closure release with ageing in pure beech (*Fagus sylvatica*) and spruce (*Picea abies*) stands. Ann For Sci 68:1105–1114.
49. Borrmann K (1996) Vierzig Jahre Naturwaldforschung im Heilige Hallen-Bestand. Allg Forst Jagdztg 23:5.
50. Christensen M, Hahn K, Mountford EP, Odor P, Standovar T, et al. (2005) Dead wood in European beech (*Fagus sylvatica*) forest reserves. For Ecol Manage 210:267–282.
51. Szwagrzyk J, Czerwczak M (1993) Spatial Patterns of Trees in Natural Forests of East-Central-Europe. J Veg Sci 4:469–476.

5

Response of Soil Respiration to Acid Rain in Forests of Different Maturity in Southern China

Guohua Liang[1,2], Xingzhao Liu[1,2], Xiaomei Chen[1,2], Qingyan Qiu[1,2], Deqiang Zhang[1], Guowei Chu[1], Juxiu Liu[1], Shizhong Liu[1], Guoyi Zhou[1]*

1 South China Botanical Garden, Chinese Academy of Sciences, Guangzhou, China, **2** University of Chinese Academy of Sciences, Beijing, China

Abstract

The response of soil respiration to acid rain in forests, especially in forests of different maturity, is poorly understood in southern China despite the fact that acid rain has become a serious environmental threat in this region in recent years. Here, we investigated this issue in three subtropical forests of different maturity [i.e. a young pine forest (PF), a transitional mixed conifer and broadleaf forest (MF) and an old-growth broadleaved forest (BF)] in southern China. Soil respiration was measured over two years under four simulated acid rain (SAR) treatments (CK, the local lake water, pH 4.5; T1, water pH 4.0; T2, water pH 3.5; and T3, water pH 3.0). Results indicated that SAR did not significantly affect soil respiration in the PF, whereas it significantly reduced soil respiration in the MF and the BF. The depressed effects on both forests occurred mostly in the warm-wet seasons and were correlated with a decrease in soil microbial activity and in fine root biomass caused by soil acidification under SAR. The sensitivity of the response of soil respiration to SAR showed an increasing trend with the progressive maturity of the three forests, which may result from their differences in acid buffering ability in soil and in litter layer. These results indicated that the depressed effect of acid rain on soil respiration in southern China may be more pronounced in the future in light of the projected change in forest maturity. However, due to the nature of this field study with chronosequence design and the related pseudoreplication for forest types, this inference should be read with caution. Further studies are needed to draw rigorous conclusions regarding the response differences among forests of different maturity using replicated forest types.

Editor: Ben Bond-Lamberty, DOE Pacific Northwest National Laboratory, United States of America

Funding: This work was financially supported by the Strategic Priority Research Program of Chinese Academy of Sciences (XDA05050205), the National Key Basic Research Development Program (973 Program) (2009CB421101) and the Guangdong Provincial Natural Science Foundation Project (No. 8351065005000001). The funders had no role in study design, data collection and analysis, decision to publish, or preparation of the manuscript.

Competing Interests: The authors have declared that no competing interests exist.

* E-mail: gyzhou@scib.ac.cn

These authors contributed equally to this work.

Introduction

Acid rain, as a result of the dissolution of atmospheric sulfur dioxide (SO_2) and nitrogen oxides (NO_x) [1,2] which originate mostly from anthropogenic activities such as industrial emission and automobile exhaust [3,4], has been recognized as a worldwide environmental problem since the 1970s [5]. Studies have affirmed that acid rain can have detrimental effects on terrestrial and aquatic ecosystems [6]. As an important component of the terrestrial ecosystems, forests have also been deteriorating widely under the chronic stress of acid rain [5]. In some forests, for example, acid rain accelerated the leaching of nutrients from plant and soil [4,7–10], stimulated Al^{3+} in soil solution which is toxic to the fine roots and the functioning of microbial community [11], and altered the forest species composition [12].

Although the emission of acidic gases in developed countries has declined in recent years, it is still increasing in many developing countries. Southern China was reported as the third largest acid rain area after Europe and the United States [13]. In this region, the annual average pH value of precipitation is generally below 4.5 in recent years, and precipitation with pH as low as 3.5 has been observed [14,15]. Soil acidification due to acid rain has also occurred during the past 32–35 years, with a decrease of soil pH ranged from 0.1 to 1.0 pH units in some forests [16,17]. One of the consequences of the fast soil acidification in this region is an alteration of the biogeochemical cycles and a challenge of the stability in the forest ecosystems [18].

Second only to gross photosynthesis (100–120 PgC yr^{-1}), CO_2 emission from soils (i.e. soil respiration, 68–100 PgC yr^{-1}) is a major pathway in the global carbon cycle [19,20]. Forest soil is an important source of CO_2 in atmosphere [21], and soil respiration in forest is therefore a key process that underlies our understanding of the terrestrial carbon cycle [22]. Also, its response to environmental changes is an increasing concern [23]. Acid rain changes the conditions of soil and plant roots in forests, which is thought to have potential effects on soil respiration [24]. Although many efforts have been devoted to investigating the response of soil respiration in forests to acid rain, the results are often inconsistent. Decreases, increases, or unchanged in soil respiration after SAR treatments have been reported [25–27]. Moreover, up to now, most studies of SAR on soil respiration have been performed in temperate forest ecosystems in developed countries such as Europe and the United States [26]. There have been very few field studies on the response of soil respiration to acid rain in subtropical forests in southern China [28,29]. Stand composition of forests in this region generally changes from coniferous to mixed coniferous and broad-leaved and to broad-leaved in the process of forest succession [30] and therefore, forms forests of different maturity. Due to the different environmental conditions such as soil properties [31] and litter layer properties [32], we hypothesized that the response of soil respiration to acid rain would be different in forests of different maturity. For these reasons mentioned above, we carried out a field experiment to investigate the effects of SAR on soil respiration in three forests of different maturity at the Dinghushan Forest Ecosystem Research Station in southern China.

The aims of this study were to (1) determine if acid rain addition affects soil respiration in subtropical forests with high acidic soil, (2) compare the differences in the response of soil respiration to SAR among forests of different maturity, (3) and to identify possible mechanisms of the observed effects.

Materials and Methods

Ethics Statement

The study site is maintained by the South China Botanical Garden, Chinese Academy of Sciences. The location is within the Dinghushan Forest Ecosystem Research Station. All necessary permits were obtained for the described field study. The field study did not involve endangered or protected species. Data will be made available upon request.

Site Description

The Dinghushan Forest Ecosystem Research Station, with an area of 1133 ha and an elevation ranging from 10 to 1000 m above sea level, is located in the middle part of Guangdong Province in southern China (112°30′–112°33′E, 23°09′–23°11′N). The area is characterized by a typical south subtropical monsoon climate with a distinct seasonal pattern. The annual mean temperature is 21°C with the maximum and minimum monthly mean temperature being 28.0°C in July and 12.6°C in January, respectively. The annual precipitation is 1927 mm, of which nearly 80% falls in the warm-wet season (April – September) and 20% in the cool-dry season (October – March) [33]. The annual average relative humidity is 82%. Acid rain was a threat in this area with an annual average pH value of precipitation proximately 4. 90 and a frequency of acid rain above 63% [34]. Bedrocks of the station are classified as Devonian sandstone and shale [35] and soils are classified as lateritic red earth (oxisol), loamy in texture, and acidic [36]. At the station, there are three types of forests of different maturity: the pine forest (PF), the mixed conifer and broadleaf forest (MF) and the broadleaf forest (BF) with age of more than 60, 110 and 400 years, respectively. They represent forests in young, transition, and old-growth stages in the region [30,37]. During natural succession, heliophytes gradually invade the pine forests to form mixed forests, and mesophytes subsequently invade the mixed forests and eventually transform them into evergreen broadleaf forests [31]. The dominant species were *Pinus massoniana* in the PF, *Castanopsis chinensis*, *pinus massoniana* and *Schima superba* in the MF and *Cryptocarya concinna*, *Machilus chinensis* and *Cryptocarya chinensis* in the BF [32]. The main characteristics of the forests are listed in Table 1.

Experimental Treatments

Four SAR treatments were established by irrigating the plots with water of different pHs: CK (pH 4.5, the local lake water), T1 (water pH 4.0), T2 (water pH 3.5) and T3 (water pH 3.0). There were three replicates for each treatment. Twelve plots were established in each forest, with 10 m×10 m for each plot and surrounded by a 3 m wide buffer strip. All treatments were arranged randomly. According to Liu *et al.* [38], the pH value of precipitation in this region ranged from 4.36 during the dry period to 5.61 during the wet one. Considering the pH value of precipitation would probably decrease in the future, we set the T1 treatment as the lowest pH value observed in the natural rain and other two pH levels 0.5 unit lower each time. To reflect the real and the tendency of mole ratio of S:N according to the previous acid rain records, acidic solutions were prepared by adding a mixture of H_2SO_4 and HNO_3 in a 1:1 mole ratio to the local lake water. SAR treatments were initiated in June 2009 and were sprayed twice a month during the soil respiration measurement period. The simulated rainfall was applied to each plot below the canopy using a gasoline engine sprayer. The amount applied to each plot was 40 L per application. During the experimental period, the total H^+ load each plot received was 9.6, 32, 96 mol ha^{-1} yr^{-1} in the T1, T2 and T3 treatments, respectively, which was equal to about 0.6, 2.0 and 6.0 times, respectively of that in the through-fall of the three forests.

Table 1. Stand characteristics of the pine forest (PF), the mixed conifer and broadleaf forest (MF) and the broadleaved forest (BF) including the elevation, soil organic carbon (SOC), litterfall and accumulated litter.

Forest	PF	MF	BF
Elevation (m)	200–300	220–300	220–300
SOC** (g kg^{-1})	39.5±14.0	25.1±4.4	15.1±3.1
Litterfall* (g m^{-2} yr^{-1})	598.3±56.1	701.2±83.8	893.2±59.2
Accumulated litter* (g m^{-2})	1058±121	686±96	575±77

*From Yan *et al.* (2006) [85].
**FromYi *et al.* (2007) [86]. The mean values of 0–15 cm soil layer.

Field Sampling and Measurements

Ten months after the initial SAR application, soil respiration measurements were made twice a month from April 5, 2010 to March 20, 2012. Soil respiration was measured between 9:00 am and 12:00 am each time which were close to the daily mean, based on a study at an adjacent site where the diurnal pattern of soil respiration was measured [31]. Soil respiration was measured using a Li-8100 Infrared Gas Analyzer (Li-Cor Inc., Lincoln, NE, USA) with attached survey chamber. Two PVC collars (20 cm in diameter) were permanently anchored 5 cm into the soil in each SAR treatment plot. To eliminate the influence of plants on soil respiration, all living plants in the collars were removed before soil respiration measurement. Soil temperature (°C) at 5 cm depth and soil moisture (volumetric water content, %) of the top 5 cm soil layer were monitored simultaneously adjacent to each PVC collar using a digital thermometer and a PMKit [31], respectively.

To determine soil pH value, soil samples were collected in all forests in June and December of each year during the study period. We also collected soil samples in June 2011 for determining soil microbial biomass carbon. Two composite samples of four cores (2.5 cm inside diameter) from the upper soil layer (0–10 cm) were collected randomly from each plot. The composite samples were gently mixed and stored at 4°C until processing. Dead roots, litter and plant residues were picked out and the samples were passed through a 2-mm-mesh sieve. The soil pH values were measured using a glass electrode (1:2.5 soil-water ratio) after shaking the samples to equilibration for approximately 30 min [39]. The soil microbial biomass carbon was estimated using DOC difference and a Kc factor of 0.33 [40,41]. To determine fine root biomass, roots of 0–10 cm soil layer was collected by using a 6.8 cm diameter stainless-steel corer in June 2011. Two composite samples of three cores were randomly collected from each plot. The fine roots (diameter ≤2 mm) were separated by washing and sieving, dried at 60°C for 48 h and weighed [42].

Statistical Analysis

Soil respiration in each pot was calculated as the mean of the two collar measurements, getting a sample size of three for each treatment in the analysis. Repeated measures ANOVA with Tukey's HSD test was performed to examine the effects of SAR treatments on soil respiration, soil temperature, soil moisture and soil pH value among treatments for the study period. Standard t-test was used to test the seasonal (i.e. warm-wet vs. cool-dry seasons) difference in means of soil respiration rate, soil temperature and soil moisture. One-way ANOVA with Tukey's HSD test was used to test the difference among treatments in means of annual soil respiration, soil pH value, soil microbial biomass carbon and fine root biomass, and among the control (CK) plots in three different forests. An exponential and a linear equation were used: $R = a\exp^{(bT)}$ (1); $R = cM+d$ (2) [43,44], where R is soil respiration rate (μmolCO$_2$ m^{-2} s^{-1}), T is soil temperature (°C), M is volumetric soil moisture (%) and a, b, c and d are constants fitted to the regression equation. The Q_{10} value which was defined as the difference in respiration rates over a 10°C interval was calculated using the exponential relationship between soil respiration and soil temperature ($Q_{10} = \exp^{(10b)}$) [45,46]. One-way ANOVA test was also used to compare the regression slopes (b values) among treatments. Data analyses were carried out using the SAS software (SAS Version 8.0, SAS Institute Inc. Cary, NC, USA). Statistical significant differences were set with $p<0.05$ unless otherwise stated.

Results

Soil Temperature and Soil Moisture

Soil temperature and soil moisture (Fig. 1 a, b) exhibited strong seasonal patterns for all treatments in the three forests. Soil was warmer and wetter from April to September (the warm-wet season) than that from October to March of the next year (the cool-dry season) ($p<0.05$ for all). Annual mean soil temperature in the control (CK) plots was in the order of PF (22.30±0.13°C)>MF (20.67±0.06°C)>BF (20.00±0.04°C) ($p<0.05$). Annual mean soil moisture in the control (CK) plots was not significantly different between the MF (26.98±0.61%) and the BF (27.68±0.71%), while the PF (13.76±0.24%) was significantly lower than the MF and the BF ($p<0.05$). There was no treatment effect on neither soil temperature nor soil moisture ($p>0.05$ for both in all forests) during the study period.

Soil Respiration

Soil respiration in different treatments also followed a clear seasonal pattern in all three forests during the study period, with significantly higher rates in the warm-wet seasons and lower rates in the cool-dry ones ($p<0.05$) (Table 2; Fig. 1 c). Mean annual soil respiration in the control (CK) plots was 4.30±0.35, 4.20±0.20 and 4.28±0.13 kgCO$_2$ m^{-2} yr^{-1} in the PF, the MF and the BF, respectively, and there was no significant difference among three forests ($p>0.05$) (Table 3; Fig. 1 c).

The sensitivity of the response of soil respiration to SAR showed an increasing trend with the progressive maturity of three forests. Compared with the CK treatment, mean annual soil respiration was 0.3, 1.7 and 4.0% lower in the T1, T2 and T3 treatments, respectively in the PF. The repeated measures ANOVA showed that SAR did not affect soil respiration in the PF ($p=0.97$), but it significantly reduced soil respiration in the MF and the BF($p=0.02$ and 0.01, respectively). Compared with the CK treatment, mean annual soil respiration in the T1, T2 and T3 treatments was changed by −0.1, 7.8 and 14.0%, respectively in the MF and 0, 8.9 and 15.8%, respectively in the BF (Table 3; Fig. 1 c). Both in the MF and the BF, there were no significant differences among the CK, T1 and T2 treatments, while the T3 treatment was significantly lower than the CK and T1 treatments ($p<0.05$ for both). These negative effects were evident in the warm-wet seasons (($p<0.01$ and $p=0.02$ in the MF and the BF, respectively), but not in the cool-dry ones ($p>0.05$ for both) (see Table 2; Fig. 1 c).

In addition, by analyzing the annual soil respiration of each year, we found that these negative effects had been strengthened over time. Compared with the CK treatment, there were −1.73–1.12, 1.37–12.86 and 1.56–12.99% lower in the acid treatment plots in the PF, the MF and the BF, respectively ($p>0.05$ for all) in the first year, and 2.55–7.21, 1.65–17.23 and 1.60–18.79% lower in the PF, the MF and the BF, respectively (significantly different in the MF and the BF with $p<0.05$ for both) in the second year (Table 3; Fig. 1 c).

In all treatments of the three forests, soil respiration exhibited significantly positive exponential relationships with soil temperature ($p<0.01$, with R^2 ranging from 0.34 to 0.41, 0.51 to 0.67 and 0.59 to 0.71 in the PF, the MF and the BF, respectively) and significantly positive linear relationships with soil moisture ($p<0.01$, with R^2 ranging from 0.10 to 0.23, 0.25 to 0.45 and 0.19 to 0.22 in the PF, the MF and the BF, respectively) (Table 4; Fig. 2). The mean temperature sensitivity (Q_{10}) values for the CK, T1, T2 and T3 treatments were 1.75, 1.75, 1.69 and 1.71 in the PF, 1.80, 1.88, 1.76 and 1.72 in the MF and 1.88, 1.87, 1.80 and 1.74 in the BF, respectively (Table 4). Q_{10} value showed a declining trend with the intensification of acidification, although it was not statistically significant.

Soil pH Value

From December 2009 to December 2011, mean soil pH values in the control (CK) plots were 3.97±0.06, 3.88±0.01 and 3.90±0.01 in the PF, the MF and the BF, respectively (Fig. 3). Mean soil pH values in the MF and the BF were significantly lower than that in the PF ($p<0.05$). The repeated measures ANOVA showed that SAR did not affect soil pH value in the PF ($p=0.997$), but it significantly reduced soil pH value in the BF ($p=0.01$) and the reduction was marginally significant in the MF ($p=0.07$). Also, we observed that these declining trends in the BF and the MF had been strengthened over time. The significant differences of soil pH value among treatments were found in the BF in June 2011 and December 2011(the pH values in the T3 and T2 treatments were significant lower than those in the CK and T1 treatments in both time with $p<0.05$) and in the MF in June 2011(the pH value in the T3 treatment was significant lower that in the CK treatment with $p<0.05$).

Soil Microbial Biomass Carbon and Fine Root Biomass

Mean microbial biomass carbon in the control (CK) plots in the MF (486.12±55.03 mg kg^{-1}) and the BF (603.76±46.18 mg kg^{-1}) was significantly higher than that in the PF (205.42±44.00 mg kg^{-1}) ($p<0.05$). On the contrary, fine root biomass in the control (CK) plots in the MF (84.03±6.00 g m^{-2})

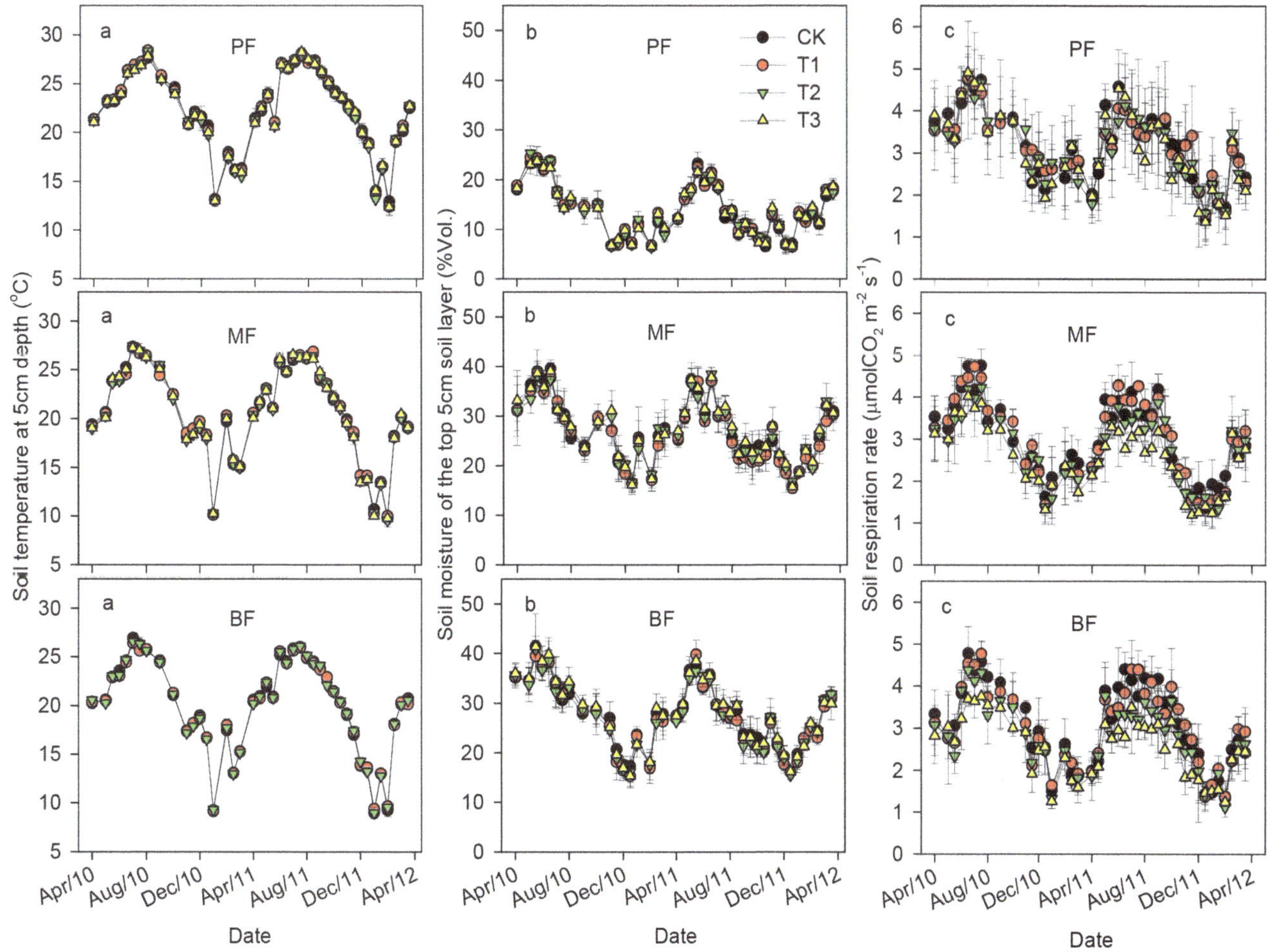

Figure 1. Seasonal dynamics of soil temperature, soil moisture and soil respiration under different SAR treatments in the pine forest (PF), the mixed conifer and broadleaf forest (MF) and the broadleaved forest (BF). (a) soil temperature at 5 cm depth; (b) volumetric soil moisture of the top 5 cm soil layer; (c) soil respiration rate. Error bars are standard errors of the mean (n = 3 for all the treatments). The treatments are: CK = control, T1 = pH 4.0, T2 = pH 3.5, T3 = pH 3.0.

and the BF (86.53±7.54 g m^{-2}) was significantly lower than that in the PF (135.53±27.60 g m^{-2}) ($p<0.05$) (Fig. 4). Compared with the CK treatment, the microbial biomass carbon was 2.5, 9.0 and 20.0% lower in the T1, T2 and T3 treatments, respectively in the MF and 6.4, 13.6 and 15.6% lower, respectively in the BF, and significant differences were found among the T3, T1 and CK treatments in the MF and between the T3 and CK treatments in the BF ($p<0.05$ for both). Although the fine root biomass had the same decreasing tendency with 3.3, 16.9 and 27.3% lower in the T1, T2 and T3 treatments, respectively compared with the CK treatment in the MF and 4.0, 8.5 and 21.0% lower, respectively in the BF, the differences among treatments were not statistically significant in both forests ($p>0.05$ for both). However, there were no significant differences among treatments in the microbial biomass carbon and the fine root biomass in the PF ($p>0.05$ for all).

Table 2. Mean soil respiration rate in the pine forest (PF), the mixed conifer and broadleaf forest (MF) and the broadleaved forest (BF) under different SAR treatments (mean ± standard deviations).

Forest	Season	CK	T1	T2	T3
PF	Wet season	3.69±0.21 a*	3.63±0.59 a*	3.60±0.55 a*	3.61±0.06 a*
	Dry season	2.46±0.23 a*	2.56±0.67 a*	2.44±0.41 a*	2.27±0.25 a*
MF	Wet season	3.72±0.16 a*	3.78±0.11 a*	3.41±0.03 b*	3.12±0.20 b*
	Dry season	2.27±0.18 a*	2.24±0.15 a*	2.12±0.28 a*	1.95±0.13 a*
BF	Wet season	3.72±0.16 a*	3.64±0.28 a*	3.29±0.32 ab*	3.03±0.05 b*
	Dry season	2.43±0.03 a*	2.49±0.06 a*	2.27±0.27 a*	2.09±0.12 a*

n = 3 for all the treatments. Mean values within a row with different lowercase letter have significant treatment differences at $p = 0.05$ level. Means values within each column indicated by the asterisk (*) show significant seasonal differences at $p = 0.05$ level. The treatments are: CK = control, T1 = pH 4.0, T2 = pH 3.5, T3 = pH 3.0. Unit: μmolCO2 m^{-2} s^{-1}.

Discussion

Soil Respiration of the Control (CK) Plots in Three Forests

The mean annual soil respiration in the control (CK) plots of our forests (Table 3) was in the same range as a previous study in adjacent forests [31] and as some similar studies worldwide [47–49]. The mean temperature sensitivity (Q_{10}) values for soil respiration in the control (CK) plots (Table 4) was similar to the

Table 3. Annual soil respiration (R) and its % decreased for each year in the pine forest (PF), the mixed conifer and broadleaf forest (MF) and the broadleaved forest (BF) under different SAR treatments (mean ± standard deviations).

		CK	T1		T2		T3	
Forest	**Year**	**R**	**R**	**% decreased**	**R**	**% decreased**	**R**	**% decreased**
PF	The 1st year	4.49±0.59 a	4.57±0.37 a	−1.73	4.54±0.42 a	−1.20	4.44±0.44 a	1.12
	The 2nd year	4.11±0.11 a	4.01±1.09 a	2.55	3.91±0.09 a	4.95	3.82±0.67 a	7.21
	Mean	4.30±0.35 a	4.29±0.72 a	0.32	4.23±0.26 a	1.74	4.13±0.55 a	4.03
MF	The 1st year	4.41±0.07 a	4.35±0.05 a	1.37	4.04±0.35 a	8.28	3.84±0.49 a	12.86
	The 2nd year	3.99±0.36 a	4.06±0.20 a	−1.65	3.70±0.11 ab	7.34	3.30±0.17 b	17.23
	Mean	4.20±0.20 a	4.20±0.12 a	−0.07	3.87±0.21 ab	7.83	3.61±0.24 b	13.99
BF	The 1st year	4.40±0.21 a	4.33±0.32 a	1.56	4.07±0.27 a	7.60	3.83±0.18 a	12.99
	The 2nd year	4.16±0.21 a	4.23±0.16 a	−1.60	3.73±0.52 ab	10.32	3.38±0.14 b	18.79
	Mean	4.28±0.13 a	4.28±0.24 a	0.01	3.90±0.37 ab	8.92	3.61±0.09 b	15.81

n = 3 for all the treatments. Mean values within a row with different lowercase letter have significant treatment differences at $p = 0.05$ level. The treatments are: CK = control, T1 = pH 4.0, T2 = pH 3.5, T3 = pH 3.0. % decreased = 100 ((R of the CK treatment − R of each treatment)/R of the CK treatment) % (in the same time period); negative values within the column means % increased. The 1st year: April 2010 to March 2011; The 2nd year: April 2011 to March 2012; Mean: April 2010 to March 2012. Unit: kgCO2 m^{-2} yr^{-1} for R.

reported values in a subtropical forest in China (1.75–2.55, [50]), but lower than those in a temperate forest (2.6–3.2, [51]). Soil respiration in the control (CK) plots was not significantly different among the three forests of different maturity, which may be related to the compensation mechanism discussed below. Soil respiration mainly consists of two components: autotrophic (root) respiration and heterotrophic (microbial) respiration, which are related to the amount of living root biomass and the decomposition of litter and soil organic matter, respectively [52,53]. Although the microbial biomass and SOC in the control (CK) plots in the PF were lower than those in the MF and the BF, the fine root biomass was higher in the PF (Table 1; Fig. 4). Therefore, the relatively lower heterotrophic respiration could be compensated by the higher autotrophic respiration in the PF.

Effect of Soil Temperature and Soil Moisture on the Seasonality of Soil Respiration

Soil respiration in all treatments of the three forests exhibited strong seasonal patterns with higher rates in the warm-wet seasons and lower rates in the cool-dry ones (Table 2; Fig. 1 c), which is consistent with the results reported in some subtropical forests [31,54–56] and in some temperate forests [21,57,58]. In addition, similar to the related studies in some warm and wet forests [42,49], positive exponential relationships between soil respiration and soil temperature, as well as positive linear relationships between soil respiration and soil moisture have been found in our forests (Table 4; Fig. 2). The dual temperature and moisture controls on soil respiration in this study may be related to the monsoon tropical climate of our studied region with high temperature occurs simultaneously with high moisture [37]. Therefore, high plant growth and soil microbial activity in the warm-wet season can stimulate greater soil respiration in this region [59].

Effects and Mechanisms of SAR on Soil Respiration

Our results demonstrated that soil respiration in all three forests were depressed after exposure to SAR. This result was consistent with several laboratory and field experimental results [29,58,60], indicating that the response of soil respiration was susceptible to acid rain in subtropical forests. However, the sensitivity of the response of soil respiration to SAR was different among the three forests. The response of soil respiration to SAR was less significant in the PF than those in the MF and the BF (4.0, 14.0 and 15.8% lower in the T3 treatment compared with the CK treatment, respectively) and the differences among treatments were significant in the MF and the BF (Table 3; Fig. 1 c).

Several mechanisms may help explain the depression of soil respiration after SAR in the MF and the BF and that mostly occurred in the warm-wet season. First, soil acidification under SAR may decrease heterotrophic respiration from the microbial community. After SAR treatment for 25 months, mean soil pH values in the T3 treatment in both forests were significantly lower than those in the CK treatment (Fig. 3), which indicated the aggravation of soil acidification. Many related studies in temperate and subtropical forests also suggest that SAR decreases soil pH value [29,61,62]. Lower soil pH value due to the toxicity of high H^+ loads can change the population, community structure and biological activity of soil decomposers [63]. In this study, we found that soil microbial biomass carbon in the warm-wet season significantly decreased in the T3 treatment both in the MF and the BF (Fig. 4), which was consistent with several studies in temperate and subtropical forests [64–66]. This suggests that microbial activity was restrained under the stress of acid rain. The reduction of microbial activity, on one hand, would slow the mineralization and decomposition rates of soil organic matter, thus inhibiting CO_2 emission from soil [62]; on the other hand, litter decomposition would be inhibited, which therefore, leads to the reduction of microbial CO_2 production from litter [67]. Many studies have also reported that the litter decomposition rates in subtropical and temperate area slowed down under the stress of SAR [68,69].

Second, autotrophic respiration from plant roots may decrease after soil acidification. It was reported that fine root biomass was significantly correlated with soil respiration rate [22], and SAR decreased fine root biomass and inhibited seedling growth [70]. In one way, cumulative effects of acid rain will lead to nutrient leaching [71–73] as the amount of H^+ cation in the soil increases and they replace the basic cation in the argilo-humic complex. A nutrient depletion will then follow in the long term and the growth of plants can be affected [38]. In another way, when the soil pH value decreases, the concentration of free moving metallic ions

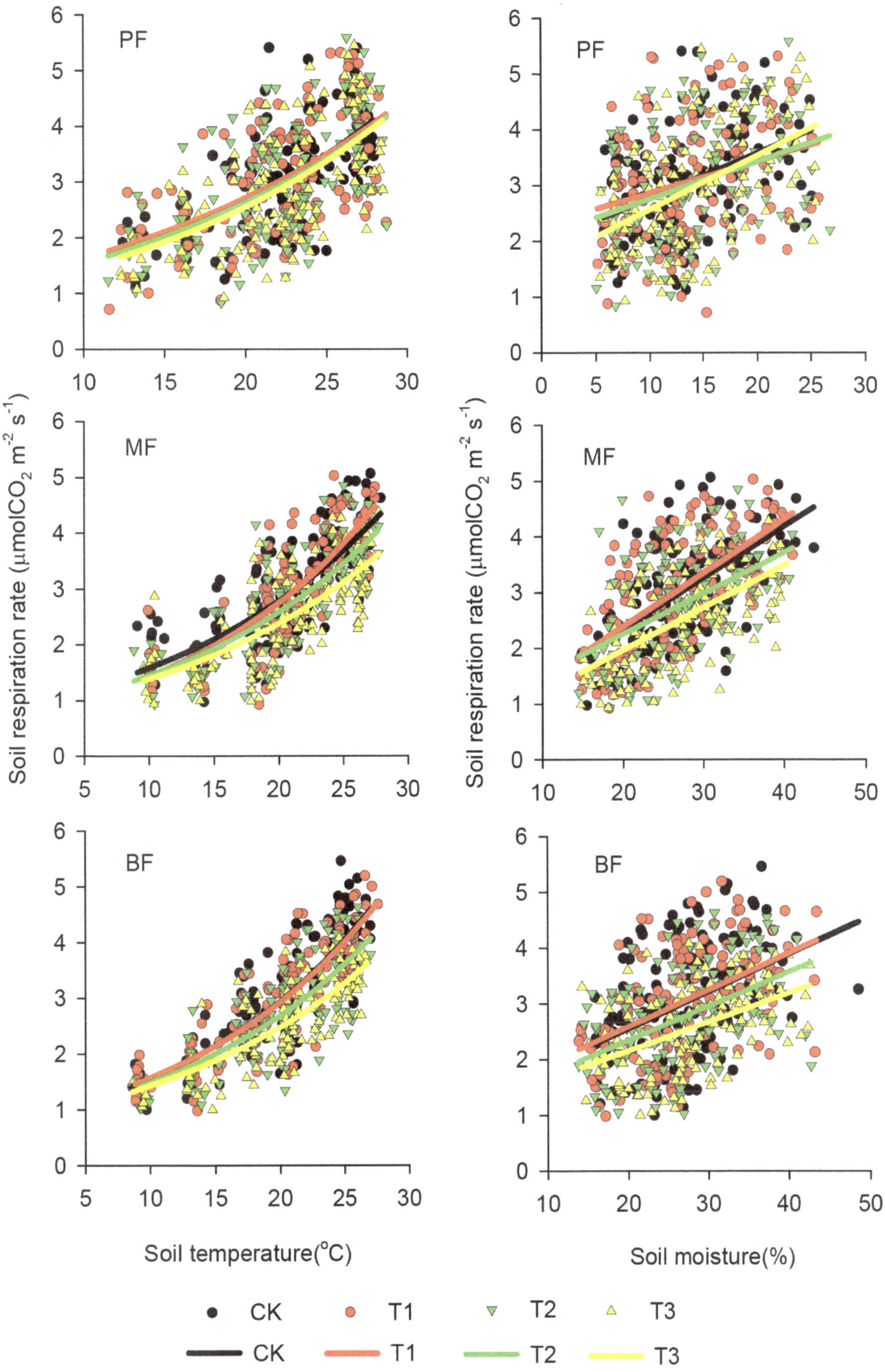
PF
PF
MF
MF
BF
BF
Soil respiration rate (μmolCO2 m-2 s-1)
Soil respiration rate (μmolCO2 m-2 s-1)
Soil temperature(°C)
Soil moisture(%)
CK
T1
T2
T3
CK
T1
T2
T3

Figure 2. Relationship of soil respiration rate with soil temperature or soil moisture under different SAR treatments in the pine forest (PF), the mixed conifer and broadleaf forest (MF) and the broadleaved forest (BF). n = 126 for all the treatments. The treatments are: CK = control, T1 = pH 4.0, T2 = pH 3.5, T3 = pH 3.0.

(e.g. Al^{3+}, Mn^{2+}) will increase [74,75]. Those metals show very high degrees of toxicity for many plants [76,77], and thus affect root respiration [14]. In this study, the fine root biomass in the warm-wet season in the MF and the BF revealed a negative response to the increasing level of SAR (Fig. 4), which was also accompanied by the decreased CO_2 loss from the soils.

Under the stress of acid rain, the soil will be acidified gradually [34,75]. The declining trends of soil pH value in the MF and the BF had been strengthened over time during the study period (Fig. 3), which also indicated that the soil acidification under SAR was a gradual process in these two forests. Therefore, the negative effects of SAR on soil respiration had been strengthened over time accordingly (Table 3). Furthermore, we found an effect of SAR on the temperature response of soil respiration, with a decline in Q_{10} in all forests (Table 4). This result suggests that SAR would decrease the temperature sensitivity of respiration. As a respiratory substrate, root biomass plays an important role in the response of soil respiration to soil temperature; temperature sensitivity of soil respiration decreases when the substrate supply is low [59]. Therefore, the reduction of root biomass in this study could result in lower temperature sensitivity. Although *b* values for the exponential curve among treatments were not significantly different in all forests, we suggest that this decrease would be more significant under the continued SAR in the future. Moreover, the decline of the temperature sensitivity of respiration was increasingly pronounced with the progressive maturity of three forests, which indirectly indicated the same trend in the sensibility of the response of soil respiration to SAR.

Mechanisms of the Different Responses of Soil Respiration among Three Forests

We suspect that the different responses of soil respiration to SAR among three forests may result from their differences in soil and in litter layer which, ultimately, are caused by the differences in maturity. First, acid buffering abilities of soil are different among the three forests. The decrease of soil pH is the most direct indicator of soil acidification [62]. At our study sites, although the soil pH values of 0–10 cm depth were highly acidic with mean pH values below 4 in the control (CK) plots of all forests, soil acidification was aggravated when exposed to an addition of H^+. This may relate to the low base saturation (BS, less than 10%, [78]) in the soil of our forests, as the BS can reflect the buffering capacity of soil to acid rain [79]. However, unlike the situations in the MF and the BF, there was no significant difference among treatments in the soil pH value in the PF (Fig. 3). Yu *et al.* [80] pointed out that soil acidification under acid rain is closely related to the degree of pH value of original soil, that is the higher the original soil pH, the smaller the affect of soil acidification by acid rain. Soil pH value in the control (CK) plots in the PF was significantly higher than those in the MF and the BF (Fig. 3), indicating that the soil in the PF was less susceptible to acid rain. In addition, the BS in soil was decrease with the progressive maturity of three forests [78], which also suggested that the acid buffering ability of soil was the highest in the PF.

Second, acid buffering abilities of litter layer are different among the three forests. The litter layer is not only a source of soil respiration but also an influence of soil respiration by indirect effects on biological processes in the underlying soil [32]. The decaying litter above the surface of soil can mitigate the effects of acid rain by the exchange between the base cation in the litter and H^+ in the acid rain [81]. Although the annual litterfall decreased

Table 4. Model for relationships between the soil respiration (*R*) and soil temperature at 5 cm depth (*T*) and volumetric soil moisture of the top 5 cm soil layer (*M*) in the pine forest (PF), the mixed conifer and broadleaf forest (MF) and the broadleaved forest (BF).

Model		$R = a\exp^{(bT)}$ *				$R = aM+b$*		
Forest	**Treatment**	*a*	*b*	Q_{10}	R^2	*a*	*B*	R^2
PF	CK	0.8375±0.3837	0.0585±0.0177 a	1.75	0.41	0.0698±0.0125	2.0850±0.1520	0.18
	T1	1.0012±0.6896	0.0588±0.0455 a	1.75	0.34	0.0577±0.0431	2.1087±0.6338	0.10
	T2	0.7604±0.1062	0.0569±0.0121 a	1.69	0.35	0.1071±0.0369	1.3543±0.7017	0.12
	T3	0.9266±0.3183	0.0518±0.0066 a	1.71	0.36	0.0504±0.0284	2.2875±0.8416	0.23
MF	CK	0.7807±0.2344	0.0595±0.0127 a	1.80	0.57	0.0873±0.0228	0.5439±0.8123	0.40
	T1	0.7462±0.1461	0.0646±0.0069 a	1.88	0.67	0.0918±0.0110	0.5101±0.3868	0.39
	T2	0.9138±0.2138	0.0557±0.0122 a	1.76	0.61	0.0770±0.0054	0.5489±0.1247	0.25
	T3	0.8523±0.1124	0.0490±0.0085 a	1.72	0.51	0.0666±0.0193	0.5011±0.4646	0.35
BF	CK	0.7340±0.1494	0.0687±0.0111 a	1.88	0.70	0.0898±0.0112	0.6235±0.4818	0.19
	T1	0.8353±0.1566	0.0630±0.0082 a	1.87	0.71	0.0870±0.0096	0.7855±0.2888	0.22
	T2	0.8724±0.1447	0.0546±0.0117 a	1.80	0.59	0.0805±0.0446	0.5894±0.8082	0.21
	T3	0.8069±0.1378	0.0539±0.0065 a	1.74	0.64	0.0632±0.0274	0.7312±0.9059	0.22

*$p<0.01$ in all treatments of all forests.
n = 126 for all the treatments. R^2 is the determination of coefficient. The treatments are: CK = control, T1 = pH 4.0, T2 = pH 3.5, T3 = pH 3.0.

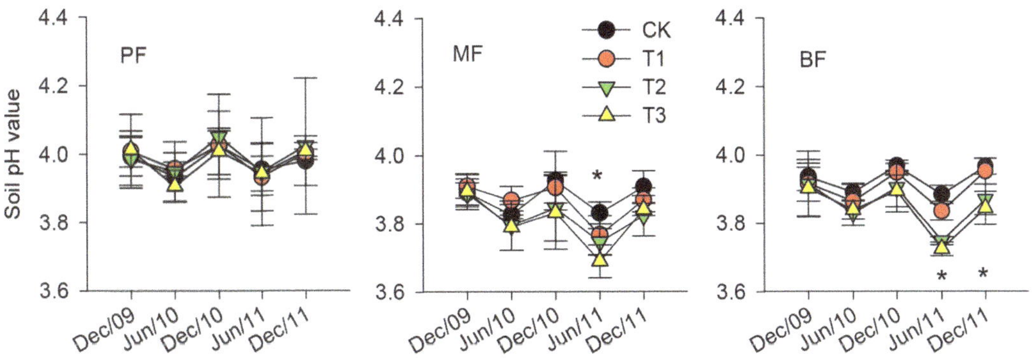

Figure 3. Dynamics of soil pH value under different SAR treatments in the pine forest (PF), the mixed conifer and broadleaf forest (MF) and the broadleaved forest (BF). Error bars are standard errors of the mean (n = 3 for all the treatments). The asterisk (*) indicates significant difference among treatments at $p = 0.05$. The treatments are: CK = control, T1 = pH 4.0, T2 = pH 3.5, T3 = pH 3.0.

with the progressive maturity of three forests, the amount of litter present was in the opposite trend (PF>MF>BF) (Table 1) due to the decreased trend of decomposition rate [82]. Therefore, the acid buffering ability of litter layer was more evident in the PF than those in the MF and the BF. The result of Liu *et al.* [83] also suggested that the acid buffering ability of litter layer was best in

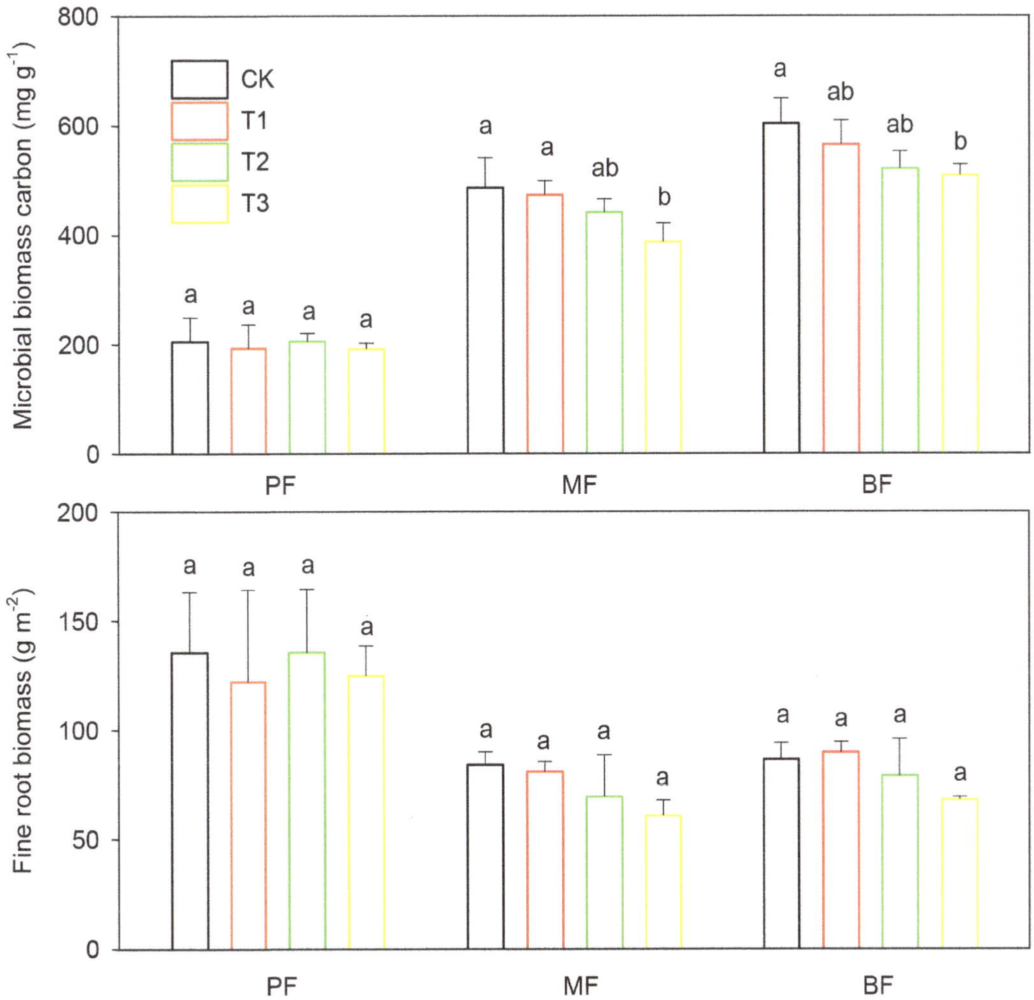

Figure 4. Soil microbial biomass carbon and fine root biomass under different SAR treatments in the pine forest (PF), the mixed conifer and broadleaf forest (MF) and the broadleaved forest (BF). Error bars are standard errors of the mean (n = 3 for all the treatments). Different lowercase letters denote significant difference ($p = 0.05$) between treatments. The treatments are: CK = control, T1 = pH 4.0, T2 = pH 3.5, T3 = pH 3.0.

the pine forest among six forest types in southwestern Subtropical region of China. The acid buffering ability of soil and litter layer in the PF mitigated the acidification of soil and the reductions of soil microbial activity and fine root biomass under SAR, thus alleviated the depression of soil respiration.

Limitation of the Study

In this study, we selected three typical forest types of different maturity in southern China, tested the effects of acid rain on soil respiration, and compared the response differences among forests of different maturity. One shortcoming of this study was that we did not have true replication for forest age; this is a common feature of many chronosequence designs [84] but means we cannot draw formal inferences about the effect of age on the interaction between soil respiration and SAR. Thus, the inferences regarding the response differences among forest types should be read with caution. Further studies are needed to draw rigorous conclusions regarding the response differences among forests of different maturity using replicated forest types.

Conclusions

SAR did not affected soil respiration in the PF during the study period, but significantly reduced soil respiration in the MF and the BF. These depressed effects on both forests occurred mostly in the warm-wet seasons and were correlated with the decreases in soil microbial biomass carbon and fine root biomass caused by soil acidification under SAR. The sensitivity of the response of soil respiration to SAR showed an increasing trend with the progressive maturity of three forests. This result indicated that the depressed effect of acid rain on soil respiration in southern China may be more pronounced in the future, as the young forests that are currently dominant in the region due to the widespread afforestation in recent years would gradually become mature. Further studies are still needed to draw rigorous conclusions regarding the response differences among forests of different maturity using replicated forest types.

Acknowledgments

We would like to thank the constructive comments from Prof. Dafeng Hui, which have greatly improved the quality of the paper.

Author Contributions

Conceived and designed the experiments: DZ GZ GL. Performed the experiments: GL XC QQ GC SL. Analyzed the data: GL XC XL DZ GZ. Wrote the paper: GL XL GZ DZ JL.

References

1. Liao BH, Larssen T, Seip HM (1998) Response of five Chinese forest soils to acidic inputs: batch experiment. Geoderma 86: 295–316.
2. Lawrence GB, David MB, Shortlel WC (1995) A new mechanism for calcium loss in forest-floor soils. Nature 378: 162–165.
3. Kita I, Sato T, Kase Y, Mitropoulos P (2004) Neutral rains at Athens, Greece: a natural safeguard against acidification of rains. Science of Total Environment 327: 285–294.
4. Zhang JE, Ouyang Y, Ling DJ (2007) Impacts of simulated acid rain on cation leaching from the Latosol in south China. Chemosphere 67: 2131–2137.
5. Blank LW (1985) A new type of forest decline in Germany. Nature 314: 311–314.
6. Reuss JO, Cosbyt BJ, Wright RF (1987) Chemical processes governing soil and water acidification. Nature 329: 27–32.
7. Fairfaxn JAW, Lepp W (1975) Effect of simulated 'acid rain' on cation loss from leaves. Nature 255: 324–325.
8. Turner DP, Tingey DT (1990) Foliar leaching and root uptake of Ca, Mg and K in relation to acid fog effects on Douglas-fir. Water, Air and Soil Pollution 49: 205–214.
9. Reddy GB, Reinert RA, Eason G (1991) Enzymatic changes in the rhizosphere of loblolly pine exposed to ozone and acid rain. Soil Biology and Biochemistry 23: 1115–1119.
10. Pabian SE, Ermer NM, Tzilkowski WM, Brittingham MC (2012) Effects of liming on forage availability and nutrient content in a forest impacted by acid rain. PLoS ONE 7(6): e39755. doi:10.1371/journal.pone.0039755.
11. Kuperman RG (1996) Relationships between soil properties and community structure of soil macroinvertebrates in oak-hickory forests along an acidic deposition gradient. Applied Soil Ecology 4: 125–137.
12. Schaberg PG, DeHayes DH, Hawley GJ (2001) Anthropogenic calcium depletion: a unique threat to forest ecosystem health? Ecosystem Health 7: 214–218.
13. Wang WX, Xu PJ (2009) Research progress in precipitation chemistry in China. Progress in Chemistry 21: 266–281.
14. Fan HB, Wang YH (2000) Effects of simulated acid rain on germination, foliar damage, chlorophyll contents and seedling growth of five hardwood species growing in China. Forest Ecology and Management 126: 321–329.
15. Cao YZ, Wang SY, Zhang G, Luo JY, Lu SY (2009) Chemical characteristics of wet precipitation at an urban site of Guangzhou, South China. Atmospheric Research 94: 462–469.
16. Pan XG (1992) Acidification of soils in Mount Lushan over the last 35 years. Pedosphere 2: 179–182.
17. Dai ZH, Liu YX, Wang XJ, Zhao DW (1998) Changes in pH, CEC and exchangeable acidity of some forest soils in southern China during the last 32–35 years. Water, Air and Soil Pollution 108: 377–390.
18. Busch G, Lammel G, Beese FO, Feichter J, Dentener FJ, et al. (2001) Forest ecosystems and the changing patterns of nitrogen input and acid deposition today and in the future based on a scenario. Environmental Science and Pollution Research 8: 95–102.
19. Musselman RC, Fox DC (1991) A review of the role of temperate forests in the global CO_2 balance. Journal of Air Waste Management Association 41: 798–807.
20. Raich JW, Schlesinger WH (1992) The global carbon dioxide efflux in soil respiration and its relationship to vegetation and climate. Tellus 44: 81–99.
21. Bowden RD, Davidson E, Savage K, Arabia C, Steudler P (2004) Chronic nitrogen additions reduce total soil and microbial respiration in temperate forest soils at the Harvard Forest. Forest Ecology and Management 196: 43–56.
22. Davidson EA, Janssens IA, Luo YQ (2006) On the variability of respiration in terrestrial ecosystems: Moving beyond Q_{10}. Global Change Biology 12: 154–164.
23. Yang YS, Dong B, Xie JS, Chen SG, Gao R, et al. (2004) Soil respiration of forest ecosystems and its respondence to global change. Acta Ecologica Sinica 24: 583–591 (in Chinese with English abstract).
24. Kuzyakov Y, Larionova AA (2005) Root and rhizomicrobial respiration: A review of approaches to estimate respiration by autotrophic and heterotrophic organisms in soil. Journal Plant Nutrition and Soil Science 168: 503–520.
25. Fritze H (1992) Effects of environmental pollution on forest soil microflora – a review. Silva Fenn 26: 37–47.
26. Salonius PO (1990) Respiration rates in forest soil organic horizon materials treated with simulated acid rain. Canadian Journal of Forest Research 20: 910–913.
27. Will ME, Graetz DA, Roof BS (1986) Effect of simulated acid precipitation on soil microbial activity in a typic quartzipsamment. Journal of vironmental Quality 15: 399–403.
28. Ouyang XJ, Zhou GY, Huang ZL, Liu JX, Zhang DQ, et al. (2008) Effect of simulated acid Rain on potential carbon and nitrogen mineralization in forest soils. Pedosphere 18: 503–514.
29. Xie XZ, Jiang H, Yu SQ, Liu YY, Yuan HY, et al. (2009) Effect of simulated acid rain on soil respiration of *Pinus massoniana* and *Cunninghamia lanceolata*. Acta Ecologica Sinica 29: 5713–5720 (in Chinese with English abstract).
30. Peng SL, Wang BS (1995) Forest succession at Dinghushan, Guangdong, China. Chinese Journal of Botany 7: 75–80.
31. Tang XL, Liu SG, Zhou GY, Zhang DQ, Zhou CY (2006) Soil-atmoshpheric exchange of CO_2, CH_4 and N_2O in three subtropical forest ecosystems in southern China. Global Change Biology 12: 546–560.
32. Yan JH, Zhang DQ, Zhou GY, Liu JX (2009) Soil respiration associated with forest succession in subtropical forests in Dinghushan Biosphere Reserve. Soil Biology and Biochemistry 41: 991–999.
33. Huang ZF, Fan ZG (1982) The climate of Ding Hu Shan. Tropical and Subtropical Forest Ecosystem 1: 11–23 (in Chinese with English abstract).
34. Liu JX, Zhou GY, Chu GW, Zhang QM (2003) Effects of soil acidity on the soil nutrients under dinghushan monsoon evergreen broad-leaved forest. Acta Pedologica Sinica 40: 763–767 (in Chinese with English abstract).
35. Wu H, Deng H, Zheng L (1982) Physico-geographical features of Dinghushan and their dynamic analyses. Tropical and Subtropical Forest Ecosystem Research 1: 1–10.
36. Ding MM, Brown S, Lugo AE (2001) A continental subtropical forest in China compared with an insular subtropical forest in the Caribbean. General Technical Report IITF-17, USDA (United States Department of Agriculture) Forest Service, International Institute of Tropical Forestry, Rıo Piedras, Puerto Rico.
37. Kong GH, Huang ZL, Zhang QM, Liu SZ, Mo JM, et al. (1997) Type, structure, dynamics and management of the lower subtropical evergreen broad-

leaved forest in the Dinghushan Biosphere Reserve of China. Tropics 6: 335–350.
38. Liu JX, Zhou GY, Zhang DQ (2006) Effects of acidic solutions on element dynamics in monsoon evergreen broad-leaved forest at Dinghushan, China Part 1: Dynamics of K, Na, Ca, Mg and P. Environmental Science and Pollution Research 14: 123–129.
39. Dick WA, Cheng L, Wang P (2000) Soil acid and alkaline phosphatase activity as pH adjustment indicators. Soil Biology and Biochemistry 32: 1915–1919.
40. Jenkinson DS (1987) Determination of microbial biomass carbon and nitrogen in soil. In: Wilson JT (ed) Advances in nitrogen cycling in agricultural ecosystems, CAB International, Wallingford 368–386.
41. Vance ED, Brookes SA, Jenkinson DS (1987) An extraction method for measuring soil microbial biomass C. Soil Biology and Biochemistry 19: 703–707.
42. Cleveland CC, Townsend AR (2006) Nutrient additions to a tropical rain forest drive substantial soil carbon dioxide losses to the atmosphere. Proceedings of National Academy of Sciences 103: 10316–10321.
43. Raich JW, Potter CS (1995) Global patterns of carbon dioxide emissions from soils. Global Biochemical Cycles 9: 23–36.
44. Luo YQ, Wan SQ, Hui DF, Wallace LL (2001) Acclimatization of soil respiration to warming in a tallgrass prairie. Nature 413: 622–625.
45. Loyd J, Taylor JA (1994) On the temperature dependence of soil respiration. Functional Ecology 8: 315–323.
46. Xu M, Qi Y (2001) Spatial and seasonal variations of Q_{10} determined by soil respiration measurements at a Sierra Nevadan forest. Global Biogeochemical Cycles 15: 687–696.
47. Giardina CP, Ryan MG (2002) Total belowground carbon allocation in a fast-growing Eucalyptus plantation estimated using a carbon balance approach. Ecosystems 5: 487–499.
48. Salimon CI, Davidson EA, Victoria RL, Melo AWF (2004) CO_2 flux from soil in pastures and forests in southwestern Amazonia. Global Change Biology 10: 833–843.
49. Sotta ED, Meir P, Malhi Y, Nobre AD, Hodnet M, et al. (2004) Soil CO_2 efflux in a tropical forest in the central Amazon. Global Change Biology 10: 601–617.
50. Huang CC, Ge Y, Chang J, Chang J, Lu R, et al. (1999) Studies on the soil respiration of three woody plant communities in the east mid-subtropical zone, China. Acta Ecologica Sinica 19: 324–328 (in Chinese with English abstract).
51. Bekku YS, Nakatsubo T, Kume A, Adachid M, Koizumid H (2003) Effect of warming on the temperature dependence of soil respiration rate in arctic, temperate and tropical soils. Applied Soil Ecology 22: 205–210.
52. Bernhardt ES, Barber JJ, Pippen JS, Taneva L, Andrews JA, et al. (2006) Long-term effects of free air CO_2 enrichment (FACE) on soil respiration. Biogeochemistry 77: 91–116.
53. Kuzyakov Y (2006) Sources of CO_2 efflux from soil and review of partitioning methods. Soil Biology and Biochemistry 38: 425–448.
54. Zhang DQ, Sun XM, Zhou GY, Yan JH, Wang YS, et al. (2006) Seaonal dynamics of soil CO_2 effluxes with responses to environmental factors in lower subtropical forest of China. Science in China Series, Dokl. Earth Sci. 49: 139–149.
55. Mo JM, Zhang W, Zhu WX, Fang YT, Li DJ, et al. (2007) Response of soil respiration to simulated N deposition in a disturbed and a rehabilitated tropical forest in southern China. Plant Soil 296: 125–135.
56. Mo JM, Zhang W, Zhu WX, Gundersen P, Fang YT, et al. (2008) Nitrogen addition reduces soil respiration in a mature tropical forest in southern China. Global Change Biology 14: 403–412.
57. Dong YS, Peng GB, Li J (1996) Seasonal variations of CO_2, CH_4 and N_2O fluxes from temperate forest soil. Acta Geographica Sinica 51: 120–128 (in Chinese with English abstract).
58. Zhang Y, Wang LX, Chen ST, Hu ZH, Shen XS, et al. (2011) Effects of simulated acid rain on soil respiration in a northern subtropical secondary forest. China Environmental Science 31: 1541–1547 (in Chinese with English abstract).
59. Deng Q, Liu SZ, Liu JX, Meng Z, Zhang DQ (2007) Contributions of litter-fall to soil respiration and its affacting factors in southern subtropical forests of China. Advances in Earth Science 22: 976–986 (in Chinese with English abstract).
60. Chen ST, Shen XH, Hu ZH, Chen HS, Shi YS, et al. (2012) Effects of simulated acid rain on soil CO_2 emission in a secondary forest in subtropical China. Geoderma 189–190: 65–71.
61. Neuvonen S, Suomela J (1990) The effects of simulated acid rain on pine needle and birch leaf litter decomposition. Journal of applied ecology 2: 857–872.
62. Liu YY, Jiang H, Li YH, Yuan HY (2010) A short-term effect of simulated acid rain on the soil respiration of the compound system of Chinese fir seedling-soil. Acta Ecologica Sinica 30: 2010–2017 (in Chinese with English abstract).
63. Falappi D, Farini A, Ranalli G, Sorlini C (1994) Effects of simulated acid rain on some microbiological parameters of subacid soil. Chemosphere 28: 1087–1095.
64. Zhou CL, Qi YC (1988) Effects of acid rain on activities of soil microganisms. Journal of Ecology 7: 21–24.
65. Greszta J (1992) Humus degradation under the influence of simulated acid rain. Water, Air and Soil Pollution 63: 51–66.
66. Pennanen T, Fritze H, Vanhala P, Kiikkila O, Neuvonen S, et al. (1998) Structure of a microbial community in soil after prolonged addition of low levels of simulated acid rain. Applied And Environmental Microbiology 64: 2173–2180.
67. Wolters V (1991) Effects of acid rain on leaf-litter decomposition in a beech forest on calcareous soil. Biology and fertility of soils 11: 151–156.
68. Hong JH, Jiang H, Ma YD, Yu SQ, Li W, et al. (2009) The influence of acid rain on the leaf litter decomposition of three dominant trees in the subtropical forests. Acta Ecologica Sinica 29: 5246–5250 (in Chinese with English abstract).
69. Ma YD, Jiang H, Yu SQ, Zhou GM, Dou RP, et al. (2010) Effects of simulated acid rain on the decomposition of *Phyllostachys pubescens*. ActaCTA Scientiarum Naturalium Universitatis Sunyatseni 49: 95–99 (in Chinese with English abstract).
70. Schaedle M, Thornton FC, Raynal DJ, Tepper HB (1989) Response of tree seedlings to aluminum. Tree Physiology 5: 337–356.
71. Oene VH (1992) Acid deposition and forest nutrient imbalance: A modeling approach. Water, Air and Soil Pollution 63: 33–50.
72. Baba M, Okazaki M, Hashitani T (1995) Effect of acidic deposition on forested andisols in the Tama hill region of Japan. Environmental Pollution 89: 97–106.
73. Renner R (1995) Scientific uncertainty scuttles new acid rain standard. Environmental science and technology 29: 464–466.
74. Mulder J, Breemen N, Eijck HC (1989) Depletion of soil aluminium by acid deposition and implications for acid neutralization. Nature 337: 247–249.
75. Liu JX, Zhou GY, Zhang DQ (2003) The cumulative effects of acid rain on the soil and responses of *Schima superba* at Dinghushan. China Environmental Science 23: 90–94 (in Chinese with English abstract).
76. Phillips RP, Yanai RD (2004) The effects of $AlCl_3$ additions on rhizosphere soil and fine root chemistry of sugar maple (Acer Saccharum). Water, Air and Soil Pollution 159: 339–356.
77. Joner EJ, Eldhuset TD, Lange H, Åsa Frostegård (2005) Changes in the microbial community in a forest soil amended with aluminium in situ. Plant Soil 275: 295–304.
78. Wen DZ, Zhou GY, Kong GH, Yu MD (2000) A status report on the effects involved with acid rain on plants, soils and surface waters of three selected terrestrial ecosystems in lower subtropical china. Chinese Journal of Ecology 19: 11–18 (in Chinese with English abstract).
79. Mcfee WW (1980) Sensitivity of soil regions to long term acid precipitation. Atmospheric Sulfur Deposition, Ann Arbor Science, Ann Arbor, Michigan 477.
80. Yu YC, Ding AF, Hu J, Meng L (2001) Effects of simulated acid rain on soil acidification and base ions transplant. Journal of Nanjing Forestry University 25: 39–42 (in Chinese with English abstract).
81. Salim R, Khalaf S (1994) A role for decaying leaves in mitigating the harmful effects as acid rain. Journal of Environmental Science and Health. Part A: Environmental Science and Engineering and Toxicology 29: 115–127.
82. Huang YH, Li YL, Xiao Y, Wenigmannc KO, Zhou GY, et al. (2011) Controls of litter quality on the carbon sink in soils through partitioning the products of decomposing litter in a forest succession series in South China. Forest Ecology and Management 261: 1170–1177.
83. Liu N, Wang YJ, Zhao ZJ, Zhao YY, Feng HM, et al. (2011) Responses of typical forest litters to acid deposition in southwestern subtropical region. Journal of Soil and Water Conservation 25: 108–120 (in Chinese with English abstract).
84. Johnson EA, Miyanishi K (2008) Testing the assumptions of chronosequences in succession. Ecology Letters 11: 419–431.
85. Yan JH, Wang YP, Zhou GY, Zhang DQ (2006) Estimates of soil respiration and net primary production of three forests at different succession stages in South China. Global Change Biology 12: 810–821.
86. Yi ZG, Wang XM, Sheng GY (2007) Soil uptake of carbonyl sulfide in subtropical forests with different successional stages in south. China Journal of Geophysical Research 112: D08302, doi:10.1029/2006JD008048.

6

Patterns of Detection and Capture Are Associated with Cohabiting Predators and Prey

Billie T. Lazenby[1,2]*, **Christopher R. Dickman**[1]

1 School of Biological Sciences, University of Sydney, Sydney, New South Wales, Australia, **2** Department of Primary Industries Parks Water and Environment, Hobart, Tasmania, Australia

Abstract

Avoidance behaviour can play an important role in structuring ecosystems but can be difficult to uncover and quantify. Remote cameras have great but as yet unrealized potential to uncover patterns arising from predatory, competitive or other interactions that structure animal communities by detecting species that are active at the same sites and recording their behaviours and times of activity. Here, we use multi-season, two-species occupancy models to test for evidence of interactions between introduced (feral cat *Felis catus*) and native predator (Tasmanian devil *Sarcophilus harrisii*) and predator and small mammal (swamp rat *Rattus lutreolus velutinus*) combinations at baited camera sites in the cool temperate forests of southern Tasmania. In addition, we investigate the capture rates of swamp rats in traps scented with feral cat and devil faecal odours. We observed that one species could reduce the probability of detecting another at a camera site. In particular, feral cats were detected less frequently at camera sites occupied by devils, whereas patterns of swamp rat detection associated with devils or feral cats varied with study site. Captures of swamp rats were not associated with odours on traps, although fewer captures tended to occur in traps scented with the faecal odour of feral cats. The observation that a native carnivorous marsupial, the Tasmanian devil, can suppress the detectability of an introduced eutherian predator, the feral cat, is consistent with a dominant predator – mesopredator relationship. Such a relationship has important implications for the interaction between feral cats and the lower trophic guilds that form their prey, especially if cat activity increases in places where devil populations are declining. More generally, population estimates derived from devices such as remote cameras need to acknowledge the potential for one species to change the detectability of another, and incorporate this in assessments of numbers and survival.

Editor: Matt Hayward, Australian Wildlife Conservancy, Australia

Funding: This project was funded by grants from the WV Scott Estate, the University of Sydney postgraduate research support scheme, and an Australian Postgraduate Award. The funders had no role in study design, data collection and analysis, decision to publish, or preparation of the manuscript.

Competing Interests: The authors have declared that no competing interests exist.

* E-mail: Billie.Lazenby@dpipwe.tas.gov.au

Introduction

The interactions between animals can be important determinants of species' distributions and abundances [1], [2], [3]. Observations of behaviour and experiments exploring animal responses to cues from co-occurring species have been an important source of information which has helped to describe and quantify relationships such as those between predators and between predators and prey (e.g., [4], [5], [6]).

Remote cameras are being used increasingly to study wildlife [7]. They can record multiple species and offer a unique opportunity to observe spatial and temporal patterns of visitation at camera sites by different taxa by time-stamping photographs, functioning for relatively long periods without user input compared to other methods such as sand padding, and operating in variable environmental conditions. Such data lend themselves to two-species occupancy modelling, which accounts for imperfect detection of different species within a maximum likelihood framework [8]. For example, if inverse spatial or temporal patterns between two species are observed, they may indicate that one species is dominant or inducing fear in the other. Asymmetrical interactions often occur between different-sized predators, predators and prey, or dominant and subordinate competitors [9], [10].

Two-species occupancy modelling is a robust method for modelling observations of patterns consistent with asymmetrical interactions; in addition to allowing for differences in detection probability between species, other covariates such as habitat, which can be important explanatory variables for observed patterns, can also be incorporated [8]. It is a good starting point for designing experiments that explicitly address the causal processes that determine patterns arising from interactions between species.

Fear, manifested as avoidance behaviour, is an important structuring component in many ecosystems [11], [12]. For example, small mammalian carnivores often fear larger species and alter their habitat use when dominants are present [13], [14], [15], leading to reduced population size and activity of the subordinate predators. Mesopredator release occurs when a subordinate predator responds to the decline or extinction of a dominant predator by increasing population size, activity, or impact on prey, and it has been observed in many ecosystems globally [16], [17]. The feral cat *Felis catus* L. provides a good example. This carnivore frequently occupies a mesopredator role and can have destructive impacts on small prey if the suppressive effects of a dominant species are removed [18], [19], [20], [21], [22].

Fear and avoidance also characterize many predator-prey relationships; in mammals, these relationships are often mediated by odour (e.g., [23]). Selection can be expected to favour individuals that recognize and avoid the odour of predators that are a significant source of mortality, hence reducing their probability of encounter with the predator itself [24], [25]. Avoidance of predator odours by small mammals is often pronounced where predators and prey have coevolved [4] but, in systems where predators are novel, avoidance responses by prey may not always occur (e.g., [26], [27]). Lack of recognition and avoidance of novel predators may explain, in part, why small mammals have fared poorly in Australia over the last 200 years [28]. Many species in the Australian region experience high rates of depredation from the recently introduced red fox *Vulpes vulpes* L. and feral cat, with at least 16 species declining to extinction in the presence of these predators [29]. However, some Australian rodents avoid the odours of novel predators, perhaps suggesting that predation already has selected for prey that can recognize and avoid them [30].

In this study we use cameras to uncover relationships between introduced and native predators and prey in Tasmania, a large (68 000 km^2) island south of the continental mainland of Australia. Feral cats were originally introduced to Tasmania some 200 years ago [31], with red foxes arriving over the last decade or more [32], [33]. A sub-species of swamp rat, *Rattus lutreolus velutinus* (Thomas), is endemic to Tasmania and is particularly useful, as a potential prey species, for field-based behavioural studies because it is widespread and highly detectable compared to other native small mammals on the island. Adult swamp rats weigh on average 100 g [34] and have an approximately equal head-body compared to tail length, and ears that are set close to the head compared to those of co-occurring small mammals. These features make them distinguishable in photographs taken by remote cameras. The world's largest extant marsupial carnivore, the Tasmanian devil *Sarcophilus harrisii* (Boitard), also is endemic to Tasmania [35]. There have been recent declines in devil populations of up to 85% in some areas as a result of a contagious and invariably fatal cancer – devil facial tumour disease (DFTD) [36]. These declines have been associated with increases in spotlight counts of feral cats, perhaps via mesopredator release [37]. Feral cats – at less than half the body mass of Tasmanian devils (adult devils weigh 8–10 kg whereas cats weigh 2–4 kg) – are highly likely to be subordinate in any encounters between the two species. As small mammals such as the swamp rat form a very large part of the diet of feral cats compared to devils [38], our goal was to better understand relationships between these predators and prey and improve our ability to manage them.

We investigated predator and prey interactions and responses to a range of predator odour cues in the cool temperate forests of southern Tasmania. Our specific aims were to 1) quantify patterns of detection at camera sites of devils and feral cats, devils and swamp rats, and feral cats and swamp rats; and 2) investigate the trap response of the swamp rat to devil and feral cat faecal odours. We hypothesized that feral cats would show reduced detection at camera sites where devils were recorded and that swamp rats would be detected less often at sites with devils and feral cats. We hypothesized further that swamp rats would avoid the faecal odours of feral cats and devils.

Materials and Methods

Ethics Statement

This study was conducted under ethics approval from the University of Sydney Animal Ethics Committee number L04/8/2008/3/4878 and scientific permits issued by the Tasmanian Department of Primary Industries Parks Water and Environment for research on native wildlife (TFA108121, TFA10046, and TFA11137).

Study Sites

Three study sites, each covering ~40 km^2, were selected in southern Tasmania based on similarities in flora, fauna and land use (Fig. 1). They are referred to henceforth as Southwest, Mt Field, and Tasman Peninsula (Fig. 2). The central coordinate for each site in longitude and latitude is: Southwest 146° 22′ 26.4″, −42° 48′ 28.8″; Mt Field 146° 41′ 16.8″, −42° 39′ 46.8″; and Tasman Peninsula 147° 54′ 36″, −43° 3′ 50.4″. The outer margins of each study site were separated by at least 15 km. Vegetation at each site was primarily cool temperate wet forest, but included minor components of other habitats such as highland treeless vegetation. More specifically, the vegetation dominating the three study sites, classified according to the TASVEG 2.0 Tasmanian mapping system [39], was:

- Dry eucalypt forest and woodland (emergent eucalypts over a relatively open ground storey compared to rainforest and wet eucalypt vegetation);
- Rainforest and related scrub (characterised by the following genera: *Nothofagus*, *Atherosperma*, *Eucryphia* and *Athrotaxis*);
- Wet eucalypt forest and woodland (emergent eucalypts over a rainforest or scrub community); and
- Agricultural exotic and urban vegetation (characterised by plantation operations including eucalypt and pine plantations). Plantations and active forestry operations occurred or had occurred recently in small parts of all study areas.

Areas that were not subjected to forestry operations were classified as 'reserved land' such as National Park. All study sites were traversed by a network of gravel roads, fire trails and other tracks.

The average annual maximum and minimum temperatures at the study sites, recorded by the Bureau of Meteorology, ranged from 7 to 16 degrees, and 1 to 9 degrees, respectively (www.bom.gov.au). Mean annual rainfall ranged from 960 to 2500 mm and there were above average rainfalls on all sites during 2009, 2010 and 2011.

Remote Cameras

Up to 18 remote cameras (a minimum of 15 and an average of 17) were set for at least one week per survey at each study site in April and June 2009 and at least two weeks per survey in April and again in June at each site in 2010 and 2011. We used DigitalEye^{TM} 7.2 (Pixcontroller Inc.) digital trail cameras set 1.0–1.5 km apart on the sides of dirt and gravel vehicular trails. They were set in a systematic grid pattern and placed in the same locations for each survey (Fig. 3). All units were set 5–10 m from a track or trail in one of the four vegetation types (Table 1) to hide them from public view.

The cameras featured an infra-red flash that is less likely to disturb animals compared to visible white light flashes [40]. They also had a passive infra-red triggering system that detects body heat and motion before triggering a photo. Each remote camera consisted of a removable 7.2 megapixel digital camera, electronics

Figure 1. Map of Australia showing Tasmania. The arrow indicates the 68 000 km^2 island State of Tasmania to the south of the continental mainland of Australia. Map courtesy of Geoscience Australia.

control board and 9 V battery, all encased in a waterproof, camouflage-painted pelican case. We set the passive infrared sensitivity (PIR) switch at medium (standard factory setting), and the switch control board to record in 'trail mode' so that photographs were taken at least once every second after the PIR sensor had been activated by an animal. The switch control board was also set such that the camera would record photographs 24 hours a day.

Cameras were fixed to solid tree trunks ~30 cm above ground using straps. They were set in positions that received full shade throughout the day to reduce the contrast in light and shadow on the ground; this reduced triggering of the PIR sensor and avoided runs of empty photographs. Cameras were tilted 5 to 10 degrees from the vertical by wedging the back of the camera unit to limit the PIR sensor range to an area that could be illuminated by the infra-red flash.

A scent lure and food reward consisting of Juro™ tuna emulsion (Juro Oz Pro tackle, Australia) and fish-based tinned cat food in jelly was placed 1.5–2.0 m from each camera unit for each survey. Two dessert spoons of tinned cat food and 50–75 ml of tuna emulsion were spread in a 0.25 m^2 area that was the focal point for the camera. Tuna emulsion was also squirted on 1–2 branches up to 2 m off the ground above the focal point of the camera to maximize the chances of the lure scent entering air streams. Additional tuna emulsion was placed in a perforated film canister that was staked into the ground after the first two standard camera surveys. Mid-trip checks were conducted during surveys lasting more than a week (i.e. all the 2010 and 2011 surveys) to re-bait, and to replace camera batteries and memory cards. Photographs from remote cameras were downloaded on completion of each survey. The time-stamp on each photograph was converted to Eastern Standard Time.

Storage of information. We stored information from over 58 500 photographs collected during the study in a Microsoft AccessTM database. This included time, date, species, confidence of species identification, 24 h survey period (from 1700 h to 1700 h the following day), study site and camera site. We also recorded whether sites were scent marked by devils (Fig. 4). Behaviours consistent with scent-marking were clearly distinguishable in photographs and included anal dragging, defecating and urinating.

We used a query that allowed us to extract occurrence matrices for each species and each survey (custom written by Studio Q, Sandy Bay, Tasmania). The matrices consisted of columns for 24 h survey periods and rows for camera numbers. A '1', '0' or '.' was recorded in each cell where '1' signified one or more occurrences within the defined 24 h period, '0' signified no recorded occurrences, and '.' denoted camera failure.

Figure 2. Map of Tasmania showing study sites. Each site is represented by a black dot.

Small Mammal Trap Response to Predator Odours

Faecal odour trials were conducted at two of the three study sites in April and June 2011 using type-A (30×10×10 cm) Elliott small mammal traps (Elliott Scientific, Upwey, Victoria). Trials were carried out away from cameras to avoid any interference. We trialed four odours: feral cat, Tasmanian devil, Tasmanian pademelon *Thylogale billardierii* (Desmarest), and clear (no scent). Faecal odours of the herbivorous pademelon were included to ensure that any avoidance by small mammals of predator-scented traps represented genuine avoidance rather than a simple preference for clean traps.

Fresh faecal samples were obtained from a variety of wild and captive sources and then frozen. Carnivore scats were placed in a freezer at −80°C for at least two weeks to kill pathogens such as *Toxoplasma gondii* and hydatid tapeworms, after which time they were stored in a normal commercial freezer at −20°C. A portion of each scat was removed and placed in a container labeled with the species of origin to defrost prior to deployment. Where possible, we included portions of scat from different individuals and sexes in defrosting containers. Following defrosting, each container was mixed into a faecal slurry with water (ratio ~5:1),

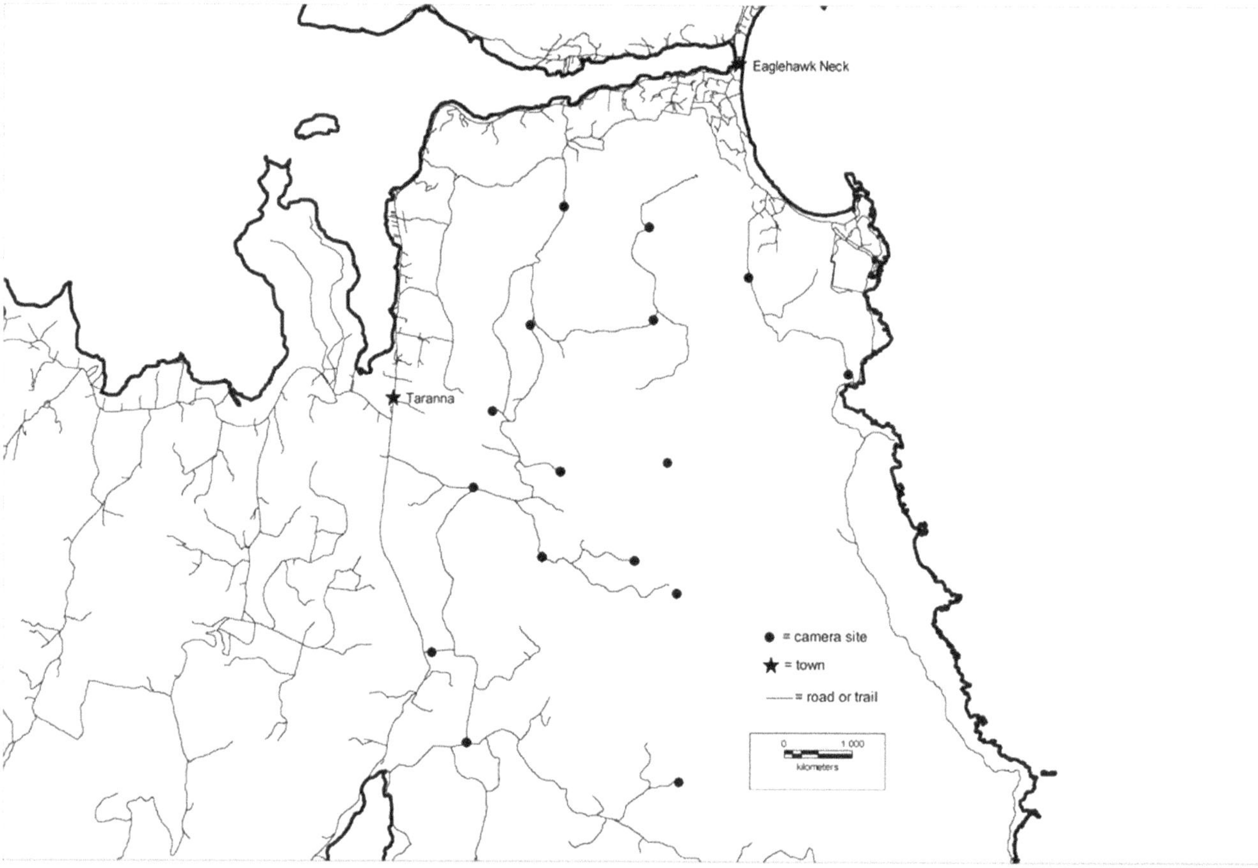

Figure 3. Map of camera placement within the Tasman Peninsula. An example of the spatial arrangement of cameras along roads and trails within a study site.

with care taken not to cross-contaminate the odours of different species.

We constructed a grid of 20 trap stations and set four Elliott traps at each station. Stations were separated by 20–30 m. A smear of slurry was placed on the door of each Elliott trap (excluding the 'clear' odour) such that there was one of each of the scented traps at each trapping station. Traps labeled with their respective scents were set in a circular pattern, ~1 m apart, opening towards the centre of the circle. This protocol follows those used in several previous investigations [25], [26], [28].

Traps were baited with a mixture of peanut butter, rolled oats and honey; a handful of commercial rabbit bedding (clean woodchip shavings with the dust extracted) also was placed at the back of each trap for insulation. A waterproof plastic sheet was placed over the back of the traps and secured to keep captured animals dry. Traps were left *in situ* for four nights, but were re-scented after two nights. Trapping stints of a maximum of four nights were separated by at least a month before faecal odour trials were carried out at the same sites. Faecal odour trials were conducted over a total of 13 nights. Traps were checked from first light every morning, and captured animals were weighed, measured, marked with an individual-specific ear clip, then released. Traps that captured animals were replaced with clean traps with new bait, bedding and fresh faecal smears. Dirty traps were scrubbed with water and air dried in sunlight off-site. Records from trap stations capturing more than one animal overnight were excluded from analyses as animals captured after the first-caught individual would experience less choice of trap odours. Captures of the same individual over successive days were included.

Table 1. Vegetation associated with camera sites.

Study area	Dry eucalypt forest and woodland	Rainforest and related scrub	Wet eucalypt forest and woodland	Agricultural exotic and urban vegetation
Southwest	4	0	5	8
Mt Field	2	8	2	6
Tasman Peninsula	3	3	0	11

The number of camera sites situated within different vegetation types within each study area.

Figure 4. A Tasmanian devil *Sarcophilus harrisii*, scent-marking a camera station with an anal drag. Note the curved tail, raised hind foot, and posterior part of the body touching the ground. Scent-marking was often observed in a series of photographs and was evidenced by anal-dragging, defecating or urinating.

Analyses

Camera surveys. We used two-species multi-season occupancy models to test whether the presence of one species altered the probability of detection of another at cameras [8], [41]. We standardized camera survey effort by restricting our analyses to the first 7 or 14 days for 2009 and 2010–2011 respectively. Models were constructed in PRESENCE 4.0 [42] for the following pairs of species: devils and feral cats; devils and swamp rats; and feral cats and swamp rats. The models were based on data from over 3570 camera trap nights. We limited our analyses to patterns of detection because our camera spacing was not distant enough to ensure spatial independence between survey units for the two larger species; the devil and feral cat, therefore comparisons of probability of site occupancy may have produced erroneous results [38].

Two-species models contain many parameters, and large data sets are necessary to achieve reliable model convergence and variance estimates. As a result, the only small mammal included in our modelling was the swamp rat because of its frequent occurrence across the study sites. We considered modelling small mammals as a group, but preliminary investigation [38] revealed that different small mammal species had different habitat associations and probabilities of detection, thus rendering them unsuitable to be treated as one group.

Three alternative set-ups can be implemented in PRESENCE 4.0 for two-species multi-season occupancy models [41]. We used parameterization number one, which is conceptually the simplest alternative. We did not experience problems with model convergence that are sometimes associated with covariates in this parameterization, probably because we limited our analyses to detection rather than occupancy and detection. Parameterization one consists of the following parameters:

PsiA - the probability of occupancy for species A, regardless of occupancy status of species B

PsiB - the probability of occupancy for species B, regardless of occupancy status of species A

pA - the probability of detecting species A during the jth survey, if only species A is present

pB - the probability of detecting species B during the jth survey, if only species B is present

rA - the probability of detecting species A, given both species are present

rB – the probability of detecting species B, given both species are present

Phi – an expression of whether two species co-occur independently at survey sites; i.e. a 'species interaction factor' (SIF); it is defined by the following equation:

$$Phi = \frac{PsiAB}{PsiA \times PsiB}$$

where PsiAB = the probability of both species being present.

Values <1 indicate that two species co-occur less often than expected, suggesting possible avoidance or competitive exclusion, while values >1 indicate a positive association.

Delta – an expression of whether two species are detected independently at survey sites, and termed a 'detection species interaction factor'; it is defined by the following equation:

Table 2. Animal occurrences at cameras.

Species or group	Common name	Small, medium or carnivore*	No. of occurrences	Southwest	Mt Field	Tasman Peninsula
Rattus lutreolus	Swamp rat	Small	753 (308)	316 (143)	17 (12)	420 (153)
Thylogale billardierii	Tasmanian pademelon	Medium	617	98	114	405
Mus musculus	House mouse	Small	409	0	83	326
Sarcophilus harrisii	Tasmanian devil	Carnivore	389 (290)	164 (126)	192 (138)	33 (26)
Trichosurus vulpecula	Brush-tailed possum	Medium	329	27	394	59
Dasyurus viverrinus	Eastern quoll	Carnivore	281	3	278	0
Pseudomys higginsi	Long-tailed mouse	Small	258	111	17	130
Felis catus	Feral cat	Carnivore	249 (187)	75 (58)	65 (45)	109 (84)
Rattus rattus	Black rat	Small	211	0	51	160
Dasyurus maculatus	Spotted-tailed quoll	Carnivore	55	54	0	1
Antechinus swainsonii	Dusky antechinus	Small	45	1	7	37
Potorous tridactylus	Long-nosed potoroo	Medium	33	7	0	26
Macropus rufogriseus	Bennett's wallaby	Medium	32	1	24	7
Vombatus ursinus	Wombat	Medium	26	6	5	15
Canis familiaris	Dog	Other	15	0	1	14
Tachyglossus aculeatus	Echidna	Medium	15	2	6	7
Bettongia gaimardi	Tasmanian bettong	Medium	8	0	8	0
Perameles gunnii	Eastern Barred Bandicoot	Medium	9	0	4	5
Isoodon obesulus	Southern Brown Bandicoot	Medium	8	3	3	2
Hydromys chrysogaster	Water rat	Small	3	3	0	0
Pseudocheirus peregrinus	Ringtail Possum	Medium	3	2	0	1
Sminthopsis leucopus	White-footed dunnart	Small	2	2	0	0
Cercartetus sp.	Pygmy possum	Small	1	1	0	0
Total small mammals			2281	1136	272	873
Unidentified small mammals			599	216	97	286
Total medium mammals			1399	170	612	617
Unidentified medium mammals (including carnivores)			165	24	55	86
Total carnivores			974	296	535	143
Total ground-foraging birds			985	428	190	367

The total numbers of occurrences (expressed as visits to a camera station – each visit with no longer than 5 minutes duration between consecutive photographs) and number of occurrences per study area (Southwest, Mt Field and Tasman Peninsula) for species, and animal groups, at camera sites set for 14 nights in 2009 and 28 nights in 2010 and 2011 in cool temperate forest in southern Tasmania. Numbers in parentheses represent the number of detections per 24 h survey period for species used in two-species occupancy models.

*Mammals were classified as small if the average adult weight of the species was 1–499 g, or medium for 500 g or above. The carnivore group included all medium to large mammalian carnivores such as feral cats, Tasmanian devils, eastern quolls and spotted-tailed quolls. It did not include domestic dogs, which were included in a separate category.

$$Delta = \frac{rAB}{rA \times rB}$$

where rAB = the probability of detecting both species.

Values <1 indicate that cameras are less likely to detect one species during a 24 h survey period if the other species was detected during the same 24 h period, and the converse if values are >1.

Gamma – the probability that an unoccupied site in season t is occupied by the species in season t+1.

Epsilon - the probability that a site occupied in season t is unoccupied by the species in season t+1.

Given that our analyses were limited to probability of detection, parameters PsiA and PsiB were modelled independently, Phi was not estimated, and gamma and epsilon were modelled as constants.

Several environmental variables, considered *a priori* to have the potential to influence probabilities of detection at camera sites, were included in the two-species models. These were:

Study area – a covariate for each of the three study sites. Study sites where one of the species in the two-species model was recorded rarely were excluded from the analyses. This included Mt Field for the devil and swamp rat model, and Mt Field for the feral cat and swamp rat model.

Time since baiting – a score for the number of days elapsed since camera sites had been baited, such that bait was modelled to decay linearly (i.e. the first survey day was scored 0, the second 0.1, the third 0.2 and so on) over each 7-day period.

Figure 5. A swamp rat, *Rattus lutreolus velutinus*, visiting a camera station. Note the distinctive equal head-body compared to tail length, and ears that are set close to the head.

Habitat – classification of camera sites into one of four vegetation groups based on groupings in the TASVEG 2.0 Tasmanian vegetation mapping system [39]. The habitat groups were dry eucalypt forest and woodland, wet eucalypt forest and woodland, rainforest and related scrub, and agricultural, exotic and urban vegetation. There was a generally similar representation by habitat groups at camera sites across the three study areas (Table 1). Vegetation structure and cover can influence species' habitat preferences and hence occupancy and detection probabilities; however, we used broad classifications of habitat based largely on floristics rather than microhabitat assessments of structure and cover. This was because selection of camera sites within habitat groups was standardized to facilitate better camera function towards microhabitats that received day-long shade and were clear of vegetation within the sensitivity range of the camera.

Season – a covariate for seasonal variation in detection probability. We were unable to model season with other covariates because the resulting models had a large number of parameters for our data set (30 parameters for a global model of season alone). Multi-season single-species site occupancy models indicated that there was little seasonal variation in detection probability for devils and feral cats but there was more seasonal variation for swamp rats [38].

Several questions were addressed regarding the probability of detection of the tested species at camera sites:

Was detection independent, irrespective of the species present within each 24 h period; i.e. did delta = 1?

Did the presence of one species at a camera site during a one or two week study site survey affect the detection of another; i.e. did pA = rA and/or pB = rB?

Models were ranked using Akaike's Information Criterion (AIC). Models with small AIC values (delta AIC ≤ 2) were considered reasonable descriptors of the data, and the above hypotheses were tested by setting various constraints on the top-ranked model such as delta to equal one, pA to equal rA, and pB to equal rB.

Faecal odour trials. Numbers of single captures in traps bearing each odour-type were tallied over both study areas, and differences in capture frequencies tested using χ^2 goodness of fit analyses.

Results

We recorded at least 23 species of mammals from 3754 visits at camera sites, as well as another 985 visits by birds (Table 2). Fifty-nine observations were made of devils scent marking during the course of the study between 2009 and 2011; 21 in Southwest, 32 at Mt Field, and six on the Tasman Peninsula (Fig. 4). There were 389, 308 and 249 records at cameras within 24 h periods of the devil, swamp rat (Fig. 5), and feral cat respectively. These data formed the basis of the two-species occupancy models. There were no photographs of the introduced red fox.

Two-Species Models

Devils and feral cats. Two-species models comparing site occupancy and detection probabilities of devils and feral cats at camera sites are shown in Table 3, and the outputs from the models within two AIC points of the top model are shown in Table 6. The results (derived from models within two AIC points of the top model) indicate that:

Table 3. Model selection statistics for Tasmanian devils and feral cats.

Model	AIC	delta AIC*	AIC wt†	Model likelihood	No. par‡	−2xLogLik**
p*Sh* (s+h), p*Fc*(bxs), r*Sh* (s+h),r*Fc* (bxs),delta(.)	3059.11	0.00	0.3801	1.0000	22	3015.11
p*Sh* (s), p*Fc*(bxs), r*Sh*(s),r*Fc* (bxs),delta(.)	3059.20	0.09	0.3633	0.9560	19	3021.20
p*Sh* (bxs), p*Fc* (bxs), r*Sh*(bxs),r*Fc* (bxs),delta(.)	3061.94	2.83	0.0923	0.2429	21	3019.94
p*Sh* (s), p*Fc*(bxs), r*Sh*(bxs),r*Fc*(s),delta(b)	3061.94	2.83	0.0923	0.2429	20	3021.94
p*Sh* (s+h), p*Fc* = r*Fc* (bxs),r*Sh* (s+h),delta(.)	3063.97	4.86	0.0335	0.0880	18	3027.97
p*Sh* (s), p*Fc* (bxs), rSh(s),rFc (bxs),delta(s)	3065.69	6.58	0.0142	0.0373	21	3023.69
p*Sh* (bxs+h), p*Fc* (bxs+h), r*Sh*(bxs+h),rFc (bxs+h),delta(.)	3066.90	7.79	0.0077	0.0203	24	3018.90
p*Sh* (s), p*Fc* = r*Fc* (bxs), r*Sh*(s),delta(.)	3067.09	7.98	0.0070	0.0185	15	3037.09
p*Sh* (s), p*Fc*(s), r*Sh*(s),r*Fc* (s),delta(.)	3067.34	8.23	0.0062	0.0163	17	3033.34
p*Sh* (bxs), p*Fc* = r*Fc* (bxs), r*Sh*(bxs),delta(.)	3069.12	10.01	0.0025	0.0067	17	3035.12
p*Sh* = r*Sh* (s+h), p*Fc*(bxs),r*Fc* (bxs),delta(.)	**3073.58**	**14.47**	**0.0003**	**0.0007**	**19**	**3035.58**
p*Sh* = r*Sh* (s), p*Fc*(bxs),r*Fc* (bxs),delta(.)	**3077.29**	**18.18**	**0.0000**	**0.0001**	**16**	**3045.29**
p*Sh*(s+h), p*Fc*(bxs), r*Sh* (s+h), r*Fc* (bxs),delta(= 1)	**3083.09**	**23.98**	**0.0000**	**0.0000**	**22**	**3039.09**
p*Sh*(s), p*Fc*(bxs), r*Sh* (s), r*Fc* (bxs),delta(= 1)	**3085.50**	**26.39**	**0.0000**	**0.0000**	**19**	**3047.50**

The ten most supported multi-season two-species occupancy models based on AIC (Akaike's Information Criterion) in program PRESENCE 4.0. Models that did not fall within the top ten that tested specific hypotheses are shown in bold below the top ten. The models were fitted to detection data from three study sites in cool temperate forests in southern Tasmania during standardized surveys from 2009–2011. The terms in parentheses represent the sources of variation in model parameters. 's' denotes study site (Southwest, Mt Field and Tasman Peninsula), 'h' denotes habitat (dry eucalypt forest and woodland, wet eucalyptus forest and woodland, rainforest and related scrub, and agricultural exotic), 'b' denotes days since baiting, and 'season' denotes a model where detection varied with season. '.' indicates a parameter set equal across species and survey times. The probabilities of site occupancy of devils and feral cats were estimated independently, and the probability of site occupancy, colonization and extinction were constrained to be constant for both species and covariates for all models. Tasmanian devil *Sarcophilus harrisii* is abbreviated *Sh* and feral cat *Felis catus* are abbreviated *Fc*.
Delta AIC is the difference in AIC values between each model and the model with the lowest AIC.
†AIC wt is the model weight.
‡Number of parameters in the model.
**Twice the negative log-likelihood.

- Devils had a lower probability of site occupancy compared to feral cats; the probability of devil detection also varied between study sites, being markedly low on the Tasman Peninsula.
- Feral cats were detected consistently less often at cameras where devils were detected (Fig. 6). This pattern was consistent across the three study sites that were tested, and the probability of detecting a feral cat at sites where devils were detected was often less than half that of sites where devils were not detected (Figs 6A,B,C).
- Devils were detected more often at cameras where feral cats had been detected. Mt Field was an exception; here, devils were detected at similar rates regardless of the detection of feral cats.
- Feral cats and devils were detected less often than expected if their detections were independent within the same 24 h period. In other words, the probability of detecting both cats and devils within the same 24 h period was less than the probability of detecting either species alone. Estimates of delta under the most supported models ranged from 0.46 to 0.47 (±0.07 SE).
- Habitat and study site were important covariates for devil detection. Baiting and site were important covariates for feral cat detection.

Devils and swamp rats. Models within two AIC points of the top model comparing site occupancy and detection probabilities of devils and swamp rats (Table 4, Table 7) indicate that:

- Swamp rats had a lower probability of site occupancy than devils, but they had a higher probability of detection.
- Patterns were inconsistent regarding the detection of swamp rats at sites where devils were detected: the probability of detection of swamp rats increased at sites where devils were detected in the Southwest according to one of the top models, but decreased according to the other top model. Swamp rat detection decreased at sites where devils were detected under both models for the Tasman Peninsula.
- The probability of detecting devils was unaffected by the detection of swamp rats.
- The two species were detected less often within the same 24 h period than expected if they were detected independently. Estimates of delta under the most supported models decreased with days since baiting at both sites and ranged from 0.66 (±0.06 SE) –0.16 (±0.08 SE) in the Southwest and 0.58 (±0.15 SE) –0.12 (±0.09 SE) on the Tasman Peninsula.
- Habitat and study site were important covariates for devils. Time since baiting, habitat and study site were important covariates for swamp rats.

Feral cats and swamp rats. Feral cat and swamp rat models within two AIC points of the top model (Table 5, Table 8) show that:

- Swamp rats had a lower probability of site occupancy than feral cats, but swamp rats had a higher probability of detection.
- The probability of detecting feral cats at cameras where swamp rats were present was similar for the Southwest, but reduced on the Tasman Peninsula.
- The probability of detecting swamp rats at sites where feral cats were detected was greater in the Southwest but less on the

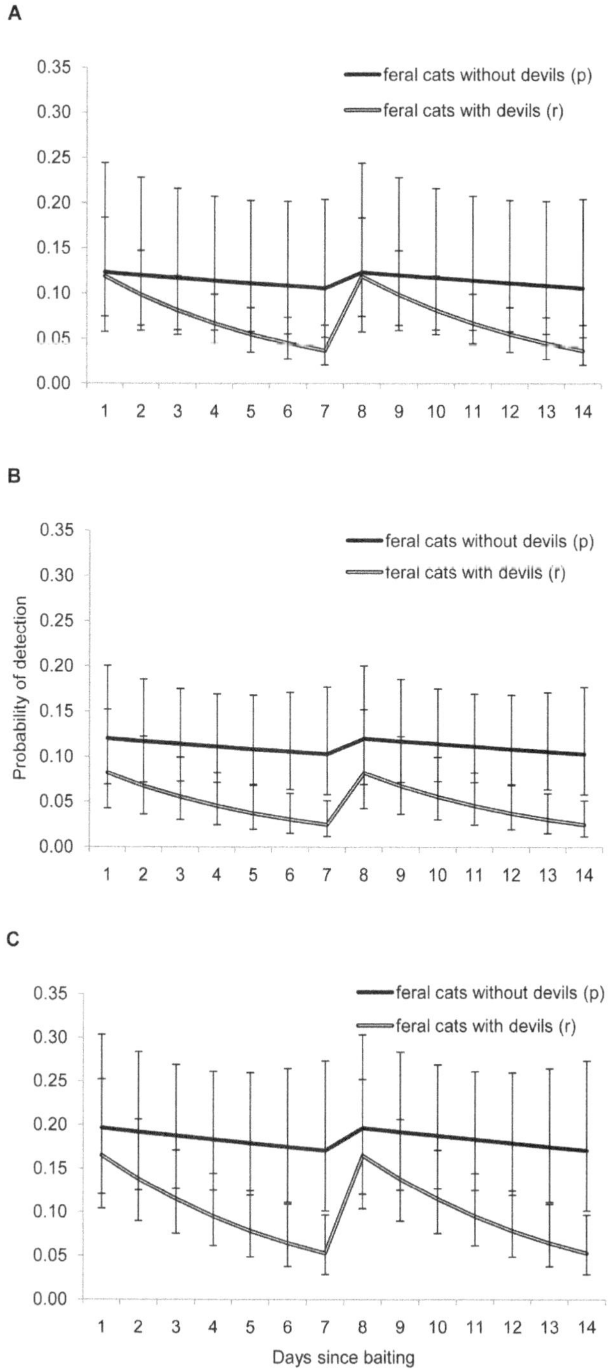

Figure 6. The probability of detecting feral cats *Felis catus* in relation to Tasmanian devils *Sarcophilus harrisii*. Study sites represented are: A. Southwest, B. Mt Field and C. Tasman Peninsula. Trends are shown as a function of time since baiting; cameras were initially baited on day one and re-baited on day seven. Estimates of probability of detection were generated in PRESENCE 4.0 under the highest-ranked AIC$_C$ model in a two-species multi-season analysis. Bars represent 95% CI.

Tasman Peninsula compared to sites where feral cats were not detected.

- The two species were detected less often within the same 24 h period than expected if they were detected independently. Estimates of delta were 0.49 (±0.06 SE) for the Southwest and 0.67–0.68 (±0.05 SE) for the Tasman Peninsula.
- Time since baiting, habitat and study site were important covariates for swamp rats. Time since baiting and study site were important covariates for feral cats.

Small Mammal Trap Responses to Predator Odours

There were 46 captures of swamp rats across the two study sites (Table 9) and fewer were captured in traps scented with feral cat odour compared to the other odours (Fig. 7). In comparison, devil-scented traps captured the most swamp rats compared to other odours. While trends were apparent, there were no significant differences in response to odour by swamp rats ($\chi^2_3 = 5.13$, $p = 0.162$).

Discussion

Fear, manifested as avoidance behaviour, can be an important factor in shaping individual behaviour and population dynamics [23], and can be interpreted as a survival adaptation. Fear responses have been observed in a wide range of taxa, including mesopredators and prey [17]. A better understanding of avoidance behaviour is a precursor to effective conservation management and to understanding ecological interactions more generally. This is especially important for invasive predators because they often have particularly adverse impacts on prey [43], [44]. In addition, mesopredator relationships between invasive predators can lead to unexpected outcomes for other species in the absence of an integrated pest management approach [45]. We investigated patterns of detection between two predators, the endangered marsupial Tasmanian devil and the introduced eutherian feral cat, and one native small mammal, the swamp rat.

We found that feral cats were detected less often at cameras where devils were detected; this was apparent at all three of our study sites, and occurred for the duration of the 1–2 week study site surveys. Devils also tended to be detected more often at cameras where feral cats had been detected. The responses of the swamp rat to devil and cat cues at cameras were less clear and depended on study site, although capture rates of swamp rats in Elliott traps tended to be less where feral cat but not devil faecal odours were present. Thus, the results of the devil and feral cat two-species occupancy analyses at cameras supported our initial hypothesis, but models incorporating the swamp rat were less clear. The results of the faecal odour trials provided very limited support for our hypothesis that swamp rats would avoid feral cat scented traps, and no support for our hypothesis that they would avoid devil scented traps.

Two-Species Occupancy Models

We used a series of two-species occupancy models to elucidate patterns of detection between the three study species. Our models showed that the probability of detecting a given species could be associated with the detection of two or more species, thus indicating that patterns of detection are at least three-way, and realistically might be far more complex given that over 20 species were recorded at cameras. In this regard our models were a considerable simplification of the real system. Modelling interactions among multiple species within a maximum likelihood framework is not yet possible with available software. A three-species model for one predator and two prey has been developed [46]; however, this model does not accommodate two predators and one prey, or hierarchical interactions between two predators

Table 4. Model selection statistics for Tasmanian devils and swamp rats.

Model	AIC	delta AIC*	AIC wt†	Model likelihood	No. par‡	−2xLogLik**
p*Sh* (s+h), p*Rl*(bxs+h), r*Sh*(s+h),r*Rl* (bxs+h),delta(bxs)	2314.33	0.00	0.2654	1.0000	20	2274.33
p*Sh* = r*Sh* (s+h), p*Rl*(bxs+h),r*Rl* (bxs+h),delta(bxs)	2314.46	0.13	0.2487	0.9371	18	2278.46
p*Sh* = r*Sh* (s+h), p*Rl*(bxs+h),r*Rl* (bxs+h),delta(s)	2316.20	2.01	0.1042	0.3962	17	2282.20
p*Sh* = r*Sh* (xbxs+h),p*Rl* (xbxs+h),r*Rl* (xbxs+h),delta(xbxs)	2316.45	2.12	0.0919	0.3465	19	2278.45
p*Sh* (bxs), p*Rl*(bxs), r*Sh*(bxs),r*Rl* (bxs),delta(bxs)	2317.63	3.30	0.0510	0.1920	19	2279.63
p*Sh* = r*Sh* (xbxs),p*Rl* (xbxs),r*Rl* (xbxs+h),delta(xbxs)	2317.81	3.48	0.0466	0.1755	16	2285.81
p*Sh* = r*Sh* (xbxs+h),p*Rl* (xbxs+h),r*Rl* (xbxs+h),delta(= 1)	2318.68	4.35	0.0301	0.1136	17	2284.68
p*Sh* (bxs), p*Rl*(bxs), r*Sh*(bxs),r*Rl* (bxs),delta(b)	2318.80	4.47	0.0284	0.1070	18	2282.80
p*Sh* (xbxs+h),p*Rl* (xbxs+h), r*Sh* (xbxs+h),r*Rl* (xbxs+h),delta(xbxs)	2319.37	5.04	0.0214	0.0805	22	2275.37
p*Sh* (xbxs+h),p*Rl* (xbxs+h), r*Sh* (xbxs+h),r*Rl* (xbxs+h),delta(xbxs+h)	2319.39	5.06	0.0211	0.0797	22	2275.39
p*Sh* (s+h), p*Rl*(bxs+h), r*Sh*(s+h),r*Rl* (bxs+h),delta(bxs)	**2320.67**	**6.34**	**0.0111**	**0.0420**	**18**	**2284.67**
p*Sh* (s+h), p*Rl* = r*Rl* (bxs+h), r*Sh*(s+h),delta(bxs)	**2345.14**	**30.81**	**0.0000**	**0.0000**	**17**	**2311.14**

The ten most supported multi-season two-species occupancy models based on AIC (Akaike's Information Criterion) in program PRESENCE 4.0. Models that did not fall within the top ten that tested specific hypotheses are shown in bold below the top ten. The models were fitted to detection data from two study sites in cool temperate forests in southern Tasmania during standardized surveys from 2009–2011. The terms in parentheses represent the sources of variation in model parameters. 's' denotes study site (Southwest and Tasman Peninsula), 'h' denotes habitat (dry eucalypt forest and woodland, wet eucalyptus forest and woodland, rainforest and related scrub, and agricultural exotic), 'b' denotes days since baiting, and 'season' denotes a model where detection varied with season. '.' indicates a parameter set equal across species and survey times. The probabilities of site occupancy of devils and swamp rats were estimated independently, and the probability of site occupancy, colonization and extinction were constrained to be constant for both species and covariates for all models. Tasmanian devil *Sarcophilus harrisii* is abbreviated *Sh* and swamp rat *Rattus lutreolus* is abbreviated *Rl*.
Delta AIC is the difference in AIC values between each model and the model with the lowest AIC.
†AIC wt is the model weight.
‡Number of parameters in the model.
**Twice the negative log-likelihood.

Table 5. Model selection statistics for feral cats and swamp rats.

Model	AIC	delta AIC*	AIC wt†	Model likelihood	No. par‡	−2xLogLik**
p*Fc* (bxs),p*Rl* (bxs+h),r*Fc* (bxs),r*Rl* (bxs+h),delta (s)	2271.95	0	0.6358	1	21	2229.95
p*Fc* (bxs),p*Rl* (bxs+h), r*Fc* (bxs),r*Rl* (bxs+h),delta(bxs)	2273.94	1.99	0.2351	0.3697	22	2229.94
p*Fc* (bxs),p*Rl* (bxs+h),r*Fc* (bxs),r*Rl* (bxs+h),delta(.)	2275.52	3.57	0.1067	0.1678	20	2235.52
p*Fc* (bxs+h),p*Rl* (bxs+h), r*Fc* (bxs+h),r*Rl* (bxs+h),delta(bxs+h)	2281.12	9.17	0.0065	0.0102	22	2237.12
p*Fc* (bxs),p*Rl* (bxs+h),r*Fc* (bxs),r*Rl* (bxs+h),delta(= 1), gam,eps	2281.17	9.22	0.0063	0.01	20	2241.17
p*Fc* (bxs),p*Rl* (bxs), r*Fc* (bxs),r*Rl* (bxs),delta(bxs), gam,eps	2282.82	10.87	0.0028	0.0044	19	2244.82
p*Fc* (bxs),p*Rl* (bxs+h), r*Fc* (bxs),r*Rl* (bxs+h),delta(bxs+h), gam,eps	2282.86	10.91	0.0027	0.0043	22	2238.86
p*Fc* = r*Fc* (bxs),p*Rl* (bxs+h),r*Rl* (bxs+h),delta(s), gam,eps	2283.26	11.31	0.0022	0.0035	18	2247.26
p*Fc* (bxs),p*Rl* (bxs),r*Fc* (bxs),r*Rl* (bxs),delta(.), gam,eps	2284.78	12.83	0.001	0.0016	17	2250.78
p*Fc* (bxs),p*Rl* (bxs+h),r*Fc* (bxs),r*Rl* (bxs+h),delta(b), gam,eps	2285.22	13.27	0.0008	0.0013	21	2243.22
p*Fc* (bxs),p*Rl* = r*Rl* (bxs+h),r*Fc* (bxs),delta(s)	**2305.43**	**33.48**	**0**	**0**	**18**	**2269.43**

The ten most supported multi-season two-species occupancy models based on AIC (Akaike's Information Criterion) in program PRESENCE 4.0. Models that did not fall within the top ten that tested specific hypotheses are shown in bold below the top ten. The models were fitted to detection data from two study sites in cool temperate forests in southern Tasmania during standardized surveys from 2009–2011. The terms in parentheses represent the sources of variation in model parameters. 's' denotes study site (Southwest and Tasman Peninsula), 'h' denotes habitat (dry eucalypt forest and woodland, wet eucalyptus forest and woodland, rainforest and related scrub, and agricultural exotic), 'b' denotes days since baiting, and 'season' denotes a model where detection varied with season. '.' indicates a parameter set equal across species and survey times. The probabilities of site occupancy of feral cats and swamp rats were estimated independently, and the probability of site occupancy, colonization and extinction were constrained to be constant for both species and covariates for all models. Feral cat *Felis catus* is abbreviated *Fc* and swamp rat *Rattus lutreolus* is abbreviated *Rl*.
*Delta AIC is the difference in AIC values between each model and the model with the lowest AIC.
†AIC wt is the model weight.
‡Number of parameters in the model.
**Twice the negative log-likelihood.

Table 6. Model outputs from two-species occupancy models for Tasmanian devils and feral cats.

p*Sh* (s+h), p*Fc*(bxs), r*Sh* (s+h),r*Fc* (bxs),delta(.)

Psi *Sh*	Psi *Fc*	p *Sh*	p *Fc*	r *Sh*	r *Fc*	delta
0.63 (0.12)	0.70 (0.13)	SW 0.07 (0.02) D	SW 0.12 (0.05) 1	SW 0.13 (0.03) D	SW 0.12 (0.03) 1	0.47 (0.07)
		SW 0.09 (0.03) W	SW 0.12 (0.04) 2	SW 0.18 (0.03) W	SW 0.10 (0.02) 2	
		SW 0.10 (0.03) R	SW 0.11 (0.04) 3	SW 0.19 (0.03) R	SW 0.08 (0.01) 3	
		MtF 0.17 (0.06) R	SW 0.11 (0.04) 4	MtF 0.15 (0.05) R	SW 0.07 (0.01) 4	
		MtF 0.12 (0.05) D	SW 0.11 (0.04) 5	MtF 0.10 (0.03) D	SW 0.05 (0.01) 5	
		MtF 0.17 (0.04) W	SW 0.11 (0.04) 6	MtF 0.14 (0.03) W	SW 0.05 (0.01) 6	
		MtF 0.28 (0.04) A	SW 0.11 (0.04) 7	MtF 0.24 (0.04) A	SW 0.04 (0.01) 7	
		TP 0.002 (0.002) D	MtF 0.12 (0.03) 1	TP 0.05 (0.02) D	MtF 0.08 (0.02) 1	
		TP 0.005 (0.004) A	MtF 0.12 (0.03) 2	TP 0.13 (0.04) A	MtF 0.07 (0.02) 2	
		TP 0.002 (0.002) W	MtF 0.11 (0.03) 3	TP 0.07 (0.02) W	MtF 0.06 (0.02) 3	
			MtF 0.11 (0.02) 4		MtF 0.05 (0.01) 4	
			MtF 0.11 (0.03) 5		MtF 0.04 (0.01) 5	
			MtF 0.11 (0.03) 6		MtF 0.03 (0.01) 6	
			MtF 0.10 (0.03) 7		MtF 0.02 (0.01) 7	
			TP 0.20 (0.05) 1		TP 0.17 (0.04) 1	
			TP 0.19 (0.04) 2		TP 0.14 (0.03) 2	
			TP 0.19 (0.04) 3		TP 0.11 (0.02) 3	
			TP 0.18 (0.03) 4		TP 0.10 (0.02) 4	
			TP 0.18 (0.04) 5		TP 0.08 (0.02) 5	
			TP 0.17 (0.04) 6		TP 0.06 (0.02) 6	
			TP 0.17 (0.04) 7		TP 0.05 (0.02) 7	

p*Sh* (s), p*Fc*(bxs), r*Sh*(s),r*Fc* (bxs),delta(.)

Psi *Sh*	Psi *Fc*	p *Sh*	p *Fc*	r *Sh*	r *Fc*	delta
0.62 (0.12)	0.76 (0.12)	SW 0.08 (0.03)	SW 0.12 (0.05) 1	SW 0.17 (0.02)	SW 0.11 (0.03) 1	0.46 (0.07)
		MtF 0.08 (0.03)	SW 0.12 (0.04) 2	MtF 0.26 (0.02)	SW 0.09 (0.02) 2	
		TP 0.002 (0.005)	SW 0.11 (0.04) 3	TP 0.07 (0.02)	SW 0.08 (0.02) 3	
			SW 0.11 (0.04) 4		SW 0.07 (0.01) 4	
			SW 0.10 (0.04) 5		SW 0.05 (0.01) 5	
			SW 0.10 (0.04) 6		SW 0.05 (0.01) 6	
			SW 0.10 (0.03) 7		SW 0.04 (0.01) 7	
			MtF 0.12 (0.03) 1		MtF 0.06 (0.02) 1	
			MtF 0.12 (0.03) 2		MtF 0.05 (0.01) 2	
			MtF 0.11 (0.02) 3		MtF 0.04 (0.01) 3	
			MtF 0.11 (0.02) 4		MtF 0.03 (0.01) 4	
			MtF 0.11 (0.02) 5		MtF 0.03 (0.01) 5	
			MtF 0.10 (0.03) 6		MtF 0.02 (0.01) 6	
			MtF 0.10 (0.03) 7		MtF 0.02 (0.01) 7	
			TP 0.21 (0.06) 1		TP 0.15 (0.04) 1	
			TP 0.20 (0.05) 2		TP 0.13 (0.03) 2	
			TP 0.20 (0.04) 3		TP 0.11 (0.02) 3	
			TP 0.19 (0.04) 4		TP 0.09 (0.02) 4	
			TP 0.18 (0.04) 5		TP 0.08 (0.02) 5	
			TP 0.18 (0.04) 6		TP 0.06 (0.02) 6	
			TP 0.17 (0.04) 7		TP 0.05 (0.02) 7	

Outputs are shown for models that were within two AIC of the most supported model. 's' represents study site (SW = Southwest, MtF = Mt Field or TP = Tasman Peninsula), 'h' represents habitat (D = dry eucalypt forest and woodland, W = wet eucalypt forest and woodland, R = rainforest, A = agricultural and/or exotic vegetation), 'b' represents days since baiting (1 = less than one day since baiting, 2 = less than two days since baiting and so on) and '.' represents constant. Numbers in brackets represent standard error.

Table 7. Model outputs from two-species occupancy models for Tasmanian devils and swamp rats.

p*Sh* (s+h), p*Rl*(bxs+h), r*Sh*(s+h),r*Rl* (bxs+h),delta(bxs)

Psi *Sh*	Psi *Rl*	p *Sh*	p *Rl*	r *Sh*	r *Rl*	delta
0.99 (0.17)	0.11 (0.06)	SW 0.16 (0.03) D	SW 0.21 (0.10) D 1	SW 0.16 (0.03) D	SW 0.46 (0.05) D 1	SW 0.66 (0.06) 1
		SW 0.15 (0.02) W	SW 0.17 (0.08) D 2	SW 0.15 (0.02) W	SW 0.44 (0.05) D 2	SW 0.57 (0.05) 2
		SW 0.11 (0.02) R	SW 0.13 (0.07) D 3	SW 0.11 (0.02) R	SW 0.41 (0.04) D 3	SW 0.47 (0.05) 3
		TP 0.04 (0.01) D	SW 0.11 (0.06) D 4	TP 0.05 (0.01) D	SW 0.40 (0.04) D 4	SW 0.38 (0.06) 4
		TP 0.02 (0.01) A	SW 0.08 (0.05) D 5	TP 0.02 (0.01) A	SW 0.36 (0.04) D 5	SW 0.29 (0.08) 5
		TP 0.04 (0.01) W	SW 0.06 (0.04) D 6	TP 0.04 (0.01) W	SW 0.34 (0.04) D 6	SW 0.22 (0.08) 6
			SW 0.05 (0.03) D 7		SW 0.32 (0.05) D 7	SW 0.16 (0.08) 7
			TP 0.78 (0.07) D 1		TP 0.26 (0.05) D 1	TP 0.58 (0.15) 1
			TP 0.73 (0.07) D 2		TP 0.24 (0.05) D 2	TP 0.48 (0.15) 2
			TP 0.67 (0.07) D 3		TP 0.22 (0.04) D 3	TP 0.39 (0.15) 3
			TP 0.61 (0.06) D 4		TP 0.21 (0.04) D 4	TP 0.30 (0.14) 4
			TP 0.54 (0.06) D 5		TP 0.19 (0.04) D 5	TP 0.23 (0.13) 5
			TP 0.47 (0.07) D 6		TP 0.17 (0.04) D 6	TP 0.17 (0.11) 6
			TP 0.40 (0.08) D 7		TP 0.16 (0.04) D 7	TP 0.12 (0.09) 7

p*Sh*=r*Sh* (s+h), p*Rl*(bxs+h),r*Rl* (bxs+h),delta(bxs)

Psi *Sh*	Psi *Rl*	p *Sh*	p *Rl*	r *Sh*	r *Rl*	delta
0.74 (0.16)	0.11 (0.06)	SW 0.15 (0.02) D	SW 0.60 (0.11) 1	SW 0.15 (0.02) D	SW 0.37 (0.05) 1	SW 0.66 (0.07) 1
		SW 0.16 (0.02) W	SW 0.55 (0.11) 2	SW 0.16 (0.02) W	SW 0.35 (0.05) 2	SW 0.58 (0.06) 2
		SW 0.12 (0.02) R	SW 0.51 (0.11) 3	SW 0.12 (0.02) R	SW 0.32 (0.04) 3	SW 0.50 (0.05) 3
		TP 0.04 (0.01) D	SW 0.46 (0.10) 4	TP 0.04 (0.01) D	SW 0.29 (0.04) 4	SW 0.42 (0.07) 4
		TP 0.02 (0.01) A	SW 0.41 (0.10) 5	TP 0.02 (0.01) A	SW 0.27 (0.04) 5	SW 0.35 (0.08) 5
		TP 0.04 (0.01) W	SW 0.37 (0.11) 6	TP 0.04 (0.01) W	SW 0.25 (0.04) 6	SW 0.28 (0.10) 6
			SW 0.33 (0.11) 7		SW 0.23 (0.04) 7	SW 0.21 (0.10) 7
			TP 0.70 (0.07) 1		TP 0.24 (0.05) 1	TP 0.57 (0.15) 1
			TP 0.66 (0.06) 2		TP 0.22 (0.04) 2	TP 0.49 (0.15) 2
			TP 0.61 (0.06) 3		TP 0.20 (0.04) 3	TP 0.41 (0.15) 3
			TP 0.57 (0.06) 4		TP 0.18 (0.03) 4	TP 0.34 (0.15) 4
			TP 0.52 (0.06) 5		TP 0.16 (0.03) 5	TP 0.27 (0.14) 5
			TP 0.48 (0.07) 6		TP 0.15 (0.03) 6	TP 0.21 (0.13) 6
			TP 0.43 (0.08) 7		TP 0.13 (0.03) 7	TP 0.16 (0.12) 7

Outputs are shown for models that were within two AIC of the most supported model. 's' represents study site (SW = Southwest or TP = Tasman Peninsula), 'h' represents habitat (D = dry eucalypt forest and woodland, W = wet eucalypt forest and woodland, R = rainforest, A = agricultural and/or exotic vegetation), 'b' represents days since baiting (1 = less than one day since baiting, 2 = less than two days since baiting and so on) and '.' represents constant. Numbers in brackets represent standard error.

such as that implied in the present study between devils and feral cats.

Despite the potential influences of multiple species, the consistent pattern of detection between devils and feral cats in three study sites is compelling evidence that devils can alter feral cat detection. There have been suggestions of a mesopredator relationship between devils and cats following observations of a marked increase in spotlight counts and anecdotal reports of feral cats in the north-east of Tasmania in recent years (Department of Primary Industries and Water, unpublished data) [37]. This latter region is where DFTD appears to have been longest established and where devil populations have suffered the most dramatic declines to date. Whether these observations truly represent mesopredator release is yet to be determined. It is also important to note that if there is a competitive relationship between devils and feral cats, the strengths and characteristics of this relationship may vary spatially and temporally. For example, at higher abundances feral cats may overwhelm prey resources or kill significant numbers of juvenile devils [47]. Also, preferential use of trails has been recorded in carnivore species in sand dune habitats, and we might have recorded different patterns of detection if we had set cameras randomly throughout the landscape rather than limiting them to roads and trails [48].

Aspects of devil biology that may have facilitated their apparent dominance at baited camera stations include efficient scavenging abilities [49] and scent-marking. We regularly recorded devils scent marking camera stations (via anal-dragging, urination and defecation), whereas none of the other species photographed

Table 8. Model outputs from two-species occupancy models for feral cats and swamp rats.

Psi *Fc*	Psi *Rl*	p *Fc*	p *Rl*	r *Fc*	r *Rl*	delta
p*Fc* (bxs),p*Rl* (bxs+h),r*Fc* (bxs),r*Rl* (bxs+h),delta (s)						
0.99 (0.15)	0.11 (0.06)	SW 0.10 (0.02) 1	SW 0.36 (0.09) D 1	SW 0.11 (0.03) 1	SW 0.57 (0.06) 1	SW 0.49 (0.06)
		SW 0.09 (0.02) 2	SW 0.30 (0.07) D 2	SW 0.09 (0.02) 2	SW 0.56 (0.06) 2	TP 0.68 (0.05)
		SW 0.09 (0.02) 3	SW 0.25 (0.06) D 3	SW 0.08 (0.02) 3	SW 0.54 (0.06) 3	
		SW 0.08 (0.01) 4	SW 0.20 (0.05) D 4	SW 0.07 (0.02) 4	SW 0.52 (0.06) 4	
		SW 0.07 (0.01) 5	SW 0.16 (0.05) D 5	SW 0.06 (0.02) 5	SW 0.50 (0.06) 5	
		SW 0.07 (0.02) 6	SW 0.13 (0.04) D 6	SW 0.05 (0.02) 6	SW 0.49 (0.06) 6	
		SW 0.07 (0.02) 7	SW 0.10 (0.04) D 7	SW 0.05 (0.02) 7	SW 0.47 (0.07) 7	
		TP 0.19 (0.03) 1	TP 0.78 (0.06) D 1	TP 0.07 (0.02) 1	TP 0.34 (0.06) D 1	
		TP 0.17 (0.03) 2	TP 0.73 (0.06) D 2	TP 0.07 (0.02) 2	TP 0.33 (0.05) D 2	
		TP 0.16 (0.02) 3	TP 0.67 (0.06) D 3	TP 0.06 (0.01) 3	TP 0.31 (0.05) D 3	
		TP 0.15 (0.02) 4	TP 0.61 (0.06) D 4	TP 0.05 (0.01) 4	TP 0.30 (0.04) D 4	
		TP 0.14 (0.02) 5	TP 0.54 (0.07) D 5	TP 0.04 (0.01) 5	TP 0.28 (0.04) D 5	
		TP 0.13 (0.02) 6	TP 0.48 (0.08) D 6	TP 0.04 (0.01) 6	TP 0.27 (0.05) D 6	
		TP 0.12 (0.03) 7	TP 0.41 (0.09) D 7	TP 0.03 (0.01) 7	TP 0.26 (0.05) D 7	
p*Fc* (bxs),p*Rl* (bxs+h), r*Fc* (bxs),r*Rl* (bxs+h),delta(bxs)						
0.99 (0.15)	0.11 (0.06)	SW 0.10 (0.02) 1	SW 0.36 (0.09) D 1	SW 0.10 (0.03) 1	SW 0.57 (0.06) 1	SW 0.49 (0.08) 1
		SW 0.09 (0.02) 2	SW 0.30 (0.07) D 2	SW 0.09 (0.02) 2	SW 0.56 (0.06) 2	SW 0.49 (0.07) 2
		SW 0.09 (0.02) 3	SW 0.25 (0.06) D 3	SW 0.08 (0.02) 3	SW 0.54 (0.06) 3	SW 0.49 (0.06) 3
		SW 0.08 (0.01) 4	SW 0.20 (0.05) D 4	SW 0.07 (0.02) 4	SW 0.52 (0.06) 4	SW 0.49 (0.06) 4
		SW 0.07 (0.01) 5	SW 0.16 (0.05) D 5	SW 0.06 (0.02) 5	SW 0.50 (0.06) 5	SW 0.49 (0.08) 5
		SW 0.07 (0.02) 6	SW 0.13 (0.04) D 6	SW 0.05 (0.02) 6	SW 0.49 (0.06) 6	SW 0.49 (0.09) 6
		SW 0.06 (0.02) 7	SW 0.10 (0.04) D 7	SW 0.05 (0.02) 7	SW 0.47 (0.07) 7	SW 0.49 (0.11) 7
		TP 0.19 (0.03) 1	TP 0.78 (0.06) D 1	TP 0.08 (0.02) 1	TP 0.34 (0.06) D 1	TP 0.67 (0.06) 1
		TP 0.17 (0.02) 2	TP 0.73 (0.06) D 2	TP 0.07 (0.02) 2	TP 0.33 (0.05) D 2	TP 0.67 (0.07) 2
		TP 0.16 (0.02) 3	TP 0.67 (0.06) D 3	TP 0.06 (0.01) 3	TP 0.31 (0.05) D 3	TP 0.68 (0.05) 3
		TP 0.15 (0.02) 4	TP 0.61 (0.06) D 4	TP 0.05 (0.01) 4	TP 0.30 (0.04) D 4	TP 0.68 (0.05) 4
		TP 0.14 (0.02) 5	TP 0.54 (0.07) D 5	TP 0.04 (0.01) 5	TP 0.28 (0.04) D 5	TP 0.68 (0.07) 5
		TP 0.13 (0.02) 6	TP 0.48 (0.08) D 6	TP 0.04 (0.01) 6	TP 0.27 (0.05) D 6	TP 0.68 (0.08) 6
		TP 0.12 (0.03) 7	TP 0.41 (0.09) D 7	TP 0.03 (0.01) 7	TP 0.26 (0.05) D 7	TP 0.68 (0.10) 7

Outputs are shown for models that were within two AIC of the most supported model. 's' represents study site (SW = Southwest or TP = Tasman Peninsula), 'h' represents habitat (D = dry eucalypt forest and woodland, W = wet eucalypt forest and woodland, R = rainforest, A = agricultural and/or exotic vegetation), 'b' represents days since baiting (1 = less than one day since baiting, 2 = less than two days since baiting and so on) and '.' represents constant. Numbers in brackets represent standard error.

showed such obvious behaviours consistent with scent marking. The function of scent marking in devils is unknown, but it is reasonable to expect that it serves as a signal to other animals, including feral cats, that a devil may return. Most mammals scent mark [50] and it is believed to play an important role in inter- and intra-specific communication [25], [51].

Patterns of detection between swamp rats, feral cats and devils were not as consistent as those between devils and feral cats. This

Table 9. Swamp Rat captures in scented traps.

Site	Survey	Trap nights	Clear	Tasmanian Pademelon	Tasmanian devil	Feral cat
Tasman Peninsula	April 2011	560	8	2	5	2
Mt Field	June 2011	480	5	7	12	5

The number of captures of the Swamp Rat *Rattus lutreolus* for each survey at two study sites in Elliott traps scented with clear (no faecal smear), Tasmanian Pademelon *Thylogale billardierii* faeces, Tasmanian Devil *Sarcophilus harrisii* faeces, and feral cat faeces. Trap nights are expressed as number of traps multiplied by nights set.

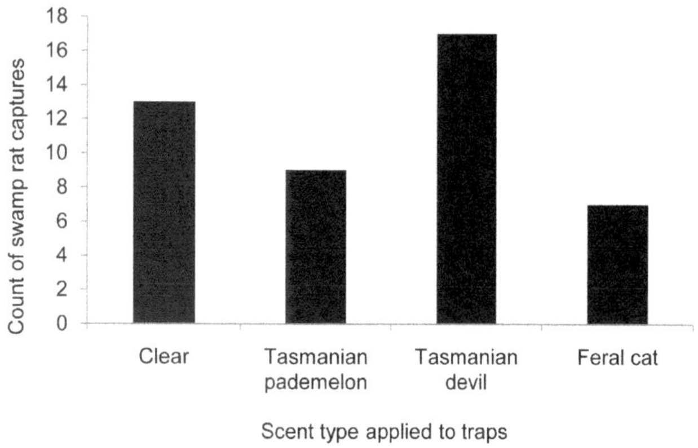

Figure 7. Captures (n=46) of swamp rat *Rattus lutreolus* in Elliott traps scented with different faecal odours. Traps were scented with one of four different odours: clear (no faecal smear), Tasmanian pademelon *Thylogale billardierii* faeces, Tasmanian devil *Sarcophilus harrisii* faeces and feral cat *Felis catus* faeces.

inconsistency may result from inadequate model structure: as noted previously, a series of two-species models was used to represent a three-way system and we were not able to fully incorporate season as a covariate in our models due to the large number of parameters involved. Single species multi-season site occupancy models indicated that there was some seasonal variability in the probability of detecting swamp rats [38]. Alternatively, the patterns of swamp rat detection that we observed may reflect reality, with animals differing in detection in different areas. Biological factors that could underlie such differences include competition between conspecifics [5], [52] different predator abundances and therefore prey vigilance [53], and different risk thresholds associated with factors such as food availability or fine-scale habitat preferences [23], [52].

Faecal Odour Trials

Faecal odour trials suggested a tendency for swamp rats to be captured less often in feral cat scented traps and more often in devil scented traps, but these trends were not significant. It is possible that our study may not have been powerful enough to detect real avoidance. If our findings were a true reflection of reality, however, then they provide partial support for a study that found swamp rats in the laboratory avoided integumental odour (odour trapped in hair from secretions of sebaceous glands on the skin) of the spotted-tailed quoll (a native carnivore in Tasmania), but did not avoid integumental odour from the red fox or feral cat [54]. Differential avoidance of odours highlights the potential for variability in measures of predator recognition depending on the predator and type of cue.

If strength of avoidance of predator cues is proportional to the intensity or duration of selection [28], feral cats could be expected to exert more predatory pressure on swamp rats than devils. In the first instance, small mammals occur much less frequently in the diet of the devil than do medium-sized mammals [55], [56], [57]. Moreover, the frequency of occurrence of small mammals is much lower in the diet of the devil compared to that in the diet of feral cats [38]. If this reasoning is correct, our failure to find stronger avoidance by swamp rats of the odour of feral cats is surprising and may suggest that selection has not been as intense as we had expected. This could reflect insufficient time for selection to have acted or, perhaps more plausibly, that devils maintained cats at such low densities prior to the arrival of DFTD that their predatory impacts on swamp rats were usually weak. Conversely, faeces may not represent reliable cues to the current risk of predation; integumental odours may arguably be more representative of impending attack and may therefore induce different and perhaps more marked anti-predator behaviour [54]. In addition, predator cues are not limited to odour, they can also include visual and acoustic cues [58], [59].

We hesitate to compare the responses by swamp rats to the predator activity recorded at cameras, and the odour cues on traps, for three reasons: there was, firstly, an inconsistent response by swamp rats at cameras; secondly, our two-species models may not represent reality to the degree necessary to reliably quantify avoidance; and thirdly, the responses of rats to odours were weak. The potential for prey avoidance to vary with cue, study site, season, and habitat structure highlights the need to conduct predator avoidance studies that account for these factors [60].

Conclusions

Patterns of detection at cameras of devils and feral cats are consistent with the interpretation of a dominant predator – mesopredator relationship, and warrant further investigation given that devils have declined dramatically as a result of devil facial tumour disease. Because small mammals constitute a higher proportion of the diet of feral cats compared with devils, there may be an increased risk to populations of small mammals if there is an increase in the activity and thus impact of feral cats resulting from devil decline. Feral cat impacts may be exacerbated further if swamp rats truly exhibit weak avoidance of predator scents. Further investigation could focus profitably on the responses of other small Tasmanian mammals to cat cues and cat predation, especially scarce endemic species that could be at particularly increased risk. More generally, we suggest that population estimates derived from devices such as remote cameras should acknowledge the potential for one species to change the detectability of another, and ensure that this potential bias is incorporated in assessments of numbers and survival.

Acknowledgments

The authors thank Ben Allen and two anonymous reviewers for their valuable comments on the draft manuscript. This work would not have been possible without logistical support from the Tasmanian Department of Primary Industries, Parks Water and Environment including the Threatened Species Section, Mt Field Parks and Wildlife Service base, and Wildlife Management Branch. Numerous landholders allowed access to their land and many volunteers assisted with fieldwork.

Author Contributions

Conceived and designed the experiments: BL CD. Performed the experiments: BL. Analyzed the data: BL. Contributed reagents/materials/analysis tools: BL CD. Wrote the paper: BL CD.

References

1. Elton CS (1967) *Animal ecology*. Great Britain: William Clowes and Sons Ltd.
2. Wootton JT, Emmerson M (2005) Measurement of interaction strength in nature. Ann Rev Ecol Evol S 36: 419–444.
3. Ritchie EG, Martin JK, Johnson CN, Fox BJ (2009) Separating the influences of environment and species interactions on patterns of distribution and abundance: competition between large herbivores. J Anim Ecol 78: 724–731.

4. Apfelbach R, Blanchard CD, Blanchard RJ, Hayes RA, McGregor IS (2005) The effects of predator odors in mammalian prey species: a review of field and laboratory studies. Neurosci Biobehav Rev 29: 1123–1144.
5. Grether GF, Losin N, Anderson CN, Okamoto K (2009) The role of interspecific interference competition in character displacement and the evolution of competitor recognition. Biol Rev 84: 617–635.
6. Nersesian CL, Banks PB, McArthur C (2012) Behavioural responses to indirect and direct predator cues by a mammalian herbivore, the common brushtail possum. Behav Ecol Sociobiol 66: 44–55.
7. Cutler TL, Swann DE (1999) Using remote photography in wildlife ecology: a review. Wildlife Soc Bull 27: 571–581.
8. MacKenzie DI, Bailey LL, Nichols JD (2004) Investigating species co-occurrence patterns when species are detected imperfectly. J Anim Ecol 73: 546–555.
9. Dickman CR (1991) Mechanisms of competition among insectivorous mammals. Oecologia 85: 464–471.
10. Letnic M, Ritchie EG, Dickman CR (2012) Top predators as biodiversity regulators: the dingo *Canis lupus dingo* as a case study. Biol Rev 87: 390–413.
11. Ripple WJ, Beschta RL (2004) Wolves and the ecology of fear: can predation risk structure ecosystems? BioScience 54: 755–766.
12. Frank DA (2008) Evidence for top predator control of a grazing ecosystem. Oikos 117: 1718–1724.
13. Palomares E, Caro TM (1999) Interspecific killing among mammalian carnivores. The Am Nat 153: 492–508.
14. Durant SM (2000) Predator avoidance, breeding experience and reproductive success in endangered cheetahs, *Acinonyx jubatus*. Anim Behav 60: 121–130.
15. Linnell JD, Strand O (2000) Interference interactions, co-existence and conservation of mammalian carnivores. Divers Distrib 6: 169–176.
16. Ritchie EG, Johnson CN (2009) Predator interactions, mesopredator release, and biodiversity conservation. Ecol Lett 12: 982–998.
17. Rogers CM, Caro MJ (1998) Song sparrows, top carnivores, and nest predation: a test of the mesopredator release hypothesis. Oecologia 116: 227–233.
18. Soulé ME, Bolger DT, Alberts AC, Wright J, Sorice M, et al. (1988) Reconstructed dynamics of rapid extinctions of chaparral-requiring birds in urban habitat islands. Conserv Biol 2: 75–92.
19. Crooks KR, Soulé ME (1999) Mesopredator release and avifaunal extinctions in a fragmented system. Nature 400: 563–566.
20. Risbey DA, Calver MC, Short J, Bradley JS, Wright IW (2000) The impacts of cats and foxes on the small vertebrate fauna of Heirisson Prong, Western Australia. II. A field experiment. Wildl Res 27: 223–235.
21. Woinarski JCZ, Legge S, Fitzsimons JA, Traill BJ, Burbidge AA, et al. (2011) The disappearing mammal fauna of northern Australia: context, cause and response. Conserv Lett 4: 192–201.
22. Brook LA, Johnson CN, Ritchie EG (2012) Effects of predator control on behaviour of an apex predator and indirect consequences for mesopredator suppression. J Appl Ecol: DOI: 10.1111/j.1365-2664.2012.02207.x.
23. Brown JS, Laundré JW, Garung M (1999) The ecology of fear: optimal foraging, game theory, and trophic interactions. J Mammal 80: 385–389.
24. Boonstra R, Rodd FH, Carleton DJ (1982) Effect of *Blarina brevicauda* on trap response of *Microtus pennsylvanicus*. Can J Zool 60: 438–442.
25. Dickman CR, Doncaster CP (1984) Responses of small mammals to red fox (*Vulpes vulpes*) odour. J Zool 204: 521–531.
26. Banks PB (1998) Responses of Australian bush rats, *Rattus fuscipes*, to the odor of introduced *Vulpes vulpes*. J Mammal 79: 1260–1264.
27. Berger J, Swenson JE, Persson I (2001) Recolonizing carnivores and naïve prey: conservation lessons from Pleistocene extinctions. Science 291: 1036–1039.
28. Banks PB, Dickman CR (2007) Alien predation and the effects of multiple levels of prey naiveté. Trends Ecol Evol 22: 229–230.
29. Johnson C (2006) *Australia's Mammal Extinctions – a 50 000 Year History*. Cambridge University Press, Cambridge, UK.
30. Kovacs EK, Crowther MS, Webb JK, Dickman CR (2012) Population and behavioural responses of native prey to alien predation. Oecologia 168: 947–957.
31. Dickman CR (1996) *Overview of the Impacts of Feral Cats on Australian Native Fauna*. Australian Nature Conservancy, Canberra, Australia.
32. Saunders GR, Gentle MN, Dickman CR (2010) The impacts and management of foxes *Vulpes vulpes* in Australia. Mammal Rev 40: 181–211.
33. DPIPWE (2012) Locations of fox activity in Tasmania. Fox Eradication Branch, Department of Primary Industries, Water and Environment, Tasmania, Australia.
34. Driessen MM (1987). A comparative dietary study of two native, sympatric murids; *Pseudomys higginsi*, and the velvet-furred rat, *Rattus lutreolus velutinus*, in southern Tasmania. Honours thesis, University of Tasmania, Hobart, Tasmania, Australia.
35. Owen D, Pemberton D (2005) *Tasmanian Devil: A Unique and Threatened Animal*. Allen and Unwin, Sydney, Australia.
36. Hawkins CE, Baars C, Hesterman H, Hocking GJ, Jones ME, et al. (2006) Emerging disease and population decline of an island endemic, the Tasmanian devil, *Sarcophilus harrisii*. Biol Conserv 131: 307–324.
37. Hollings T, Hocking G, Mooney N, Jones M, McCallum H (2011) Ecosystem impact following disease-induced top predator decline: the Tasmanian devil and DFTD. Abstract Ecological Society of Australia Conference, Hobart, Tasmania.
38. Lazenby BT (2012) Do feral cats affect small mammals? A case study from the forests of southern Tasmania. PhD thesis, University of Sydney, Sydney, Australia.
39. DPIPWE (2009) TASVEG 2.0: Tasmanian vegetation monitoring and mapping program. Department of Primary Industries, Parks, Water and Environment, Tasmania, Australia.
40. Long RA, Mackay P, Ray J, Zielinski W (2008) *Non-invasive Survey Methods for Carnivores*. Island Press, Washington, DC, USA.
41. MacKenzie DI, Nichols JD, Royle JA, Pollock KH, Bailey LL, et al. (2006) *Occupancy Estimation and Modeling: Inferring Patterns and Dynamics of Species Occurrence*. Elsevier, London, UK.
42. Hines JE (2006) PRESENCE2– Software to estimate patch occupancy and related parameters. USGS-PWRC. http://www.mbr_pwrc.usgs.gov/software/presence.html Accessed 31 January 2012.
43. King CM (1984) Immigrant killers. Oxford University Press, Auckland, New Zealand.
44. Salo P, Korpimäki E, Banks PB, Nordström M, Dickman CR (2007) Alien predators are more dangerous than native predators to prey populations. Proc Roy Soc B 274: 1237–1243.
45. Ruscoe WA, Ramsey DSL, Pech RP, Sweetapple PJ, Yockney I, et al. (2011) Unexpected consequences of control: competitive vs. predator release in a four-species assemblage of invasive mammals. Ecol Lett 14: 1035–1042.
46. Waddle JH, Dorazio RM, Walls SC, Rice KG, Beauchamp J, et al. (2010) A new parameterization for estimating co-occurrence of interacting species. Ecol Appl 20: 1467–1475.
47. Atickem A, Bekele A, Williams SD (2010) Competition between domestic dogs and Ethiopian wolf (*Canis simensis*) in the Bale Mountains National Park, Ethiopia. Afr J Ecol 48: 401–407.
48. Mahon PS, Banks PB, Dickman CR (1998) Population indices for wild carnivores: a critical study in sand-dune habitat, south-western Queensland. Wildl Res 25: 11–22.
49. Pemberton D (1990). Social organisation and behaviour of the Tasmanian devil, *Sarcophilus harrisii*. PhD, University of Tasmania, Hobart, Tasmania.
50. Ralls K (1971) Mammalian scent marking. Science 171: 443–449.
51. Thiessen D, Rice M (1976) Mammalian scent gland marking and social behaviour. Psychol Bull 83: 505–539.
52. Monamy V, Fox BJ (1999) Habitat selection by female *Rattus lutreolus* drives asymmetric competition and coexistence with *Pseudomys higginsi*. J Mammal 80: 232–242.
53. Dickman CR (1992) Predation and habitat shift in the house mouse, *Mus domesticus*. Ecology 73: 313–322.
54. McEvoy J, Sinn DL, Wapstra E (2008). Know thy enemy: Behavioural response of a native mammal (*Rattus lutreolus velutinus*) to predators of different coexistence histories. Austral Ecol 33: 922–931.
55. Taylor RJ (1986) Notes on the diet of the carnivorous mammals of the upper Henty River region, western Tasmania. Pap Proc Roy Soc Tasmania 120: 7–10.
56. Jones ME, Barmuta LA (1998) Diet overlap and relative abundance of sympatric dasyurid carnivores: a hypothesis of competition. J Anim Ecol 67: 410–421.
57. Pemberton D, Gales S, Bauer B, Gales R, Lazenby B, et al. (2008) The diet of the Tasmanian devil, *Sarcophilus harrisii*, as determined by analysis of scat and stomach contents. Pap Proc Roy Soc Tasmania 142: 13–21.
58. Blumstein DT, Daniel JC, Griffin AS, Evans CS (2000) Insular tammar wallabies (*Macropus eugenii*) respond to visual but not acoustic cues from predators. Behav Ecol 11: 528–535.
59. Jones ME, Smith GC, Jones SM (2004) Is anti-predator behaviour in Tasmanian eastern quolls (*Dasyurus viverrinus*) effective against introduced predators? Anim Conserv 7: 155–160.
60. Hayes A, Nahrung HF, Wilson JC (2006) The responses of native Australian rodents to predator odours varies seasonally: a by-product of life history variation? Anim Behav 71: 1307–1314.

7

The Distribution and Abundance of an Island Population of Koalas (*Phascolarctos cinereus*) in the Far North of Their Geographic Range

Denise C. McGregor, Sarah E. Kerr, Andrew K. Krockenberger*

Centre for Tropical Environmental and Sustainability Science, and School of Marine and Tropical Biology, James Cook University, Cairns, Queensland, Australia

Abstract

Koalas are an iconic species of charismatic megafauna, of substantial social and conservation significance. They are widely distributed, often at low densities, and individuals can be difficult to detect, making population surveys challenging and costly. Consequently, koala population estimates have been limited and the results inconsistent. The aims of this study were to estimate the distribution, relative abundance and population size of the koalas on Magnetic Island, far north Queensland. Population densities were estimated in 18 different vegetation types present on the island using a Fecal Standing Crop Method. Koala density ranged from 0.404 ha^{-1}, recorded in forest red gum and bloodwood woodland, to absence from eight of the vegetation types surveyed. The second highest density of 0.297 koalas ha^{-1} was recorded in mixed eucalypt woodland, which covers 45% of the island. The total abundance of koalas on Magnetic Island, not including those present in urban areas, was estimated at 825±175 (SEM). The large variation in koala density across vegetation types reinforces the need for sampling stratification when calculating abundance over large areas, as uniformity of habitat quality cannot be assumed. In this context, koala populations also occur in low densities in areas generally regarded as poor quality koala habitat. These results highlight the importance of protecting vegetation communities not traditionally considered to have high conservation value to koalas, as these habitats may be essential for maintaining viable, widespread, low-density populations. The results from this study provide a baseline to assess future trends in koala distribution, density and abundance on Magnetic Island.

Editor: Marco Festa-Bianchet, Université de Sherbrooke, Canada

Funding: The study was not supported under any external grant funding, but relied on in-kind support (provision of accommodation) from the Magnetic Island Community Development Association and internal funds from James Cook University. The funders had no role in study design, data collection and analysis, decision to publish, or preparation of the manuscript.

Competing Interests: The authors have declared that no competing interests exist.

* E-mail: Andrew.Krockenberger@jcu.edu.au

Introduction

Koalas (*Phascolarctos cinereus*) are arboreal folivores that occur in the eucalypt forests of eastern Australia. Their current distribution is widespread, covering approximately one million square kilometers and 30 biogeographical regions, from the tropical forests of northern Queensland to the temperate forests of the Victoria coast [1], [2]. However, their distribution is not continuous, but patchy, composed of many separate populations isolated from other groups by unsuitable habitat [3]. This is primarily a reflection of forest fragmentation [4] as the distribution and density of koalas is limited by the presence of *Eucalyptus* and *Corymbia* species that comprise the koala diet [3], [5]. Due to their sedentary nature [3], koalas can be difficult to detect. This combination of factors makes estimating koala abundance difficult and costly [6–8], and has resulted in a limited number of estimates at regional, state and national levels [6], [9], [10].

The distribution of the koala has contracted by more than 50% from pre-European distribution [11], [12], with much of the reduction attributed to extensive fragmentation of koala habitat in Queensland [13]. The current distribution of koalas within their reduced range faces ongoing threats from clearing, fragmentation, expanding urbanization, disease, vehicular traffic, domestic dogs and bushfire [6], [12], [14], [15]. Koalas are also vulnerable to climatic extremes, particularly prolonged periods of unusually high temperatures and droughts [16], [17]. These conditions can lead to extensive leaf fall, subsequently affecting nutrient quality and moisture content available to koalas, resulting in population crashes [16–18]. As hotter and drier conditions continue, as is expected with climate change, koala populations will be adversely affected, with reduced populations restricted to diminishing riparian habitats [17]. This conclusion is supported by past observations of koala population crashes of as much as 80% in just 14 years associated with drought in southwestern Queensland [17], [18].

Comprehensive, reliable estimates of distribution and abundance are fundamental to the successful long-term conservation and management of a species [5], [19]. However, koala population studies have been performed predominantly in areas where koalas are known to occur in high densities [20–22], restricting the understanding of widespread, low density populations in many areas of northern, western and central Queensland [13], [23]. This increases the potential for inaccurate state and national population estimates. The extrapolation of broadscale koala density measurements from local studies is problematic in the absence of appropriate stratification as densities can range from 0.001 to

8.9 koalas ha^{-1} [20], [23]. Ignoring low-density populations of koalas over large areas can lead to underestimates [24]. This was the case on Kangaroo Island where the initial population estimate was calculated from surveys within 1400 hectares of high quality habitat while the remaining ~200,000 hectares of medium and poor quality habitats were erroneously disregarded as having insignificant koala abundance [24]. This error resulted in a population size underestimate of 22,000 koalas [22], [24] and ultimately rendered the $1.25 million management scheme ineffective [22]. Many researchers agree that until additional, consistent and robust estimates of local, regional and national abundance can be made, and population trends clarified, inconsistencies will continue to hinder conservation efforts [6], [25], [9], [10].

Survey Methods

Nationally recognized standards for the assessment of koala distribution and abundance have not been established. Abundance and density estimations have primarily been obtained from transect counts, with fixed boundaries [20], [23], [26–30]. Inference from community surveys [11], [14]; distance sampling [8]; mark-resight [19], [22], [31] and fecal pellet surveys [25], [17] have also been used. However, abundance surveys of low density, patchy koala populations, scattered over large regions, such as those typical of Queensland, have been challenging for conventional methods, such as direct counts, where even extensive efforts can lead to limited data or overlooked animals and inaccurate results [6].

Counting indirect animal signs (e.g., scat, nests, calls, tracks) provides an alternative to surveying elusive, low-density animals [32] that occupy dense or widespread habitats [33] such as the koala [34]. The Fecal Standing Crop Method (FSCM) [35], [36] can be used to calculate absolute abundance of animals from fecal pellet abundance estimated from transect searches [37]. This method requires pellet abundance to be divided by two additional parameters; the daily rate of pellet production of the species and the decomposition rate or maximum age of pellets collected from transects [34]. Koalas are highly suited to this method as the required parameters can be accurately estimated. Koalas have distinctive fecal pellets that are easily found under trees they have occupied [7], [38]. This contrasts sharply with the difficulty of locating these elusive animals [34]. While defecation rates are difficult to establish for most species, the sedentary nature and roosting behavior of koalas [3] facilitates this estimation [25].

Given that a large proportion of koala populations occur in Queensland [11], accurate estimations of their abundance within the state are an essential component to conservation management of the species [19]. Magnetic Island, a popular tourist destination in north Queensland, is economically and environmentally affected by the presence of a population of koalas introduced in the 1930's as a conservation response to population crashes and reductions in distribution throughout the koala's range [3], [16]. Unlike koala populations in isolates in the south of their range, there is no evidence that the Magnetic Island population has experienced extreme boom and bust cycles of population growth [3], [39], [40], [41]. As this population is located at the northern limit of the koalas' range it may represent a sentinel population for monitoring abundance trends under future climates. The aims of this study were to determine the distribution, relative abundance and population size of the koalas on Magnetic Island. This information had not previously been assessed and will be instrumental for future koala conservation and management and will provide future opportunities to monitor population trends in this closed population of koalas over time.

Materials and Methods

Ethics Statement

This study complies with the Australian National Health and Medical Research Council's Code of Practice for Care and Use of Animals for Scientific Purposes (2004) and Queensland State legislation and was approved by the James Cook University Animal Ethics Committee (A1343) and QLD NPWS Scientific Purposes permit WITK05490308.

Magnetic Island Study Area and Climate

Magnetic Island is situated 5 kilometers off the north Queensland coast of Australia, near Townsville (19°08′S 148°50′E). The island is approximately 5184 hectares in area and includes Magnetic Island National Park (2716 hectares). The island's rugged peaks rise sharply from sea level to its summit, Mt Cook (497 meters above sea level). A plateau, divided by higher peaks with slopes exceeding 40°, occurs above 200 m [42]. The island is dominated by massive, exposed granite boulders. There is no permanent watercourse except a small tributary to Gustav Creek (pers. obs.). Urban settlements are restricted to the relatively flat areas of Horseshoe, Geoffrey, Nelly, Picnic, Bolger, and Young Bays. The human population is estimated at just over 2100 [43].

The climate of Magnetic Island is characterized by warm, dry winters and hot wet summers. July is usually the coldest month (mean daily minimum of 13.6°C) and December is typically the warmest month (mean daily maximum of 31.5°C) [44]. The wet season occurs from December to March and includes 75% of the 1196 mm average annual rainfall [44]. On average, September is the driest month of the year (mean monthly minimum rainfall of 10.8 mm) with February typically the wettest (mean maximum rainfall of 307.1 mm) [44]. Tropical cyclones occur in this area, with the most recent being cyclone 'Yasi' that struck Magnetic Island in January 2011, bringing torrential rain and destructive winds with gusts of 135 km/hr [45].

This study was conducted between August and October, 2011, during the dry season. During this period, the monthly mean maximum temperatures were 26.0°C for August, 27.7°C for September and 29.4°C for October and the monthly mean minimum temperatures were 14.7°C for August, 17.4°C for September and 20.7°C for October [44]. The total monthly rainfall on Magnetic Island for the same period was recorded at 0 mm in July, August and September and 15.8 mm in October [44]. Rainfall in October primarily occurred on two days; 3.4 mm on October 17 and 11.2 mm on October 15 [44].

Vegetation Patterns and stratification of study sites

Geology, landforms and soil patterns were used by Sandercoe [42] to categorize Magnetic Island into five landform classifications with differing vegetation communities: I) Foreshore unconsolidated sediments, II) Coastal lowlands and sands and piedmont deposits, III) Granite hills and Lithosols and talus slopes, IV) Plateau and hills of Mt Cook, V) Agglomerate hills of the West Point area. These five landforms were further divided into 23 vegetation types (Table 1) [42].

Eighteen of the 23 vegetation types contain at least one of the 15 *Eucalyptus* or *Corymbia* species found on the island [42]. Mixed eucalypt forest (vegetation type 17) covers 45% of the island and contains the greatest diversity of eucalypts with 12 species [42]. The most predominant species are yellow stringybark (*Eucalyptus acmenoides*), narrow-leafed ironbark (*Eucalyptus drepanophylla*), pink bloodwood (*Corymbia intermedia*), ghost gum (*Corymbia aparrerinja*), and Carbeen (*Eucalyptus tessellaris*) [42]. Koalas in the mixed eucalypt forest of Magnetic Island have previously been found to

Table 1. Vegetation types of Magnetic Island grouped by major landform divisions.

Magnetic Island Vegetation Types			Area (hectares)	Area (% of island)
I	**Type**	**Foreshore unconsolidated sediments**		
	1.	Stilted mangrove forest	98.2	2.0
	2.	Grey mangrove forest	2.9	0.6
	3.	Mixed mangrove shrubland	82.3	1.6
	4.	Saltmarsh and samphire flats	51.9	1.0
II		**Coastal lowlands on sands and piedmont**		
	5.	Coastal sheoak woodland	27.9	0.6
	6.	Weeping teatree and bulkuru swamp	19.9	0.4
	7.	Moreton bay ash flats	182.2	3.7
	8.	Forest red gum and bloodwood woodland	25.4	0.5
	9.	Poplar gum and bloodwood woodland	31.6	0.6
	10.	Littoral scrub	34.6	0.7
III		**Granite hills of lithosois and talus slopes**		
	11.	Araucaria forest	42.0	0.8
	12.	Mixed lowland coastal forests	127.5	2.6
	13.	Low vine forest amongst boulders	263.0	5.3
	14.	Vine forest	94.3	1.9
	15.	Mixed semi-deciduous woodland	319.4	6.4
	16.	Mixed semi-deciduous low open woodland	116.1	2.3
	17.	Mixed eucalypt woodland	2232.9	44.8
	18.	Acacia scrubland	152.8	3.1
	19.	Grassland +/− sparse trees and scrubs	197.6	4.0
	20.	Mallee brush box forests	145.5	2.9
IV		**Plateau and hills of Mt. Cook**		
	21.	Cabbage tree palm and forest sheoak forest	73.8	1.5
	22.	Forest sheoak and grass tree shrubland	31.9	0.6
V		**Agglomerate hills of the West Point area**		
	23.	Mixed open low scrub	87.26	1.7
	24.	Disturbed (urban) areas	548.43	11.0

(Adapted from Sandercoe 1990).

prefer *Corymbia intermedia*, *Corymbia erythrophloia* and *Eucalyptus drepanophylla* (Tindall pers. comm.).

Koala densities can vary significantly between differing habitats [22], [23] so the survey was stratified using the 23 vegetation types identified by Sandercoe [42]. Vegetation areas 1–5 were not surveyed as they were classified as mangroves, saltmarshes, samphire flats and sand dunes. Eucalypts do not occur in these vegetation types [42], nor had koala sightings been documented in these areas by local rangers (Petersen pers. comm.). Approximately 1% of the total area of each vegetation type was surveyed, or a minimum of six transects, whichever was greater. Transects were chosen from at least two areas from each vegetation type, except vegetation types that occurred only in area (vegetation types 9, 21 and 23). Areas within each vegetation type were selected for maximum size and distance from other surveyed areas of the same vegetation type. Urban areas were not surveyed as they are continuously disturbed by human activities such as supplemental watering, mowing, and gardening. These activities could result in the removal of pellets or an increase in their decay rate; compromising the accuracy of the FSCM.

Fecal Standing Crop Method

Fecal pellet searches were conducted between August 31 and October 27, 2012, across a total of 385, 2×100 m strip transects (Fig. 1). The first transect in each survey area was established by locating a randomized starting point, with additional transects located systematically, with a distance of 10–20 m separating transects. A GPS (Garmin Etrex) was used to locate the designated random starting point in each area and subsequent start and finish points of each adjacent transect. Transect lines were walked by a single observer in one direction, who searched and collected fecal pellets. A GPS and compass were used to navigate the transect line while a one meter stick was used to measure the distance from the line. Substrate was thoroughly searched in each transect. All vertebrate fecal pellets found, including those deposited by koalas, wallabies, possums and rodents, were collected in sealable plastic bags. Each cluster of koala fecal pellets was placed into a separate plastic bag, while all non-koala fecal pellets were placed in one bag for each transect. All fecal pellets were assigned to species by DM at the end of each day. Search times per transect ranged from 45 minutes to three hours depending on the substrate type, density of vegetation and quantity of pellets to be collected. The total

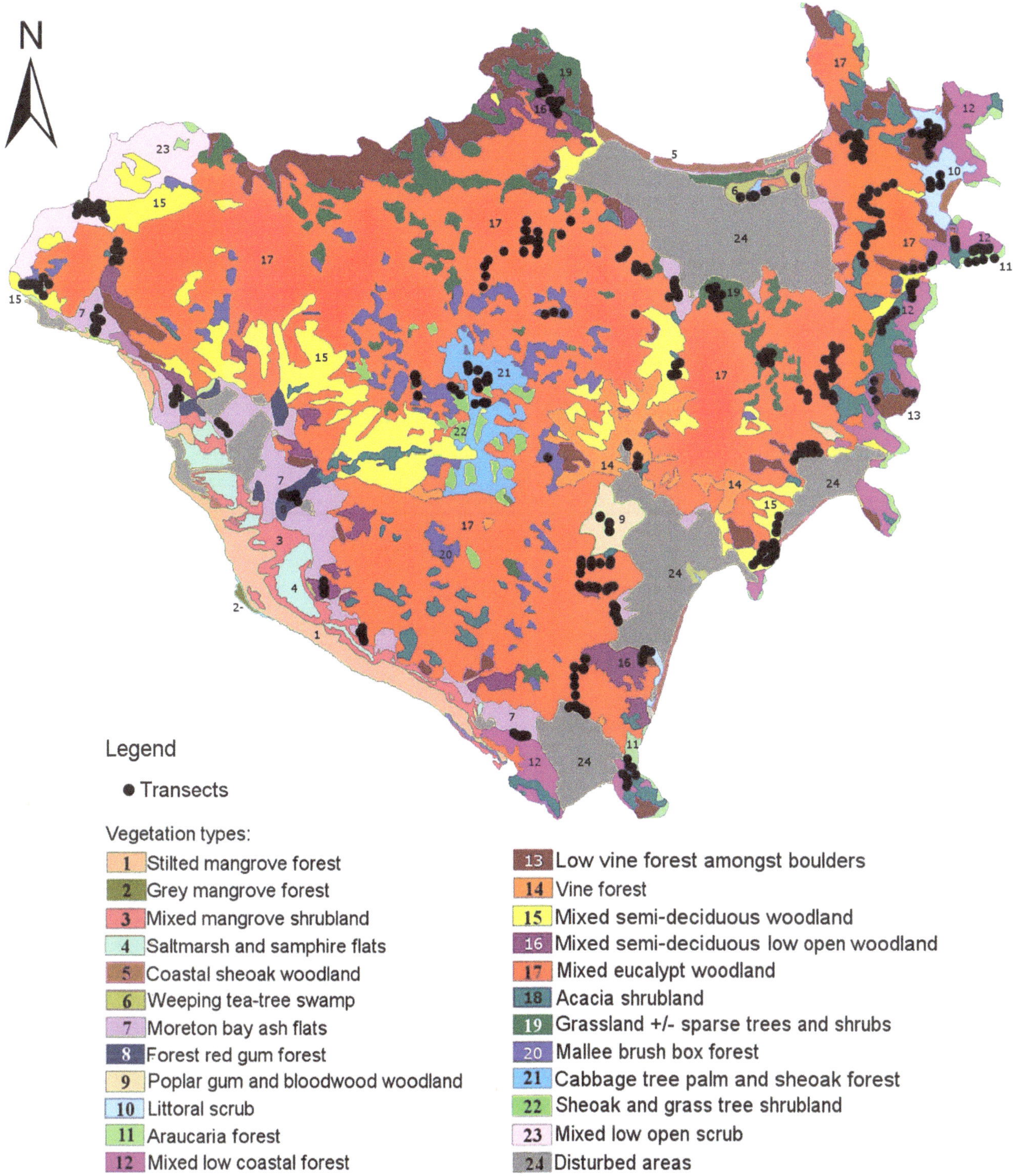

Figure 1. Map of Magnetic Island depicting vegetation types, urban areas and starting locations of transects searched. (Adapted from Sandercoe 1990).

search time across all 385 transects was approximately 656 hours. The raw data from transect searches, including transect name, search date, start and end coordinates, transect area and number of pellets found is available online from the Tropical Data Hub at James Cook University.

Daily pellet production rates were estimated in 15 free-ranging koalas (7 adult females, 5 adult males and 3 subadults). Plastic sheeting was placed under the entire canopy of each tree with a koala present, and secured in place with rocks or fallen branches, to separate previously deposited pellets and assist in the collection of fresh pellets [45], [46]. Pellets deposited were counted as long as

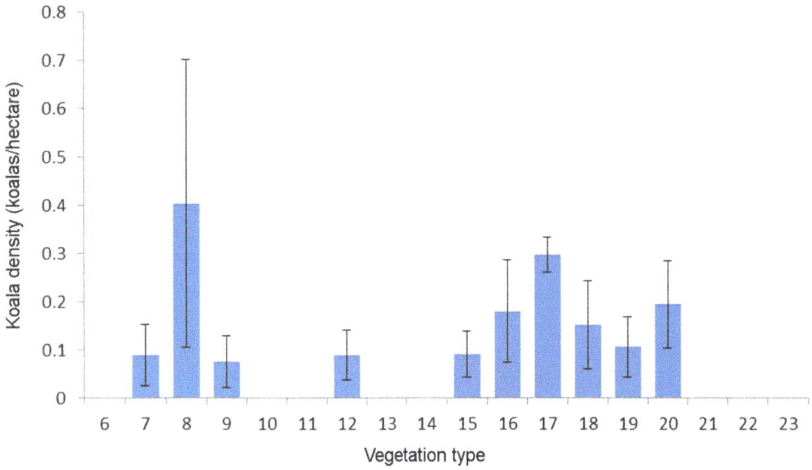

Figure 2. Density of koalas (koalas ha^{-1}) across 18 vegetation types on Magnetic Island. The mean density of koalas ± SE are shown for each vegetation type. Density estimates were derived using the fecal standing crop method. The 18 vegetation types surveyed were classified as per Sandercoe (1990), and included: 6: weeping tea-tree swamp; 7: Moreton bay ash flats; 8: forest red gum forest; 9: poplar gum and bloodwood woodland; 10: littoral scrub; 11: aracaria forest; 12: mixed low coastal forest; 13: low vine forest amongst boulders, 14: vine forest; 15: mixed semi-deciduous woodland; 16: mixed semi-deciduous low open woodland; 17: mixed eucalypt woodland; 18: acacia shrubland; 19: grassland ± sparse trees and shrubs; 20: mallee brush box forest; 21: cabbage tree palm and sheoak forest; 22: sheoak and grass tree shrubland; 23: mixed low open scrub.

each koala remained up the tree, and the occupancy time was recorded in minutes. The minimum occupancy period was 11 hours and the maximum was 23 hours. Daily estimates were then calculated from the mean number of pellets produced per minute.

The maximum age of pellets in abundance counts was estimated using a method developed by Sullivan et al. [7] based on the volatile essential oil component of eucalypt leaves, the primary component of the koala diet [47], which steadily diminishes as pellets age [7]. Fecal pellets were categorized into age classes based on their level of eucalypt odor with 'new' pellets classified as those having any eucalypt odor and "old" pellets as having no odor. A single researcher (DM) determined the presence or absence of odor in all koala pellets collected from transect searches to control for variations in individual subjective sensitivity to detect odors. Color differences were noted between fresh pellets collected from various geographic areas on the island; therefore color was not used in determining age, as was done in other studies [7], [25].

Trials to determine the number of days for which koala fecal pellets maintained their eucalypt odor were conducted concurrently with fecal pellet searches in transects. Twenty groups of 70 fresh pellets were collected over five weeks, from August 29-September 30, from defecating koalas across a variety of locations within the study area. Pellets were then allowed to age in a leafy substrate under eucalypt trees in the study area and one pellet from each group was tested for odor each day until no internal eucalypt odor was detected for three consecutive days. In this way the maximum age of 'new' pellets was established.

Based on the parameters established for the daily pellet production rate and the maximum pellet age in transect searches, the FSCM [25], [35] was used to calculate the mean koala density (± SEM) in each of the 18 vegetation types surveyed. This density was then extrapolated over the total area occupied by each vegetation type to derive an estimated absolute abundance of koalas in each vegetation type and an island-wide population estimate.

Results

Koala fecal pellets maintained their eucalypt odor for a mean of 57 days ±1 day, (SEM, n = 20). This estimate was used as the maximum age of 'new' pellets in the application of the FSCM. The mean daily production of koala fecal pellets for free-ranging koalas was 141±11 (95% CI, n = 15; Range 108–168 pellets). A total of 11,073 'new' koala fecal pellets were collected from 138 of the 385 transects searched (36%). Other fecal pellets collected from transects included 8775 'old' koala, 19,290 allied rock wallaby (Petrogale assimilis), 8799 brushtail possum (Trichosurus vulpecula), 4308 agile wallaby (Macropus agilis) and 747 unidentified rodent pellets.

Koala fecal pellets were found in 10 of the 18 surveyed vegetation types. These vegetation types contained significantly different densities of koala pellets ($\chi^2 = 72.15$, df = 17, p = 0001). Koala density ranged from 0.404 ha^{-1} to zero koalas ha^{-1} (Fig. 2; Fig. 3). The highest density was recorded in forest red gum and bloodwood woodland (vegetation type 8). The total abundance of koalas on Magnetic Island was estimated at 825±175 (SEM; Fig. 4). There were no koala fecal pellets, and hence no koalas, in eight of the vegetation types surveyed. The vegetation type with the highest koala density covers only ~25 hectares or 0.05% of Magnetic Island, so, despite the high density, only 1% of the island's koalas occurred there (Fig. 4). The second highest density of 0.297±0.036 koalas ha^{-1} was recorded in mixed eucalypt woodland (vegetation type 17), which includes 2233 hectares or 45% of the island and supports 80% of the island's koala population.

Discussion

This study recorded koala densities of 0–0.40 ha^{-1} using the FSCM. These results fall within the range of koala densities from other Queensland area studies using various methods including the FSCM (0–2.51 koalas ha^{-1}) [25]; distance sampling (0–0.76 koalas ha^{-1}) [8]; and direct counts (0.1–2.0 koalas ha^{-1}) [16], 0.4 koalas ha^{-1} [28], (0.02–0.4 koalas ha^{-1}) [29]. Although

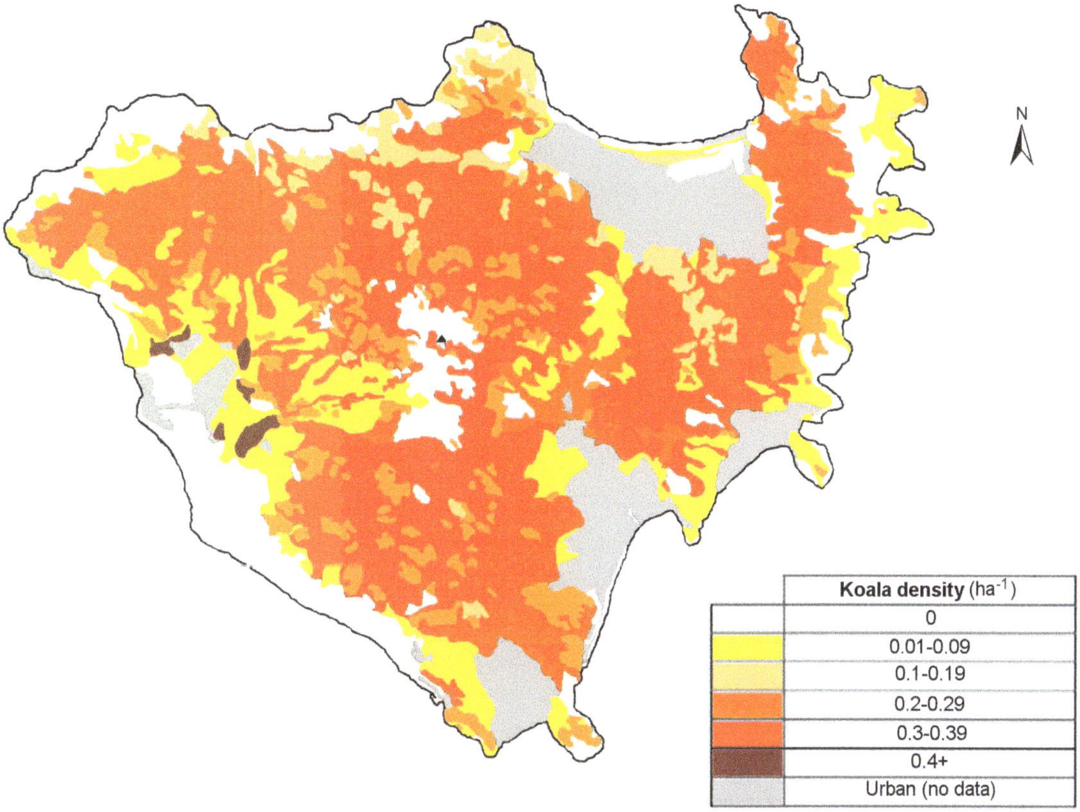

Figure 3. Map of Koala density (koalas ha^{-1}) across Magnetic Island as determined using the fecal standing crop method, with stratification by vegetation type.

the density from this study is broadly consistent with other studies from Queensland, it is much lower than abundances on southern islands [41]. For example, koala densities reach 6–8.9 koalas ha^{-1} on French Island [48], and 5 koalas ha^{-1} on Kangaroo Island [22]. These islands have a history of population spikes followed by overbrowsing, koala starvation and dramatic population crashes

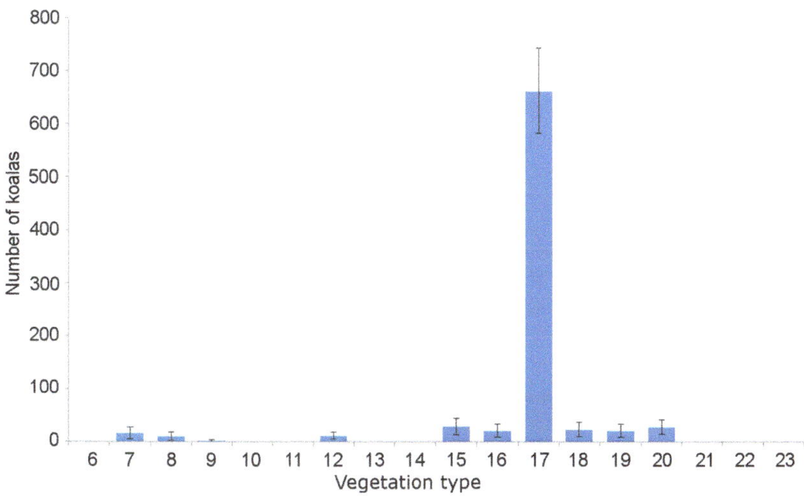

Figure 4. Abundance of koalas across 18 vegetation types on Magnetic Island. The mean abundance of koalas ± SE are shown for each vegetation type. Abundance estimates were derived using the fecal standing crop method. The 18 vegetation types surveyed were classified as per Sandercoe (1990), and included: 6: weeping tea-tree swamp; 7: Moreton bay ash flats; 8: forest red gum forest; 9: poplar gum and bloodwood woodland; 10: littoral scrub; 11: aracaria forest; 12: mixed low coastal forest; 13: low vine forest amongst boulders, 14: vine forest; 15: mixed semi-deciduous woodland; 16: mixed semi-deciduous low open woodland; 17: mixed eucalypt woodland; 18: acacia shrubland; 19: grassland ± sparse trees and shrubs; 20: mallee brush box forest; 21: cabbage tree palm and sheoak forest; 22: sheoak and grass tree shrubland; 23: mixed low open scrub.

[41]. Despite similar numbers of koalas being introduced to those islands over a similar time period, koala density on Magnetic Island has remained considerably lower than southern islands, without documented population crashes. However, direct comparisons of density between these extreme limits of the koala range are difficult due to wide variations in habitats and the unknown sustainable carrying capacities on northern islands.

Pellet age

Pellets 56.70±1 day old or less were classified as 'new' according to odor. This is considerably longer than the 'new' classification determined previously [7] and likely the result of the absence of rainfall during this study as well as differences in the aging method. Sullivan et al. [7] classified pellets with a strong eucalypt odor as 'new' and pellets with a weak eucalypt odor as old. In the current study, pellets were considered 'new' until there was complete absence of eucalypt odor. However, because the categorization was consistent within this study it is unlikely that this difference in methodology between studies resulted in any impact on the estimations of koala density.

Even within the same species, pellet aging has been shown to be highly variable based on differences in habitat and climatic events [7], [49], [50]. To minimize errors associated with pellet age, measurement of pellet aging should occur in the time period just prior to and concurrent to pellet sampling [25]. Longer decay rates in koala fecal pellets have been associated with dry climatic conditions [50], whereas increased rainfall and substrate moisture has been suggested to increase the breakdown and relocation of feces [32], [35]. Consequently, the dry season is the most appropriate survey period for implementing the FSCM in the tropics of Queensland.

We observed that throughout the nine weeks of aging trials, pellets did not show cracking or signs of disintegration. However, once past the critical ageing threshold, pellets quickly lost their eucalypt odor. The slow decay rate of koala pellets minimize the risk of underestimation but may increase the risk of overestimation if pellets cannot be aged accurately [32], [50]. For example, large amounts of 'old' pellets occur along dry creek beds where they appear to have accumulated during the wet season (pers. obs.). Counting only 'new' pellets, as opposed to aging pellets from defecation to decay, limited the variability of prolonged environmental and biotic factors on the breakdown and relocation of pellets.

Pellet deposition rates

Defecation rates can be difficult to establish as it is essential to know the exact amount of time animals spent in the area and the amount of dung accumulated in that time [32]. One of the disadvantages commonly identified with the FSCM is the evasive movement of animals [51] that prevents accurate measurements of defecation. Consequently, for most animals, estimating this parameter requires confining them in a small area, often necessitating artificial feeding, which may alter defecation rates [52]. However, as koalas are sedentary by nature [3], defecation rates can be easily estimated. Our estimate of koala daily pellet production of 140.60±10.85 (95% CI) is consistent with the other two published estimates of 150.75±12.55, in free-ranging koalas, [25] and 174±29 pellets day in captive koalas [46].

Due to the low nutrient, sclerophyllous diet of koalas it has been suggested that their digestive processes are likely to be regular instead of episodic, and generally uniform across habitats [53]. However, Ellis et al. [45] found koala pellets deposited disproportionally over a diurnal cycle, with higher numbers of pellets at peak activity times from 1800–2400 hours and therefore suggested pellets be collected over a 24 hour period to avoid bias from potential circadian activity patterns. In this study, pellets were collected from 14 of the koalas for approximately 12 hours from 600–1800 hours (one koala remained in the tree for 23 hours). Sullivan et al. [25] collected pellets over a 24 hour period and reported a 9% higher mean daily pellet production rate in free-ranging koalas then we found. If we have underestimated defecation rates due to measurement largely during daylight hours, then applying the pellet production rate measured by Sullivan et al. [25] suggests our study could have overestimated the population by 9% (74±16 koalas), well within our estimate of error.

Sullivan et al. [25] found their upper estimates to underestimate koala density by <20% when compared with direct counts of koalas in the same area. To avoid possible underestimation of koala density in the current study, the entire area of each transect was thoroughly searched, as opposed to only under the canopies of eucalypt trees [17], [25]. While non-eucalypt trees contribute only a minor part of the koala diet, tree species not preferred for food, including non-eucalypt genera, are used opportunistically by koalas for roosting and sleeping [11], [54], [55]. Consequently, we do not expect the same overall underestimate in this study.

Conservation Implications

This study recorded the distribution of koalas in significantly different densities across differing vegetation types, with the highest density occurring in the habitat with the most desirable food species. However, the majority of koalas within this population do not occur in this high quality habitat, as it covers only 5% of the island. Witt and Pahl [56] were first to record significant koala populations within low-quality habitat. Prior to this it was presumed that koalas in Queensland were largely restricted to riverine communities whose predominant vegetation was river red gum (*Eucalyptus camaldulensis*) and coolabah (*Eucalyptus coolabah*) [25]. Recent results from the Mulgalands also suggested a wider variety of vegetation communities utilized by koalas [25]. The failure of management due to underestimation of the koala population on Kangaroo Island [24] highlights the need to stratify population surveys by vegetation types, surveying in a complete range of habitats. While high quality habitats warrant conservation efforts, these results advocate efforts to protect vegetation communities that might traditionally have been considered to have low fauna conservation value, as they may still be essential for maintaining viable, widespread, low-density koala populations [22], [56].

There is general agreement that habitat destruction poses the greatest proximate threat to the conservation of koala populations [11], [57]. Unfortunately, determining which tree species are most preferred by koalas, and therefore should be protected, has been difficult [11]. A complex set of factors has been associated with koala habitat quality. These factors vary widely across regions and include floristic composition, water availability, leaf nutrients, soil type, topography, land use and fire regimes [16], [58], [59]. Given the broad-scale distribution of koalas, investigating differing vegetation communities and the koala densities they support may be a more useful approach to management and conservation planning.

Urban areas

Urban areas, particularly in southeast Queensland, can support substantial populations of koala [60], [61]. Dique et al. [8] estimated koala density in the urban habitats of the Pine Rivers Shire, Queensland, to be between 0.06 and 0.42 ha^{-1}, or 25% of the regional population. Urban areas account for 548 hectares, or 11%, of the total area of Magnetic Island [42].

This study did not assess the density of koalas in the urban areas, as the FSCM would be inappropriate within an urban setting. Transect searches could be difficult due to restricted access on private property, but more importantly, human interference from landscape maintenance and irrigation would result in inaccuracies when using the FSCM. However, multiple sightings of koalas and their signs (pellets and scratches) were seen or reported within urban areas during the study period. Based on anecdotal reports (Petersen pers. comm), it is possible that koala abundance in the urban habitats of Magnetic Island is relatively high. The omission of these areas means that the true island-wide population of koalas is greater than the estimated 825±175 SE reported in this study. In the interest of an estimation of the total population size, if we assume that koalas occur within urban areas on Magnetic Island at the same average density as across the other habitats on the island and the proportional uncertainty remains constant, then we might expect that another 102±21 koalas occur in urban areas, making the total koala population of Magnetic island around 927±195 (SE). Community-response surveys have been used in other regions of Queensland in the past, primarily for determining distribution; however, their accuracy in estimating abundance is controversial [6], [10], [11]. A further survey designed to determine koala population in urban areas would improve the quality of this population estimate.

Conclusion

The most advantageous time to enact conservation management is before a population has been reduced to a point where opportunities become limited [62]. There appears to be sufficient evidence to conclude that the broad-scale distribution of koalas has decreased by at least 50%, with abundance declining as much as 80% in some areas. However, vigorous populations in diverse locations offer unique possibilities for future conservation [6]. Sixteen koalas were originally introduced to Magnetic Island in 1931–32 as part of an attempt to provide island sanctuaries in response to drastic declines in mainland populations. If recent alarming declines in isolated koala populations in New South Wales [63] and in Queensland [17] are representative of koala population trends across their range, then island populations such as that on Magnetic Island may fulfill their original intent as island sanctuaries. Although there currently is no data on population trends on Magnetic Island, the results from this study provide a baseline to assess future trends in koala distribution, density and abundance. There is little ongoing deforestation on the island, with development limited to the bay areas in this otherwise steep and rocky terrain [42]. Therefore, if declines occur here it is unlikely to be due to broadscale habitat destruction and more likely they are the result of other factors, such as climate change, drought or disease.

Divergence in survey results estimating koala abundance is a key factor in uncertainty regarding conservation status and management, directly retarding conservation efforts for the species. Until robust estimates of abundance can be achieved, inconsistent estimates will continue to hinder conservation efforts [6], [7], [9], [10]. The advantages of using the FSCM to estimate koala abundance include the ability to readily locate koala fecal pellets, as opposed to the more elusive koalas, it is non-invasive [7], [32] and fecal pellets can be used for a variety of additional investigations including genetics and disease analysis [32]. Application of the FSCM across broader areas of the koala's geographic range has the potential to substantially improve our understanding of koala population dynamics.

Acknowledgments

We thank Magnetic Island Community Development Association (and the various members who donated accommodation), Lorna Hempstead, Queensland Park and Wildlife Rangers Jo Petersen, Nathan Winn and Patrick Centurino, SeaLink, Adam Fakes, Tuff Trax Adventure Tours, Alex Feck, Kristie Ouellette and the many student and community volunteers.

Author Contributions

Conceived and designed the experiments: DCM SEK AKK. Performed the experiments: DCM SEK. Analyzed the data: DCM SEK AKK. Contributed reagents/materials/analysis tools: DCM SEK AKK. Wrote the paper: DCM SEK AKK.

References

1. Handasyde KA, Krockenberger AK, Martin R (2008) Family Phascolarctidae Koala. In: Van Dyck S, Strahana R. editors. Mammals of Australia. Reed New Holland: Sydney.
2. McAlpine CA, Rhodes JR, Bowen D, Lunney D, Callaghan JG, et al. (2008) Can multiscale models of species' distribution be generalized from region to region? Case study of the koala. J Appl Ecol 45: 558–567.
3. Martin R, Handasyde KA (1999) Population dynamics of the koala (Phascolarctos cinereus) in south-eastern Australia. In: Lee AK, Handasyde KA, Sanson GD, editors. Biology of the Koala. Chipping Norton (Australia): Surrey Beatty. 75–84.
4. McAlpine CA, Rhodes JR, Callaghan JG, Bowen ME, Lunney D, et al. (2006) The importance of forest area and configuration relative to local habitat factors for conserving forest mammals: A case study of koalas in Queensland, Australia. Biol Conserv 132: 153–65.
5. Dique DS, Preece HJ, Thompson J, Villiers D (2004) Determining the distribution and abundance of a regional koala population in south-east Queensland for conservation management. Wildl Res 31: 109–17.
6. Melzer A, Carrick F, Menkhorst P, Lunney D, John BS (2000) Overview, Critical Assessment, and Conservation Implications of Koala Distribution and Abundance. Conserv Biol 14: 619–28.
7. Sullivan BJ, Baxter GS, Lisle AT (2002) Low-density koala (*Phascolarctos cinereus*) populations in the mulgalands of south-west Queensland. I. Faecal pellet sampling protocol. Wildl Res 29: 455–62.
8. Dique DS, de Villiers DL, Preece HJ (2003) Evaluation of line-transect sampling for estimating koala abundance in the Pine Rivers Shire, south-east Queensland. Wildl Res 30: 127–33.
9. Cork SJ, Clark TW, Mazur N (2000) Conclusions and recommendations for koala conservation. Conserv Biol 14: 702–704.
10. Phillips SS (2000) Population trends and the koala conservation debate. Conserv Biol 14: 650–9.
11. Phillips B (1990) Koalas the little Australians we'd all hate to lose. Canberra: Australian Government Publishing Service.
12. Maxwell S, Burbidge AA, Morris K (1996) The 1996 action plan for Australian marsupials and monotremes. Project 500 for the IUCN/SSC Australasian Marsupial and Monotreme Specialist Group of the World Conservation Union/ Species Survival Commission. Canberra: Environment Australia.
13. Patterson R (1996) The distribution of koalas in Queensland. In: Gordon G, editor. Koalas: Research for Management-Proceedings of the Brisbane Koala Symposium. Brisbane: Australian Government Publishing Service, World Koala Research Incorporated. 75–81.
14. Reed PC, Lunney D, Walker P (1990) A 1986–1987 survey of the koala *Phascolarctos cineretus* (Goldfuss) in New South Wales and an ecological interpretation of its distribution. In: Lee AK, Handasyde KA, Sanson GD, editors. Chipping Norton (Australia): Surrey Beatty. 55–74.
15. Australian and New Zealand Environment and Conservation Council (1998) National koala conservation strategy. Canberra: Environment Australia.
16. Gordon G, McGreevy DG, Lawrie CL (1990) Koala populations in Queensland: major limiting factors. In: Lee AK, Handasyde KA, Sanson GD, editors. Biology of the Koala. Chipping Norton (Australia): Surrey Beatty. 85–95.
17. Seabrook L, McAlpine C, Baxter G, Rhodes J, Bradley A, et al. (2011) Drought-driven change in wildlife distribution and numbers: a case study of koalas in south west Queensland. Wildl Res 38: 509–524.
18. Gordon G, Brown AS, Pulsford T (1988) A koala (*Phascolarctos-cinereus*) Population crash during drought and heat-wave conditions in southwestern Queensland. Austral Ecol 13: 451–461.
19. Caughley G, Sinclair ARE (1994) Wildlife Ecology and Management. Cambridge: Blackwell Science.

20. Mitchell P, Martin R (1990) The structure and dynamics of koala populations-French Island in perspective. In: Lee AK, Handasyde KA, Sanson GD, editors. Biology of the Koala. Chipping Norton (Australia): Surrey Beatty. 97–108.
21. Smith AP, Andrews S (1997) Koala habitat, abundance and distribution in the Pine Creek study area. State Forests of New South Wales, Coffs Harbour. Report.
22. Masters P, Duka T, Berris S, Moss G (2004) Koalas on Kangaroo Island: from introduction to pest status in less than a century. Wildl Res 31: 267–272.
23. Melzer A, Lamb D (1994) Low density populations of the koala (*Phascolarctos cinereus*) in central Queensland. Proceedings of the Royal Society of Queensland 104: 89–93.
24. St. John B (1997) Risk assessment and koala management in South Australia. Aust Zool Australian Biologist 10: 47–56.
25. Sullivan BJ, Baxter GS, Lisle AT, Pahl L, Norris WM (2004) Low-density koala (Phascolarctos cinereus) populations in the mulgalands of south-west Queensland. IV. Abundance and conservation status. Wildl Res 31: 19–29.
26. Gall BC (1980) Aspects of the ecology of the koala *Phascolarctos cinereus* (Goldfuss), in Tucki Tucki Nature Reserve, New South Wales. Wildl Res 7: 167–176.
27. Every KR (1986) Evaluation of a decline in population of the koala *Phascolarctos cinereus* (Goldfuss) in Ventor Reserve, Phillip Island Victoria by means of a triple count technique. Wildl Res 13: 517–525.
28. White NA, Kunst ND (1990) Aspects of the ecology of the koala in south-eastern Queensland. In: Lee AK, Handasyde KA, Sanson GD, editors. Biology of the Koala. Chipping Norton (Australia): Surrey Beatty. 109–116.
29. Melzer A (1995) Aspects of the ecology of the koala, *Phascolarctos cinereus* (Goldftiss, 1817), in the sub-humid woodlands of central Queensland. Ph.D. thesis. Brisbane: University of Queensland, St. Lucia.
30. Dique DS, Penfold G, Thompson J, Leslie R, Preece H (2001) Koala distribution and density in southeast Queensland: The accuracy and precision of koala surveys. In Lyons K, Melzer A, Carrick F, Lamb D, editors. The Research and Management of Non-urban Koala Populations. Rockhampton: Koala Research Center of Queensland. 105–121.
31. Hasegawa M (1995) Habitat utilisation by koalas (*Phascolarctos cinereus*) at Point Halloran, Queensland. M.S. thesis. Brisbane: University of Queensland, St. Lucia.
32. Putman RJ (1984) Fact from feces. Mamm Rev 14: 79–97.
33. Laing SE, Buckland ST, Burn RW, Lambie D, Amphlett A (2003) Dung and nest surveys: estimating decay rates. J Appl Ecol 40: 1102–1111.
34. Munks S, Corkrey R, Foley W (1996) Characteristics of arboreal marsupial habitat in the semi-arid woodlands of northern Queensland. Wildl Res 23: 185–95.
35. Johnson CN, Jarman PJ (1987) Macropod studies at Wallaby Creek: A validation of the use of dung-pellet counts for measuring absolute densities of populations of macropodids. Wildl Res 14: 139–145.
36. Latham J, Staines BW, Gorman ML (1996) The relative densities of red (*Cervus elaphus*) and roe (*Capreolus capreolus*) deer and their relationship in Scottish plantation forests. J Zool 240: 285–299.
37. Bennett A, Ratcliffe P, Jones E, Mansfield H, Sands R (2005) In:Hill d, Fasham M, Tucker G, Shewry M, Shaw P, editors. Handbook of Biodiversity Methods: Survey, Evaluation and Monitoring. Cambridge: Cambridge University Press. 450–471.
38. Triggs B (1996) Tracks, scats and other traces: a field guide to Australian mammals. South Melbourne: Oxford University Press.
39. Martin RW (1985) Overbrowsing, and decline of a population of the koala, Phascolarctos cinereus, in Victoria. I. Food preference and food tree defoliation. Wildl Res 12: 355–365.
40. Todd CR, Forsyth DM, Choquenot D (2008) Modelling the effects of fertility control on koala–forest dynamics. J Appl Ecol 45: 568–578.
41. Martin R, (1997) Managing over-abundance in koala populations in south-eastern Australia-future options. Aust Biol 10: 57–63.
42. Sandercoe C (1990) Vegetation of Magnetic Island. Technical report. Queensland National Parks and Wildlife Services.
43. Queensland Government Census (2006). Queensland Treasury's Office of Economic and Statistical Research website. Canberra: Office of Economic and Statistical Research. Available: http://www.oesr.qld.gov.au/our-services/coordination/census-2006/index.php. Accessed 2013 Feb 22.
44. Bureau of Meteorology (2012) Commonwealth of Australia Bureau of Meteorology website. Canberra, ACT. Available: www.bom.gov.au. Accessed 2013 Feb 22.
45. Ellis WAH, Sullivan BJ, Lisle AT, Carrick FN (1998) The spatial and temporal distribution of koala faecal pellets. Wildl Res 25: 663–8.
46. Krockenberger AK, Hume ID (2007) A flexible digestive strategy accommodates the nutritional demands of reproduction in a free-living folivore, the Koala (*Phascolarctos cinereus*). Funct Ecol 21: 748–756.
47. Hume ID, Esson C (1993) Nutrients, antinutrients and leaf selection by captive koalas (*Phascolarctos-cinereus*). Aust J Zool 41: 379–392.
48. Hindell MA (1984) The feeding ecology of the koala, Phascolarctos cinereus in a mixed Eucalyptus forest. M.Sc. Thesis, Melbourne: Monash University.
49. Telfer WR, Griffiths AD, Bowman D (2006) Scats can reveal the presence and habitat use of cryptic rock-dwelling macropods. Aust J Zool 54: 325–334.
50. Rhodes J, Lunney D, Moon C, Matthews A, McAlpine CA (2011) The consequences of using indirect signs that decay to determine species' occupancy. Ecography 34: 141–150.
51. Marques FFC, Buckland ST, Goffin D, Dixon CE, Borchers DL, et al. (2001) Estimating deer abundance from line transect surveys of dung: sika deer in southern Scotland. J Appl Ecol 38: 349–63.
52. Smith AD (1964) Defaecation rates of mule deer. J Wildl Manage 28: 435–444.
53. Cork SJ, Sanson GD (1990) Digestion and nutrition in the koala: a review. In: Lee AK, Handasyde KA, Sanson GD, editors. Biology of the Koala. Chipping Norton (Australia): Surrey Beatty. 129–144.
54. Lee A, Martin R (1988) The koala: a natural history. Sydney: New South Wales University Press.
55. Hindell MA, Lee AK (1987) Habitat use and tree preference of koalas in mixed eucalypt forest. Wildl Res 12: 349–360.
56. Witt GB, Pahl LI (1995) Mulgaland communities of south-west Queensland as habitat for koalas. In Page M, Beutel T, editors. Ecological Research and Management in the Mulgalands. Brisbane: University of Queensland, Gatton College. 91–95.
57. Gordon G, Hrdina F, Patterson R (2006) Decline in the distribution of the koala *Phascolarctos cinereus* in Queensland. Aust Zool 33: 345–358.
58. Cork SJ, Braithwaite LW (1996) Resource availability, eucalypt chemical defenses, and habitat quality for leaf-eating marsupials. In Gordon G, editor. Koalas: Research for Management. Brisbane: World Koala Research Inc. 9–16.
59. Melzer A, Houston W (2001) An overview of the understanding of koala ecology: how much more do we need to know? In: Lyons K, Melzer A, Carrick F, Lamb D, editors. The Research and Management of Non-urban Koala Populations. Rockhampton: Koala Research Center of Queensland. 105–121.
60. Knight RL (1999) Private lands: the neglected geography. Conserv Biol 13: 223–224.
61. Lunney D, Matthews A, Moon C, Ferrier S (2000) Incorporating habitat mapping into practical koala conservation on private lands. Conserv Biol 14: 669–80.
62. Penn A, Sherwin WB, Gordon G, Lunney D, Melzer A, et al. (2000) Demographic forecasting in koala conservation. Conserv Biol 14: 629–38.
63. Department of Environment and Resource Management (2009) Decline of the Koala Coast Population: Population Status in 2008. Queensland Government.

8

Mapping the Global Emergence of *Batrachochytrium dendrobatidis*, the Amphibian Chytrid Fungus

Deanna H. Olson[1*¶], David M. Aanensen[2¶], Kathryn L. Ronnenberg[1], Christopher I. Powell[2], Susan F. Walker[2], Jon Bielby[3], Trenton W. J. Garner[4], George Weaver[1], The *Bd* Mapping Group¶, Matthew C. Fisher[2¶]

1 Pacific Northwest Research Station, U.S. Forest Service, Corvallis, Oregon, United States of America, **2** Department of Infectious Disease Epidemiology, Imperial College, London, United Kingdom, **3** Institute of Zoology, Zoological Society of London, London, United Kingdom, **4** Department of Statistics, Oregon State University, Corvallis, Oregon, United States of America

Abstract

The rapid worldwide emergence of the amphibian pathogen *Batrachochytrium dendrobatidis* (*Bd*) is having a profound negative impact on biodiversity. However, global research efforts are fragmented and an overarching synthesis of global infection data is lacking. Here, we provide results from a community tool for the compilation of worldwide *Bd* presence and report on the analyses of data collated over a four-year period. Using this online database, we analysed: 1) spatial and taxonomic patterns of infection, including amphibian families that appear over- and under-infected; 2) relationships between *Bd* occurrence and declining amphibian species, including associations among *Bd* occurrence, species richness, and enigmatic population declines; and 3) patterns of environmental correlates with *Bd*, including climate metrics for all species combined and three families (Hylidae, Bufonidae, Ranidae) separately, at both a global scale and regional (U.S.A.) scale. These associations provide new insights for downscaled hypothesis testing. The pathogen has been detected in 52 of 82 countries in which sampling was reported, and it has been detected in 516 of 1240 (42%) amphibian species. We show that detected *Bd* infections are related to amphibian biodiversity and locations experiencing rapid enigmatic declines, supporting the hypothesis that greater complexity of amphibian communities increases the likelihood of emergence of infection and transmission of *Bd*. Using a global model including all sampled species, the odds of *Bd* detection decreased with increasing temperature range at a site. Further consideration of temperature range, rather than maximum or minimum temperatures, may provide new insights into *Bd*-host ecology. Whereas caution is necessary when interpreting such a broad global dataset, the use of our pathogen database is helping to inform studies of the epidemiology of *Bd*, as well as enabling regional, national, and international prioritization of conservation efforts. We provide recommendations for adaptive management to enhance the database utility and relevance.

Editor: Jason E. Stajich, University of California Riverside, United States of America

Funding: Funds were provided by the UK Natural Environmental Research Council (E006701/1), The ERAnet BIODIVERSA project RACE, The Wellcome Trust (WT08972), and the US Forest Service, Pacific Northwest Research Station. The funders had no role in study design, data collection and analysis, decision to publish, or preparation of the manuscript.

Competing Interests: The authors have declared that no competing interests exist.

* E-mail: dedeolson@fs.fed.us

¶ These authors contributed equally to this work.

¶ Membership of The *Bd* Mapping Group is provided in the Acknowledgments.

Introduction

The specter of multiple-host infectious diseases leading to the rapid devastation of entire communities is alarming but, so far, rare. However, a handful of highly virulent multi-host pathogens are known, and these have had profound impacts across populations and species (e.g., the rinderpest Morbillivirus; West Nile Virus; bat *Geomyces destructans*). Rapid aggregation, synthesis, and analysis of disease data are needed to advance both science and management of such diseases. While technological advances have aided our ability to respond to emerging disease, such as by the use of proactive modeling [1], our ability to rapidly detect, assess, and report globally-emerging pathogens has lagged for wild animal diseases that are not zoonotic or are not transmitted to livestock or companion animals. For these diseases, infection and mortality are often cryptic and research studies may take years to be completed before being published.

Herein we report on a community surveillance effort addressing the emergence of the amphibian chytrid fungus *Batrachochytrium dendrobatidis* (*Bd*). Approximately one-third of global amphibian species have imperiled status, and the emergence of *Bd* is known to be a proximate driver of amphibian species declines and extinctions [2–8]. *Bd* has an unusually wide host-range: it has been detected infecting hundreds of species. Cascading effects owing to declines of multiple host species as a result of chytridiomycosis could potentially undermine ecosystem stability and function [9,10], although some studies have not supported key functional roles of amphibians in local ecosystems [11]. Hence, although caution is necessary because amphibians' ecological roles will vary with location, as a worst case scenario, the potential of *Bd* to act synergistically with other anthropogenic drivers may catastrophically disturb biological communities and substantially contribute to the ongoing 6th mass extinction event [8].

Although infectious diseases are a normal component of wildlife ecology, only recently are diseases being widely recognized as critical conservation concerns, notably for amphibians [12]. *Bd* was described as a species in the late 1990s [13], and our basic biological knowledge of this novel pathogen is still accruing. After over a dozen years of research, many aspects of its ecology, epidemiology, and pathogenicity remain uncertain. For example, vectors and pathways of transmission of *Bd* across spatial scales are little understood, as are the factors affecting species-specific susceptibility to the clinical disease, chytridiomycosis [14,15], although reservoir amphibian hosts have been identified, including some involved in commercial trade of amphibians [16–20]. *Bd* genotype affects *Bd* virulence, and population genomics studies have shown that the worldwide emergence of chytridiomycosis has occurred as a consequence of 20th-century emergence of an aggressive lineage [16]. Concern has heightened about *Bd* transmission in various arms of the commercial trade in amphibians which annually involves millions of individuals [15,21]. Transmission pathways may include migratory waterbirds [22], and some regional studies have reported an association of *Bd* occurrence with amphibian species' proximity to human development, suggesting uncharacterised pathways of introduction and transmission [23]. *Bd* has been detected in water samples [24], supporting transmission via fomites. Due to the recent recognition of *Bd* as an emerging infectious disease and an invasive species, *Bd* was listed as a notifiable disease by the World Organization of Animal Health (OIE) in 2009, resulting in international recommendations to forestall further spread via anthropogenic activities [25]. The need for effective surveillance, including rapid data accrual and interpretation in order to identify emerging patterns and processes, is integral to the progression of both *Bd* science and management objectives. In particular, our incomplete understanding of the global distribution of *Bd* is key information that is required to assess the importance of various mechanisms that may have contributed to the rapid emergence of chytridiomycosis [26] or continue to contribute to the ongoing spread of *Bd*.

We report on the outputs of a multi-phase project for compiling worldwide *Bd* data and maintaining an updated system for global-scale assessment of the pathogen, the Global *Bd* Mapping Project [27]. A web-based system (www.Bd-maps.net) for collation of *Bd* incidence and associated metadata was produced [28], with the aims of providing new insights into both *Bd* occurrence patterns and the development of hypotheses for the study of *Bd* ecology and epidemiology, and to inform species- and land-management planning efforts. Using data gleaned from the published literature or submitted as primary data to *Bd*-maps.net by members of the *Bd* Mapping Group over a four year period, we report: 1) spatial and taxonomic patterns of infection, including amphibian families that appear over- and under-infected; 2) relationships between *Bd* occurrence and declining amphibian species, including an analysis of *Bd* occurrence, species richness, and occurrence of enigmatic population declines; and 3) patterns of environmental correlates, including climate metrics for all species and for three families (Hylidae, Bufonidae, Ranidae), at both the global scale and the U.S.A. scale (which has higher spatial resolution of climate metrics allowing downscaled analyses). Our analyses contribute new insights to hypotheses of geographic and climatic relationships with *Bd* occurrence [29–37].

Results

Bd-maps.net

Bd-maps.net contains a web-accessible database for data storage, facilities for data uploading, basic summary statistics, and visualisation of spatial data utilising Google Maps and Google Earth. Data are submitted in a number of ways, through: 1) literature review with subsequent geo-coding and direct database submission; 2) direct online data entry by community users; and 3) a smartphone interface for the direct submission of location and metadata using the data gathering tool, EpiCollect [38]. Summaries of *Bd* data can be viewed globally or by country, and all data are now available for community downloading for further off-line analyses such as those presented here (see Materials and Methods).

Cautions expressed in Materials and Methods for use of *Bd*-maps.net data should be reiterated here; in particular, results of analyses should be used to inform predictions for further examination, and are not conclusive due to several forms of potential data bias. Importantly, distribution maps derived from the *Bd*-maps database do not represent the distribution of chytridiomycosis, but locations where *Bd* sampling has occurred, where the fungus has been detected, or where it was not detected. These locations have a suite of potential biases as discussed in Materials and Methods, and we have addressed these biases to some extent by the data aggregations that we have conducted. Type I errors indicate false *Bd*-positive findings, and may occur in the dataset if *Bd* were detected at a site but the assay to determine this was incorrectly run, or was subject to cross-contamination. A Type II error is a false *Bd*-negative finding. This might occur if *Bd* occurs at a site, but was not detected. A low-level *Bd* infection, an insufficient sample size, or the presence of PCR inhibitors, could result in a false negative report [39]. We did not conduct quality assurance procedures to examine data for the use of acceptable swabbing, histology or PCR techniques, or for sample sizes. These types of errors have been increasingly recognized over the last decade, and are likely reduced for surveillance efforts reported in peer-reviewed publications. Nevertheless, in our analyses, by aggregating all data at a site level and by conducting analyses at the 0.5 degree latitude and longitude grid cell size, we may have reduced the chance of both false-positive and false-negative findings by the compilation of results among different studies (and their different sampling and analytical approaches), researchers, species, seasons, or years. An additional consideration is the sample sizes used in our various analyses and summary statistics; the larger samples may represent more robust datasets with less bias. Nevertheless, we offer our results as hypotheses for further testing, and temper our conclusions accordingly.

Spatial and taxonomic patterns of *Bd* detection

Bd has been detected in 56 of 82 (68%) countries (Figure 1), and in 516 of 1,240 (42%) species, determined using a data set of more than 36,000 individuals (*Bd*-maps.net data snapshot August 2012). *Bd* was detected at 1,814 of 4,281 (48%) sites tallied for our analyses (data snapshot July 2010). Broad-scale distribution patterns are evident, with *Bd* widely detected in the Americas, and detected only patchily in Africa, Asia, and Europe. Spatial biases and gaps in sampling across the globe contribute to the clusters of *Bd* locations on the world map. In particular, Asia had been largely unsampled at the time of our data analyses; however, recently, Swei et al. [26] sampled in 15 Asian countries and found only 2.35% *Bd* prevalence.

Bd was detected in 41 of 50 (82%) sampled families of amphibians (Table 1; data snapshot January 2011). Anurans and caudates were comparably sampled: 935 of 6,177 (15%) extant

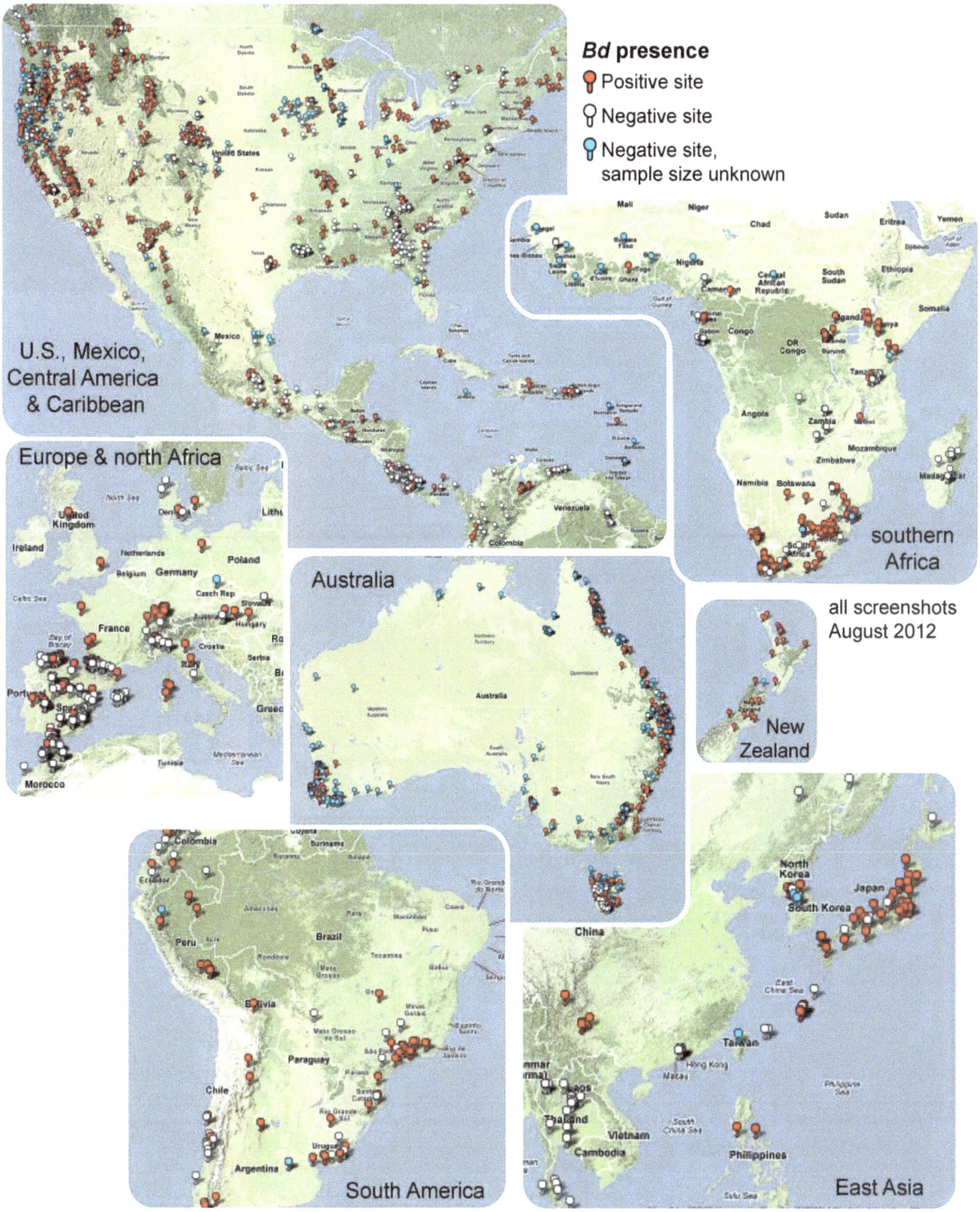

Figure 1. Global distribution of the amphibian chytrid fungus, *Batrachochytrium dendrobatidis* (*Bd*). Maps downloaded from www.Bd-maps.net (15 August 2012); *Bd*-positive (red) and *Bd*-negative (white, blue) sites are shown.

and recognized anuran species were included in our study, while 114 of 629 (18%) caudates were also included (Table 1). Caecilians remained relatively unsampled, with only 6 of 190 (3%) species registered in the database (Table 1, Table 2) (total species counts from AmphibiaWeb.org, accessed 25 July 2012). Three speciose anuran families, Hylidae, Ranidae, and Bufonidae, accounted for 30% of the species in which *Bd* was detected (Table 2, Table S1).

We examined which families deviated from the expected level of infection given the overall background prevalence in our data set (508/1,055; 48%), and found broad taxonomic patterns (Table 2). Microhylidae, Arthroleptidae, and Mantellidae were under-

Table 1. Numbers of amphibian species and families with *Batrachochytrium dendrobatidis* detections as of January 2011.

	Species				Families			
Order	***Bd* Detected**	***Bd* Not Detected**	**Total**	**Percent Detected**	***Bd* Detected**	***Bd* Not Detected**	**Total**	**Percent Detected**
Anura	449	486	935	48.0	35	6	40	87.5
Caudata	59	55	114	51.8	6	2	8	75
Gymnophiona	0	6	6	0	0	2	2	0
Total	**508**	**547**	**1055**	**48.2**	**41**	**9**	**50**	**82**

infected anurans: these families are principally terrestrial, and many species are direct developers (no tadpoles), which would reduce their exposure to the aquatic infective stage of *Bd*. Furthermore, the majority of tadpoles of microhylids that exhibit indirect development lack keratinized mouthparts, the target of infection in larval anurans [17]. Although the Bufonidae were also under-infected, some of the most dramatic *Bd*-driven declines have been seen in bufonids, such as *Anaxyrus boreas* (formerly *Bufo*) in North America and *Atelopus* spp. in Central America. Caudates of the family Hynobiidae were also under-infected, perhaps due to endemicity in Asia, where *Bd* prevalence is low and presence is patchy. Over-infected families (Table 2) were quite diverse and included those with relatively localized to broad distributions, with both aquatic and terrestrial life histories, including the direct-developing Craugastoridae. These patterns may be used to generate testable hypotheses, yet we note that family-scale infection patterns also may reflect a sampling bias in our non-randomly collected data, especially for families with taxonomically or geographically restricted sampling efforts.

Bd and declining amphibian populations

The Global Amphibian Assessment, GAA [7], reported that 47% of rapid amphibian species declines were 'enigmatic'. While *Bd* has been suggested as a potential driver of these declines [40], data and analyses in support of this hypothesis have been lacking. Using the GAA data layer of world occurrences of enigmatic declines, we indirectly addressed this with data submitted to *Bd*-maps.net. We found that the average number of amphibian species experiencing enigmatic declines at a site was greater at sites where *Bd* was detected than where it was not detected ($p<0.001$; 3,961 sites used in analyses; Figure 2). The occurrence of enigmatic declines increased with increasing species richness, however the rate of increase of enigmatic decline occurrence with species richness did not differ between sites with and without *Bd* ($p=0.32$; Figure 2). We repeated the analysis using data for the 3 families of amphibians [Bufonidae (908 sites), Hylidae (881 sites), and Ranidae (1,223 sites)] that dominated the database. As with the global data set, the average number of enigmatic declines was greater for sites with *Bd* than for sites without *Bd* for the Bufonidae and Ranidae, but the rate of increase in the relationship between the occurrence of enigmatic declines and species richness did not differ with *Bd* occurrence. The increase in enigmatic declines with richness was higher for sites without *Bd* for sites with hylids ($p<0.001$). Although such analyses may be biased, we report them because they may provide insights for follow-up studies. Due to these associations, we included both occurrences of enigmatic declines and species richness in our multivariate environmental models of *Bd* occurrence, below.

For species in which *Bd* infection was detected, 106 of 449 (24%) anurans and 12 of 60 (20%) caudates were listed in the three Threatened categories of the IUCN Red List (Critically Endangered, Endangered, Vulnerable; Table S1). It is important to note that these are counts of species that are recognized as being threatened without reference to what threats, according to the IUCN Threats Classification Scheme, are responsible for their classification. Animals under threat from *Bd* would be classified as 8.1 'Invasive non-native/alien species/diseases, 8.2 'Problematic native species/diseases' or 8.4 'Problematic species/diseases of unknown origin'. Although *Bd* infection does not equate to it being a threat in every instance, upon inspection of the Red List, *Bd* may be under-reported as a concern, given the frequency of its occurrence as we report it here. For example, of the 12 caudates known to be infected with *Bd* that are currently ranked as threatened, *Bd* infection was listed as a threat for only one species, and disease, generally, was listed for a second species. To assess the relative impact of threats globally, an appropriate analysis would involve comparing the number of threatened species that are Red Listed because of a given process, versus the number of species that are exposed to that process. Currently, we are unable make such a comparison in the context of *Bd* because the only data we have are the infection status of a relatively small subset of the total number of amphibian species (1,240/6,996 species, 18%; species count from AmphibiaWeb.org, accessed 25 July 2012) and we lack conclusive data on symptoms of the disease chytridiomycosis for most of these species.

Bd and environmental correlates: World and U.S.A. scales

Our full model of odds of *Bd* occurrence by environmental parameters at sites using all world data (data snapshot from July 2010) included biomes, latitude, occurrence of enigmatic declines, temperature range, and average annual precipitation as significant explanatory variables (Table S2). Removing latitude from our analysis, which we considered warranted due to the overwhelming focus of *Bd*-surveillance in the northern hemisphere (Figure S1, Figure S2), resulted in only minor changes in the other coefficients, and slight changes to the order of biome entry into the model (Table 3; N = 3,733 sites used in analysis). After accounting for differences in odds of *Bd* detection between biomes, we found that the odds of *Bd* detection at discrete sites; 1) increased by 14% with each known increase in occurrence of enigmatic declines at a site; 2) decreased by 8.8% with each degree increase in temperature range at a site; and 3) decreased by 0.05% with each mm increase in average annual precipitation at a site (Table 3, Figure 3, Figure 4). Among biomes, sites in Montane Grasslands and Shrublands (N = 56 sites) had the highest odds of *Bd* occurrence. To better understand the inverse relationship of *Bd* presence with world temperature range values at sites, it is important to

Table 2. Numbers of species with *Batrachochytrium dendrobatidis* (*Bd*) detections by amphibian family, and results of randomisation tests (see Materials and Methods section) to determine whether each sampled amphibian family was over- or under-infected, compared to what we would expect by chance given an overall prevalence of 508/1055 species.

Family	**Species**			**Prev.**	**p-value**	**Prev. pattern**
	***Bd* Detected**	***Bd* Not Detected**	**Total Sampled**			
Anura						
Heleophrynidae	3	0	3	1	0.9652	NA
Pelobatidae	2	0	2	1	0.9278	NA
Calyptocephalellidae	1	0	1	1	0.8494	NA
Ceratophryidae	11	1	12	0.92	**0.9988***	**Over**
Hylodidae	8	1	9	0.89	**0.9936***	**Over**
Hemiphractidae	7	2	9	0.78	0.9641	NSD
Leiuperidae	6	2	8	0.75	0.9386	NSD
Aromobatidae	6	2	8	0.75	0.9386	NSD
Alytidae	5	2	7	0.71	0.8928	NSD
Pyxicephalidae	11	5	16	0.69	0.9510	NSD
Craugastoridae	17	8	25	0.68	**0.9782***	**Over**
Dendrobatidae	14	7	21	0.67	0.9542	NSD
Limnodynastidae	12	6	18	0.67	0.9442	NSD
Cycloramphidae	4	2	6	0.67	0.8137	NSD
Pipidae	10	6	16	0.63	0.8761	NSD
Hylidae	89	59	148	0.6	**0.9992***	**Over**
Myobatrachidae	20	14	34	0.59	0.8980	NSD
Ranidae	55	43	98	0.56	0.9513	NSD
Eleutherodactylidae	15	12	27	0.56	0.7858	NSD
Centrolenidae	9	7	16	0.56	0.7361	NSD
Hyperoliidae	27	27	54	0.50	0.6162	NSD
Bombinatoridae	2	2	4	0.50	0.5309	NSD
Strabomantidae	30	33	63	0.48	0.4610	NSD
Leptodactylidae	8	10	18	0.44	0.3776	NSD
Leiopelmatidae	2	3	6	0.40	0.3572	NSD
Bufonidae	43	68	111	0.39	**0.0178***	**Under**
Ptychadenidae	5	8	13	0.38	0.2439	NSD
Phrynobatrachidae	5	8	13	0.38	0.2418	NSD
Dicroglossidae	6	15	21	0.29	0.0333	NSD
Rhacophoridae	5	14	19	0.26	0.0286	NSD
Petropedetidae	1	5	6	0.17	0.0643	NSD
Scaphiopodidae	1	5	6	0.17	0.0608	NSD
Microhylidae	4	21	25	0.16	**0.0006***	**Under**
Megophryidae	1	6	7	0.14	0.0344	NSD
Arthroleptidae	4	26	30	0.13	**6.19E-05***	**Under**
Mantellidae	0	48	48	0	**1.05E-11***	**Under**
Ceratobatrachidae	0	3	3	0	0.04688	NA
Brevicipitidae	0	3	3	0	0.0456	NA
Pelodytidae	0	1	1	0	0.1676	NA
Hemisotidae	0	1	1	0	0.1652	NA
Total Anura	**449**	**486**	**935**			
Caudata						
Cryptobranchidae	3	0	3	1	0.9646	NA
Salamandridae	14	4	18	0.78	**0.9947***	**Over**
Ambystomatidae	13	5	18	0.72	**0.9809***	**Over**

Table 2. Cont.

Family	Species			Prev.	p-value	Prev. pattern
	Bd Detected	*Bd* Not Detected	Total Sampled			
Proteidae	1	1	2	0.50	0.5264	NA
Sirenidae	1	1	2	0.50	0.5191	NA
Plethodontidae	27	34	61	0.44	0.2647	NSD
Hynobiidae	0	9	9	0	**0.0018***	**Under**
Amphiumidae	0	1	1	0	0.1683	NA
Total Caudata	**59**	**55**	**114**			
Gymnophiona						
Caeciliidae	0	5	5	0	**0.0165***	**Under**
Ichthyophiidae	0	1	1	0	0.1624	NA

Prev. = Prevalence: *Bd* detected/Total sampled. * denotes a significant deviation from the expected level of infection in that family. 'NSD' = not significant. 'NA' denotes that insufficient species had been sampled in that family to detect a significant deviation from a random level of infection. Power-analyses using binomial tests indicated that with six species sampled in a Family, there was sufficient power to detect a deviation in either direction. '*Bd* not detected' data have limitations, as indicated in Supplemental Information. Notes on taxonomic names used are provided in Supplemental Information. Detections in table updated as of January 2011.

recognize that these temperature ranges represented the *differences* in average monthly values, and spanned 4.8 to 20.8°C. Large temperature ranges were found in: 1) Deserts and Xeric Shrublands; 2) Tropical and Subtropical Coniferous Forests; and 3) Temperate Coniferous Forests. Sites in montane areas of Mexico and Chile had moderately large temperature ranges. Sites with small temperature ranges in our analysis occurred in Tasmania, Italy, Denmark, Alaska, and South Africa.

Family-scale analyses were first conducted using all data to explore potential *Bd*-associations between higher-order taxonomy (family) and geographic patterns. Our analyses included the three main amphibian families in our world sample (Ranidae, Bufonidae, Hylidae: 78% of sites; Table S3). In our Bufonidae model (N = 888 sites; Figure 4), the odds of *Bd* detection increased 44.7% with each 1,000-m increase in elevation, and were highest in Boreal Forest and Taiga, Temperate Coniferous Forests, Montane Grasslands and Shrublands, and Tropical and Subtropical Dry Broadleaf Forests. In Hylidae (N = 871 sites; Figure 4), the odds of detecting *Bd*: 1) increased by 10% with each incremental increase in occurrence of enigmatic declines; 2) increased by 1.9% with each degree increase in average annual temperature; and 3) decreased by 14.9% with each degree increase in average temperature range. The highest odds of *Bd* detection for Hylidae were in Montane Grasslands and Shrublands, Tropical and Subtropical Dry Broadleaf Forests, and Temperate Grasslands, Savannahs, and Shrublands. In Ranidae (N = 1,184 sites; Figure 4), the odds of *Bd* detection increased by 6.4% with each degree increase in average temperature range, decreased by 4% with each incremental increase in species richness, and decreased by 20% with each 1000-m increase in elevation. The highest odds of *Bd* detection in ranids occurred in two biomes, Tropical and Subtropical Grasslands, and Deserts and Xeric Shrubland. Several biomes were dropped from the ranid analysis due to data gaps.

Half of our compiled sites are located in the U.S.A., where we also were able to use environmental data with finer spatial resolution in our analyses. U.S.A. models of *Bd* occurrence were developed for all species combined, and for the three main families (83% of total sites sampled in U.S.A.), similar to the world-scale analyses. The full U.S.A. model with all taxa included (1,880 sites), showed an increase in odds of *Bd* detection with species richness, minimum temperature, an interaction between richness and minimum temperature, and biome. Due to the interaction term, interpretation of this model was more complex. For sites with species richness>15, the highest odds of *Bd* detection were at sites

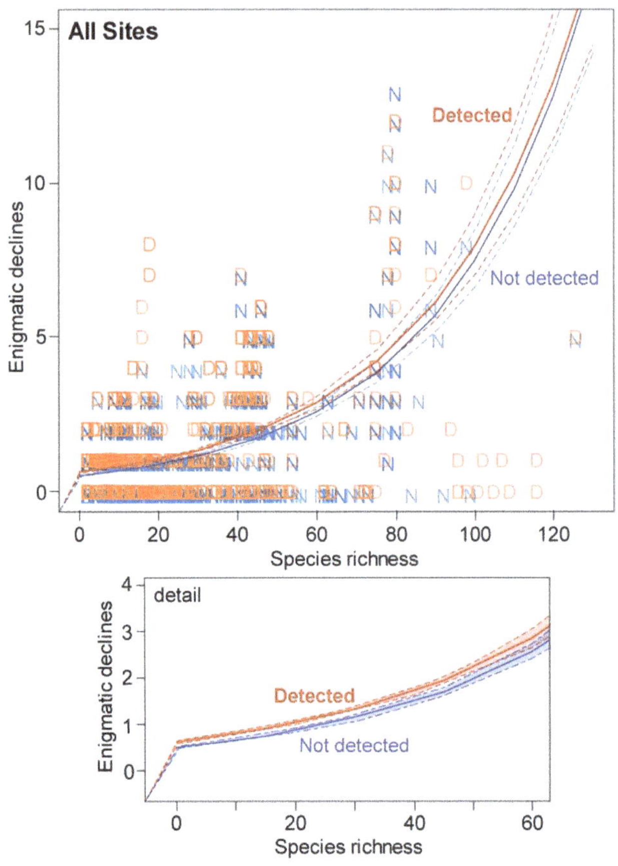

Figure 2. Global locations with enigmatic amphibian declines were positively associated with amphibian species richness (data from GAA [7]), and this relationship increased with the occurrence of the amphibian chytrid fungus, *Batrachochytrium dendrobatidis*, as modeled with logistic regression. Dashed lines indicate 95% confidence interval.

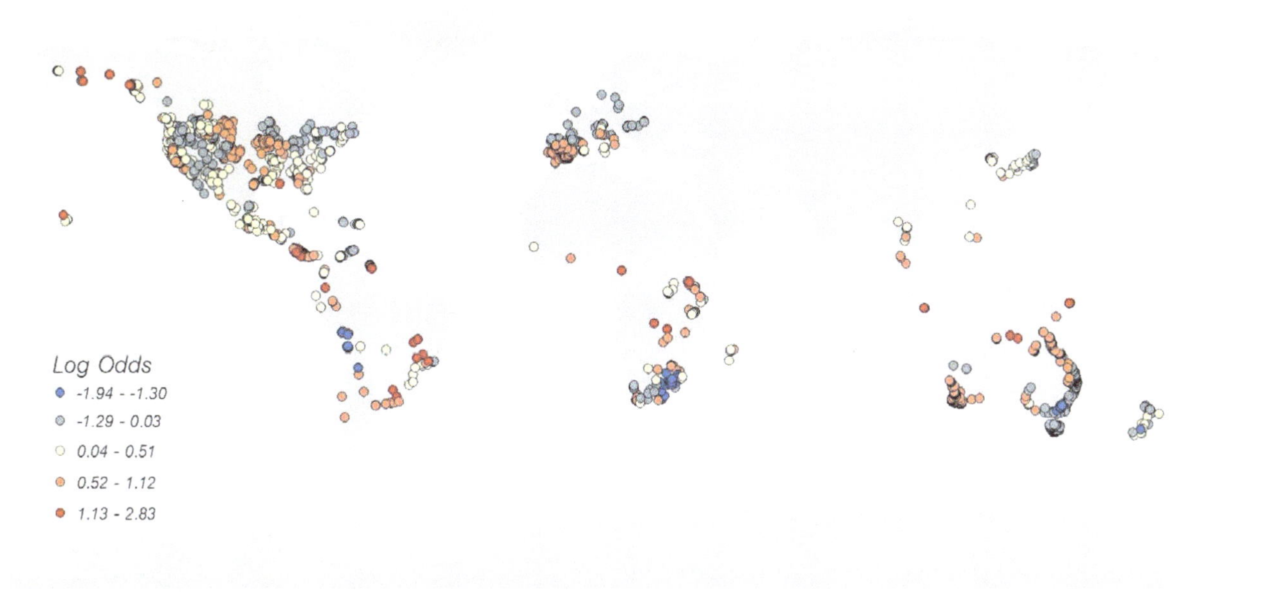

Figure 3. Odds of *Batrachochytrium dendrobatidis* (*Bd*) detection in amphibians derived from a regression model using data from *Bd*-maps.net from all species combined (data snapshot July 2010; N = 3,733 locations).

with average annual minimum temperatures<5°C. Sites with richness<10 species had roughly constant odds of *Bd* detection. For sites with average minimum temperature above 5°C, there was little change in odds of detection. *Bd* detection likelihood was greatest in Temperate Broadleaf and Mixed Forests. The U.S.A. Bufonidae model showed that the odds of *Bd* detection increased by 6.7% with each additional species at a site, and decreased 12.5% with each degree of average annual temperature. Two biomes were positively associated with bufonid *Bd* odds of detection: Deserts and Xeric Shrublands, and Temperate Coniferous Forests. The U.S.A. Hylidae model included only species richness and occurrence of enigmatic declines: when enigmatic declines = 1, there was an estimated 29% increase in odds of *Bd* detection when species richness increased by one. The map of this model was not informative due to the few enigmatic declines known for U.S.A. hylids. The U.S.A. Ranidae model included only biome. The odds of detecting *Bd* in U.S.A. ranids were highest in Desert and Xeric Shrublands, Temperate Broadleaf and Mixed Forests, and Temperate Coniferous Forests.

Discussion

Preliminary analyses of data gathered over a four-year period support broad-scale differences in patterns of infection between amphibian families. These results provide insights to family-specific environmental traits associated with *Bd* infection. Down-scaled environmental hypotheses and assessments are warranted to gauge relationships that may help explain infection patterns at local to regional scales.

Largely due to efforts of the *Bd* Mapping community, we show that the spatial distribution of *Bd* is highly heterogeneous. It appears that a proximal driver of this heterogeneity is the rapid, continuing spread of *Bd* into naïve regions, and it appears that *Bd* has not yet reached a global equilibrium. Within this increasing pattern of spread, we identified nested variables that may influence the ecology and epidemiology of *Bd* following potential introduction into a region; these variables contain taxonomic (see under- and over-infected families), environmental (e.g., temperature range and precipitation at a site), and community-level (e.g., species richness) determinants. While models that combine spatial-spread dynamics with stationary site-specific equilibria are required for understanding and predicting the current and further impact of *Bd*, model clarity will only be achieved when we have a better understanding of the fine-scale epidemiological processes that reflect amphibian diversity, habitat, and community-level interactions.

Bd relationships with temperature are supported by experimental work showing that *Bd* does not grow well at temperatures≥28°C [13,41], field work describing increased infection prevalence and mortality under cooler conditions [30,31,42,43], and other *Bd*-occurrence models, such as the downscaled model using the software Maxent which was applied to data from Australia [44]. Although altitude was not a significant predictor in the current analysis (Figure S3), the highest odds of detecting *Bd* were in the biome Montane Grasslands and Shrublands (Figure S4). This suggests that altitude *per se* is not a globally-uniform predictor of *Bd*-presence, and that the risk of chytridiomycosis to montane-associated species may need to be assessed using regional, spatially restricted data sets. Species-specific and site-specific approaches may yield a better understanding of the dynamics of *Bd*-occurrence among such montane amphibians.

Our analyses support the associations of *Bd* with global-level temperature metrics, supporting hypotheses that the future occurrence of chytridiomycosis will respond to changes in global climate conditions [32,45]. However, there appears to be little support at this time that climate change is associated with chytridiomycosis, within at least parts of its range where this has been studied [35]. Our global data and the model created for Australia [44] suggest that both temperature range and precipitation may be of particular importance for future investigations. Our model predicts an almost 9% decrease in the odds of *Bd* detection with each degree-C increase in temperature range at a site, with Hylidae contributing significantly to this result and Ranidae having an opposing tendency. This aspect of our study will be particularly informative for predicting the effects of warmer or more variable weather patterns on the worldwide incidence of

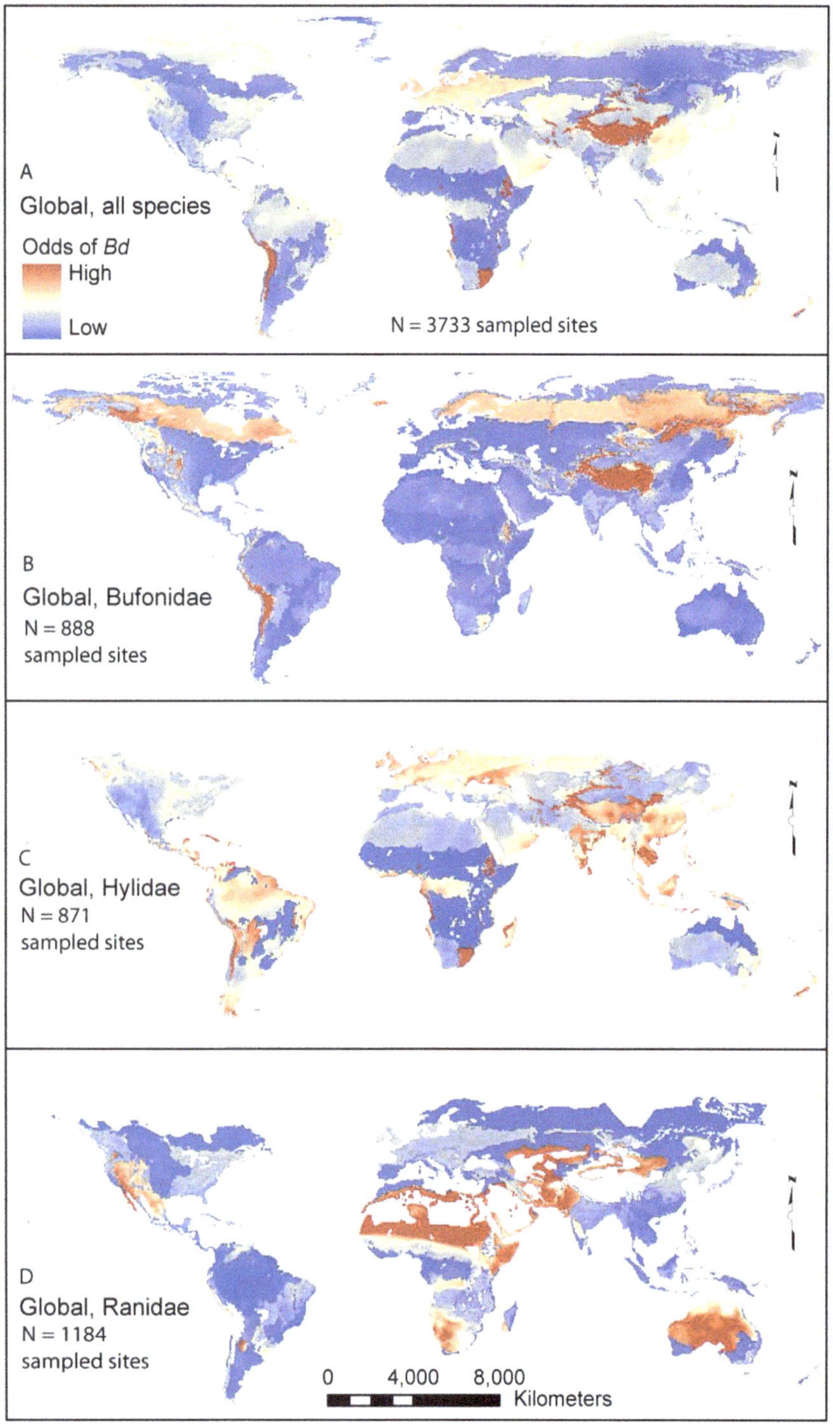

Figure 4. Extrapolated global maps of the odds of *Batrachochytrium dendrobatidis* (*Bd*) detection in amphibians derived from regression models. The global model using data from all species combined is shown (A), as well as subsets of the world data for species within the amphibian families Bufonidae (B), Hylidae (C), and Ranidae (D). Although these map projections depict the odds of *Bd* detection at the world scale, it should be noted that amphibians do not occur everywhere in the world, and in particular, the three amphibian families (B,C,D) are not native to all regions world-wide. Significant model parameters differed among models and included landscape-scale site attributes including climate metrics and biotic factors, such as biome or amphibian species richness (see text). Gaps in the mapped models are due to a lack of species richness data, or amphibian absence, for certain regions.

chytridiomycosis. A caveat here is that such broad-brush analyses inevitably miss fine-scale relationships between amphibians, their microhabitats, and the little-understood relationship between temperature and the ectothermic immune-response: these factors are all likely key to determining the local outcome of infection [46].

Our finding of increased occurrence of *Bd* with species richness supports the hypothesis that greater complexity of amphibian

Table 3. Significant parameters in the regression model using *Batrachochytrium dendrobatidis* (*Bd*) occurrence and data from wild-occurring amphibians having precise locations in the global *Bd* database (N = 3,733 locations).

Parameter	Coefficient	SE	P value
Intercept	1.186	0.262	--
Enigmatic Declines	0.141	0.028	<0.001
Temperature Range	−0.089	0.017	<0.001
Average Annual Precipitation	−0.0005	0.000	<0.001
Biome	--	--	<0.001
Tundra (included in intercept)	--	--	
Montane Grasslands & Shrublands	1.929	0.366	
Mediterranean Forest, Woodlands & Scrub	−0.598	0.126	
Tropical & Subtropical Dry Broadleaf Forests	−0.346	0.406	
Boreal Forests/Taiga	−0.649	0.340	
Mangroves	0.251	0.773	
Deserts & Xeric Shrublands	0.471	0.186	
Flooded Grasslands & Savannas	0.323	0.576	
Temperate Broadleaf & Mixed Forests	0.397	0.122	
Temperate Grasslands, Savannas & Shrublands	−0.346	0.172	
Temperate Coniferous Forests	0.229	0.124	
Tropical & Subtropical Grasslands	−0.582	0.198	
Tropical & Subtropical Moist Broadleaf Forests	0.224	0.172	
Tropical & Subtropical Coniferous Forests	0.503	0.278	

Individual biome coefficients represent additive shifts in the *Bd*-occurrence odds ratio, whereas other coefficients represent multiplicative changes in odds of detecting *Bd*. Logistic regression with likelihood ratio test statistic (Chi^2) was used.

communities may amplify the transmission of *Bd*. Theoretical community epidemiology shows that when between-species transmission rates ($\beta_{between}$) increase, both the prevalence of infection and the proportion of time that hosts are infected increases [47]. Where heterogeneity exists between species in their transmission potential, then even poor transmitters of the pathogen ($\beta_{within} \rightarrow 0$) may suffer repeated exposures in multi-host systems. Additionally, the number of secondary infections caused by a single infected individual (R_0) may increase in relation to increasing species (reservoir) density [48]. For amphibians, it is known that mortality as a consequence of exposure to *Bd* is both dose-dependent [15,49,50] and species-specific [17,51]; some species succumb to very low levels of infection while others never exhibit clinical signs of disease [14,52]. It is possible that the strong positive relationship between species richness and enigmatic decline detailed in Figure 2 reflects the epidemic 'forcing' of infection in heterogeneous speciose systems following the introduction of *Bd*. However, the concurrence of rare species with restricted geographic ranges in areas with high species richness also may contribute to this pattern, if those rare species are also more likely to be tied to Red List criteria for enigmatic declines. Nevertheless, the risk that chytridiomycosis poses to a single amphibian species may be strongly tied to location, as well as being species-specific, with the epidemiological outcome dependent on a potentially very large number of inter-related biotic and abiotic variables. Mechanistic multi-species epidemiological models incorporating taxonomic selectivity in risk alongside community composition and local-scale environmental drivers will provide the key to unlocking the site-specific dynamics of *Bd* and chytridiomycosis, and thus assessing the risk to particular species and amphibian communities. However, the spatial modeling of an infectious disease at this level of complexity, with such broad host-numbers and parameter values, has never been attempted and will present a considerable challenge to mathematical epidemiologists. To aid this process, the modeling of amphibian-*Bd* dynamics in simple systems with few hosts, aided by controlled laboratory experiments for accurate parameter assessment, will prove highly valuable [52,53].

The impact of the *Bd* mapping project is still unfolding. First, this project demonstrates the value of social networking among scientists, with data compiled from hundreds of contributors who have assessed the occurrence of *Bd* at over 4,000 sites worldwide. *Bd*-maps.net allows an ongoing synergy among researchers and managers as global data are communicated or exported for further uses. Second, the project has led to a number of new studies and programs. As a consequence of the project, biologists have taken the initiative to: 1) initiate *Bd* monitoring and assessment in their region and to leverage funding for projects with expanded scope (e.g., the pan-European chytridiomycosis risk assessment RACE; www.bd-maps.eu); 2) use *Bd*-maps.net to direct work toward spatial and taxonomic gaps (e.g., U.S.A. states, Southeast Asia; e.g., *Bd* surveillance papers published in the Amphibian Diseases section of *Herpetological Review*); 3) to instigate disease mitigation programs in focal systems (e.g., Mallorca and *Alytes muletensis*; [54]); 4) include *Bd*-occurrence data in aquatic invasive species alerts for land management (e.g., for water-draw decisions for western US wildfire management; Figure S5); and 5) to facilitate novel investigations into the complex ecological patterns associated with *Bd* infection [55].

To enhance the utility of *Bd*-maps.net for the global science and management community, dedicated curation and community deposition of data are needed. Currently, part-time efforts of individuals at two institutions, the U.S. Forest Service, Pacific Northwest Research Station, Oregon, U.S.A., and Imperial College, London, U.K., allow limited technical assistance for data contributors. Broader institutional support from the global community could tie data reporting to permit processes and publications for broader community participation and uses. For example, infection trend monitoring could easily be conducted for local jurisdictions, would be more effective with surveillance reporting tied to existing permitting processes, and could lead to insights for understanding transmission vectors or infectivity correlates. Molecular typing of *Bd* isolates [16] is continuing and is planned to be integrated with mapping data to understand the global spread of infection within a population genomics framework [56]. The addition of molecular data to *Bd*-maps.net, alongside rapid next-generation genotyping techniques, may aid in identifying the major sources and sinks of infection, and the dominant global vectors of *Bd*. We are applying our lessons-learned in the development of Bd-maps to the current development of a global Ranavirus Reporting System [57,58]. This developing system includes data fields to assess data certainty and potential bias. Understanding of the vectors for both *Bd* and Ranaviruses will be key for regional biosecurity, with subsequent emphasis on management of the human-mediated transmission pathways. An expected outcome is the need for increased transparency of disease transmission pathways in amphibian trade within the context of the OIE legislation, as well as a call for all countries contributing to the OIE infrastructure to contribute to the detection and containment of novel wildlife diseases. These approaches are synergistic with those proposed to contain the emergence of newly emerging human diseases, and while the challenges are great, the imperative to curb the spread of these globally-devastating infections has never been greater.

Materials and Methods

Data compilation

Bd location data were compiled from published literature and unpublished studies (Table S1). We defined 'site' as a geographically distinct location having specific coordinates and/or locality description. A 'record' was a database entry for a single species for a single year, at a particular site. A given site may have a number of records for different species or years, some for which *Bd* was found, others for which it was not found. A 'country centroid' is a point designation for a type of site that placed sampled species in the country of origin, where finer-scale location data were not available. Country centroid data were compiled generally from early years of *Bd* surveillance before Global Positioning Systems were frequently used to determine site coordinates.

Site-level data were used in analyses, and if both positive and negative *Bd* records were available for a site, the site was designated *Bd*-positive. Analyses did not include data from captive animals, vague locations, or country centroids. For analyses, a snapshot of site-level data was taken in July 2010, including all data compiled in our database up to that time. Analyses were conducted using a global grid spatial resolution of 0.5 degrees latitude and longitude, which was the finest scale available for the global climate data. Finer-scale habitat patterns will not be apparent from our analyses, such as those with more restricted temperature regimes. Parameters used in our models are listed in Table S3.

To provide a more updated summary of *Bd* occurrence (Table 1, Table 2, Table S1), we report tallies of *Bd*-positive and *Bd*-negative data compiled up through a data snapshot taken on 30 January 2011. These summaries were used for the over/under-infection randomization analysis. Our data are available via a world *Bd* mapping database and interactive web application that we developed for the *Bd* mapping project, which has merged and extended pre-existing data compilation efforts (http://www.Bd-maps.net). The *Bd*-maps.net website facilitates automatic surveillance by country or species at several spatial scales which can be user-determined by an interactive mapping tool using Google Maps technology. Hence, real-time counts by country and species are available online, and recent data snapshots of these tallies were taken on 25 July 2012 for this paper. Some sensitive data were not released for precise mapping, and hence our hierarchical mapping method enables their mapping at world scale but not at finer scales. However, sensitive data were included in our data snapshots for both analyses and more recent summaries. For the purposes of this paper, a data snapshot was undertaken for our analyses (June 2010), and we have provided additional time-sensitive snapshots for more updated information in Figure 1 (August 2012) and Table S1 (January 2011). Data updates and analyses are time-intensive procedures to ensure duplicate records are screened from our counts, accounting for different snapshot dates reported here. The most recent (July 2012) counts of *Bd*-infected countries and species is the only realtime data summary available at this time.

Caution is needed due to several potential data constraints when interpreting our maps and analyses. Limited quality assurance was conducted, such that if inaccurate methods were used or results reported, those data may not have been screened out of this compilation (e.g., false positives and negatives may be included in the data). *Bd* prevalence in a population may be low, such that a small sample size of individuals screened for *Bd* may not detect it, and consequently could yield a false negative report ('no detection' does not indicate absence of *Bd*; see [39]). This could be a particular concern for data collected before 2008, when the recommendation for sample sizes>59 be collected to assess *Bd* occurrence when prevalence is low [39]. In addition, it is possible that negative data are not consistently reported, especially from early exploratory studies or due to the challenge of publishing negative data, and hence negative data may be under-represented in our compilation. To conduct adequate quality assurance of the global data set to reduce potential biases, individual studies would need to be screened, which would require each data source to be revisited and assessed. Our aggregation of data at the 0.5 degree latitude and longitude scale likely reduces these biases by compiling data among different sampling events, including separate studies (and their methods), researchers, species, seasons, and years. Finally, our world map of *Bd* detections (Figure 1) includes samples taken from living or dead animals—appearing healthy or symptomatic, native and introduced animals, and wild and captive animals. We do not map the disease chytridiomycosis wherein animals showed symptoms of the disease. This last point should be emphasized; we report *Bd* occurrence and hence the threat of disease and potential taxonomic or geographic sources of disease transmission, but we have not distinguished locations with mortality attributed to the disease and our maps do not represent declining amphibian populations. In addition, our logistic regression models should not be used for direct prediction, and coefficients from the fitted models should only be interpreted in terms of relative (not absolute) contribution to the odds of *Bd* detection. Because sites were not randomly selected, the inference of each model is restricted to those sites used in the analysis. However, with 3,733 sites used in our full model, we feel there is useful predictive capacity of those data to be used as hypotheses for future investigation and later downscaled analyses, especially for the better-sampled regions of the world and more heavily-sampled amphibian families.

During modeling, all of our 3,961 *Bd* sites during that data snapshot did not fall cleanly within available environmental coverages, and hence some were not included in regression analyses. For example, the biome coverage did not overlay cleanly with the *Bd* site data. For 227 sites, site location did not correspond to an environmental coverage after the overlay was conducted, and these were eliminated from the model. These sites were distributed around the world and often were near coastlines, suggesting the coastal accuracy of the landscape-scale environmental boundaries was poor; hence, these sites had very low elevations. However, there were often neighboring locations that were sorted into a biome, so locations with clusters of samples were not generally lost from analyses. We also deleted the 'lake' biome from analyses (1 site) because lentic habitats within other biome types were not distinguished.

Statistical Analysis

Using logistic regression (programmed in R software [59]), we examined associations between the *Bd* occurrence at a site (world scale: all data and three well-sampled families separately – Bufonidae, Hylidae, Ranidae) and: 1) latitude; 2) elevation [60]; 3) biome [61]; 4) amphibian species richness [7]; and 5) global temperature and precipitation metrics ([62]; Climatic Research Unit, University of East Anglia, Norwich, UK; www.cru.uea.ac.uk) (Table S2). These parameters were chosen to represent world-scale ecological factors that may be correlated with *Bd*. Our intent was to use our robust data set to address existing hypotheses of *Bd*-occurrence correlations, and to contribute to new ecological research and surveillance directions. We did not include other factors that may better represent a human footprint signal from modern-day disturbances (e.g., proximity to roads, cities, agricultural areas) largely because these data layers are inconsistent at the world scale.

Additionally, because the United States was a relatively well-sampled large-scale region, we conducted similar analyses for all U.S.A. data and well-sampled families in the U.S.A. (Bufonidae, Hylidae, Ranidae). More precise temperature and precipitation data were used in the U.S.A. analyses (Parameter-elevation Regressions on Independent Slopes Model [PRISM], C. Daley, PRISM Climate Group, http://prism.oregonstate.edu/). Altogether, eight *Bd*-occurrence models were generated: world scale, all data and data from three families; U.S.A. scale, all data and data from three families.

Logistic regression also was used to examine the relationship between locations with enigmatic population declines [7] and both *Bd* detection and species richness (decline and richness data sources; [7]) detailed in Table S3. Again, we conducted eight analyses: world scale with all world data, and world data from three families; U.S.A. scale with all U.S.A. data, and U.S.A. data from three families.

Randomisation tests were used to determine whether each sampled amphibian family was over- or under-infected, compared to what we would expect by chance given an overall prevalence of 508/1055 species (using the January 2011 data snapshot reflected in Table 1, Table 2, and Table S1). Ten thousand unconstrained randomizations were performed using a computer program written in the statistical language R [59]; infection data was randomised across all species in the data set, and the number of species in the focal family that were randomly assigned as infected was counted. The null distribution generated was then compared to the observed number of infected species within the focal family, and the null hypothesis (that infection was random) was rejected if the observed value lay in the 2.5% for either tail of the null distribution, and statistical power was sufficient (no. species>6). Power-analyses using binomial tests indicated that with six species sampled in a family, there was sufficient power to detect a deviation in either direction.

Supporting Information

Figure S1 Occurrence of *Batrachochytrium dendrobatidis* (*Bd*) at sites by latitude. The high number of sample sites in the Northern Hemisphere, particularly the United States and Spain, gives a pronounced skew to this distribution.

Figure S2 Taxonomic patterns of *Batrachochytrium dendrobatidis* (*Bd*) occurrence by latitude. The extraordinary peak in species richness in the 8°N range highlights the overlap of many families of amphibians in Central America and also the intensive sampling represented by Lips et al. (2003, 2006). Of additional interest are the very broad latitudinal ranges of the families Hylidae, Bufonidae, and Ranidae. This range explains, in part, why they were so widely sampled, and *Bd* was so widely detected among them, and therefore also why they were chosen as the three families we modeled. Note also how the number of families thins out at very high and very low latitudes.

Figure S3 Occurrence of *Batrachochytrium dendrobatidis* at sites by elevation, with annotation by ecoregion of site. Vertical black lines indicate the range of elevations covered by a particular ecoregion. For the most part, only ecoregions or aggregates of ecoregions with more than 20 sites represented are shown in the annotation. Low elevation sites (below 1000 m) were broadly distributed across the world, and across a wide variety of ecoregions. 'bsl' = 'below sea level', for a site in the Coachella Valley, California, U.S.A. Country abbreviations: AR = Argentina, CA = Canada, CH = Switzerland, CL = Chile, ES = Spain, FR = France, IT = Italy, MX = Mexico, PE = Peru, US = United States.

Figure S4 Occurrence of *Batrachochytrium dendrobatidis* (*Bd*) at sites among 15 world biomes. 'Lake' (98) was dropped from our analyses, as there was only one site so classified owing to the coarse scale of the bioregional data relative to our site locations. Among biomes, the highest odds of detecting *Bd* were in Montane Grasslands and Shrublands (Biome 10, reported from Australia, New Zealand, South Africa, and the Chilean and Peruvian Andes), and the lowest odds of detection were in Mediterranean Forests, Woodlands, and Scrub (Biome 12, based on sites in Australia, Spain, Italy, South Africa, and California (U.S.A.), and in Tundra (Biome 11, based on sites in the interior and Kenai Peninsula of Alaska, U.S.A.).

Figure S5 *Batrachochytrium dendrobatidis* (*Bd*) occurrence by 6th-field Hydrologic Units (HU; watershed) for the U.S.A. Natural resource planning and management decisions often occur by watershed in the US. For example, disease disinfection protocols often stipulate disinfection between drainages, and 6th-field watershed delineations are used for water draw decisions during wildfire season in parts of the US. Aquatic invasive species including *Bd* are mapped by watershed to inform decision-makers about risk of transmission during water draw decisions. Map reflects data as of January 2011.

Table S1 *Bd* detections by species, with references and countries of detection. Family-level taxonomy is shown according to Frost et al. (2006), Grant et al. (2006), Frost (2009), and Hedges et al. (2008). Species name is shown as given in the report of the *Bd* occurrence. Where assignments to genus or species have changed since the species was reported as being found with *Bd*, the older name is given in parentheses. For further information on taxonomy, see Taxonomic Notes. The abbreviation (cap.) after a species name indicates an infected captive animal or animals. Status is Conservation Status according to the IUCN Red List (IUCN 2010). Categories are defined as: EX = extinct; EW = Extinct in the Wild; CR = Critically Endangered; EN = Endangered; VU = Vulnerable (the previous three categories are considered Threatened by the IUCN); NT = Near Threatened; LC = Least Concern; DD = Data Deficient; NE = Not Evaluated (newly discovered or newly recognized species). Last updated from published literature in March 2011; references are given in Supplemental Information text, File S1.

Table S2 Parameters used in regression analyses. Numbers in parentheses after biomes indicate number of sites in the full analysis (N = 3733 total sites). Sites with no biome assigned (n = 227) were dropped from the analysis. Lake biome (N = 1 site) was also dropped due to inconsistency of lentic habitat designation.

Table S3 Significant parameters in three family-scale logistic regression models using *Batrachochytrium dendrobatidis* (*Bd*) occurrence data from wild-occurring amphibians having exact and approximate locations in

the global *Bd* database. Individual biome coefficients represent additive shifts in the *Bd*-occurrence odds ratio, whereas other coefficients represent multiplicative changes in odds of detecting *Bd*. Logistic regression with likelihood ratio test statistic (Chi^2) was used.

File S1 Supplemental Information: Taxonomic Notes; References for Supplemental Material.

Acknowledgments

The *Bd* Mapping Group contributed data and assisted in early project development: M. Adams, L. D. Acevedo Cendales, T. Allison; Arizona Game and Fish Department, R. Bakal, K. Beard, B. Bell, J. Bettaso, J. Bielby, M. Bingham, P. Bishop, A. Blaustein, W. Bolívar-G., J. Bosch, S. Bovero, D. Bradford, C. Brown, I. Bruce-Miller, E. Bull, P. Burrowes, M. Byrne, D. Caldwell, Canadian Cooperative Wildlife Health Centre, F. Castro, J. Charbonneau, T. Chestnut, S. Chinnadurai, J. Collins, C. Cortez, F. Clare, A. A. Cunningham, R. Cutter, R. Dankovics, E. W. Davidson, I. De la Riva, A. de Silva, D. Dombrowski, K. Duncan, L. du Preez, L. Ellenburg, K. Field, P. Fisher, J. Fisher, S. Fisher, R. Flitcroft, J. Foster, S. Fox, L. Friis, E. Garcia, T. W. J. Garner, S. Germaine, C. Goldberg, M. Goode, P. Govindarajulu, R. Grasso, A. Gray, D. E. Green, E. Greenbaum, M. Hahr, A. Haigh, C. Harland, T. Hawley, M. Hayes, V. Hemingway, J.-M. Hero, M. I. Herrera-M, A. Hettyey, F. Hock, C. Ivanyi, T. James, J. Jensen, R. Joglar, J. Johnson, P. Johnson, T. Jones, K. Kaiser, N. Karraker, K. Kendell, J. Kielgast, M. Kluber, R. Knapp, J. Kolby, K. Kriger, S. Kupferberg, M. Lampo, M. Lannoo, R. Lehtinen, D. Lesbarrères, A. Lind, K. Lips, A. Litvintseva, M. Lo, J. Loda, J. Longcore, A. Longo, S. Lötters, R. Lovich, J. Lowe, R. Maneyro, B. Maxell, M. McDowell, D. McGriff, J. McKay, J. Mendelson, K. Milliron, R. Moore, S. Muskopf, E. Muths, J. Nachtmann, P. Nanjappa-Mitchell, N. Nieto, Northern Leopard Frog Recovery Team, L. O'Donnell, S. Okada, G. Padgett-Flohr, G. Parra Olea, C. Pearl, A. Pessier, C. Peterson, D. Pflugmacher, D. Pilliod, K. Pope, J. Reaser, K. Reed, M. Reeves, U. Reyer, D. Rödder, A. Rodriguez, M.-O. Roedel, P. C. Rosen, B. Rothermel, J. Rowley, V. St-Amour, A. Salzberg, B. Schmidt, G. Scott, S. Shaw, L. Shirose, R. Solis, E. Sos, M. Sredl, D. Sumerlin, N. Suzuki, T. Tadevosayan, C. Tait, F. Toledo, B. Turner, U. S. Forest Service Sierra Nevada Amphibian Monitoring Program, K. Vance-Borland, B. E. Velásquez, M. Vences, V. Vredenberg, J. Vörös, S. Wagner, J. Ware, L. Watts, C. Weldon, H. Welsh, M. Whitney, L. Wilkinson, W.-M. Wong, D. Woodhams, and Partners in Amphibian and Reptile Conservation. We thank K. Christiansen for assisting with Geographic Information Systems analyses.

Author Contributions

Conceived Global Bd Mapping Project: DHO. Conceptually designed interactive website: DMA MCF DHO. Led website development: DMA. Assisted website development: SFW CIP. Led compilation of world data set and summarisation: KLR. Analyzed the data: GW JB DHO MCF KLR SFW. Contributed reagents/materials/analysis tools: DHO DMA KLR CIP SFW JB TWJG GW MCF. Wrote the paper: DHO DMA KLR JB TWJG MCF SFW GW CIP.

References

1. Ferguson NM, Donnelly CA, Anderson RM (2001) Transmission intensity and impact of control policies on the foot and mouth epidemic in Great Britain. Nature 413: 542–548.
2. Berger L, Speare R, Daszak P, Green DE, Cunningham AA, et al. (1998) Chytridiomycosis causes amphibian mortality associated with population declines in the rain forests of Australia and Central America. Proc Natl Acad Sci U S A 95: 9031–9036.
3. Johnson PTJ (2006) Amphibian diversity: Decimation by disease. Proc Natl Acad Sci U S A 103: 3011–3012.
4. Gascon C (2007) Amphibian Conservation Action Plan. IUCN/SSC Amphibian Specialist Group, Gland, Switzerland and Cambridge, UK.
5. McCallum ML (2007) Amphibian decline or extinction? Current declines dwarf background extinction rate. J Herpetol 41: 483–491.
6. Skerratt LF, Berger L, Speare R, Cashins S, McDonald KR, et al. (2007) Spread of chytridiomycosis has caused the rapid global decline and extinction of frogs. EcoHealth 4: 125–134.
7. Stuart SN, Chanson JS, Cox NA, Young BE, Rodrigues AS, et al. (2004) Status and trends of amphibian declines and extinctions worldwide. Science 306: 1783–1786.
8. Wake DB, Vredenburg VT (2008) Are we in the midst of the sixth mass extinction? A view from the world of amphibians. Proc Natl Acad Sci U S A 105: 11466–11473.
9. Colón-Gaud C, Whiles MR, Kilham SS, Lips KR, Pringle CM, et al. (2009) Assessing ecological responses to catastrophic amphibian declines: patterns of macroinvertebrate production and food web structure in upland Panamanian streams. imnol Oceanogr 54: 331–343.
10. Colón-Gaud C, Whiles MR, Kilham SS, Lips KR, Pringle CM, et al. (2010) Macroinvertebrate community responses to catastrophic amphibian declines in neotropical streams. J North Amer Benthol Soc 2: 1185–1198.
11. Connelly S, Pringle CM, Kilham S, Whiles MR, Lips KR, et al. (2011) Do tadpoles affect leaf decomposition in neotropical streams? Freshwater Biol 56: 1863–1875.
12. Fisher MC, Henk DA, Briggs C, Brownstein JS, Madoff L, et al. (2012) Emerging fungal threats to animal, plant and ecosystem health. Nature 484: 186–194.
13. Longcore JE, Pessier AP, Nichols DK (1999) *Batrachochytrium dendrobatidis* gen et sp nov, a chytrid pathogenic to amphibians. Mycologia 91: 219–227.
14. Fisher MC, Garner T, Walker J (2009) Global emergence of *Batrachochytrium dendrobatidis* and amphibian chytridiomycosis in space, time and host. Ann Rev Microbiol 63:291–310.
15. Garner TW, Walker S, Bosch J, Leech S, Rowcliffe M, et al. (2009) Life history trade-offs influence mortality associated with the amphibian pathogen *Batrachochytrium dendrobatidis*. Oikos 118:783–791.
16. Farrer RA, Weinert LA, Bielby J, Garner TWJ, Balloux F, et al. (2011) Multiple emergences of genetically diverse amphibian-infecting chytrids include a globalized hypervirulent recombinant lineage. Proc Natl Acad Sci U S A. doi:10.1073/pnas.1111915108
17. Reeder NMM, Pessier AP, Vredenburg VT (2012) A reservoir species for the emerging amphibian pathogen *Batrachochytrium dendrobatidis* thrives in a landscape decimated by disease. PLoS ONE 7(3): e33567. doi:10.1371/journal.pone.0033567
18. Daszak P, Strieby A, Cunningham AA, Longcore JE, Brown CC, et al. (2004) Experimental evidence that the bullfrog (*Rana catesbeiana*) is a potential carrier of chytridiomycosis, an emerging fungal disease of amphibians. Herpetol J 14: 201–207.
19. Garner TW, Perkins M, Govindarajulu P, Seglie D, Walker S, et al. (2006) The emerging amphibian pathogen *Batrachochytrium dendrobatidis* globally infects introduced populations of the North American bullfrog, *Rana catesbeiana*. Biol Lett 2: 455–459.
20. Fisher MC, Garner TWJ (2007) The relationship between the introduction of *Batrachochytrium dendrobatidis*, the international trade in amphibians and introduced amphibian species. Fungal Biol Rev 21: 2–9.
21. Picco AM, Collins JP (2008) Amphibian commerce as a likely source of pathogen pollution. Conserv Biol 22: 1582–1589.
22. Garmyn A, Van Rooij P, Pasmans F, Hellebuyck T, Van Den Broeck W, et al. (2012) Waterfowl: Potential environmental reservoirs of the chytrid fungus *Batrachochytrium dendrobatidis*. PLoS ONE 7(4): e35038. doi:10.1371/journal.pone.0035038
23. El Mouden EH, Slimani T, Donaire D, Fernández-Beaskoetxea S, Fisher MC, et al. (2011) First record of the chytrid fungus *Batrachochytrium dendrobatidis* in North Africa. Herpetol Rev 42: 71–75.
24. Kirshtein JD, Anderson CW, Wood JS, Longcore JE, Voytek MA (2007) Quantitative PCR detection of *Batrachochytrium dendrobatidis* DNA from sediments and water. Dis Aquat Org 77:11–15.
25. OIE (2008) Report of the meeting of the OIE Aquatic animal health standards commission, Paris, 3–7 March 2008. 76th General Session.
26. Swei A, Rowley JJL, Rödder D, Diesmos MLL, Diesmos AC, et al. (2011) Is chytridiomycosis an emerging infectious disease in Asia? PLoS ONE 6(8): e23179. doi:10.1371/journal.pone.0023179
27. Olson DH, Ronnenberg KL (2008) http://www.parcplace.org/bdmap2008update.html.
28. Aanensen DM (2009) The development and use of bioinformatic web applications for Infectious Disease Microbiology. Thesis. https://unicorn.lib.ic.ac.uk/uhtbin/ckey/825195.
29. Woodhams DC, Alford RA (2005) Ecology of chytridiomycosis in rainforest stream frog assemblages of tropical Queensland. Conserv Biol 19: 1449–1459.
30. Berger L, Speare R, Hines H, Marantelli G, Hyatt AD, et al. (2004) Effect of season and temperature on mortality in amphibians due to chytridiomycosis. Aust Vet J 82: 31–36.
31. Ron S (2005) Predicting the distribution of the amphibian pathogen *Batrachochytrium dendrobatidis* in the New World. Biotropica 37: 209–221.

32. Bosch J, Carrascal LM, Duran L, Walker S, Fisher MC (2007) Climate change and outbreaks of amphibian chytridiomycosis in a montane area of Central Spain; is there a link? Proc R Soc Lond [Biol] 274: 253–260.
33. Kriger KM, Hero JM (2007) Large-scale seasonal variation in the prevalence and severity of chytridiomycosis. J Zool 271: 352–359.
34. Lips KR, Brem F, Brenes R, Reeve JD, Alford RA, et al. (2006) Emerging infectious disease and the loss of biodiversity in a Neotropical amphibian community. Proc Natl Acad Sci U S A 103: 3165–3170.
35. Lips KR, Diffendorfer J, Mendelson III JR, Sears MW (2008) Riding the wave: Reconciling the roles of disease and climate change in amphibian declines. PLoS Biol 6: 441–454.
36. Muths E, Pilliod DS, Livo LJ (2008) Distribution and environmental limitations of an amphibian pathogen in the Rocky Mountains, USA. Biol Cons 141: 1484–1492.
37. Murray KA, Skerratt LF, Speare R, McCallum H (2009) Impact and dynamics of disease in species threatened by the amphibian chytrid fungus, *Batrachochytrium dendrobatidis*. Conserv Biol 23: 1242–1252.
38. Aanensen DM, Huntley DM, Feil EJ, al-Own F, Spratt BG (2009) EpiCollect: Linking smartphones to web applications for epidemiology, ecology and community data collection. PLos ONE 4(9): e6968. doi:10.1371/journal.-pone.0006968.
39. Skerratt LF, Berger L, Hines HB, McDonald KR, Mendez D, et al. (2008) Survey protocol for detecting chytridiomycosis in all Australian frog populations. Dis Aquat Org 80: 85–94.
40. Bielby J, Cooper N, Cunningham AA, Garner TW, Purvis A (2008) Predicting susceptibility to future declines in the world's frogs. Conserv Lett 1: 82–90.
41. Piotrowski JS, Annis SL, Longcore JE (2004) Physiology of *Batrachochytrium dendrobatidis*, a chytrid pathogen of amphibians. Mycologia 96: 9–15.
42. Retallick RWR, McCallum H, Speare R (2004) Endemic infection of the amphibian chytrid fungus in a frog community post-decline. PloS Biol 2: 1965–1971.
43. Forrest MJ, Schlaepfer MA (2011) Nothing a hot bath won't cure: Infection rates of amphibian chytrid fungus correlate negatively with water temperature under natural field settings. PLoS ONE 6(2):e28444. doi:10.1371/journal.pone.0028444.
44. Murray KA, Retallick RWR, Puschendorf R, Skerratt LF, Rosauer D, et al. (2011) Assessing spatial patterns of disease risk to biodiversity: implications for the management of the amphibian pathogen, *Batrachochytrium dendrobatidis*. J Anim Ecol 48:163–173.
45. Pounds AJ, Bustamante MR, Coloma LA, Consuegra JA, Fogden MPL et al. (2006) Widespread amphibian extinctions from epidemic disease driven by global warming. Nature 439: 161–167.
46. Lovich R, Ryan MJ, Pessier A, Claypool B (2008) Infection with the fungus *Batrachochytrium dendrobatidis* in a non-native *Lithobates berlandieri* below sea level in the Coachella Valley, California, USA. Herpetol Rev 39.
47. Fenton A, Pedersen AB (2005) Community epidemiology framework for classifying disease threats. Emerg Infect Dis 11: 1815–1821.
48. de Castro F, Bolker B (2005) Mechanisms of disease-induced extinction. Ecol Lett 8: 117–126.
49. Vredenburg VT, Knapp R, Tunstall T, Briggs CJ (2010) Dynamics of an emerging disease drive large-scale amphibian population extinctions. Proc Natl Acad Sci U S A 107: 9689–9694.
50. Kinney VC, Heemeyer JL, Pessier AP, Lannoo MJ (2011) Seasonal pattern of *Batrachochytrium dendrobatidis* infection and mortality in *Lithobates areolatus*: Affirmation of Vredenburg's '10,000 Zoospore Rule'. PLoS ONE 6: e16708.
51. Woodhams DC, Ardipradja K, Alford RA, Marantelli G, Reinert LK, et al. (2007) Resistance to chytridiomycosis varies among amphibian species and is correlated with skin peptide defenses. Anim Conserv 10: 409–417.
52. Briggs CJ, Vredenburg VT, Knapp RA, Rachowicz LJ (2005) Investigating the population-level effects of chytridiomycosis: An emerging infectious disease of amphibians. Ecology 86: 3149–3159.
53. Mitchell KM, Churcher TS, Garner TWG, Fisher MC (2008) Persistence of the emerging pathogen *Batrachochytrium dendrobatidis* outside the amphibian host greatly increases the probability of host extinction. Proc R Soc [Biol] 275: 329–334.
54. Lubick N (2010) Emergency medicine for frogs. Nature 465, 680–681.
55. Bancroft B, Han B, Searle C, Michael L, Olson DH, et al. (2011) Environmental and biological correlates of species susceptibility to *Batrachochytrium dendrobatidis* in amphibians. Biol Conserv 20:1911–1920.
56. James TY, Litvintseva A, Vilgalys R, Morgan JA, Taylor JW, et al. (2009) Rapid global expansion of the fungal disease chytridiomycosis into declining and healthy amphibian populations. PLoS Pathogens 5: e1000458. doi:10.1371/journal.ppat.1000458.
57. Duffus A, Olson D (2011) The establishment of a global Ranavirus reporting system. Froglog 96: 37.
58. Marschang RE, Miller D (2011) 2011 International Ranavirus symposium. J Herpetol Med Surgery 21: 1–2.
59. R Development Core Team (2009) R: A language and environment for statistical computing. R Foundation for Statistical Computing, Vienna, Austria ISBN 3-900051-07-0, URL http://www.R-project.org.
60. USGS [United States Geological Survey] (2006) Global digital elevation model GTOPO30. [US Geological Survey] global digital elevation model GTO-PO30<http://edc.usgs.gov/products/elevation/gtopo30/gtopo30.html>(accessed 28 October 2008).
61. Olson DM (2001). BioScience 933–938. <http://www.worldwildlife.org/science/data/item1875.html>[accessed 28 October 2008)].
62. Mitchell TD (2004) Tyndall Centre Working Paper No. 55.<http://www.tyndall.ac.uk/publications/working_papers/wp55_summary.shtml>(Accessed 4 November 2008)

9

Soil Temperature Triggers the Onset of Photosynthesis in Korean Pine

Jiabing Wu, Dexin Guan*, Fenhui Yuan, Anzhi Wang, Changjie Jin

State Key Laboratory of Forest and Soil Ecology, Institute of Applied Ecology, Chinese Academy of Sciences, Shenyang, P.R. China

Abstract

In forest ecosystems, the onset of spring photosynthesis may have an important influence on the annual carbon balance. However, triggers for the onset of photosynthesis have yet to be clearly identified, especially for temperate evergreen conifers. The effects of climatic factors on recovery of photosynthetic capacity in a Korean pine forest were investigated in the field. No photosynthesis was detectable when the soil temperature was below 0°C even if the air temperature was far beyond 15°C. The onset of photosynthesis and sap flow was coincident with the time of soil thawing. The rates of recovery of photosynthetic capacity highly fluctuated with air temperature after onset of photosynthesis, and intermittent frost events remarkably inhibited the photosynthetic capacity of the needles. The results suggest that earlier soil thawing is more important than air temperature increases in triggering the onset of photosynthesis in Korean pine in temperate zones under global warming scenarios.

Editor: Carl J. Bernacchi, University of Illinois, United States of America

Funding: This study was jointly financed by State Key Development Program of Basic Research (SKDPBR) (2009CB421101), and the Youth Fund for Creative Research Groups, Institute of Applied Ecology, Chinese Academy of Sciences. The funders had no role in study design, data collection and analysis, decision to publish, or preparation of the manuscript.

Competing Interests: The authors have declared that no competing interests exist.

* E-mail: bjfu_1999@hotmail.com

Introduction

In temperate forests, the onset of spring photosynthesis may have an important influence on the carbon assimilation of trees [1,2] and hence influence the carbon balance of forest ecosystems [3]. The onset of photosynthesis is a crucial process in plant phenology. Accurate evaluation of the starting date of photosynthesis and its connections with future climatic factors are essential for understanding and modeling the ecosystem carbon balance. For example, enhanced warming at high northern latitudes has been reported to result in the earlier onset of plant photosynthesis and therefore an increase in forest productivity [4–5].

The photosynthetic capacity of deciduous trees in boreal and temperate forests varies considerably with the annual cycle of presence and absence of foliage. Therefore, for deciduous trees, the onset of photosynthesis in spring is linked to the easily observable growth of leaves. Several studies report that air temperature triggers the recovery of boreal and temperate deciduous forest photosynthesis in spring (e.g., [6–7]), and the effects of air temperature on the onset of photosynthesis have been well modeled [7]. Accordingly, an earlier onset of photosynthesis in spring for these forests is reported following global warming [8–9].

Evergreen coniferous trees growing in boreal and temperate zones have a clear annual cycle of growth and dormancy period, considering that the photosynthetic capacity of the needles is generally inhibited during the wintertime by low air temperature [10]. However, variations in the timing of onset of photosynthesis as well as its determining factors, are not easily detectable. Tanja et al. (2003) [11] found that air temperature is the main trigger of the onset of plant photosynthesis at five evergreen boreal coniferous field stations. Sevanto et al. (2006) [12] found that the photosynthesis of *Pinus sylvestris* begins when the air temperature reaches 3°C to 4°C, even though the soil remains frozen, and argued that water for needle photosynthesis is possibly available from stored stem water. However, several temperature treatment experiments and field investigations on evergreen conifers have also shown that the onset of photosynthesis cannot be observed even after several days of exposure to optimal air temperature when the soil is frozen [8,13–14]. Some studies have reported that soil temperature is the primary control of boreal conifer photosynthesis and related this effect to the inhibition of water uptake from frozen soils (e.g., [15,16]). A literature overview shows that there are extensive studies on the seasonal development of tree photosynthesis in temperate forests. However, further studies are still need to improve the understanding of the factors that trigger the onset of photosynthetic recovery for temperate evergreen coniferous trees.

The global temperature is expected to rise in future decades, and warming has been predicted to be especially pronounced during spring in regions located at mid-and high latitudes [17]. To further our knowledge of the effects of climate and climate change on plant phenology, existing worldwide CO_2 flux measurements from the micrometeorological tower sites provide a valuable resource to investigate the effects of climate on photosynthetic phenology across biome and climate zones. Using long-term measurements of net ecosystem carbon exchange from an eddy covariance flux tower, and concurrent measurements of needle photosynthesis, sap flow in trees, and associated meteorological variables, we attempt to determine the environmental factors initiating the onset of photosynthesis for Korean pine in the temperate forest of Northeast China. In particular, we evaluate the importance of soil temperature versus air temperature in determining the onset of photosynthesis in temperate evergreen coniferous trees.

Methods

Ethics Statement

The field studies were conducted at the Changbai Mountains National Nature Reserve, in northeastern China. It is a practice base for the researchers of the Chinese Academy of Sciences. Thus we could conduct experiments there without specific permits. The experiments conducted in this study do not involve or impact endangered or protected species.

Observation Sites and Climatic Background

The experimental site is located within the National Natural Conservation Park of Changbai Mountains, northeast of China (42° 24′ 10″ N, 128° 05′ 46″ E, with an elevation of 740 m). The study area lies in a monsoon-influenced, temperate continental climate zone. Its annual mean temperature is 3.6°C, and its annual mean precipitation is 690 mm. The site is vegetated with virgin old-growth Korean pine (*Pinus koraiensis*) as dominant species and interspersed with tuan linden (*Tilia amurensis*), Mono maple (*Acer mono*) and larch (*Olgensis* var.). The mean canopy height is about 27 m and its maximum leaf area index is 6.0 [18].

Observations and Data Processing

A flux tower was located in the Korean pine forest in Changbai Mountains. Fluxes of CO_2 and H_2O were measured using eddy covariance methodology at a 60.0 m high micrometeorological tower. Air temperatures and three-dimensional wind speeds were measured with a three-dimensional sonic anemometer (CSAT3, Campbell Sci., Inc., USA). A fast-response, open-path infrared gas analyzer (LI-7500, LI-Cor Inc., USA) was used to measure CO_2 and H_2O concentrations. Analog signals from the sonic anemometer and the infrared gas analyzer were sampled by a data logger (CR3000, Campbell Sci., Inc., USA) at 10 Hz.

Fluxes of CO_2 were calculated at 30 min intervals. Linear trends in the 10 Hz raw data of air temperature and CO_2 concentrations were removed. The wind components measured by the sonic anemometer were rotated so that the mean transverse velocity and the mean vertical velocity were reduced to zero for each 30 min interval [19]. Transfer functions were applied to correct the high-frequency CO_2 flux losses caused by the sensor separation, and the frequency response of the data acquisition system [20]. CO_2 fluxes were corrected for density fluctuations arising from air temperature and humidity [21]. Finally, the flux data were screened using an integral turbulence characteristics test following Foken and Wichura (1996) [22], which removed data points associated with disturbed and underdeveloped turbulence. The percentages of the remaining flux data in the study period for 2006, 2007, 2008, and 2009 were 57.4%, 63.8%, 57.0%, and 67.1%, respectively. Data gaps were filled by using standardized methods: linear interpolation for small gaps shorter than 2.0 hr and mean diurnal variation method for longer gaps [23].

CO_2 flux measurements spanning from April to May (DOY 91 to DOY 151) in the years 2006 to 2009 were used. These periods fully covered the stages of soil thawing, onset of needle photosynthesis, and full development of foliage photosynthesis. The onset of photosynthesis in trees is actually a gradual process but we treated it as an on-off phenomenon in this study for simplicity. In the current study, we defined the onset of photosynthesis as the point when the three-day moving average of integrated net ecosystem carbon exchange during daytime (NEE_d) first falls below zero. Daytime was defined as the period in which net radiation is greater than 10 $W \cdot m^{-2}$.

Soil temperatures were measured using multilevel thermocouple probes, with two sets of sensors (105T, Campbell Sci., Inc., USA) placed at 5, 10, 20, and 50 cm below the soil surface under the canopy of pine trees. Air temperature was measured with one set of HMP45C probes (Vaisala, Helsinki, Finland) placed at 2, 16, 22, 26, 32, and 60 m height of the flux tower. Net radiation was measured with CNR-1 net radiometer (Kipp & Zonen, Delft, Netherlands) at 40 m height. These ancillary temperature measurements responded to frequencies up to 0.5 Hz. Six Granier type thermal dissipation probes (TDP30, Rainroot Scientific Limited, Beijing, China) were used to measure the sap flux density of tree stems. Two sensors were inserted into each of the three pine stems at 1.5 m height. One sensor was located on the eastern side and another was placed on the western side of the stem.

Light response curves were performed using a portable photosynthetic analyzer (LI-6400; LI-Cor Inc., USA) equipped with a standard 3 cm×2 cm leaf chamber and a red/blue light source (6400–02B). Three pine trees were chosen and photosynthesis measurements on each tree were carried out daily for 15 consecutive days during the soil thawing period. Measurements were performed on branches (about 2 cm in stem diameter) that were cut from the mid-portion of the canopy and were immediately placed in water. Pine needles were allowed to adapt to each light level for 3 minutes before each point was recorded.

Results

Environmental Information

Meteorological information on the Korean pine forest is presented in Table 1. In general, remarkable annual variations in the monthly mean values of temperature and precipitation in April and May were observed, especially in April of 2006, which showed the lowest air temperature (Fig. 1A) and precipitation. Although statistically insignificant, possibly due to the limited observed time range, a trend of warmer springs (0.97°C/yr, $r^2 = 0.84$, $p<0.1$) was observed in the study area, as suggested by the linear fitting of the mean air temperature during April and May over the four-year period. Hence, a trend of earlier end of the soil thawing period (from DOY 117 to DOY 98) was observed, as indicated by measured soil temperatures at a depth of 5 cm.

Details on the variation in temperature are shown in Fig. 1. Air temperature varied widely during spring photosynthesis recovery period. The monthly air temperature in April over four years of observation averaged 5.8°C. The minimum air temperature, which was observed in 2006, was 3.2°C, and the maximum air temperature, which was observed in 2008, was 9.4°C. Earlier soil thawing corresponding to warmer springs was demonstrated by the annual variation in soil temperature. During the observation period, 26 m height air temperature showed significant variations from −6.5°C to 20.8°C before the onset of photosynthesis, while 5 cm depth soil temperatures remained constant at around 0°C (isothermal stage), and the period of isothermal stage generally lasted for around three weeks each year. The earliest start of soil thawing at 5 cm depth was observed in early April 2008 (DOY 98), and the latest start of soil thawing was observed in April 2006 (DOY 117). Soil temperatures rose rapidly after the end of soil thawing, reaching a maximum of 13.0°C by the end of May in 2006.

Table 1. The basic information on meteorological factors, monthly cumulative net ecosystem carbon exchange rates (NEE_d) and the dates of onset of photosynthesis investigated from Changbai Mountains forest site during study period.

Year	Month	Air temperature (°C)	Precipitation (mm)	Cumulative NEE_d ($gC \cdot m^{-2}$)	Date of soil unfrozen (DOY)	Onset of photosynthesis (DOY)
2006	April	3.2	23	5.6	117	117
	May	13.1	50.6	−36	–	–
2007	April	3.9	33.9	1.1	109	110
	May	11.9	93.3	−46.4	–	–
2008	April	9.4	33.0	−3.2	98	99
	May	10.6	72.5	−75.2	–	–
2009	April	6.8	36.5	−9.9	100	100
	May	14.6	46.7	−63.9	–	–

Patterns of Photosynthesis and its Controls

The timing of annual onset of photosynthesis ranged from DOY 99 to DOY 117, with an average starting date of DOY 107. The earliest start of photosynthesis appeared in April 2008 (Tab. 1), coinciding with the highest monthly mean air temperature. The latest signs of photosynthesis appeared in April 2006, coinciding with the appearance of the lowest monthly mean air temperature. However, data analysis suggests that air temperature is not the main trigger of the onset of tree photosynthesis, as observed from seasonal evolutions of air and soil temperature and synchronous variations in the daytime net ecosystem carbon exchange (Fig. 1). The onset of photosynthesis coincided with the end of soil thawing, and the maximum lag of onset of photosynthesis related to the end of soil thawing was generally within one day. After onset of photosynthesis, photosynthetic rates obviously fluctuated with daily and seasonal evolutions of air temperature throughout the study period.

Further data analysis suggests the existence of a threshold value of accumulated growing degree days (GDD, which is defined as a cumulative sum of daily mean air temperatures above 0°C from the beginning of the soil isothermal stage period) related to the timing of onset of photosynthesis. As observed from Fig. 2, photosynthesis began when the GDD had accumulated to 105°C in all of the observed years. The GDD was significantly higher in 2008 and 2009 than in 2006 and 2007, obviously contributing to the earlier onset of photosynthesis in the latter two years.

In situ measurements of photosynthetic light response curves with a LI-6400 portable photosynthesis system also suggested that the pine needles could not begin photosynthesis when soil is frozen. As can be seen from Fig. 3, after 20 min of adaptation to saturated light, photosynthetic carbon uptake was not detected in one-year-old pine needles even though the air temperature was far beyond 15°C while soil temperature T_s was below 0°C. Comparably, photosynthetic activity was detected immediately at the end of the isothermal period (T_s reached 0.6°C).

The effects of soil thawing on tree photosynthesis were also well reflected by measurements of sap flow. As observed from Fig. 4, early in the spring, when the pine needles were still dormant, air temperatures rose above freezing during the day but dropped back below freezing at night. Sap flow was not initiated in the tree until the soil temperature in the upper rooting zone (0–20 cm depth) soil had reached a certain temperature above 0°C, even when the

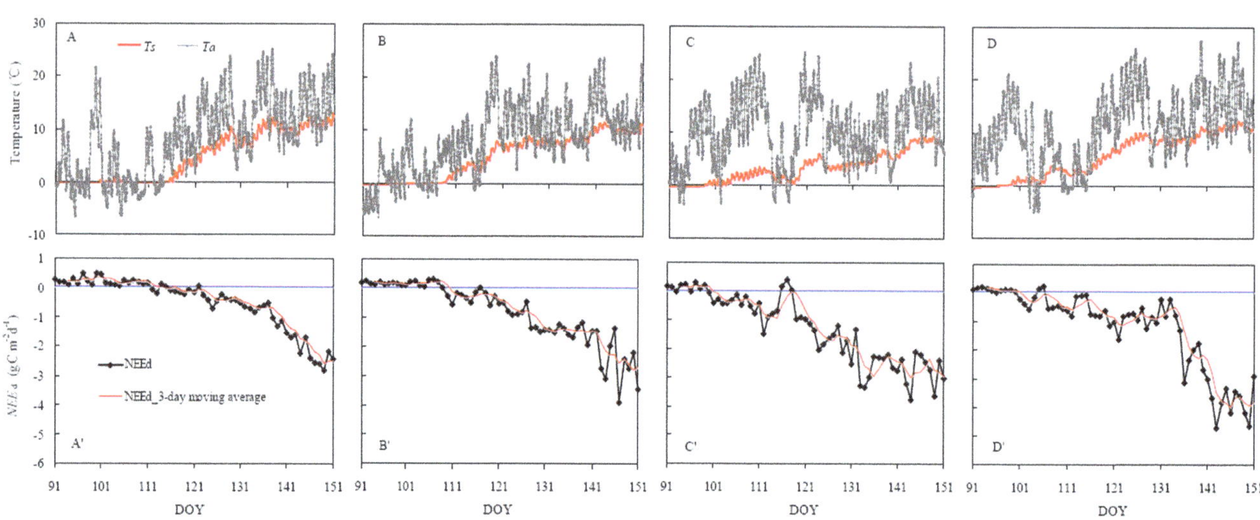

Figure 1. Seasonal courses in three-day moving average of NEE_d (daily integrated net ecosystem carbon exchange), the 26-m height air temperature and 5-cm depth soil temperature at the Changbai Mountains Korean pine forest from DOY 91 to DOY 151 (1st of April to 31st of May) in year of 2006 (A–A'), 2007 (B–B'), 2008 (C–C') and 2009 (D–D').

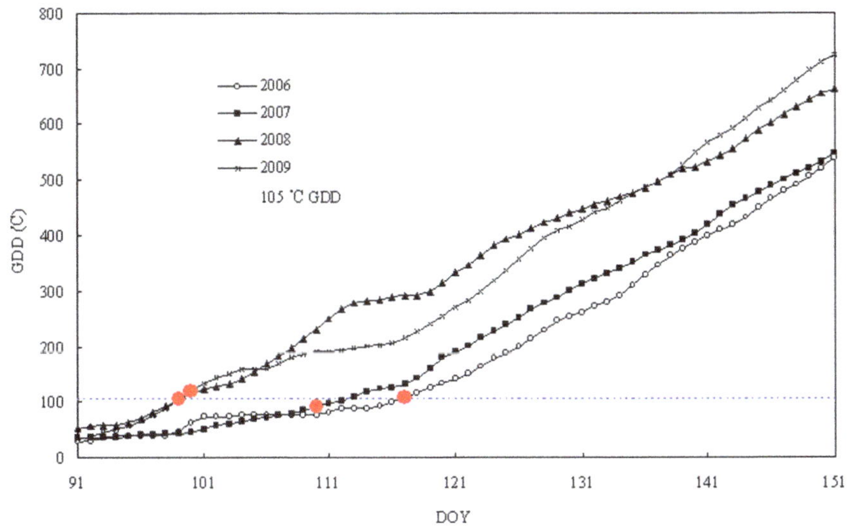

Figure 2. Onset of Korean pine photosynthesis as denoted by accumulated growing degree days (GDD) measured at canopy height (26 m). Horizontal lines mark the 105°C isotherm of the GDD threshold. The circles denote the timing of onset of photosynthesis in each year.

atmospheric water demand was likely large under high air temperature conditions. Pines with favorable air temperature (up to 13.2°C) did not take up water until the end of the thawing period of the surface soil.

Further analysis suggests that the low air temperature obviously inhibits photosynthetic recovery after onset of photosynthesis in Korean pine. An extreme case of low temperature inhibition was the occurrence of frost. We defined a "frost event" as an average temperature decrease of at least 10°C within 24 hours, and simultaneously the minimum temperature drops down to 0°C or lower. Frost events occurred during recovery of photosynthetic capacity in all of the observed years except in 2006, as shown in Table 2. Generally, frost events caused declines of over 80% in net photosynthetic rates and even caused transitions from net carbon assimilation to net carbon emission in the studied forest in the daytime. Frost-induced inhibition of photosynthesis was more severe in 2008 than in other years because frost events occurred comparatively late in the recovery period. As we can see from Fig. 1C, a continuous three-day net carbon emission was observed from DOY 116 to DOY 118 even though photosynthesis had already fully developed by this time. Recovery of photosynthesis to pre-frost levels was completed during the subsequent two to three days, and the recovery duration was dependent on the severity of frost.

Discussion

The timing and rate of recovery of photosynthetic capacity in spring influence the potential carbon balance of boreal and temperate forests. Several approaches have been used to define the seasonal start of ecosystem photosynthesis. Tanja et al. (2003) [11] defined the onset of photosynthesis in five boreal coniferous forests as the date when the half-hourly *NEE* first falls below 20% of the maximum summer *NEE*. Moore et al. (2006) [24] defined the onset of photosynthesis in a temperate bog as the date when the estimated daily photosynthesis (GPP) first becomes greater than zero. In the current study, the onset of photosynthesis in Korean pine forest was defined as a cumulative daytime *NEE* less than zero. This definition avoids the uncertainties in GPP estimations, considering potential errors in estimating daytime ecosystem respiration fluxes because of the light inhibition of leaf respiration [25]. Besides, without consideration of the high nighttime ecosystem respiration of forests with high biomass, NEE_d can more sensitively reflect the development of plant photosynthesis. It

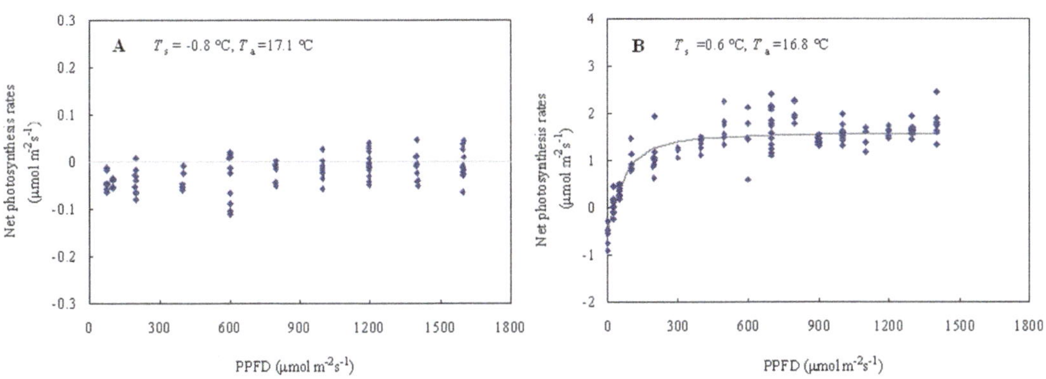

Figure 3. Light response curves for needles of Korean pine measured one day before soil thawing (A) and two days after the soil temperature was above freezing (B). PPFD is photosynthetic photon flux density, T_s is soil temperature and T_a is air temperature.

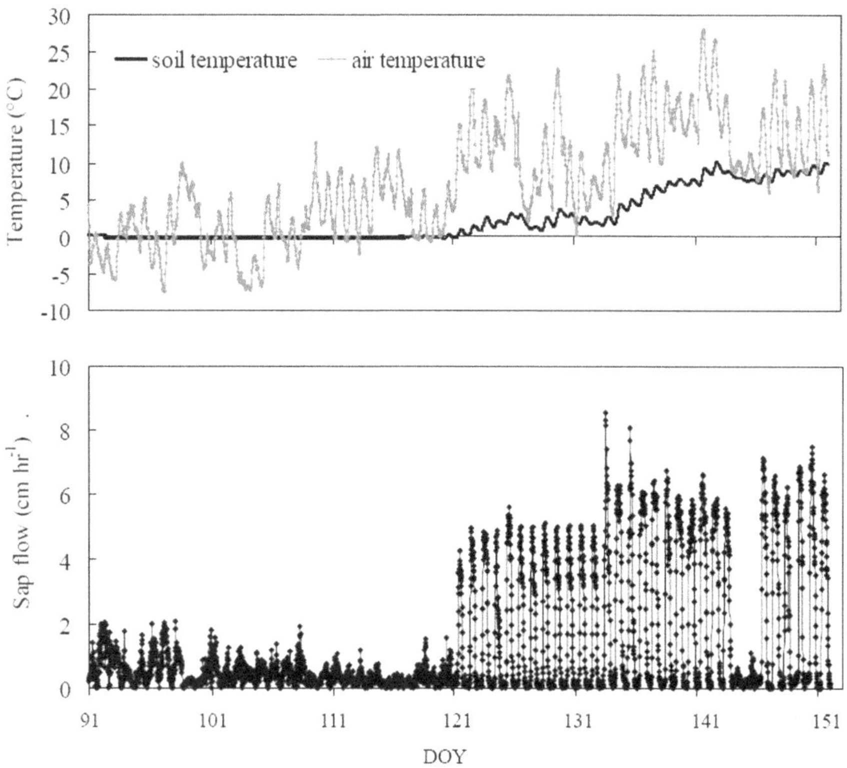

Figure 4. Seasonal evolution of air and soil temperature and sap flow rates measured in Korean pine stems in the early spring of 2010.

has to been mentioned that for this proposed method, the onset of photosynthesis will be detected only when the daytime photosynthesis will exceed daytime ecosystem respiration, which could lead to biases, depending on the strength of ecosystem respiration in the spring season. However, from another perspective, this method can effectively prevent the misjudgment caused by the disturbance of earlier onset of photosynthesis of early-spring herbs growing on the forest floor. According to the investigation conducted on April 21, 2005 [26], there were 12 herb species (including *Anemone amurensis*, *Adonis amurensis*, *Corydalis repens*, and so on) growing on the floor of this pine forest and their biomass was 588.7 kg·hm^{-2}.

Based on global scale correlations between NDVI and environmental factors, Schultz and Halpert (1993) [27] found that onset of spring photosynthesis is mainly influenced by temperature in cold regions, by both precipitation and temperature in temperate regions, and by precipitation in regions where the amplitude of the annual temperature cycle is small. Such responses have been verified by investigations at the ecosystem scale and individual plant level [24]. Overall, the site-based research results are generally consistent with regional scale findings, but each dataset presents special properties. Suni et al. (2003) [9] reported that air temperature triggers the recovery of evergreen boreal forest photosynthesis in spring. Ensminger et al. (2006) [28] also considered that air temperature is responsible for the onset of photosynthesis in spring. However, onset of photosynthesis in plants has been mostly attributed to soil temperature-related triggers. For example, by using eddy covariance technique, Hollinger et al. (1999) [29] found that significant ecosystem carbon uptake of an evergreen coniferous in the Howland Forest site began with the thawing of the soil in early April, and hence concluded that the threshold for the regulation of photosynthetic carbon assimilation is the spring thawing of forest soil. Control experiments on soil temperature also suggest that the timing of onset of photosynthesis is delayed by delayed soil thawing [30] and advanced by earlier soil thawing [31].

Table 2. The inhibitory effects of frost on recovery of photosynthetic capacity in Korean pine in the Changbai Mountains region over the four observation years.

Year	Date of occurrence of frost (DOY)	Time duration of onset of photosynthesis (days)	2 days pre-frost mean NEE_d ($gC \cdot m^{-2} \cdot d^{-1}$)	2 days post-frost mean NEE_d ($gC \cdot m^{-2} \cdot d^{-1}$)	Recovery duration (days)
2006	-	-	-	-	-
2007	116	6	−0.42	−0.08	2
2008	115	16	−0.72	0.25	3
2009	104	4	−0.49	0.19	2

In the current study, measurements from the eddy covariance and sap flow systems showed that no photosynthetic activity (Fig. 1) or root water supply for photosynthesis (Fig. 4) was available in Korean pine while the soil was frozen despite the daily mean air temperatures have reached and maintained above 10°C for several weeks. The sap flow measurements suggest that water uptake of roots is restricted when soil is frozen, limiting water flow into the trees and, ultimately, photosynthesis. Whilst photosynthetic measurements of pine needles with the LI-6400 portable photosynthesis system suggested that no photosynthesis occurred while the soil was frozen despite the air maintaining a temperature of over 15°C (Fig. 3). Immediate photosynthetic responses to soil thawing verified that soil temperature is essential for the onset of photosynthesis in temperate coniferous trees. This study case also suggests that once air temperatures permit photosynthesis, the availability of unfrozen soil water triggers the onset of pine photosynthesis. Similarly, Havranek and Tranquillini (1995) and Monson et al. (2005) [32–33] found that photosynthesis could not begin recovering until liquid water is available in the soil. Jarvis and Linder (2000) [16] argued that photosynthesis couldn't commence in evergreen trees until the soil has thawed because the onset of plant photosynthesis must be accompanied by water supply to leaves and it would be impossible for trees to take up water from frozen soil.

Fig. 1 shows that forests can carry out photosynthesis with air temperature at canopy height below 0°C after onset of needle photosynthesis. For example, the observed data on April 22, 2009 showed that the maximum air temperature was –0.42°C, while the daily NEE_d value was close to, but smaller than zero (Fig. 1D). Substantial reductions but not total inhibited in needles photosynthesis in early spring indicated that subfreezing daytime temperatures had a large impact on gas exchange because soil temperatures decreased slightly or did not change during this period. This finding further demonstrates that the importance of air temperatures on gas exchange after onset of photosynthesis is remarkable. However, from another perspective, it also suggests that it is soil temperature, rather than air temperature, that determines the onset of photosynthesis under natural field conditions. Reversible inhibitions of net photosynthesis have been observed in Scots pine seedlings in the field with temperatures below the freezing point of needles [34]. Monson et al. (2005) have noted that in evergreen pine and spruce forests the spring recovery can be reversible, after a cold spell [33]. However several studies also reported that temperate evergreen coniferous forest maintains photosynthetic carbon assimilation in winter [29,35–36], even in cool-temperature regions with cold continental winters [37]. Currently less is known about the factors trigger the switch on or off of tree photosynthesis in winter or early spring. Our results contribute to the limited literature on the photosynthetic phenology of temperate trees by showing that soil temperature is a primary factor determining the timing of onset of temperate conifer photosynthesis and that the rate of photosynthetic recovery is generally limited by low air temperatures.

One of the most significant findings from this study is that the recovery of photosynthetic capacity can be reversed by a frost event in evergreen coniferous forests. Such events frequently occur at the stage of spring photosynthetic recovery in boreal and temperate forests. Even a short-term frost event during the transitional periods from dormancy to growth in early spring could result in remarkable damage to the pine forests [38]. Recovery of net ecosystem carbon exchange rates of pine forest from frost is completed about two to three days after a temperature increase. Davidson et al. (2004) [39] found that recovery of photosynthesis to pre-frost levels of evergreen eucalyptus trees required around three consecutive frost-free nights and depended on the severity of frost. Both cases of recovery duration for evergreen trees are faster than the recovery of deciduous forests. For example, Oksanen et al. (2005) [40] reported that recovery of photosynthesis of birch forests from frost has not been completed even after 14 days. This finding is an interesting phenomenon worthy of further study in the future. A possible explanation for this different photosynthesis recovery ability is that structural injuries to the foliage of deciduous trees are more severe than that of evergreen trees.

Similar to other studies [41,42], the timing of onset of photosynthesis is a function of the accumulated temperature sum above a given threshold (Fig. 2), which suggests that the accumulation of heat controls onset of photosynthesis. The GDD is a good predictor of the onset of photosynthesis suggesting that the accumulation of heat controls onset of photosynthetic activity. However, it is not a general principle that plant photosynthetic recovery responses to a threshold GDD, considering that an arbitrary threshold temperature (e.g., 0, 5, and even 10°C) is used for calculating GDD in different studies. In addition, frost can reverse recovery in photosynthetic capacity [43,44], and, therefore, stochastic frost events can confound the simple relationship between photosynthetic recovery and GDD. In the current study, we chose to calculate GDD based on the beginning of the soil isothermal stage because this time corresponds to the possible availability of liquid water from thawing soil.

The timing of the spring thaw varies between years. In the current study, the difference of 19 days between the earliest (year of 2008) and the latest (year of 2006) soil thaw represents an increase of almost 13% in season length. It is consistent with the statement that the length of the northern hemisphere growing season may be increasing and this may affect carbon uptake of terrestrial ecosystem [45].

Conclusions

In the current study, the effects of soil and air temperature on the spring recovery of photosynthesis in pine forest were assessed in the field. The onset of photosynthesis in Korean pine was closely related to the end of soil thawing, and the rate of recovery of photosynthetic capacity was largely dependent on air temperature. The pine needles could start photosynthesis immediately after soil water became available for the roots. Intermittent frost events remarkably inhibited recovery of photosynthetic capacity in evergreen conifers during spring; this effect was reversible immediately after the temperature was raised back to former values. The results suggested that soil temperature is more important than air temperature in determining the onset of photosynthesis in Korean pine stands in temperate zones. The biochemistry of photosynthesis can apparently tolerate temperatures below 0°C. The findings verified that soil temperatures, rather than foliar biochemistry, is more important to the onset of photosynthesis of pine trees.

Author Contributions

Conceived and designed the experiments: JW DG. Performed the experiments: JW FY AW. Analyzed the data: JW CJ. Wrote the paper: JW DG.

References

1. Goulden ML, Wofsy SC, Harden JW, Trumbore SE, Crill PM, et al. (1998) Sensitivity of boreal forest carbon balance to soil thaw. Science 279: 214–217.
2. Ham JM, Knapp AK (1998) Fluxes of CO_2, water vapor, and energy from a prairie ecosystem during the seasonal transition from carbon sink to carbon source. Agric For Meteorol 89: 1–14.
3. White MA, Thornton PE, Running SW (1997) A continental phenology model for monitoring vegetation responses to interannual climatic variability. Global Biogeochem Cycles 11: 217–234.
4. Idso CD, Idso SB, Kimball BA, Park HS, Hoober JK, et al. (2000) Ultra-enhanced spring branch growth in CO_2 enriched trees: can it alter the phase of the atmospheres seasonal CO_2 cycle? Environ Exp Bot 43: 91–100.
5. Saxe H, Cannell MGR, Johnsen B, Ryan MG, Vourlitiset G (2001) Tree and forest functioning in response to global warming. New Phytol 149: 369–399.
6. Badger MR, Mjorkman O, Armond PA (1982) An analysis of photosynthetic response and adaptation to temperature in higher plants: Temperature acclimation in the desert evergreen Nerium oleander L. Plant Cell Environ 5: 85–99.
7. Hunter AF, Lechowicz MJ (1992) Predicting the timing of budburst in temperature trees. J Appl Ecol 29: 597–604.
8. Ensminger I, Schmidt L, Lloyd J (2008) Soil temperature and intermittent frost modulate the rate of recovery of photosynthesis in Scots pine under simulated spring conditions. New Phytol 177: 428–442.
9. Suni T, Berninger F, Vesala T, Markkanen T, Hari P, et al. (2003) Air temperature triggers the recovery of evergreen boreal forest photosynthesis in spring. Global Change Biol 9: 1410–1426.
10. Bergh J, McMurtrie RE, Linder S (1998) Climatic factors controlling the productivity of Norway spruce: a model-based analysis. For Ecol Manage 110: 127–139.
11. Tanja S, Berninger F, Vesala T, Markkanen T, Hari P, et al. (2003) Air temperature triggers the recovery of evergreen forest photosynthesis in spring. Global Change Biol 9: 1410–1426.
12. Sevanto S, Suni T, Pumpanen J, Grönholm T, Kolari P, et al. (2006) Wintertime photosynthesis and water uptake in a boreal forest. Tree Physiol 26: 749–757.
13. Ottander C, Öquist G (1991) Recovery of photosynthesis in winter stressed Scots pine. Plant Cell Environ 14: 345–349.
14. Ottander C, Campbell D, Öquist G (1995) Seasonal changes in photosystem II organisation and pigment composition in Pinus sylvestris. Planta 197: 176–183.
15. Schwarz PA, Fahey TJ, Dawson TE (1997) Seasonal air and soil temperature effects on photosynthesis in red spruce (*Picea rubens*) saplings. Tree Physiol 17: 187–194.
16. Jarvis P, Linder S (2000) Constraints to growth of boreal forests. Nature 405: 904–905.
17. IPCC (2007) Contribution of Working Group I to the Fourth Assessment Report of the Intergovernmental Panel on Climate Change, 2007. Eds: Solomon S, Qin D, Manning M, Chen Z, Marquis M, Averyt KB, Tignor M, Miller HL. Cambridge: Cambridge University Press.
18. Wu J, Guan D, Yuan F, Yang H, Wang A, et al. (2012) Evolution of atmospheric carbon dioxide concentration at different temporal scales recorded in a tall forest. Atmos Environ 61: 9–14.
19. Wilczak JM, Oncley SP, Sage SA (2001) Sonic anemometer tilt correction algorithms. Bound-Layer Meteorol 99: 127–150.
20. Moore CJ (1986) Frequency response corrections for eddy correlation systems. Bound-Lay Meteorol 37: 17–35.
21. Webb EK, Pearman GI, Leuning R (1980) Correction of flux measurements for density effects due to heat and water vapor transfer. Q J R Meteorol Soc 106: 85–100.
22. Falge E, Baldocchi D, Olson R, Anthoni P, Aubinet M, et al. (2001) Gap filling strategies for defensible annual sums of net ecosystem exchange. Agric For Meteorol 107: 43–69.
23. Foken T, Wichura B (1996) Tools for quality assessment of surface based flux measurements. Agric For Meteorol 78: 83–105.
24. Moore TR, Lafleur PM, Poon DMI, Heumann BW, Seaquist JW, et al. (2006) Spring photosynthesis in a cool temperate bog. Global Change Biol 12: 2323–2335.
25. Janssens IA, Lankreijer H, Matteucci G, Kowalskil AS, Buchmann N, et al. (2001) Productivity overshadows temperature in determining soil and ecosystem respiration across European forests. Global Change Biol 7: 269–278.
26. Zhou Y, Yin Y, Yu F, Hao Z, Ye J (2007) Nutrient uptake and return of early-spring herbs in broadleaved Korean pine forest in Changbai Mountain. Chin J Ecol 26: 793–796.
27. Schultz PA, Halpert MS (1993) Global correlation of temperature, NDVI and precipitation. Adv Space Res 13: 277–280.
28. Ensminger I, Busch F, Huner N (2006) Photostasis and cold acclimation: sensing low temperature through photosynthesis. Physiol Plant 126: 28–44.
29. Hollinger DY, Goltz SM, Davidson EA, Lee JT, Tu K, et al. (1999) Seasonal patterns and environmental control of carbon dioxide and water vapour exchange in an ecotonal boreal forest. Global Change Biol 5: 891–902.
30. Strand M., Lundmark T, Söderbergh I, Melander PE (2002) Impacts of seasonal air and soil temperatures on photosynthesis in Scots pine trees. Tree Physiol 22: 839–847.
31. Bergh J, Linder S (1999) Effects of soil warming during spring on photosynthetic recovery in boreal Norway spruce stands. Global Change Biol 5: 245–253.
32. Havranek WM, Tranquillini WT (1995) Physiological processes during winter dormancy and their ecological significance. In: Ecophysiology of Coniferous Forests. Eds: Smith WK, and Hinckley TM. Academic Press, New York, 95–124.
33. Monson RK, Sparks JP, Rosenstiel TN, Scott-Denton LE, Huxman TE, et al. (2005) Climatic influences on net ecosystem CO_2 exchange during the transition from wintertime carbon source to springtime carbon sink in a high-elevation, subalpine forest. Oecologia 146: 130–147.
34. Lundmark T, Hällgren JE, Hedén J (1988) Recovery from winter depression of photosynthesis in pine and spruce. Trees 2: 110–114.
35. Ellsworth DS (2000) Seasonal CO_2 assimilation and stomatal limitations in a Pinus taeda canopy. Tree Physiol 20: 435–445.
36. Yuan F, Arain MA, Barr AG, Black TA, Bourque CPA, et al. (2008) Modeling analysis of primary controls on net ecosystem productivity of seven boreal and temperate coniferous forests across a continental transect. Global Change Biol 14: 1765–1784.
37. Saitoh TM, Tamagawa I, Muraoka H, Lee NY, Yashiro Y, et al. (2010) Carbon dioxide exchange in a cool-temperate evergreen coniferous forest over complex topography in Japan during two years with contrasting climates. J Plant Res 123(4): 473–483.
38. Zheng Y, Yang Q, Xu M, Chi Y, Shen R, et al. (2012) Responses of Pinus massoniana and Pinus taeda to freezing in temperate forests in central China. Scand J Forest Res 27: 520–531.
39. Davidson N, Battaglia M, Close D (2004) Photosynthetic responses to overnight frost in Eucalyptus nitens and E. globules. Trees 18: 245–252.
40. Oksanen E, Freiwald V, Prozherina N, Rousi M (2005). Photosynthesis of birch (*Betula pendula*) is sensitive to springtime frost and ozone. Can J For Res 35: 703–712.
41. Hannerz M (1999) Evaluation of temperature models for predicting bud burst in Norway spruce. Can J For Res 29: 1–11.
42. Linkosalo T (2000) Mutual regularity of spring phenology of some boreal tree species: predicting with other species and phenological models. Can J For Res 30: 667–673.
43. Delucia EH, Smith WK (1987) Air temperature limitations on photosynthesis in Englemann spruce during summer. Can J For Res 17: 527–533.
44. Ensminger I, Sveshnikov D, Cambell DA, Funks C, Jansson S, et al. (2004) Intermittent low temperatures constrain spring recovery of photosynthesis in boreal Scots pine forests. Global Change Biol 10: 995–1008.
45. Myneni RB, Keeling CD, Tucker CJ, Asrar G, Nemani RR (1997) Increased plant growth in the northern high latitudes from 1981 to 1991. Nature 386: 698–702.

Interactive Effects of Nitrogen and Phosphorus on Soil Microbial Communities in a Tropical Forest

Lei Liu[1,2], Tao Zhang[1], Frank S. Gilliam[3], Per Gundersen[4], Wei Zhang[1], Hao Chen[1], Jiangming Mo[1]*

1 Key Laboratory of Vegetation Restoration and Management of Degraded Ecosystems, South China Botanical Garden, Chinese Academy of Sciences, Guangzhou, China, **2** State Key Laboratory of Urban and Regional Ecology, Research Center for Eco-Environmental Sciences, Chinese Academy of Sciences, Beijing, China, **3** Department of Biological Science, Marshall University, Huntington, West Virginia, United States of America, **4** Danish Centre for Forest, Landscape and Planning, University of Copenhagen, Copenhagen, Denmark

Abstract

Elevated nitrogen (N) deposition in humid tropical regions may exacerbate phosphorus (P) deficiency in forests on highly weathered soils. However, it is not clear how P availability affects soil microbes and soil carbon (C), or how P processes interact with N deposition in tropical forests. We examined the effects of N and P additions on soil microbes and soil C pools in a N-saturated old-growth tropical forest in southern China to test the hypotheses that (1) N and P addition will have opposing effects on soil microbial biomass and activity, (2) N and P addition will alter the composition of the microbial community, (3) the addition of N and P will have interactive effects on soil microbes and (4) addition-mediated changes in microbial communities would feed back on soil C pools. Phospholipid fatty acid (PLFA) analysis was used to quantify the soil microbial community following four treatments: Control, N addition (15 g N m^{-2} yr^{-1}), P addition (15 g P m^{-2} yr^{-1}), and N&P addition (15 g N m^{-2} yr^{-1} plus 15 g P m^{-2} yr^{-1}). These were applied from 2007 to 2011. Whereas additions of P increased soil microbial biomass, additions of N reduced soil microbial biomass. These effects, however, were transient, disappearing over longer periods. Moreover, N additions significantly increased relative abundance of fungal PLFAs and P additions significantly increased relative abundance of arbuscular mycorrhizal (AM) fungi PLFAs. Nitrogen addition had a negative effect on light fraction C, but no effect on heavy fraction C and total soil C. In contrast, P addition significantly decreased both light fraction C and total soil C. However, there were no interactions between N addition and P addition on soil microbes. Our results suggest that these nutrients are not co-limiting, and that P rather than N is limiting in this tropical forest.

Editor: Ben Bond-Lamberty, DOE Pacific Northwest National Laboratory, United States of America

Funding: This study was funded by National Natural Science Foundation of China (No 41273143) and the National Key Basic Research 973 Programme (2010CB833502). P. Gundersen received support from the Water Research Initiative (ViVa), the Faculty of Life Sciences, University of Copenhagen. The funders had no role in study design, data collection and analysis, decision to publish, or preparation of the manuscript.

Competing Interests: The authors have declared that no competing interests exist.

* E-mail: mojm@scib.ac.cn

Introduction

Biogeochemical cycling of nitrogen (N) is greatly altered by anthropogenic activities, with global cycling rates estimated to have increased by approximately 100% since mid-1900s [1]. In Asia, emissions of reactive N have increased dramatically [2], leading to deposition of 30–73 kg N ha^{-1} yr^{-1} in some subtropical forests of southern China [3]. Such high rates of N deposition are comparable to the highest levels of N deposition occurring in forests of North America and Europe [4,5] where adverse effects of excess N, including soil acidification, nutrient imbalance, nitrate leaching, loss of biodiversity, and even forest decline, have been reported in some forests [6–9].

In contrast to temperate forests, which are often N-limited under natural conditions, tropical forests more typically exhibit phosphorus (P) limitation, with soils often highly acidic and low in base cation availability [10,11]. Most fertilization experiments conducted in tropical forests have demonstrated a stronger response of plant biomass to added P than to added N, supporting the paradigm that tropical ecosystems on old soils are predominantly P limited [1,12]. Cleveland *et al.* (2011) conducted a meta-analysis of carbon (C)-nutrient-climate relationships in 113 sites across the tropical forest biome and found that P availability regulated net primary production and other ecosystem processes in lowland tropical forests [13]. In contrast to these long-standing ideas, however, recent studies have demonstrated that N, P and K or base cations all limit forest plants growth in tropical forests [14,15].

Elevated atmospheric N deposition may further enhance such P limitation in tropical forests. Some tropical forests of the high deposition region in southeastern China are indeed N saturated and exhibit high rates of nitrate leaching [3], particularly an old-growth forest of the Dinghushan Biosphere Reserve (DHSBR), which supports representative natural vegetation in the region and receives high N deposition. We have reported that up to 60 kg N $ha^{-1}yr^{-1}$ is leached from the old-growth forest, and that Al^{3+} concentrations increased and the soil pH decreased significantly in high N addition plots [16,17,18]. Further work demonstrated that the herbaceous layer at this site is quite sensitive to N deposition [19]. A 30-yr time series of plant chemistry and production in this forest revealed signs of progressive P limitation, including decreases in available soil P, increases in N/P ratios in leaves and litterfall and in litterfall amount, and decreases in aboveground primary production [20]. Another study in this forest indicated an increase in the soil organic carbon (SOC) over the same period [21].

The occurrence of opposite trends in plant production and in SOC stock at DHSBR may only be explained by major changes in soil organic matter (SOM) quality and turnover, i.e., that increased N availability and decreased P availability have reduced the decomposition rate of SOM. Although work at DHSBR demonstrated decreases in litter decomposition, soil respiration, and soil microbial biomass with experimental N addition [22–25], less is known about the effects of P and particularly about N/P interactions on these processes.

Soil microorganisms are the drivers of decomposition processes, but soil nutrient availability may influence soil microbial growth and activity [26]. Nutritional constraints linked to scarcity of available P can shape soil microbial community composition in highly weathered soils [27]. Gallardo and Schlesinger (1994) suggested that microbial P limitation may be common in highly weathered soils in which P tends to be bound in iron or aluminum sesquioxides [28]. Experimental P additions in tropical rain forests have revealed that microbial utilization of at least labile fractions of soil organic carbon (SOC) was P limited [29]. Moreover, N- and P-induced shifts in microbial communities should cause corresponding shifts in the functional and metabolic potentials of the communities, resulting in a change in decomposition rates [30]. Recently, Cusack *et al.* (2011a) reported links between microbial responses to N deposition and shifts in SOM quality and quantity in two tropical forests [31].

As we known, understanding nutrient limitation is a key to predicting how the C cycle will respond to environmental change [32]. However, identifying nutrient limitation in tropical forests and resolving its importance therein is a complex undertaking. Especially, the effects of nutrient interactions on C cycle in tropical forests are poorly known. Understanding of the effects of N and P as well as their interactions on C processes and soil microbial communities in tropical forests, we established a full factorial N and P addition experiment at DHSBR in 2007 from which some short-term results for the P addition was presented by Liu *et al.* [33].

In this paper we report the long-term (>4 year) changes in microbial biomass, activity and community structure to the N-P treatments. The purpose of this study is to explore potential feedback mechanisms and microbial shifts due to N-P interactions that may explain why SOC accumulates at DHSBR. Our hypotheses are that:i) N and P addition will have opposite effects (decrease and increase, respectively) on soil microbial biomass and activity; ii) the addition of N and P will have interactive effects on soil microbes; iii) the composition of the microbial community will change after N and/or P addition and iv) such changes in microbial communities would feed back on soil carbon pools.

Materials and Methods

Ethics statement

No specific permits were required for the described field studies. This research station (Dinghushan Biosphere Reserve) belongs to South China Botanical Garden, Chinese Academy of Sciences. This study also supported by this institute. We confirmed that the location is not privately-owned or protected in any way. We also confirmed that the field studies did not involve endangered or protected species.

Site description

This study was conducted in the 1200 ha Dinghushan Biosphere Reserve (DHSBR), located in the central part of Guangdong Province, southeastern China (112°10′ E, 23°10′ N). Among the forest types at DHSBR is an old-growth monsoon evergreen broadleaf forest, which is about 250–300 m above sea level and occupies 20% of the reserve. This forest is typical of undisturbed forests of tropical China, having been protected by monks for more than 400 years and experiencing minimal direct human impacts [34]. Major plant species include *Castanopsis chinensis* Hance, *Schima superba* Chardn. & Champ., *Cryptocarya chinensis* (Hance) Hemsl., *Cryptocarya concinna* Hance, *Machilus chinensis* (Champ. Ex Benth.) Hemsl., *Syzygium rehderianum* Merr. & Perry in the tree layer and *Calamus rhabdicladus* Burret, *Ardisia quinquegona* Bl. and *Hemigramma decurrens* (Hook.) Copel in the understory layer [19]. Mean annual litter biomass production is ~8.3 Mg ha^{-1} yr^{-1} [35]. Stem density, tree height and mean diameter at breast height are summarized in Table 1.

The reserve has a typical monsoon and humid climate (*sensu* Holdridge, 1967), with an average annual precipitation of 1927 mm that exhibits a distinct seasonal pattern with 75% falling from March to August and only 6% falling from December to February [36]. The mean annual temperature is 21 °C, with minimum monthly mean temperature of 12.6 °C in January and maximum of 28.0 °C in July; annual mean relative humidity is 80% [36]. Inorganic N deposition measured in throughfall was 33 kg N $ha^{-1}yr^{-1}$, with an additional input as dissolved organic N at 15–20 kg N ha^{-1} yr^{-1} [16].

Soil in the reserve is lateritic red earth formed from sandstone [34]. The soil depth in the old-growth forest is more than 60 cm to the top of the C horizon [34]. The forest stands used in the experiment are situated on mountain slopes ranging from 15°–35°. General soil chemical properties are listed in Table 2.

Experimental treatment

In 2007, four treatments were established in five replicates each: Control, N-addition (15 g N m^{-2} yr^{-1}), P-addition (15 g P m^{-2} yr^{-1}), and NP-addition (15 g N m^{-2} yr^{-1} plus 15 g P m^{-2} yr^{-1}). Each of the 20 plots were 5 m×5 m were established and surrounded by a 5-m wide buffer strip. Plots size and fertilizer level were similar to those in the experiment in Costa Rica by Cleveland and Townsend [37]. Field plots and treatments were laid out randomly. Applications of N and P were made as NH_4NO_3 and NaH_2PO_4 solutions sprayed in two monthly portions below the canopy with a backpack sprayer starting from January 2007 and continuing through June 2011. Fertilizer was weighed and mixed with 5 L of water for each plot. Each control plot received 5 L of water without fertilizer.

Field sampling and measurements

Soil sampling was conducted in June 2011. From each plot, 5 soil cores (2.5 cm inner diameter) were collected randomly from a 10-cm soil depth and combined to one composite sample. The litter layer was carefully removed before sampling. After removing stones and coarse roots, soils were sieved to 2 mm mesh size and divided into two parts, one retained for measuring soil chemical parameters and the other for analysis of microbial biomass and community structure. At the same time, an *in situ* soil-core technique was used to estimate net nitrification rates [38].

Soil moisture content was measured gravimetrically using 10 g of field moist soil sample oven dried at 105 °C for 24 h. Soil pH was measured in a 1:2.5 soil/water suspension. Soil light-fraction carbon (C) and heavy-fraction C were separated by gravity following the method described by Janzen *et al.* [39] and modified

Table 1. Indices of the old-growth tropical forest at Dinghushan Biosphere Reserve.

Species	Stem density (tree ha^{-1})	Mean height (m)	Mean diameter at breast height (cm)	Basal area (m^2 ha^{-1})
Castanopsis chinensis	268	9.8	26.0	18.7
Machilus chinensis	131	9.0	14.8	4.0
Schima superba	185	9.9	18.3	6.4
Cryptocarya chinensis	270	8.3	14.3	4.4
Syzygium rehderianum	185	8.5	12.9	1.2
Other plants	1587	4.3	4.4	3.5
Total	2625			38.2

Survey was conducted in February 2007 (before the start of N and P fertilization).

by Izaurralde *et al.* [40]. Soil organic C (SOC) and C content of the light-fraction and heavy-fraction were determined by dichromate oxidation and titration with ferrous ammonium sulfate. Dissolved organic carbon (DOC) in filtered 0.5 M K_2SO4-extracts of fresh soil sample was measured with a TOC analyser (TOC-VCPH Shimadzu Corp., Japan). NH_4^+-N and NO_3^--N in filtered 2 M KCL-extracts of fresh soil sample were measured with a flow injection autoanalyzer (FIA, Lachat Instruments, USA). Available P concentration was analyzed colorimetrically after acidified ammonium persulfate digestion [41].

Soil microbial biomass C (MBC) and microbial biomass N (MBN) were estimated by chloroform fumigation-extraction [42]. Soil respiration was measured using the static chamber and gas chromatography techniques [33]. Soil microbial biomass and community structure was characterized using phospholipid fatty acid (PLFA) analysis as described by Bossio and Scow [43]. The abundance of individual fatty acids was determined as nmol per g of dry soil and standard nomenclature was used [44]. Concentrations of each PLFA were calculated based on the 19:0 internal standard concentrations. Frostegård and Bååth [45] chose a set of fatty acids to represent bacterial PLFAs, out of which i14:0, 15:0, i15:0, a15:0, i16:0, 16:1ω7c, 17:0, a17:0, i17:0, cy17:0, 18:1ω7 and cy19:0 were present in our samples. We calculated the sum of i14:0, i15:0, a15:0, i16:0, a17:0 and i17:0 as an indicator of gram-positive bacteria. In our study, Gram-negative bacteria were identified by the PLFAs: 16:1ω7c, cy17:0, 18:1ω7 and cy19:0 [46]. The fungi were identified by the PLFA 18:2ω6,9c [47], and PLFAs 16:1ω5c were used as a marker for arbuscular mycorrhizal fungi (AMF) [48]. The actinomycetes were identified by the PLFAs 10Me 16:0, 10Me 17:0 and 10Me 18:0 [49]. Other PLFAs such as 14:0, 16:0, 16:1 2OH, 16:1ω9c, 17:1ω8c, 18:1ω9c, 10Me 19:0, 18:3ω6c and 20:1ω9c were also used to analyze the composition of microbial community. The ratio of 18:2ω6,9c to total bacterial PLFAs was used to estimate the ratio of fungal to bacterial biomass (F: B) in soils [45,50].

Statistical analysis

Two-way ANOVA [Analysis of Variance, PROC GLM from SAS for Windows version 8] was used to examine the difference in soil chemical characteristics, microbial biomass, and F: B ratios among treatments. Twenty-six PLFAs were detected, identified, and included in principal component analysis (PCA) after standardisation for equal unit variance. Redundancy analysis (RDA) was used to test the relationship between soil microbial community (26 PLFAs) and environmental variables. Statistical significance tests for PCA and RDA were run using CANOCO software for Windows 4.5 (Microcomputer Power, Ithaca, NY, USA). Forward selection was based on Monte Carlo permutation (n = 499). Statistically significant differences were identified as $P<0.05$ unless otherwise stated.

Results

Soil chemical properties

By the time of sampling in June 2011, N and P treatments had significantly altered soil nutrients relative to the control (Table 2). Available P and soil pH were significantly elevated in the P-amended plots. Soil NH_4^+ concentrations did not differ among

Table 2. Soil chemical properties after fertilization treatments, measured in June 2011.

Treatments	Control	N	P	NP	Two-way anova		
					N	P	N*P
pH (H_2O)	3.69 (0.02) a	3.66 (0.03) a	4.00 (0.03) c	3.86 (0.04) b	**	***	ns
NH_4^+-N (mg kg^{-1})	10.7 (1.0) a	9.0 (0.6) a	9.2 (1.3) a	8.1 (0.7) a	ns	ns	ns
NO_3^--N (mg kg^{-1})	3.6 (0.4) a	4.6 (0.7) a	1.9 (0.2) b	3.1 (0.6) ab	*	**	ns
Available P (mg kg^{-1})	2.1 (0.1) a	4.2 (0.8) a	28.0 (2.8) b	21.3 (1.6) c	ns	***	ns
SOC (%)	5. 0(0.3) a	4. 6 (0.3) ab	3.9 (0.2) b	4.5(0.2) ab	ns	*	ns
DOC (mg kg^{-1})	209 (12) a	183 (7) a	185 (7) a	184 (5) a	ns	ns	ns

Notes: SOC (%) and DOC (mg kg^{-1}) stand for soil organic carbon and dissolved organic carbon, respectively. Values are means with SE in parentheses (N = 5). Values followed by different letters are significantly different among treatments with $p<0.05$. *, **, *** significance at the level of 0.05, 0.01 and 0.001, respectively.

treatments irrespective of the N-addition. Moreover, soil NO_3^- concentrations decreased in the P-addition plots. In addition, SOC decreased significantly in the P-addition plots, but not in the N-addition plots. There were no significant interactions between N and P additions. Light-fraction C in soils was significantly lower in both N-addition and P-addition plots than in the control plots (Fig. 1). Soil nitrification rate was significantly lower in the N-amended plots (1.6 and 0.6 mg N kg^{-1} mo^{-1} in N-addition plots and NP-addition plots) than in the control plots (4.4 mg N kg^{-1} mo^{-1}).

Soil microbial biomass and activity

Although both MBC and MBN decreased significantly by N addition after 18-month, there were no significant differences in MBC between treatments over the long-term, 52-month period (Fig. 2). Moreover, there were no interactions between N addition and P addition on soil microbial biomass (Appendix S1). Total microbial biomass, bacterial biomass and fungal biomass were also not significantly different among treatments (Fig. 3). However, MBN was significantly lower in the N-addition plots than in the P-addition plots in June 2011 (Fig. 2). Fertilization with P significantly increased soil respiration, and there was no effect of N fertilization (Fig. 4).

Soil microbial community structure

Mean abundances of gram-negative bacteria and arbuscular mycorrhizal (AM) PLFAs were significantly higher in the P-addition plots, whereas the mean abundances of fungal PLFAs was significantly lower in the control plots compared with the N-amended plots (Fig. 5). Fungal: bacterial ratios were also significantly lower in the control plots than in the N-addition plots (Fig. 3). Mean abundance of gram-positive bacteria and actinomycetes PLFAs did not differ among treatments, although relative abundances of individual PLFA 10Me 16:0, 10Me 17:0, and 10Me 18:0 (all actinomycete markers) were significantly different among treatments. Relative abundance of 10Me 17:0 was higher in the N-addition plots than in the P-addition plots, whereas 10Me 18:0 was lower in the N-addition plots compared with the P-addition plots. Relative abundance of 10Me 16:0 was higher in the NP-addition plots than in the P-addition plots (Fig. 5).

The first two axes produced by principal components analysis accounted for 63.6% of the total variation in the PLFA profile, with PC1 explaining 37.1% of the variation and separating the P addition versus non-P addition plots ($P<0.001$, Appendix S1, S2). Loadings of individual PLFAs showed that the samples in the P-addition plots were characterized mainly by high concentrations of the monounsaturated PLFAs 16:1w7c, 16:1w9c, 18:1w7c (indicated Gram-negative bacteria), and 16:1w5c (indicated AM fungi) (Appendix S3). These results were also confirmed by PLFA relative abundance data.

The correlations between soil microbial community and soil chemical properties revealed that six soil variables, including pH, soil organic carbon (SOC), NO_3^-, NH_4^+, available P, soil moisture content (SMC), explained 45.4% of the variation in soil microbial community composition. Soil microbial community structure was significantly related to NO_3^-, available P and SOC, with axis 1 explaining 29.3% of the variance and axis 2 explaining 16.1% (Appendix S4).

Discussion

Soil microbial biomass and community structure

Effects of N addition. Nitrogen additions decreased soil microbial biomass in the short term but returned to pre-treatment levels over the long term. This was an unexpected effect of long term N addition and was likely due to microbial populations adapting to the higher N conditions created by 52 months of N additions. One possible cause of decreased microbial biomass C is soil acidification resulting from NH_4^+ uptake by plants, nitrifica-

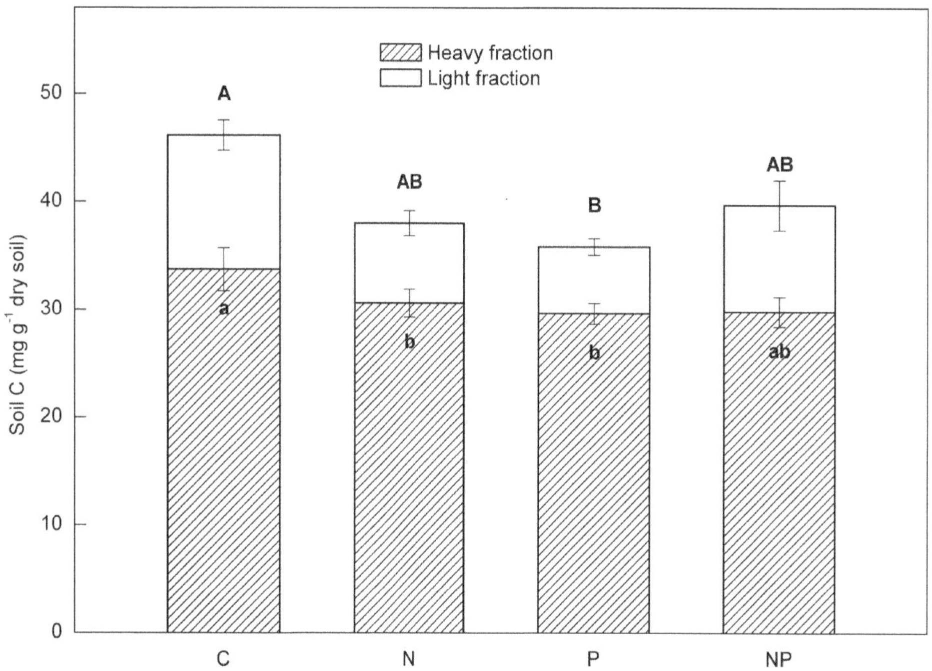

Figure 1. The contribution of two soil fractions to total soil C (Jun. 2011). C: control, N: nitrogen addition, P: phosphorus addition, NP: nitrogen and phosphorus addition. Significant differences ($p<0.05$) among treatments are indicated by different letters. Error bars show SE ($n=5$).

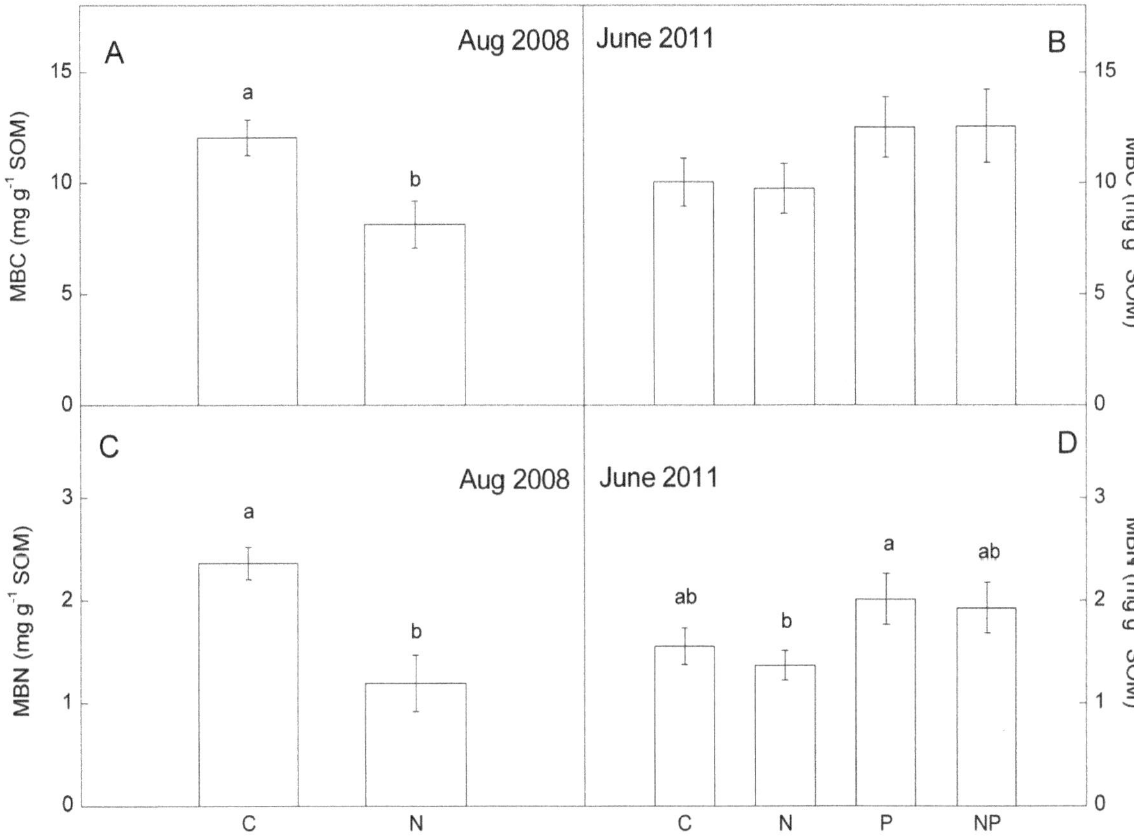

Figure 2. Microbial biomass in soils after nitrogen and phosphorus addition. Data from Aug. 2008 and June 2011. MBC: microbial biomass carbon; MBN: microbial biomass nitrogen. C: control, N: nitrogen addition, P: phosphorus addition, NP: nitrogen and phosphorus addition. Significant differences ($p<0.05$) among treatments are indicated by different letters. Error bars show SE ($n=5$).

tion of NH_4^+ in soils, and NO_3^- leaching [51]. In the present study, soil pH was already very low in the control plots and there was no significant difference between control and long term N addition plots (Table 2), suggesting that soil pH may not be affected by N addition under this N-saturated system. Balser (2001) also found that long term N addition had no effect on microbial biomass in three Hawaiian forest soils [52]. In contrast, N addition in temperate forests typically increases microbial biomass at a short term basis [53] but over the longer term, biomass generally decreases [51,54–56].

Unlike soil microbial biomass, effects of experimental inputs of N to forests on soil microbial communities are inconsistent in the literature. Nilsson *et al.* (2007) reported that N input had no effect on total fungal biomass in the soils of oak forests along a natural N deposition gradient [57]. Several studies have shown that, fungal biomarkers were reduced in soils of N-fertilized plots [58,59]. In contrast, increases in fungal biomarkers in N manipulation experiments have been reported in north temperate forest soils [60]. A common element to these studies is that they were carried out in temperate forests. Less is known, however, about how N deposition alters microbial community in tropical regions. In the present study, although there was no effect of long term N addition on soil microbial biomass, microbial community structure contrasted notably in the control versus N-addition plots, largely due to N-mediated increases in relative abundance of fungal PLFAs. Consequently, F: B ratios were significant higher in the N-addition plots comparing to the control plots (Figs.3–5). This is consistent with results reported by Balser (2001), which showed that microbial community composition was affected by long term N addition in three Hawaiian forest soils [52].

The induced changes in microbial communities was also supported by the fact that Gram negative bacteria (e.g., 16:1w7 and 18:1w7), which include several nitrifying species, decreased significantly in N-addition plots compared to the control plots, corresponding to lower net nitrification in the N-addition plots comparing to the control plots (Figs. 4,5). Gilliam *et al.* (2011) found that the Gram-negative PLFA 18:1n7c (Gram negative bacteria) was predominant in soils with highest rates of net nitrification soils at Fernow Experimental Forest, a central Appalachian hardwood forest of West Virginia, USA [26]. Billings and Ziegler (2008) observed an increase in the activity of Gram negative bacteria for pine forest soil fertilized with NH_4NO_3 [61]. Notably, N addition actually decreased net nitrification and relative abundance PLFA 16:1w7c and 18:1w7c, which include NH_4-oxidizing and NO_2-oxidizing Gram negative bacteria.

Effects of P addition. In contrast to effects of N, addition of P increased soil microbial biomass, indicating that P availability is a limiting factor for microbial growth in this old-growth tropical forest. This stimulating effect of P on microbial biomass, however, was also transient [33] (Fig. 2). Soil respiration was significantly increased after long-term P addition (Fig 4), suggesting that microbial activity is higher in the presence of addition P, allowing for more rapid transformation of soil organic matter. In a previous study, we found a similar response, concluding that it arose from enhanced P-mediated increases in C availability to microbes in

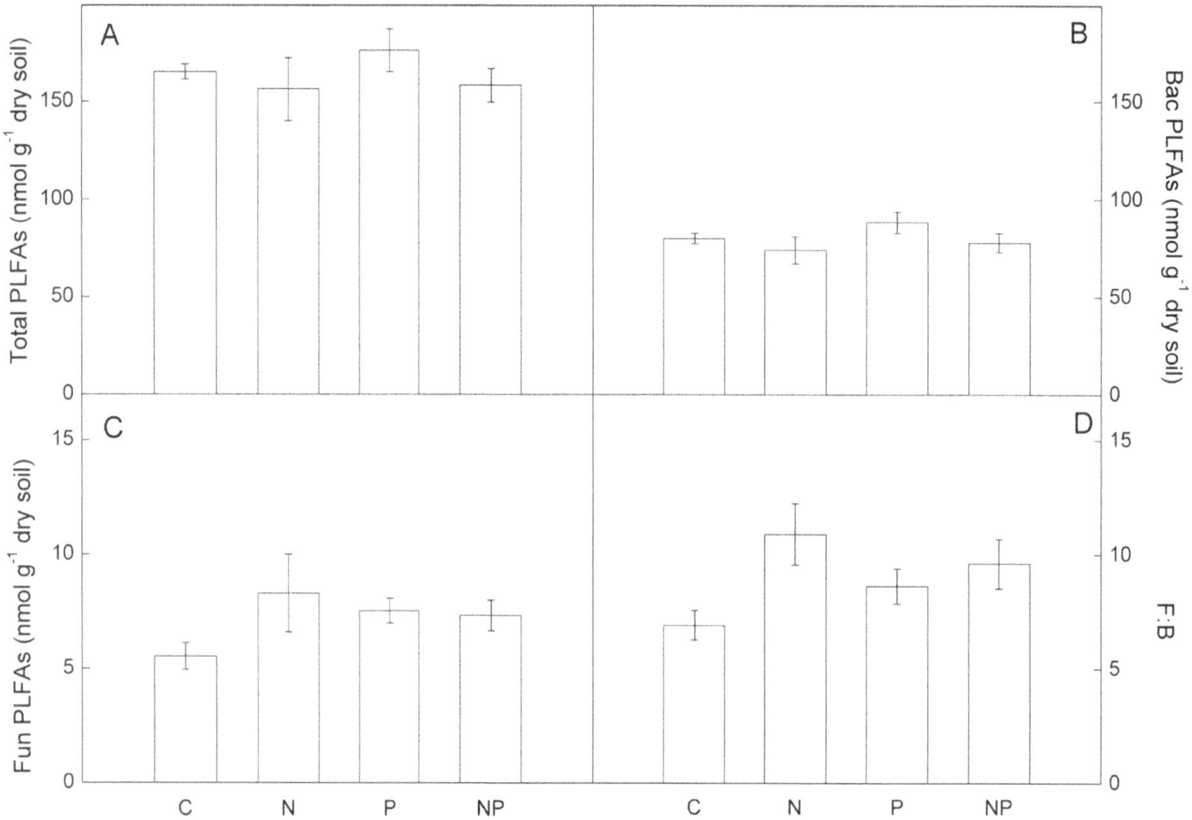

Figure 3. Comparisons of soil microbial PLFAs between treatments. Data from June 2011. C: control, N: nitrogen addition, P: phosphorus addition, NP: nitrogen and phosphorus addition. Bac PLFAs: Bacterial PLFAs; Fun PLFAs: Fungal PLFAs; F:B: the ratio of fungal to bacterial PLFAs. Significant differences ($p<0.05$) among treatments are indicated by different letters. Error bars show SE ($n=5$).

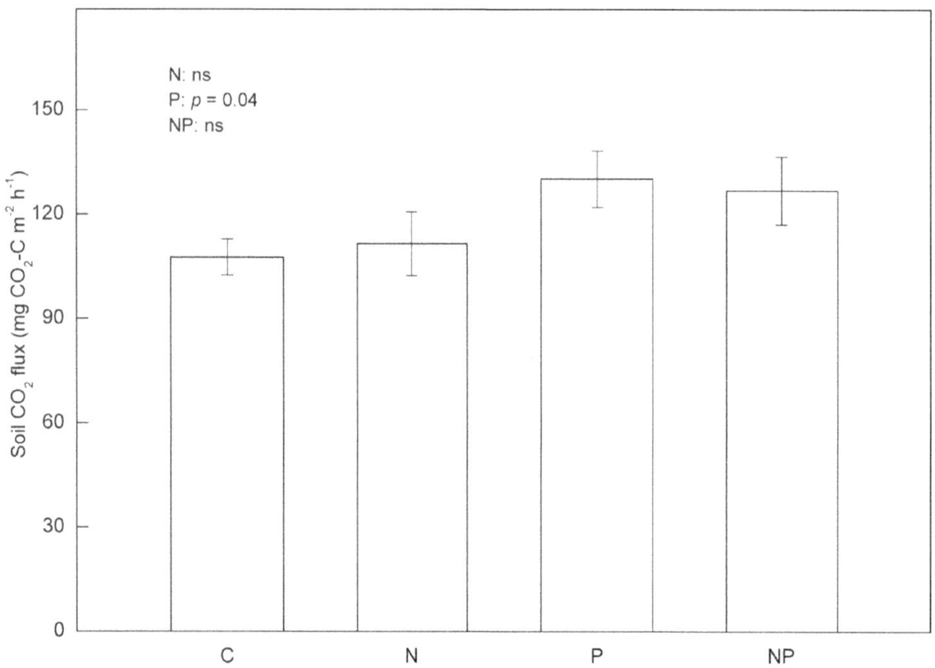

Figure 4. Comparisons of soil respiration between treatments. Values are means for three months. (Data from May-July 2011). C: control, N: nitrogen addition, P: phosphorus addition, NP: nitrogen and phosphorus addition. Error bars show SE ($n=5$).

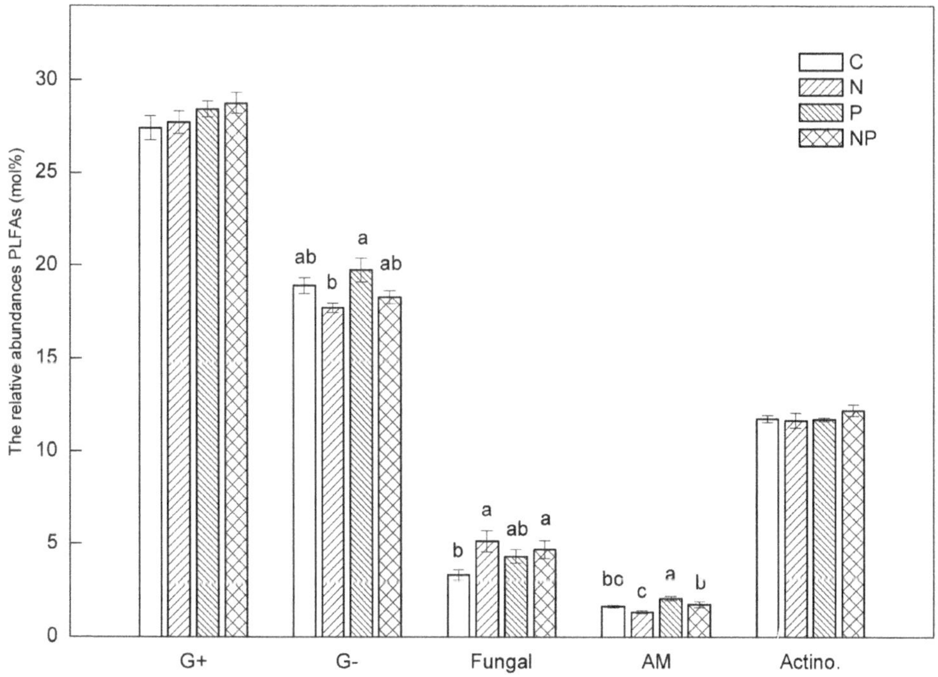

Figure 5. The relative abundances of the individual PLFAs (mol %) in soil samples. G^{+}: the proportion of gram-positive bacterial PLFAs; G^{-}: the proportion of gram-negative bacterial PLFAs; Fungi: the proportion of fungal PLFAs; AM: the proportion of AM fungal PLFAs; Actino.: the proportion of actinomycetes PLFAs. Significant differences ($p<0.05$) among treatments are indicated by different letters. Error bars show SE ($n=5$).

soils of this forest [33], thus constraining the microbial processes that relate to SOM processing.

Addition of P altered microbial community structure by significantly increasing the relative abundance of AM fungi, which obtain C from their host plants in return for mineral nutrients [62]. With N or P limiting to plant growth, plant will invest more C to AM fungi in exchange for nutrients [63]. Conversely, if N or P availability rises, a decline in AM abundance is expected [64]. However, our result is inconsistent with the expectation that P abundance suppresses plant investment in the mycorrhizal symbiosis. One possible reason was that AM fungi were initially nutrient-limited. Because this old-growth forest is N saturated and with low P availability. P additions may increase mycorrhizal growth. Since mycorrhizal fungi are more efficient scavengers for nutrients from the soil than are plant roots, the threshold for nutrient limitation may be lower for mycorrhizal fungi than for plants [65,66]. Another possible reason was increased in soil pH after P addition, as this is often associated with increased AM biomass in soil (PLFA 16:1ω5c) [67,68].

Several studies also reported that apatite addition had a significant positive influence on fungal growth in P-poor forests, but not in forests with sufficient P [69,70]. However, Groffman and Fisk (2011) found that P addition had no effect on microbial biomass and activity in a northern hardwood forest (Hubbard Brook Experimental Forest, New Hampshire, USA), concluding that P did not limit microbial biomass and respiration [71]. In contrast, the whole-watershed Ca treatment at Hubbard Brook enhanced AM colonization [72]. Kaspari *et al.* (2008) and Wright *et al.* (2011) have reported that K-limitation of microbial decomposition and N, P and K all limit a suite of ecosystem processes in a tropical forest in Panama [13,73]. These studies reveal that tropical forests may possess a greater complexity of multiple nutrient limitations than what has previously been considered.

Interactive effects of N and P addition. We expected that N and P additions would have negative and positive effects, respectively, on microbial biomass, and that microbial community composition would vary with fertilization. Furthermore, as interactive effects of combined N and P enrichment are common in many terrestrial ecosystems [1], we expected that simultaneous N and P addition would produces notable interactive effects on soil microbes in this tropical forest; to the contrary, however, none were found (Appendix S1). We suggest that this occurs because this old-growth forest is N saturated, exhibiting net loss of 8–16 kg N $ha^{-1}yr^{-1}$ from the soil [3,16]. Thus, P rather than N is limiting factor for microbe growth.

Soil organic carbon

Nutrient enrichments may feed back on ecosystem C through the effects on microbial biomass and community composition. Microbial biomass, which represents an important labile pool of nutrients in soil, plays a significant role in nutrient cycling in ecosystems [74]. Changes in the size of the microbial biomass pool may indicate changes in the soil organic matter pool. Reciprocally, microbial biomass also depends on soil organic matter [75]. In the present study, soil microbial biomass C was not significantly different among treatments (Fig. 3), suggesting that microbial biomass was not responsible for the N- and P-induced changes in soil C pool in this tropical forest.

However, we found that soil microbial community structure was significantly different among treatments. N addition increased fungal abundance, with detectable decreases in labile C compounds in this tropical forest. Previous results at this site have shown that experimental N addition significantly decreased litter decomposition [22,24]. Several studies in temperate systems have demonstrated that fertilization with N may stimulate decay of more labile C, suppress degradation of lignin, and stimulate formation of recalcitrant material [76–78]. Cusack *et al.* (2011)

reported that N fertilization reduced C content in the light fraction and increased C content in the heavy fraction in two tropical forests [79]. Our results support this conclusion in part: N addition had a negative effect on C content in the light fraction, but not in the heavy fraction. This finding may be related to the higher relative abundance of fungal PLFAs in N-addition plots compared to control plots (Figs. 3–5). However, Ramirez *et al.* (2012) found that N addition depresses soil microbial activity by shifting soil bacterial communities, yielding communities that are less capable of decomposing more recalcitrant soil carbon pools [30]. Strickland *et al.* (2009) observed no relationship between microbial community structure and C mineralization processes in soil [80]. Those studies indicated that C cycling is changing in response to N addition, driven by complex interactions among microbial composition, enzymatic capability and soil C chemistry.

Cleveland *et al.* (2002) reported that in ferralsols of tropical rainforests microbial C use was strongly constrained by P availability and suggested that more P-poor forests may display longer-term soil storage of novel inputs of C [29]. Our results that light fraction SOM and SOC significantly decreased after P-addition support this suggestion. Zhou *et al.* (2006) reported an accumulation of soil C (0–20 cm depth) at about 54 g C m-2 yr-1 over two decades in the old-growth forest [21]. One of the possible reasons is that the P-limited microbial community may increase the rates of belowground C storage [33]. Results from our study further indicated that a P-limited microbial community may actually help retain more labile forms of C in these tropical forests.

Our results have demonstrated that both the microbial community and chemistry of the SOM pool can vary greatly in response to N versus P additions. Although C content in the light fraction decreased significantly for both treatments, total soil C decreased significantly only in the P-addition plots, suggesting that soil C changes induced by fertilization may be more sensitive to P addition than to N addition in these tropical forests. Not only was soil microbial community structure significantly related to NO_3^-, available P, and SOC (Appendix S4), but soil NO_3^- concentrations significantly decreased and soil available P increased in the P-addition plots (Table 2). Previous studies have shown that litterfall production was not affected by N addition, but increased by P addition, supporting the idea that P rather than N availability may be a limiting factor for plant growth, such that P addition may increase the supply of labile plant C to microbes [23,33]. At the same time, we found that AM fungal abundance was significantly increased in the P-addition plots compared to the N-addition plots. This result indicated that abundance of AM fungi may play a larger role in organic matter decomposition and C cycling.

Supporting Information

Appendix S1 Effects of N addition, P addition and two-way interactions of N addition and P addition on soil properties and microbial characteristics.

Appendix S2 The relative abundances of the individual PLFAs (mol %) in soil samples.

Appendix S3 The phospholipid fatty acid (PLFA) pattern in soil samples.

Appendix S4 Redundancy analysis of PLFA profiles used 26 PLFAs as species and six environmental parameters.

Acknowledgments

We wish to thank Shaofeng Dong, Chuanying Xiang for their skilful assistance in field work, Jianping Wu for his assistance in laboratory work, and Drs. Sandra Brown, Shenglei Fu, Lixia Zhou and Weixing Zhu for their invaluable suggestions in earlier version of the manuscript.

Author Contributions

Conceived and designed the experiments: JMM LL PG FSG. Performed the experiments: LL TZ WZ HC. Analyzed the data: LL. Contributed reagents/materials/analysis tools: LL HC. Wrote the paper: LL.

References

1. Elser JJ, Bracken MES, Cleland EE, Gruner DS, Harpole WS, et al. (2007) Global analysis of nitrogen and phosphorus limitation of primary producers in freshwater, marine and terrestrial ecosystems. Ecology Letters 10: 1135–1142.
2. Zheng XH, Fu CB, Xu XK, Yan XD, Huang Y, et al. (2002) The Asian nitrogen cycle case study. Ambio 31: 79–87.
3. Fang YT, Yoh M, Koba K, Zhu WX, Takebayashi Y, et al. (2011) Nitrogen deposition and forest nitrogen cycling along an urban-rural transect in southern China. Global Change Biology 17: 872–885.
4. Gilliam FS, Adams MB (1996) Wetfall deposition and precipitation chemistry for a central Appalachian forest. Journal of the Air & Waste Management Association 46: 978–984.
5. MacDonald JA, Dise NB, Matzner E, Armbruster M, Gundersen P, et al. (2002) Nitrogen input together with ecosystem nitrogen enrichment predict nitrate leaching from European forests. Global Change Biology 8: 1028–1033.
6. Aber J, McDowell W, Nadelhoffer K, Magill A, Berntson G, et al. (1998) Nitrogen saturation in temperate forest ecosystems - Hypotheses revisited. Bioscience 48: 921–934.
7. Gundersen P, Emmett BA, Kjonaas OJ, Koopmans CJ, Tietema A (1998) Impact of nitrogen deposition on nitrogen cycling in forests: a synthesis of NITREX data. Forest Ecology and Management 101: 37–55.
8. Gilliam FS (2006) Response of the herbaceous layer of forest ecosystems to excess nitrogen deposition. Journal of Ecology 94: 1176–1191.
9. Bobbink R, Hicks K, Galloway J, Spranger T, Alkemade R, et al. (2010) Global assessment of nitrogen deposition effects on terrestrial plant diversity: a synthesis. Ecological Applications 20: 30–59.
10. Hall SJ, Matson PA (1999) Nitrogen oxide emissions after nitrogen additions in tropical forests. Nature 400: 152–155.
11. Hall SJ, Matson PA (2003) Nutrient status of tropical rain forests influences soil N dynamics after N additions. Ecological Monographs 73: 107–129.
12. Walker TW, Syers JK (1976) Fate of phosphorus during pedogenesis. Geoderma 15: 1–19.
13. Cleveland CC, Townsend AR, Taylor P, Alvarez-Clare S, Bustamante MMC, et al. (2011) Relationships among net primary productivity, nutrients and climate in tropical rain forest: a pan-tropical analysis (vol 14, pg 939, 2011). Ecology Letters 14: 1313–1317.
14. Wright SJ, Yavitt JB, Wurzburger N, Turner BL, Tanner EVJ, et al. (2011) Potassium, phosphorus, or nitrogen limit root allocation, tree growth, or litter production in a lowland tropical forest. Ecology 92: 1616–1625.
15. Baribault TW, Kobe RK, Finley AO (2012) Tropical tree growth is correlated with soil phosphorus, potassium, and calcium, though not for legumes. Ecological Monographs 82: 189–203.
16. Fang YT, Gundersen P, Mo JM, Zhu WX (2008) Input and output of dissolved organic and inorganic nitrogen in subtropical forests of South China under high air pollution. Biogeosciences 5: 339–352.
17. Lu X-K, Mo J-M, Gundersern P, Zhu W-X, Zhou G-Y, et al. (2009) Effect of Simulated N Deposition on Soil Exchangeable Cations in Three Forest Types of Subtropical China. Pedosphere 19: 189–198.
18. Zhang W, Mo JM, Zhou GY, Gundersen P, Fang YT, et al. (2008) Methane uptake responses to nitrogen deposition in three tropical forests in southern China. Journal of Geophysical Research-Atmospheres 113.
19. Lu X, Mo J, Gilliam FS, Zhou G, Fang Y (2010) Effects of experimental nitrogen additions on plant diversity in an old-growth tropical forest. Global Change Biology 16: 2688–2700.
20. Huang WJ, Zhou GY, Liu JX (2011) Nitrogen and phosphorus status and their influence on aboveground production under increasing nitrogen deposition in three successional forests. Acta Oecologica (in press).
21. Zhou GY, Liu SG, Li ZA, Zhang DQ, Tang XL, et al. (2006) Old-growth forests can accumulate carbon in soils. American Association for the Advancement of Science.pp.1417–1417.
22. Mo JM, Brown S, Xue J, Fang Y, Li Z (2006) Response of litter decomposition to simulated N deposition in disturbed, rehabilitated and mature forests in subtropical China. Plant and Soil 282: 135–151.

23. Mo J, Zhang W, Zhu W, Gundersen P, Fang Y, et al. (2008) Nitrogen addition reduces soil respiration in a mature tropical forest in southern China. Global Change Biology 14: 403–412.
24. Fang H, Mo JM, Peng SL, Li ZA, Wang H (2007) Cumulative effects of nitrogen additions on litter decomposition in three tropical forests in southern China. Plant and Soil 297: 233–242.
25. Wang H, Mo JM, Lu XK, Xue JH, Li J, et al. (2008) Effects of elevated nitrogen deposition on soil microbial biomass carbon in the main subtropical forests of southern China. Acta Ecologica Sinica 28: 470–478.
26. Gilliam FS, McCulley RL, Nelson JA (2011) Spatial Variability in Soil Microbial Communities in a Nitrogen-Saturated Hardwood Forest Watershed. Soil Science Society of America Journal 75: 280–286.
27. Cruz AF, Hamel C, Hanson K, Selles F, Zentner RP (2009) Thirty-seven years of soil nitrogen and phosphorus fertility management shapes the structure and function of the soil microbial community in a Brown Chernozem. Plant and Soil 315: 173–184.
28. Gallardo A, Schlesinger WH (1994) FACTORS LIMITING MICROBIAL BIOMASS IN THE MINERAL SOIL AND FOREST FLOOR OF A WARM-TEMPERATE FOREST. Soil Biology & Biochemistry 26: 1409–1415.
29. Cleveland CC, Townsend AR, Schmidt SK (2002) Phosphorus limitation of microbial processes in moist tropical forests: Evidence from short-term laboratory incubations and field studies. Ecosystems 5: 680–691.
30. Ramirez KS, Craine JM, Fierer N (2012) Consistent effects of nitrogen amendments on soil microbial communities and processes across biomes. Global Change Biology 18: 1918–1927.
31. Cusack DF, Silver WL, Torn MS, Burton SD, Firestone MK (2011) Changes in microbial community characteristics and soil organic matter with nitrogen additions in two tropical forests. Ecology 92: 621–632.
32. Townsend AR, Cleveland CC, Houlton BZ, Alden CB, White JWC (2011) Multi-element regulation of the tropical forest carbon cycle. Frontiers in Ecology and the Environment 9: 9–17.
33. Liu L, Gundersen P, Zhang T, Mo J (2012) Effects of phosphorus addition on soil microbial biomass and community composition in three forest types in tropical China. Soil Biology & Biochemistry 44: 31–38.
34. Mo J, Brown S, Peng S, Kong G (2003) Nitrogen availability in disturbed, rehabilitated and mature forests of tropical China. Forest Ecology and Management 175: 573–583.
35. Zhang DQ, Ye WH, Yu QF, Kong GH, Zhang YC (2000) The litter-fall of representative forests of successional series in Dinghushan. Acta Ecologica Sinica 20: 938–944 (in Chiense with English abstract).
36. Huang ZF, Fan ZG (1982) The climate of Dinghushan. Tropical and subtropical Forest Ecosystem 1:11–23 (in Chinese with English abstract).
37. Cleveland CC, Townsend AR (2006) Nutrient additions to a tropical rain forest drive substantial soil carbon dioxide losses to the atmosphere. Proceedings of the National Academy of Sciences of the United States of America 103: 10316–10321.
38. Raison RJ, Connell MJ, Khanna PK (1987) METHODOLOGY FOR STUDYING FLUXES OF SOIL MINERAL-N INSITU. Soil Biology & Biochemistry 19: 521–530.
39. Janzen HH, Campbell CA, Brandt SA, Lafond GP, Townleysmith L (1992) LIGHT-FRACTION ORGANIC-MATTER IN SOILS FROM LONG-TERM CROP ROTATIONS. Soil Science Society of America Journal 56: 1799–1806.
40. Izaurralde RC, Nyborg M, Solberg ED, Janzen HH, Arshad MA, et al. (1998) Carbon storage in eroded soils after five years of reclamation techniques; Lal R, Kimble JM, Follett RF, Stewart BA, editors. 369–385 p.
41. JM A (1993) Tropical soil biology and fertility: A handbook of methods, Second edition; Anderson JM, Ingram JSI, editors.xv+221p–xv+221p p.
42. Vance ED, Brookes PC, Jenkinson DS (1987) AN EXTRACTION METHOD FOR MEASURING SOIL MICROBIAL BIOMASS-C. Soil Biology & Biochemistry 19: 703–707.
43. Bossio DA, Scow KM (1998) Impacts of carbon and flooding on soil microbial communities: Phospholipid fatty acid profiles and substrate utilization patterns. Microbial Ecology 35: 265–278.
44. Tunlid A, Hoitink HAJ, Low C, White DC (1989) CHARACTERIZATION OF BACTERIA THAT SUPPRESS RHIZOCTONIA DAMPING-OFF IN BARK COMPOST MEDIA BY ANALYSIS OF FATTY-ACID BIOMARKERS. Applied and Environmental Microbiology 55: 1368–1374.
45. Frostegard A, Baath E (1996) The use of phospholipid fatty acid analysis to estimate bacterial and fungal biomass in soil. Biology and Fertility of Soils 22: 59–65.
46. Zelles L (1999) Fatty acid patterns of phospholipids and lipopolysaccharides in the characterisation of microbial communities in soil: a review. Biology and Fertility of Soils 29: 111–129.
47. Frostegard A, Baath E, Tunlid A (1993) SHIFTS IN THE STRUCTURE OF SOIL MICROBIAL COMMUNITIES IN LIMED FORESTS AS REVEALED BY PHOSPHOLIPID FATTY-ACID ANALYSIS. Soil Biology & Biochemistry 25: 723–730.
48. Olsson PA (1999) Signature fatty acids provide tools for determination of the distribution and interactions of mycorrhizal fungi in soil. Fems Microbiology Ecology 29: 303–310.
49. Zak DR, Ringelberg DB, Pregitzer KS, Randlett DL, White DC, et al. (1996) Soil microbial communities beneath Populus grandidentata crown under elevated atmospheric CO2. Ecological Applications 6: 257–262.
50. Bardgett RD, Hobbs PJ, Frostegard A (1996) Changes in soil fungal:bacterial biomass ratios following reductions in the intensity of management of an upland grassland. Biology and Fertility of Soils 22: 261–264.
51. Wallenstein MD, McNulty S, Fernandez IJ, Boggs J, Schlesinger WH (2006) Nitrogen fertilization decreases forest soil fungal and bacterial biomass in three long-term experiments. Forest Ecology and Management 222: 459–468.
52. Balser TC (2001) The impact of long-term nitrogen addition on microbial community composition in three Hawaiian forest soils. TheScientificWorldJOURNAL 1: 500–504.
53. Zhang QH, Zak JC (1998) Effects of water and nitrogen amendment on soil microbial biomass and fine root production in a semi-arid environment in West Texas. Soil Biology & Biochemistry 30: 39–45.
54. Arnebrant K, Baath E, Soderstrom B, Nohrstedt HO (1996) Soil microbial activity in eleven Swedish coniferous forests in relation to site fertility and nitrogen fertilization. Scandinavian Journal of Forest Research 11: 1–6.
55. Corre MD, Beese FO, Brumme R (2003) Soil nitrogen cycle in high nitrogen deposition forest: Changes under nitrogen saturation and liming. Ecological Applications 13: 287–298.
56. DeForest JL, Zak DR, Pregitzer KS, Burton AJ (2004) Atmospheric nitrate deposition and the microbial degradation of cellobiose and vanillin in a northern hardwood forest. Soil Biology & Biochemistry 36: 965–971.
57. Nilsson LO, Baath E, Falkengren-Grerup U, Wallander H (2007) Growth of ectomycorrhizal mycelia and composition of soil microbial communities in oak forest soils along a nitrogen deposition gradient. Oecologia 153: 375–384.
58. Demoling F, Nilsson LO, Baath E (2008) Bacterial and fungal response to nitrogen fertilization in three coniferous forest soils. Soil Biology & Biochemistry 40: 370–379.
59. Fraterrigo JM, Balser TC, Turner MG (2006) Microbial community variation and its relationship with nitrogen mineralization in historically altered forests. Ecology 87: 570–579.
60. Gallo M, Amonette R, Lauber C, Sinsabaugh RL, Zak DR (2004) Microbial community structure and oxidative enzyme activity in nitrogen-amended north temperate forest soils. Microbial Ecology 48: 218–229.
61. Billings SA, Ziegler SE (2008) Altered patterns of soil carbon substrate usage and heterotrophic respiration in a pine forest with elevated CO2 and N fertilization. Global Change Biology 14: 1025–1036.
62. Smith SE, Read DJ (2008) Mycorrhizal Symbiosis. ,San DiegoAcademic Press,Third edition.
63. Treseder KK (2004) A meta-analysis of mycorrhizal responses to nitrogen, phosphorus, and atmospheric CO2 in field studies. New Phytologist 164: 347–355.
64. Read DJ (1991) MYCORRHIZAS IN ECOSYSTEMS - NATURES RESPONSE TO THE LAW OF THE MINIMUM; Hawksworth DL, editor. 101–130 p.
65. Allen MF (1991) The ecology of mycorrhizae. Cambridge,UK:Cambridge University Press.
66. Treseder KK, Allen MF (2002) Direct nitrogen and phosphorus limitation of arbuscular mycorrhizal fungi: a model and field test. New Phytologist 155: 507–515.
67. Baath E, Anderson TH (2003) Comparison of soil fungal/bacterial ratios in a pH gradient using physiological and PLFA-based techniques. Soil Biology & Biochemistry 35: 955–963.
68. Rousk J, Brookes PC, Baath E (2010) The microbial PLFA composition as affected by pH in an arable soil. Soil Biology & Biochemistry 42: 516–520.
69. Hagerberg D, Thelin G, Wallander H (2003) The production of ectomycorrhizal mycelium in forests: Relation between forest nutrient status and local mineral sources. Plant and Soil 252: 279–290.
70. Nilsson LO, Wallander H (2003) Production of external mycelium by ectomycorrhizal fungi in a Norway spruce forest was reduced in response to nitrogen fertilization. New Phytologist 158: 409–416.
71. Groffman PM, Fisk MC (2011) Phosphate additions have no effect on microbial biomass and activity in a northern hardwood forest. Soil Biology & Biochemistry 43: 2441–2449.
72. Juice SM, Fahey TJ, Siccama TG, Driscoll CT, Denny EG, et al. (2006) Response of sugar maple to calcium addition to Northern Hardwood Forest. Ecology 87: 1267–1280.
73. Kaspari M, Garcia MN, Harms KE, Santana M, Wright SJ, et al. (2008) Multiple nutrients limit litterfall and decomposition in a tropical forest. Ecology Letters 11: 35–43.
74. Maithani K, Tripathi RS, Arunachalam A, Pandey HN (1996) Seasonal dynamics of microbial biomass C, N and P during regrowth of a disturbed subtropical humid forest in north-east India. Applied Soil Ecology 4: 31–37.
75. Wardle DA (1992) A COMPARATIVE-ASSESSMENT OF FACTORS WHICH INFLUENCE MICROBIAL BIOMASS CARBON AND NITROGEN LEVELS IN SOIL. Biological Reviews of the Cambridge Philosophical Society 67: 321–358.
76. Fog K (1988) THE EFFECT OF ADDED NITROGEN ON THE RATE OF DECOMPOSITION OF ORGANIC-MATTER. Biological Reviews of the Cambridge Philosophical Society 63: 433–462.
77. Agren G, Bosatta E, Magill AH (2001) Combining theory and experiment to understand effects of inorganic nitrogen on litter decomposition (vol 128, pg 94, 2001). Oecologia 128: 464–464.
78. Keeler BL, Hobbie SE, Kellogg LE (2009) Effects of Long-Term Nitrogen Addition on Microbial Enzyme Activity in Eight Forested and Grassland Sites:

Implications for Litter and Soil Organic Matter Decomposition. Ecosystems 12: 1–15.
79. Cusack DF, Silver WL, Torn MS, McDowell WH (2011) Effects of nitrogen additions on above- and belowground carbon dynamics in two tropical forests. Biogeochemistry 104: 203–225.
80. Strickland MS, Lauber C, Fierer N, Bradford MA (2009) Testing the functional significance of microbial community composition. Ecology 90: 441–451.

Separating the Effects of Environment and Space on Tree Species Distribution: From Population to Community

Guojun Lin[1,2], Diana Stralberg[3], Guiquan Gong[1,2], Zhongliang Huang[1]*, Wanhui Ye[1], Linfang Wu[1]

1 Key Laboratory of Vegetation Restoration and Management of Degraded Ecosystems, South China Botanical Garden, Chinese Academy of Sciences, Guangzhou, People's Republic of China, 2 University of Chinese Academy of Sciences, Beijing, People's Republic of China, 3 Department of Biological Sciences, University of Alberta, Edmonton, Alberta, Canada

Abstract

Quantifying the relative contributions of environmental conditions and spatial factors to species distribution can help improve our understanding of the processes that drive diversity patterns. In this study, based on tree inventory, topography and soil data from a 20-ha stem-mapped permanent forest plot in Guangdong Province, China, we evaluated the influence of different ecological processes at different spatial scales using canonical redundancy analysis (RDA) at the community level and multiple linear regression at the species level. At the community level, the proportion of explained variation in species distribution increased with grid-cell sizes, primarily due to a monotonic increase in the explanatory power of environmental variables. At the species level, neither environmental nor spatial factors were important determinants of overstory species' distributions at small cell sizes. However, purely spatial variables explained most of the variation in the distributions of understory species at fine and intermediate cell sizes. Midstory species showed patterns that were intermediate between those of overstory and understory species. At the 20-m cell size, the influence of spatial factors was stronger for more dispersal-limited species, suggesting that much of the spatial structuring in this community can be explained by dispersal limitation. Comparing environmental factors, soil variables had higher explanatory power than did topography for species distribution. However, both topographic and edaphic variables were highly spatial structured. Our results suggested that dispersal limitation has an important influence on fine-intermediate scale (from several to tens of meters) species distribution, while environmental variability facilitates species distribution at intermediate (from ten to tens of meters) and broad (from tens to hundreds of meters) scales.

Editor: Jean Thioulouse, CNRS - Université Lyon 1, France

Funding: This study was funded by the National Natural Science Foundation of China (30970544).URL is http://www.nsfc.gov.cn/Portal0/default166.htm. The funders had no role in study design, data collection and analysis, decision to publish, or preparation of the manuscript.

Competing Interests: The authors have declared that no competing interests exist.

* E-mail: huangzlscbg@163.com

Introduction

Understanding the forces driving species distributions is a fundamental goal in ecology, especially with respect to explaining community composition and the maintenance of species diversity [1]–[3]. Both environmental heterogeneity and dispersal limitation have frequently been cited as primary determinants of species distribution [4]–[6]. According to niche theory, species partition resources along environmental gradients and species distributions may therefore be predicted based on environmental factors. There is also increasing support for the importance of non-environmental factors such as dispersal and other population processes in structuring observed species distributions [7], [8]. The expectation in a community driven by dispersal limitation is that the difference in species composition should increase with the distance between communities [9], [10], with species distributions exhibiting spatial aggregation proportionate to dispersal limitation [11].

Recent research has evaluated the relative importance of these two driving forces by comparing the variation in species distribution explained by environmental factors (niche dimensions) to that explained by spatial factors (often assumed to be driven by dispersal limitation). Results have largely shown both types of factors to be important, suggesting that these two processes are both important in determining species' distributions and facilitating species coexistence within a community [9], [12], [13]. However, depending on issues such as data quality and site characteristics, the relative importance of environment and space to species distribution varies considerably across communities and regions. Gilbert and Lechowicz [3] found that species distribution in a North American temperate forest was organized mainly by environmental factors and only secondarily by dispersal events, while Wang et al. [14] found that spatial processes explained much more of the variation in species distribution than environmental processes in a Chinese temperate forest. Results from tropical forest studies also vary substantially [15]. For example, in western Amazonian, Tuomisto et al. [9] found that environmental determinism had a higher explanatory power than dispersal limitation, while Valencia et al. [16] found dispersal limitation to be the main driver of plant species dissimilarity. Clearly, more research is needed to make generalizations about relative contributions of environment and space to species distribution patterns.

Most studies that have quantified the relative importance of environment and space in determining species distributions have only worked at the community level, even though ecological processes act on individuals and populations. Coexisting plant species with different forms and life-history strategies may be differentially constrained by dispersal. For example, overstory species enjoy a height advantage that may help them disperse their propagules more widely, whereas understory species may be more dispersal-limited due to their low height and dispersal resistance [17], [18]. Therefore, understanding the contributing ecological processes at the species level can provide additional insights about community assembly rules.

Understanding the relative effects of environmental and spatial processes on species distribution depends on our ability to decompose the sources of observed variability [12]. The widely used variation partitioning method decomposes total variance of species distribution into the fractions explained by different ecological processes [19] [20]. Although the fraction explained by pure space is usually linked to dispersal processes [3], [21], other spatially structured environmental factors that are not included in the analysis may also contribute [12], [22], leading to an overestimation of the purely spatial fraction [23]. Therefore, further work is needed to identify the relative influence of dispersal limitation [23], [24]. Partitioning the variance of species distributions according to dispersal ability can provide insight into the importance of dispersal processes, especially when environmental factors are strongly structured by space so that they are difficult to decompose completely [9], [25], [26]. In general, species with light propagules or special dispersal organs (wings, hooks, barbs, sticky substances, or fat pulp) can diffuse longer distances and are less dispersal-limited. If dispersal limitation is an important ecological determinant of species distribution, we would expect species with low dispersal ability to have a higher proportion of their variation explained by pure space, compared to species with high dispersal ability.

Environmental conditions are commonly represented by topographic variables alone [16], but edaphic characteristics such as soil moisture and nutrient content also limit plant species distributions. Several landscape-scale studies (>1 km^2) have demonstrated the importance of soil properties in controlling tree species distributions [27], [28]. However, at local scales (<1 km^2), the role of edaphic variables in structuring the spatial distribution of trees has not been well studied in natural forests [25], especially in subtropical forests [29]. Furthermore, soil properties are strongly affected by topography [30]. Therefore, further evaluation and partitioning of the effects of soil properties and topography on tree species distribution is an important part of understanding environmental processes.

We aimed to evaluate the relative contributions of niche partitioning and dispersal limitation on species distributions using plant data from a 20 ha stem-mapped subtropical forest plot in Dinghushan, China. Based on plant inventories combined with topographic, edaphic and spatial data, we attempted to disentangle the contributions of environmental and spatial factors at the community and species levels across a range of grain (grid cell) sizes (10-, 20-, 25-, 50-, and 100-m grid cell size). Our specific objectives were 1) to quantify the relative and combined importance of environmental and spatial processes in determining species distribution at the community level and the distributions of species with different life forms; 2) to test the effects of dispersal limitation in determining species distribution, and 3) to further evaluate the role of critical environmental factors (soil and topography) in explaining species distribution. Our goal was to contribute to a larger understanding of the important ecological processes that influence species distribution and promote species coexistence in subtropical forest systems.

Materials and Methods

Ethics Statement

No specific permits were required for the described field studies. Our study site (Dinghushan Nature Reserve) is owned by the Chinese government and managed by South China Botanical Garden, Chinese Academy of Sciences. We can do our research works freely in Dinghushan under the Regulations of the People's Republic of China on Nature Reserves. Our field studies did not involve endangered or protected species.

Study site

Dinghushan Nature Reserve (23°09′21″–23°11′30″N, 112°30′39″–112°33′41″E, in Guangdong Province, China) (Fig. 1) was established in 1956 as the first nature reserve in China. It contains 1155 ha of subtropical forest. The Dinghushan climate regime is classified as south subtropical monsoon, with a mean annual temperature of 20.9°C, and monthly temperature means varying between 12.6°C and 28°C (January and July, respectively). Mean annual precipitation is 1929 mm, with most precipitation occurring between April and September, while mean annual evaporation is 1115 mm, and relative humidity averages 82%. A permanent 20-ha (400-m×500-m) plot (DHS hereafter) was established in 2005 in the Dinghushan Reserve. The plot is located in rugged terrain and includes three ridges and four valleys, with elevation ranging from 240 to 470 m, as well as complex soil properties. Furthermore, the plot is characterized by an old (>400 yrs), sheltered monsoon evergreen broadleaved forest (Fig. 1). DHS is part of the Chinese Forest Biodiversity Monitoring Network (CForBio), which consists of twelve large (5–25 ha) stem-mapped plots along a latitudinal gradient from tropical to subtropical to temperate forest (http://www.cfbiodiv.org/english/).

Species data

Following the sampling protocols promoted by the Center for Tropical Forest Science (CTFS) [31], all stems in DHS with diameter at breast height (DBH) no smaller than 1 cm were tagged, georeferenced and identified at the species level. As a result, 71,617 stems belonging to 210 species, 119 genera, and 56 families were recorded (Table S1). Among all species, over 50% (110) can be considered rare in the plot (i.e., with fewer than 20 stems) [32]. *Castanopsis chinensis (Ca.ch), Schima superba (Sc.su), Engelhardtia roxburghiana (En.ro), Syzygium rehderianum (Sy.re), Craibiodendron kwangtungense (Cr.kw), Aidia canthioides (Ai.ca), Cryptocarya chinensis (Cr.ch), Cryptocarya concinna (Cr.co), Aporosa yunnanensis (Ap.yu)* and *Ardisia quinquegona (Ar.qu)* were the ten species with the largest importance values ((relative dominance)/3+(relative frequency)/3+(relative abundance)/3. Relative dominance refers to the basal area of the focal species as a proportion of the total basal area for all species; relative frequency refers to the frequency of the focal species as a proportion of the summed frequency values for all species; relative abundance refers to the abundance of focal species as a proportion of the summed abundance values for all species). Among them, *Ca.ch* was the most dominant overstory species, with 2311 individuals and a maximum DBH of 175 cm and *Cr.ch* was the most dominant species in the midstory, with 2557 individuals and a maximum DBH of 51 cm. Understory species composition in the plot is rich and complex. We counted the number of individuals of each species in each cell for grids of five different height/width dimensions: 10-, 20-, 25-, 50-, and 100-m.

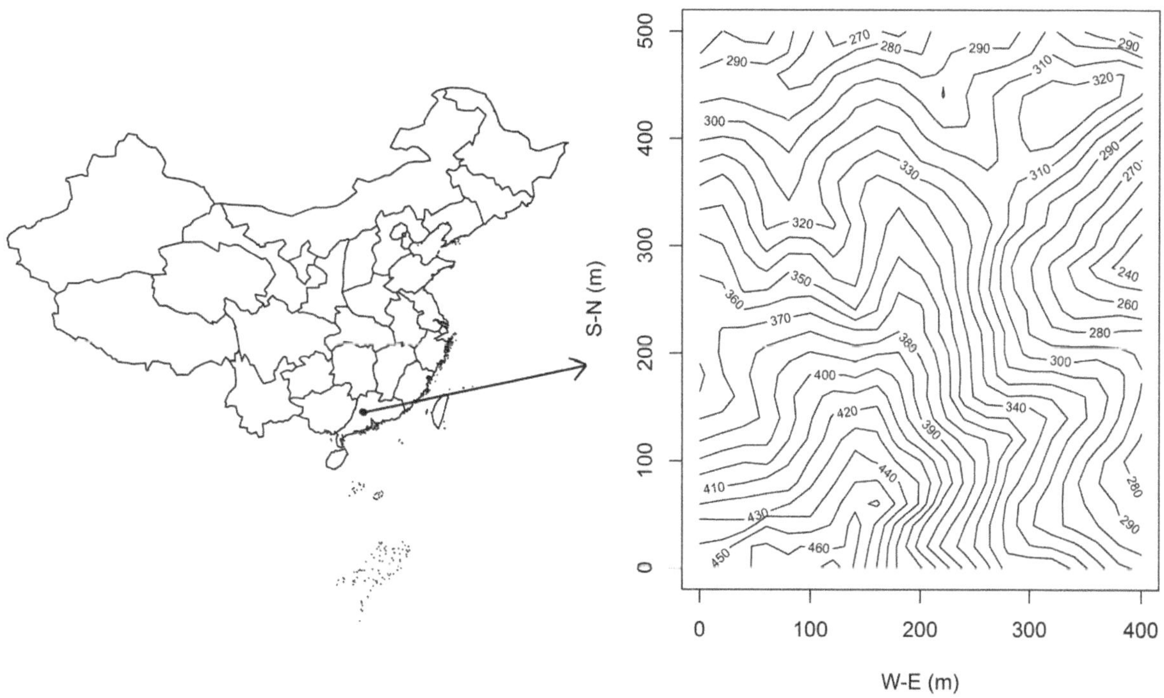

Figure 1. Location and relief map of Dinghushan and the 20 ha stem-mapped permanent forest plot. Values on the relief map represent elevations in meters, which ranges from 230 m to 470 m.

Environmental characteristics

The topography of the Dinghushan plot was quantified by measuring elevation at the four corners of each cell of a 20-m grid. Elevation values at the 5-m cell size was interpolated by ordinary kriging from 20-m data, while the values for larger cell sizes (i.e., 10-, 25-, 50-, and 100-m) were based on averages of the 5-m cells. For each cell size, we calculated the mean elevation (E), slope (S), convexity (C) and aspect (A) of each grid cell following Harms et al. [26] and Valencia et al. [16]. In order to allow non-linear relationships between topographic factors and response variables, we calculated second-order and third-order polynomials of elevation, slope and convexity. Aspect (degrees from north) was decomposed into east-west and north-south orientation, using sin(aspect) (SA) and cos(aspect) (CA) respectively.

We collected surface-soil samples (0–10 cm) by using a regular grid of points every 30 m. Each alternate grid point was paired with two additional sample points at 2, 5, or 15 m in a random compass direction from the grid to capture fine scale variation in soil properties [25]. In total, 710 samples (four of 714 samples were excluded because they fell on rocks or in creeks) were collected and ten edaphic parameters (moisture (sm), density (sd), pH, organic carbon (tc), available (an) and total nitrogen (tn), available (ap) and total phosphorus (tp), and available (ak) and total potassium (tk)) were measured on each sample. Edaphic variable values at the 5-m cell size was interpolated by ordinary kriging from these measured data, while the values for larger cell sizes (i.e., 10-, 20-, 25-, 50-m and 100-m) were based on averages of the 5-m cell.

Spatial eigenfunctions

Regardless of their origin, spatial structures cannot be neglected in ecosystems [33], [34]. Principal coordinates of neighbor matrices (PCNM) analysis is a particular case of Moran's eigenvector maps (MEM) analysis [35], an approach that has been proven to better represent spatial patterns than Euclidean distances matrices, geographic coordinates and cubic trend-surface equation [15], [21]. PCNM eigenfunctions are generated by performing a principal coordinate analysis (PCoA) on a truncated Euclidean distance matrix [35]. In our analyses, all distances larger than the distance between the centers of diagonally adjacent cells were replaced by four times that value before PCoA [12]. The axes (eigenvectors) obstained from the PCoA comprise a spectral decomposition of the spatial relationships among the sampling sites, which can be used directly as spatial variables [33]. PCNM eigenfunctions with low ranks represent large scale spatial variation and those with high ranks represent fine-scale spatial variation. More details about how to produce PCNM eigenfunctions can be found in Borcard and Legendre [33] or Dray et al. [35]. We created five sets of PCNM eigenfunctions, each corresponding to a different grid cell size (10-, 20-, 25-, 50-, and 100-m, respectively).

Variation partitioning analyses

At the community level, we conducted variation partitioning based on canonical redundancy analysis (RDA). The overall variation in community composition was divided into fractions attributable to topographic, edaphic and spatial variables, as well as combinations of these explanatory factors. We used the Hellinger-transformed abundance values of species within grid cells as response variables [36]. As partitioning results depend on the scale of analysis, we repeated the same analysis on the five different grid cell sizes (10-, 20-, 25-, 50-m and 100-m). Before conducting variation partitioning, each set of topographic and edaphic variables was standardized and selected at the 95% confidence level using forward selection [37]. Meanwhile, each set of PCNM eigenfunctions was also selected at the 95% confidence level using forward selection [37]. The total variation of species distribution was decomposed into eight fractions (see Fig. 2). Variation partitioning fractions top-unique (purely topographic),

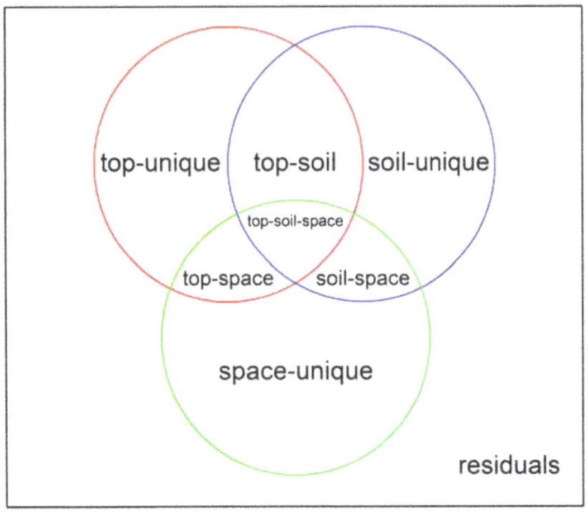

Figure 2. Eight fractions without overlap resulting from variation partitioning using topographic, edaphic and spatial explanatory tables: purely topographic (top-unique), purely edaphic (soil-unique), purely spatial (space-unique), topographically structured edaphic (top-soil), spatially structured edaphic (soil-space), spatially structured topographic (top-space), spatially and topographically structured edaphic (top-soil-space), and unexplained (residuals). Several interpretable quantities can be calculated by adding up fraction values, for example, top-all (= [purely topographic]+[topographically structured edaphic]+[-spatially structured topographic]+[spatially and topographically structured edaphic]) represents the proportion of total variation explained by topography, soil-all (= [purely edaphic]+[topographically structured edaphic]+[spatially structured edaphic]+[spatially and topographically structured edaphic]) represents the proportion of total variation explained by soil, and space-all (= [purely spatial]+[spatially structured edaphic]+[spatially structured topographic]+[spatially and topographically structured edaphic]) represents the importance of spatial variables.

soil-unique (purely edaphic) and space-unique (purely spatial) were tested using permutation tests with 999 permutations [19].

Based on RDA analysis, we further analyzed the importance of particular topographic and edaphic factors on species distribution at the community level at the 20-m cell size. We used the Hellinger-transformed abundance values of species at the 20-m cell size as response variables [36]. Standardized topographic and edaphic variables at the 20-m cell size were considered as explanatory variables respectively. For each analysis, we only selected the first three canonical axes to represent the topographic or edaphic variables, because these axes explained more than 80% of the variation. Each canonical axis was tested using a permutation test with 999 permutations [38].

At the species level, we conducted variation partitioning based on multiple linear regressions. For each of the ten species with largest importance value, the overall variation in species distribution was divided into fractions attributable to topographic, edaphic and spatial variables, as well as combinations of these explanatory factors. We used square-root-transformed abundance values of each species within grid cells as response variables [39]. Spatial variables (PCNM eigenfunctions), topographic and edaphic variables were conducted as explanatory variables in the same procedures with that in community level in our analysis.

Dispersal limitation

We studied dispersal limitation for common species (i.e., those with more than 20 stems in the plot) only, which we divided into three categories according to their dispersal ability (Table S1). Species dispersed by wind and birds were classified as high dispersal (HD), those dispersed by rodents or large mammals (e.g., wild boar) were classified as medium dispersal (MD), and those dispersed by ants or gravity were classified as low dispersal (LD). Species with mixed dispersal modes were classified according to their dominant mode. There were 46, 45, and 9 species in the HD, MD and LD groups, respectively. In order to assess the effect of dispersal limitation, we took the results of variation partitioning obtained for each group at the 20-m cell size and calculated the purely spatial fraction as a proportion of the fraction explained by environment and space combined.

We used the Hellinger-transformed abundance values of each group's species within grid cells as response variables. Other procedures for variation partitioning were the same as for the previous community level analyses in our analysis. We assumed that if dispersal limitation is important in controlling species distributions in Dinghushan plot, the distributions of species with low dispersal ability would be more dependent on spatial variables than the distributions of species with high dispersal ability. That is, the purely spatial fraction would be expected to be higher.

All analyses were conducted using R 2.11.1 [40]. Ordinary kriging was completed using package "gstat" [41]. Forward selection and PCNM eigenfunctions were computed using package "packfor" [42] and "vegan" [38], respectively. Variation partitioning and permutation test were computed using the "vegan" package [38].

Results

Variation partitioning by grid-cell size

The proportion of compositional variation explained increased monotonically with cell size, mainly due to an increase in the environmental fraction. Variation explained by topography and soil increased from 23.7% at the 10-m cell size to 54.1% at the 100-m cell size (Fig. 3). In contrast, the purely spatial (space-unique) fraction was highest at the 20-m cell size (24.1%) (Fig. 3).

Results of variation partitioning on the abundances of each of the ten most important species were generally similar to those obtained for community composition, in that environmental explanatory power generally increased with cell size (Fig. 4).

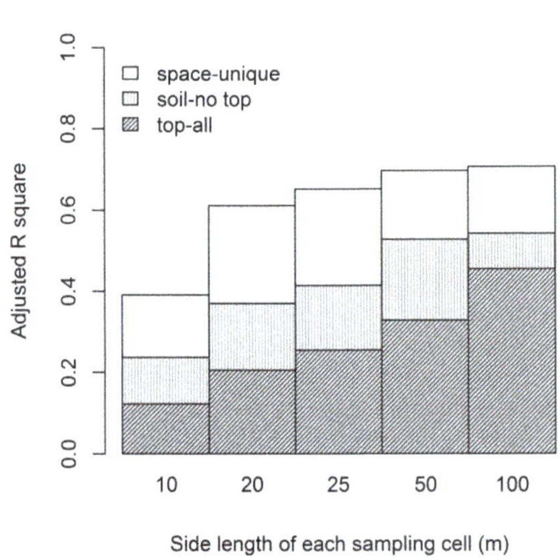

Figure 3. Overall partitioning results for species distribution at the community level. Adjusted R square (Ra^2) values corresponding to topographic (top-all), edaphic but not topographic (soil-no top), and purely spatial (space-unique) fractions.

Nevertheless, these ten species were naturally separated into three groups according to their partitioning results. For species of group A (*Ca.ch*, *Sc.su* and *En.ro*), total explanatory power increased gradually with cell size, and the unexplained variation was large in small and intermediate cell sizes. For species of group C (*Ai.ca*, *Cr.co* and *Ar.qu*), total explanatory power was fairly constant across cell size, whereas the purely spatial fraction (space-unique) fraction was large in small and intermediate cell sizes and then decreased gradually with cell size. Finally, species of group B (*Cr.ch*, *Ap.yu*, *Sy.re* and *Cr.kw*) showed patterns that were intermediate between those of groups A and C. Each of these groups represents a different set of life forms. Group A contains only overstory species, whereas group C consists solely of understory species and group B is entirely comprised of midstory species (Fig. 4).

Environmental fraction

We examined the roles of topography and soil factors in greater depth using the partitioning results obtained at the 20-m cell size. At both the species and community levels, the edaphic (soil-all) fraction was generally larger than the topographic (top-all) fraction (Table 1). Moreover, non-spatial environmental fractions of explained variation (top-unique, soil-unique and top-soil) were relatively small while those with a spatial signature (top-space, soil-space and top-soil-space) provided most of the explanatory power.

At the community level, RDA conducted using topographic variables at the 20-m cell size resulted in a first canonical axis negatively related to elevation. The second axis mainly represented slope, which was negatively correlated. The third axis mainly represented convexity, which was positively correlated (Fig. 5). Aspect had little influence. Using edaphic variables, the first canonical axis was positively correlated with levels of soil nutrients (e.g., carbon, nitrogen, phosphorus and potassium), except to the total phosphorus. The second axis was mainly related to physical characters of soil (e.g., soil density), whereas the third axis was primarily correlated with pH (Fig. 5).

Assessment of dispersal limitation

We evaluated the influence of dispersal limitation using the partitioning results obtained at the 20-m cell size (Fig. 6). Total variance explained by the topographic, edaphic and spatial variables was nearly equal across the three species groups, whereas the purely spatial fraction increased gradually as species dispersal abilities decreased. That is the purely spatial fraction as a proportion of the total explained fraction was much higher for species with low dispersal ability than for those with high dispersal ability.

Discussion

Scale-dependent ecological processes

At the community level, our variation partitioning analyses revealed scale-dependent controls on species distribution, reflecting the combination of ecological processes that operate at different spatial scales [22], [43]. In the subtropical forest that we studied, environmental variation (i.e., niche partitioning) had strong effects on species distribution at large (from tens to hundreds of meters) and intermediate (from ten to tens of meters) scales, while purely spatial processes, presumed to be driven by dispersal limitation, had non-negligible effects at intermediate scales. At fine scales (from several to ten meters), where unexplained variability was high, direct biotic interactions were presumed to exert an important influence on species distribution.

The importance of environmental variation at large and intermediate scales was consistent with findings from other studies [12], [43], and was not unexpected given the high level of environmental heterogeneity within the site. However, most of the

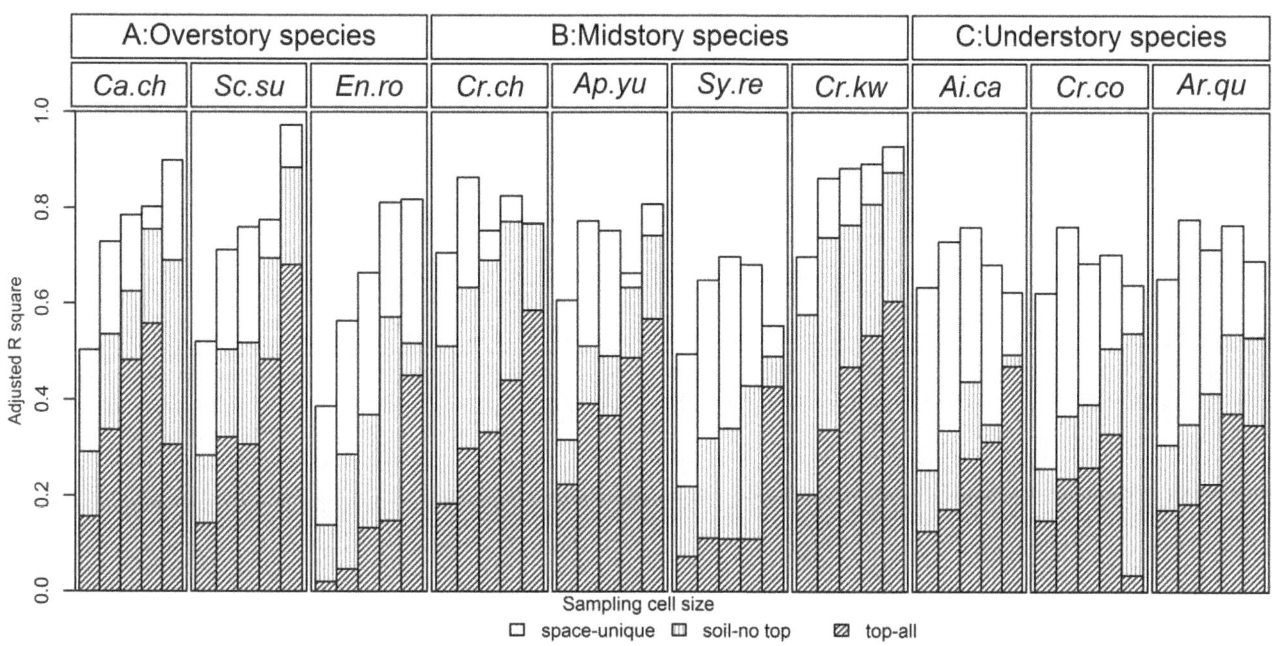

Figure 4. Life forms and variation partitioning results for the ten most abundant species in Dinghushan. The first heading indicates species life forms. The second heading indicates species codes. Ca.ch = *Castanopsis chinensis*; Sc.su = *Schima superba*; En.ro = *Engelhardtia roxburghiana*; Sy.re = *Syzygium rehderianum*; Cr.kw = *Craibiodendron kwangtungense*; Ai.ca = *Aidia canthioides*; Cr.ch = *Cryptocarya chinensis*; Cr.co = *Cryptocarya concinna*; Ap.yu = *Aporosa yunnanensis*; Ar.qu = *Ardisia quinquegona*. X-axes for each species are the same as in Fig. 3 with five different sampling cell sizes (10-, 20-, 25-, 50-m and 100-m).

Table 1. Fractions of variation explained by environmental and spatial variables at the 20-m cell size.

Species	Variation partitioning fractions								
	top-unique	soil-unique	soil-all/top-all	top-unique	soil-unique	top-soil	soil-space	top-space	top-soil-space
CC	0.205	0.289	1.41	0.005	0.003	0.001	0.162	0.076	0.123
Ca.ch	0.337	0.466	1.383	0.000	0.014	(0.001)	0.184	0.069	0.267
Sc.su	0.321	0.437	1.358	(0.004)	0.013	(0.001)	0.171	0.064	0.252
En. ro	0.047	0.280	6.006	(−0.002)	0.006	0.000	0.234	0.008	0.040
Sy.re	0.307	0.602	1.962	(0.001)	0.004	(−0.001)	0.322	0.031	0.275
Cr.kw	0.392	0.376	0.960	(0.001)	(0.001)	(−0.002)	0.119	0.135	0.258
Ai.ca	0.172	0.159	0.926	(−0.002)	(0.001)	0.001	0.164	0.179	−0.006
Cr.ch	0.111	0.279	2.498	(0.002)	0.011	(0.002)	0.197	0.040	0.068
Cr.co	0.236	0.309	1.309	(−0.001)	(0.003)	0.000	0.128	0.058	0.179
Ap.yu	0.338	0.663	1.963	(0.003)	(0.007)	(0.001)	0.394	0.072	0.262
Ar.qu	0.183	0.307	1.677	(0.006)	0.000	0.000	0.167	0.038	0.140

Notes: Fractions are the same with that in Fig. 2. top-unique = purely topographic; soil-unique = purely edaphic; top-soil = topographically structured edaphic; soil-space = spatially structured edaphic; top-space = spatially structured topographic; top-soil-space = spatially and topographically structured edaphic; top-all = [purely topographic]+[topographically structured edaphic]+[spatially structured topographic]+[spatially and topographically structured edaphic]; soil-all = [purely edaphic]+[topographically structured edaphic]+[spatially structured edaphic]+[spatially and topographically structured edaphic]. Values in the parentheses for fraction top-unique and soil-unique indicated adjusted R square values (Ra^2) not significantly different from zero. CC = species distribution at the community level; Ca.ch = *Castanopsis chinensis*; Sc.su = *Schima superba*; En.ro = *Engelhardtia roxburghiana*; Sy.re = *Syzygium rehderianum*; Cr.kw = *Craibiodendron kwangtungense*; Ai.ca = *Aidia canthioides*; Cr.ch = *Cryptocarya chinensis*; Cr.co = *Cryptocarya concinna*; Ap.yu = *Aporosa yunnanensis*; Ar.qu = *Ardisia quinquegona*.

environmental variation was spatially structured (i.e., not possible to separate from spatial factors).

The role of spatial variation, primarily at intermediate scales, was elaborated by our separate evaluation of dispersal limitation, which indicated that species with strong dispersal abilities (i.e., bird- and wind-dispersed species) had lower purely spatial variance components than species with more limited modes of dispersal. Dispersal mode is an important component of dispersal limitation. Thus, there was a large difference between the HD (high dispersal) and LD (low dispersal) groups. However, other aspects such as plant height [17] and pulp-seed attachment [44] may also affect dispersal abilities. In our analysis, we inferred dispersal limitation based on correlational evidence; the difference in the importance of space observed between the high dispersal and medium dispersal groups, while small, was notable. Analysis of plant height and seed rain data may be used to assess the influence of dispersal limitation on species distribution more directly.

Unexplained variation was highest when the grid cell size was small, suggesting the existence of other importance processes for species distribution at fine scales. Li et al. [45] found that self-thinning may influence species' spatial distribution patterns at the scale of 0–10 m. Zhang et al. [46] found evidence of strong interspecific competition at a fine scale (10 m) in temperate forest. We also infer the biotic interactions exert an important influence on fine-scale species distribution patterns in DHS plot. More direct evidence would be helpful, and may be obtained via point pattern analysis or controlled experiments.

The influence of life form

At the species level, we found that variation partitioning results varied by life form. Overstory and understory species exhibited large differences in the purely spatial fraction at small and intermediate cell sizes. For overstory species, the purely spatial fraction was low, implying that dispersal limitation had little effect on species' distributions at small and intermediate cell sizes. All three overstory species in the DHS plot have high or intermediate dispersal abilities (see Table S1). Therefore, it is reasonable that the distributions of these species are not highly influenced by dispersal limitation. However, the effect of plant height on dispersal limitation cannot be neglected [17]. Distributions of all three understory species were strongly affected by purely spatial factors, implying strong dispersal limitation. Among these three understory species, one (*Ai.ca*) is dispersed by birds. However, its low height and strong dispersal path resistance (i.e., the seed has to cross many obstacles before landing in a suitable area) counteract its dispersal mode advantage. For midstory species, the relative effects of dispersal mode, plant height and other factors could not be identified clearly.

Relative importance of soil and topography

Compared to topography, we found edaphic factors to explain a greater portion of the variance in species distribution, both at the community and species level. Soil properties directly affect plant growth, and may have stronger explanatory power than topography at local scales. In support of this notion, John et al. [25] found that the distribution patterns of 40% of tree species had strong associations with soil properties at Barro Colorado Island (BCI), and we also found that the percent of variation explained by soil (15%) was higher than that explained by topography (10%) at the 20-m scale within the BCI plot. In another study, Gleason et al. [47] found that most species (59%) showed significant fidelity to specific soil types in a diverse tropical rainforest.

We should also note that edaphic factors could not explain much additional variation in species distribution beyond topographic and spatial information. This does not mean that edaphic and topographic factors are highly correlated (topography only explained 31% of variation in soil properties at the 20-m scale) but that topographic and edaphic variables have similar spatial structuring.

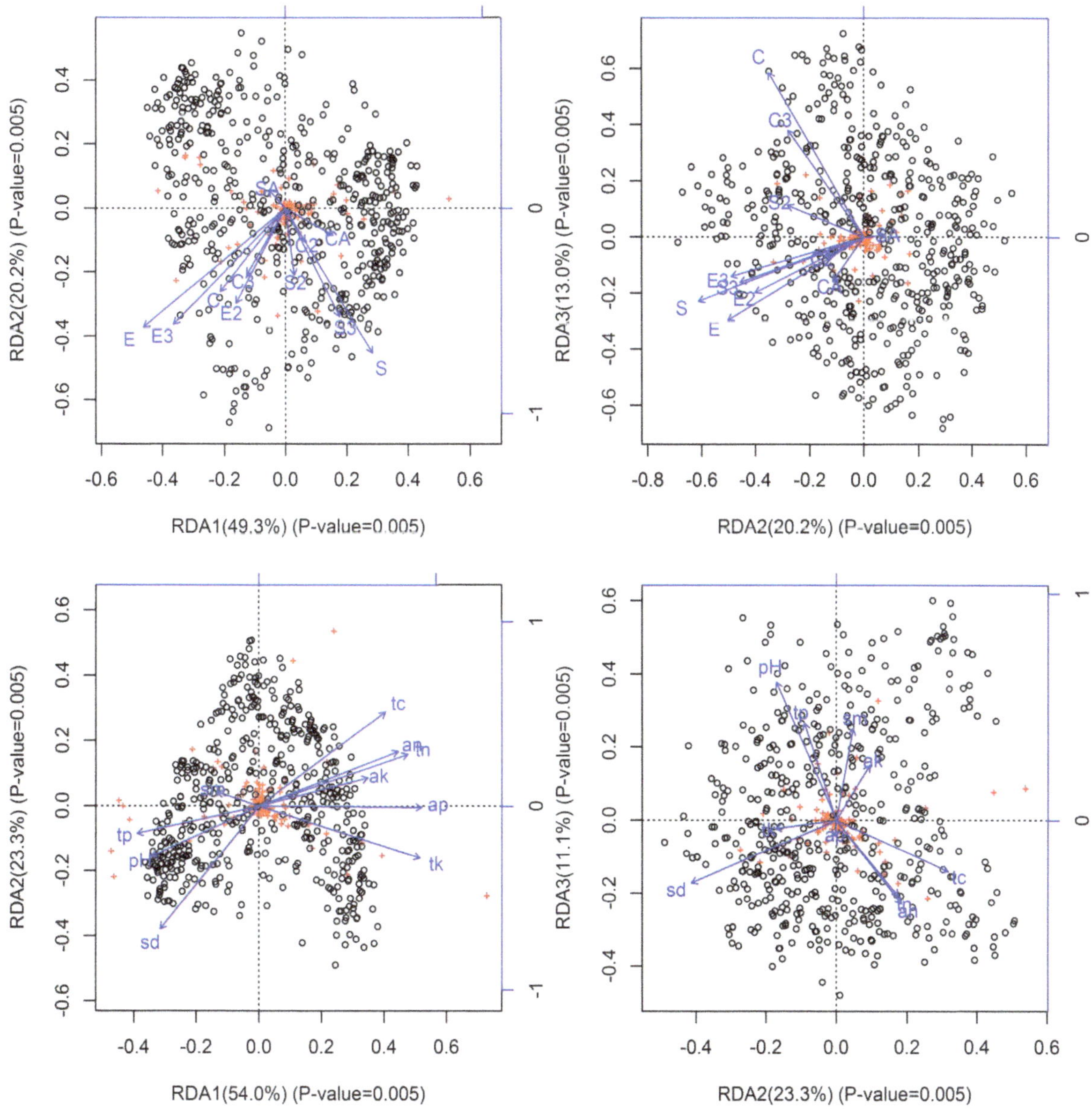

Figure 5. Canonical redundancy analysis (RDA) results at the 20-m cell size for species distribution and environmental factors at the community level. The two graphs in the top panel show relationships between species distribution and topographic factors. E = elevation, S = slope, C = convexity, SA = sin(aspect) and CA = cos(aspect).Associated numbers indicate the power. For example, E2 is the square of elevation. The two graphs in the bottom panel show relationships between species distribution and edaphic factors. sm = soil moisture; sd = soil density; pH = pH; tc = organic carbon; an = available nitrogen; tn = total nitrogen; ap = available phosphorus; tp = total phosphorus; ak = available potassium; tk = total potassium. Values on the coordinate axes represent the proportion of RDA axes. Red crosses in these graphs represent species scores, while black circles represent site scores.

Edaphic effects

At the community level, soil nutrients explained more of the variation in species distribution than that did other edaphic factors included in our analyses. Pärtel et al. [48], [49] investigated 163 case studies on habitat-productivity-plant-diversity relationships. They found that positive productivity-diversity relationships were common in tropical areas, whereas unimodal (hump-shaped) relationships dominated in the temperate zone. In the subtropical forest plot we studied, as in tropical areas, we found that levels of most nutrients were positively correlated with species distribution. The DHS plot has three salient soil features: (1) nitrogen saturation due to nitrogen deposition and an undisturbed state [50], [51]; (2) limited available phosphorus due to high precipitation, but normal levels of total phosphorus (130–260 mg/kg) [50], [51]; and (3) strong acidification (pH = 3.75±0.20). The low pH value further reduces the availability of potassium and phosphorus, particularly at pH<5 [52]. As a result, potassium and available phosphorus were positively correlated with community composition, while total phosphorus exhibited a negative relationship with community composition. Considering the low pH value in our study area, it is not surprising that pH values were positively correlated with species distribution. A striking finding was that nitrogen was also positively correlated with community composition. We speculate that mature trees are not sensitive to

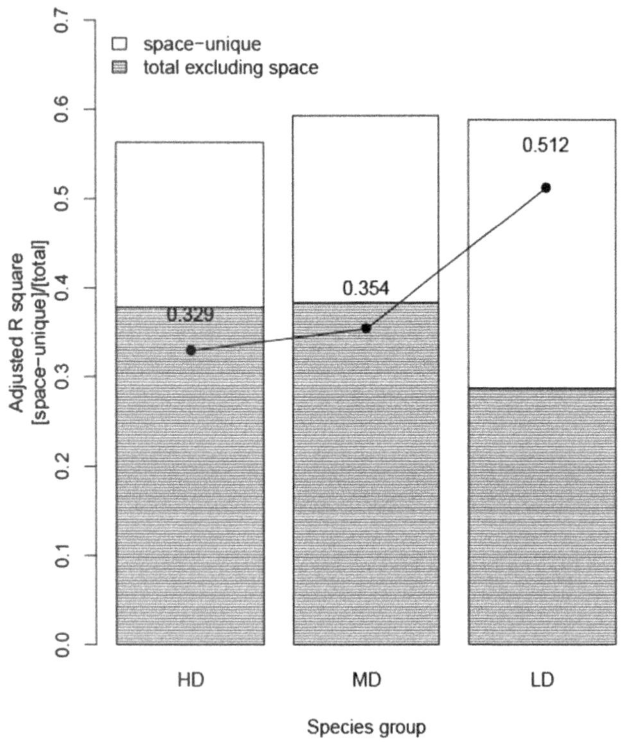

Figure 6. Variation partitioning results at the 20-m cell size for the three species groups in Dinghushan. The X-axis represents different species groups (HD: high dispersal abilities; MD: medium dispersal abilities; LD: low dispersal abilities.). Bars represent adjusted R square values for each partition: purely spatial (space-unique) and [total]-[purely spatial] (total excluding space). Points/lines represent the purely spatial fraction as a proportion of the total fraction explained.

high nitrogen levels, while Lu et al. [51] found that understory diversity decreased with high nitrogen additions (150 kg.ha^{-1}.ya^{-1}). Soil organic carbon mainly comes from plants, and soil density is correlated with quantity and quality of plant fine roots, so it's not surprising that soil organic carbon and soil density were strongly correlated with species distribution. The DHS forest has ample precipitation. Therefore, compared to soil nutrients, soil moisture did not affect species distribution strongly. According to John et al. [25], trace elements, such as boron, calcium, magnesium, iron and aluminum, may be important for species distribution. We predict that the inclusion of such information would improve the explanatory power of soil characteristics.

Topographic effects

Elevation, slope and convexity were important topographic factors structuring communities in DHS. Unlike edaphic conditions, the topography of a site influences the growth of plants only indirectly. With increasing elevation, temperature and moisture decrease in the microenvironment and may limit the richness and abundance of plant species. Five general patterns (monotone decrease, single convex peak, single concave, monotone increase and no obvious relationship) have been proposed to describe the relationships between species distribution and elevation [53]; our results support the prevalent notion that species richness and abundance decreases with elevation. Slope limits plant colonization rates, with steep slopes experiencing low rates of diaspore arrival and colonization. Although steeper slopes also include more surface area per plot, potentially accommodating more individual plants, we found more evidence of slope-limited colonization within the DHS plot, in agreement with Wang et al. [30] but not Bin et al. [54]. In DHS, species distribution was also positively correlated with convexity. It is known that convexity affects air humidity, which may explain the positive relationship between species distribution and convexity. A surprising finding was that aspect contributed little to the variance in species distribution in the DHS plot. Aspect mainly corresponds with light requirements; the high abundance of canopy species weakens the selection for light, which may be responsible for this phenomenon.

Conclusions

In this study, we found evidence that environmental heterogeneity, dispersal and biotic interaction processes worked together to determine species distributions, although their relative importance changed depending on the scale of analysis. Species life form strongly influenced variation partitioning results, with spatial explanatory power highest for understory species and lowest for overstory species. Although soil properties had higher explanatory power than topography, edaphic variables did not contribute much explanatory power beyond topographic and spatial information, due to spatial structuring of these variables. Soil nutrient availability, elevation, slope and convexity were most explanatory of species distributions at the community level.

It is not easy to separate the effects of environment and space on species distribution, even with sophisticated variation partitioning methods. However, our study provided three important new contributions. First, we separated the effects of environment and space on species distribution at the population level, and related it to species life form. We also demonstrated the influence of dispersal limitation on variation partitioning results, rather than merely speculating based on the explanatory power of purely spatial factors. Finally, we explored the role of environmental factors further by comparing the influence of topographic and edaphic factors on species distribution. Our findings improved understanding of species coexistence mechanisms in subtropical forests.

Some limitations of our study should be noted, however. We focused on a single forest plot, and therefore our findings have limited generalizability. Although many studies have attempted to separate the effects of environmental and spatial processes on species distribution patterns, many of them cannot be compared directly due to differences in methods. Thus we advocate the development of integrated research methods in order to produce a unified global dataset from which to study species coexistence mechanisms along elevational, latitudinal, and land-use gradients, to name a few.

Furthermore, as with many variation partitioning studies, our ability to separate environmental and spatial influences was limited. A large portion of the variation in species distribution was explained by spatial and environmental factors together, likely due to highly spatial autocorrelation of environmental factors. We have considered all spatially structured environmental fractions (i.e. top-space, soil-space and top-soil-space fractions in Fig. 2) as niche effects for simplicity, which may lead to an underestimation of spatial processes [55]. However, based on simulated data, Gilbert and Bennett [24] found that ordination using PCNM might under-fit the environmental component and produce inflated R^2 statistics. They further showed that all of the most commonly employed variation partitioning methods, including the analytical approaches we used, failed to accurately model the spatial and environmental components of variation. Although we conducted an independent supporting analysis (evaluating the influence of dispersal limitation) as suggested by Gilbert and Bennett [24], and we have cautiously interpreted our results, more powerful partitioning techniques should also be developed. In some simple systems, experimental manipulation of ecological processes may help illuminate the relative importance of environmental and dispersal processes on species distribution.

Supporting Information

Table S1 Species list for Dinghushan 20 ha permanent forest plot. The top 100 species are non-rare species and were used in the analysis of dispersal limitation. Importance value (IV) = ((relative dominance)/3+(relative frequency)/3+(relative abundance)/3). Species life forms were classified as overstory, midstory and understory. Species dispersed by wind and birds were classified as high dispersal (HD), those dispersed by mammals (rodents and large mammals such as wild boar and deer) were classified as medium dispersal (MD), and those dispersed by ants or gravity were classified as low dispersal (LD). Dispersal mode was identified by field staff with more than thirty years of experience in the study area.

Acknowledgments

We appreciate the invaluable work of all individuals who contributed to the field surveys of the DHS plot. We also thank Miquel De Caceres for initial language revision and Fangliang He for advice on data analysis.

Author Contributions

Analyzed the data: GL GG. Conceived and designed the experiments: GL GG ZH WY . Performed the experiments: GL GG LW . Wrote the paper: GL DS.

References

1. Chesson P (2000) Mechanisms of maintenance of species diversity. Annual Review of Ecology and Systematics 31: 343–366.
2. Chave J (2004) Neutral theory and community ecology. Ecology Letters 7: 241–253.
3. Gilbert B, Lechowicz MJ (2004) Neutrality, niches, and dispersal in a temperate forest understory. Proceedings of the National Academy of Sciences of the United States of America 101: 7651–7656.
4. Laliberté E, Paquette A, Legendre P, Bouchard A (2009) Assessing the scale-specific importance of niches and other spatial processes on beta diversity: a case study from a temperate forest. Oecologia 159: 377–388.
5. Smith TW., Lundholm JT (2010) Variation partitioning as a tool to distinguish between niche and neutral processes. Ecography 33:648–655.
6. Leibold MA, McPeek MA (2005) Coexistence of the niche and neutral perspectives in community ecology. Ecology, 87:1399–1410.
7. Legendre P, Borcard D, Peres-Neto PR (2005) Analyzing beta diversity: Partitioning the spatial variation of community composition data. Ecological Monographs 75: 435–450.
8. Freestone AL, Inouye BD (2006) Dispersal limitation and environmental heterogeneity shape scale-dependent diversity patterns in plant communities. Ecology 87: 2425–2432.
9. Tuomisto H, Ruokolainen K, Yli-Halla M (2003) Dispersal, environment, and floristic variation of Western Amazonian forests. Science 299: 241–244.
10. Condit R, Pitman N, Leigh EG, Chave J, Terborgh J, et al. (2002) Beta-diversity in tropical forest trees. Science 295: 666–669.
11. Seidler TG, Plotkin JB (2006) Seed dispersal and spatial pattern in tropical trees. PLoS Biology 4: e344.
12. Legendre P, Mi X, Ren H, Ma K, Yu M, et al. (2009) Partitioning beta diversity in a subtropical broad-leaved forest of China. Ecology 90: 663–674.
13. Cottenie K (2005) Integrating environmental and spatial processes in ecological community dynamics. Ecology Letters 8: 1175–1182.
14. Wang X, Hao Z, Ye J, Zhang J, Li B, et al. (2008) Spatial pattern of diversity in an old-growth temperate forest in Northeastern China. Acta Oecologica 33: 345–354.
15. Jones MM, Tuomisto H, Borcard D, Legendre P, Clark DB, et al. (2008) Explaining variation in tropical plant community composition: influence of environmental and spatial data quality. Oecologia 155: 593–604.
16. Valencia R, Foster RB, Villa G, Condit R, Svenning J-C, et al. (2004) Tree species distributions and local habitat variation in the Amazon: large forest plot in eastern Ecuador. Journal of Ecology 92: 214–229.
17. Thomson FJ, Moles AT, Auld TD, Kingsford RT (2011) Seed dispersal distance is more strongly correlated with plant height than with seed mass. Journal of Ecology 99: 1299–1307.
18. Cain ML, Damman H, Muir A (1998) Seed dispersal and the holocene migration of woodland herbs. Ecological Monographs 68: 325–347.
19. Peres-Neto PR, Legendre P, Dray S, Borcard D (2006) Variation partitioning of species data matrices: estimation and comparison of fractions. Ecology 87: 2614–2625.
20. Borcard D., Legendre P, Drapeau P. (1992) Partialling out the spatial component of ecological variation. Ecology 73:1045–1055.
21. Borcard D, Legendre P, Avois-Jacquet C, Tuomisto H (2004) Dissecting the spatial structure of ecological data at multiple scales. Ecology 85: 1826–1832.
22. Boulangeat I, Gravel D, Thuiller W (2012) Accounting for dispersal and biotic interactions to disentangle the drivers of species distributions and their abundances. Ecology Letters 15: 584–593.
23. Diniz-Filho JAF, Siqueira T, Padial AA, Rangel TF, Landeiro VL, et al. (2012) Spatial autocorrelation analysis allows disentangling the balance between neutral and niche processes in metacommunities. Oikos 121: 201–210.
24. Gilbert B, Bennett JR (2010) Partitioning variation in ecological communities: do the numbers add up? Journal of Applied Ecology 47: 1071–1082.
25. John R, Dalling JW, Harms kE, Yavitt JB, Stallard RF, et al. (2007) Soil nutrients influence spatial distributions of tropical tree species. Proceedings of the National Academy of Sciences of the United States of America 104: 864–869.
26. Harms KE, Condit R, Hubbell SP, Foster RB (2001) Habitat associations of trees and shrubs in a 50-ha neotropical forest plot. Journal of Ecology 89: 947–959.
27. Tuomisto H, Poulsen AD, Ruokolainen K, Moran RC, Quintana C, et al. (2003) Linking floristic patterns with soil heterogeneity and satellite imagery in Ecuadorian Amazonia. Ecological Applications, 13:352–371.
28. Phillips OL, Vargas PN, Monteagudo AL, Cruz AP, Zans MC, et al. (2003) Habitat association among Amazonian tree species: a landscape-scale approach. Journal of Ecology, 91:757–775.
29. Zhang L, Mi X, Shao H, Ma K (2011) Strong plant-soil associations in a heterogeneous subtropical broad-leaved forest. Plant and Soil, 347:211–220.
30. Wang Z, Ye W, Cao H, Huang Z, Lian J, et al. (2009) Species-topography association in a species-rich subtropical forest of China. Basic and Applied Ecology 10: 648–655.
31. Condit R (1998) Tropical forest census plots. Springer-Verlag and R. G. Landes Company, Berlin, Germany, and Georgetown, Texas. pp. 37–62.
32. Ye W, Cao H, Huang Z, Lian J, Wang Z, et al. (2008) Community structure of a 20 ha lower subtropical evergeen broadleaved forest plot in Dinghushan, China. Journal of Plant Ecology (Chinese Version) 32: 274–286.
33. Borcard D, Legendre P (2002) All-scale spatial analysis of ecological data by means of principal coordinates of neighbour matrices. Ecological Modelling 153: 51–68.
34. Peres-Neto PR, Legendre P (2010) Estimating and controlling for spatial structure in the study of ecological communities. Global Ecology and Biogeography 19: 174–184.
35. Dray S, Legendre P, Peres-Neto PR (2006) Spatial modelling: a comprehensive framework for principal coordinate analysis of neighbour matrices (PCNM). Ecological Modelling 196: 483–493.
36. Legendre P, Gallagher E (2001) Ecologically meaningful transformations for ordination of species data. Oecologia 129: 271–280.
37. Blanchet FG., Legendre P., Borcard D. (2008) Forward selection of explanatory variables. Ecology 89:2623–2632.
38. Oksanen J, Kindt R, Legendre P, O'Hara R-B (2007) Vegan: community ecology package (version 1.8–6). http://cran.r-project.org/.
39. O'Hara RB., Kotze DJ. (2010) Do not log-transform count data. Methods in Ecology and Evolution 1:118–122.
40. R development core team (2010) R: a language and environment for statistical computing. R Foundation for Statistical Computing, Vienna, Austria. http://www.r-project.org/.
41. Pebesma EJ (2010) Geostatistical modelling, prediction and simulation (vesion 0.9–69). http://cran.r-project.org/.
42. Dray S, Legendre P, Blanchet G (2007) Forward selection with permutation (Canoco p.46) (Version 0.0–7). http://r-forge.r-project.org/R/?group_id = 195.
43. Jombart T, Dray S, Dufour A-B (2009) Finding essential scales of spatial variation in ecological data: a multivariate approach. Ecography 32: 161–168.
44. Stevenson P (2011) Pulp–seed attachment is a dominant variable explaining legitimate seed dispersal: a case study on woolly monkeys. Oecologia 166: 693–701.
45. Li L, Huang Z, Ye W, Cao H, Wei S, et al. (2009) Spatial distributions of tree species in a subtropical forest of China. Oikos 118: 495–502.
46. Zhang J, Hao Z, Song B, Li B, Wang X, et al. (2009) Fine-scale species co-occurrence patterns in an old-growth temperate forest. Forest Ecology and Management 257: 2115–2120.

47. Gleason SM, Read J, Ares A, Metcalfe DJ (2010) Species-soil associations, disturbance, and nutrient cycling in an Australian tropical rainforest. Oecologia 162: 1047–1058.
48. Pärtel M, Laanisto L, Zobel M (2007) Contrasting plant productivity-diversity relationships across latitude: the role of evolutionary history. Ecology 88: 1091–1097.
49. Pärtel M, Zobel K, Laanisto L, Szava-Kovats R, Zobel M (2010) The productivity–diversity relationship: varying aims and approaches. Ecology 91: 2565–2567.
50. Mo J, Brown S, Xue J, Fang Y, Li Z (2006) Response of litter decomposition to simulated N deposition in disturbed, rehabilitated and mature forests in subtropical China. Plant and Soil 282: 135–151.
51. Lu X, Mo J, Gilliam FS, Zhou G, Fang Y (2010) Effects of experimental nitrogen additions on plant diversity in an old-growth tropical forest. Global Change Biology 16: 2688–2700.
52. Brady NC, Weil RR (2002) The nature and properties of soils (13th ed). Prentice Hall, Inc, Upper Saddle River, NJ. pp. 363–409.
53. He J, Chen W (1997) A review of gradient changes in species in diversity of land plant coummunity. Acta Ecologica Sinica (Chinese version) 17: 91–99.
54. Bin Y, Wang Z, Wang Z, Ye W, Cao H, et al. (2010) The effects of dispersal limitation and topographic heterogeneity on beta diversity and phylobetadiversity in a subtropical forest. Plant Ecology 209: 237–256.
55. Tuomisto H, Ruokolainen L, Ruokolainen K (2012) Modelling niche and neutral dynamics: on the ecological interpretation of variation partitioning results. Ecography 35:961–971.

12

Biomass Allocation Patterns across China's Terrestrial Biomes

Limei Wang[1], Longhui Li[1]*, Xi Chen[1], Xin Tian[2]*, Xiaoke Wang[3], Geping Luo[1]

1 State Key Laboratory of Desert and Oasis Ecology, Xinjiang Institute of Ecology and Geography, Chinese Academy of Sciences, Urumqi, China, **2** Institute of Forest Information Resource Techniques, Chinese Academy of Forestry, Beijing, China, **3** Beijing Urban Ecosystem Research Station, State Key Lab of Urban and Regional Ecology Research Center for Eco-Environmental Sciences, Chinese Academy of Sciences, Beijing, China

Abstract

Root to shoot ratio (RS) is commonly used to describe the biomass allocation between below- and aboveground parts of plants. Determining the key factors influencing RS and interpreting the relationship between RS and environmental factors is important for biological and ecological research. In this study, we compiled 2088 pairs of root and shoot biomass data across China's terrestrial biomes to examine variations in the RS and its responses to biotic and abiotic factors including vegetation type, soil texture, climatic variables, and stand age. The median value of RS (RSm) for grasslands, shrublands, and forests was 6.0, 0.73, and 0.23, respectively. The range of RS was considerably wide for each vegetation type. RS values for all three major vegetation types were found to be significantly correlated to mean annual precipitation (MAP) and potential water deficit index (PWDI). Mean annual temperature (MAT) also significantly affect the RS for forests and grasslands. Soil texture and forest origin altered the response of RS to climatic factors as well. An allometric formula could be used to well quantify the relationship between aboveground and belowground biomass, although each vegetation type had its own inherent allometric relationship.

Editor: Han Y.H. Chen, Lakehead University, Canada

Funding: This study was funded by the National Natural Science Foundation of China (grant number 41171094), the "Hundred Talent" Project of the Chinese Academy of Sciences (grant number Y174051001), one of National Basic Research Program of China (grant number 2013CB733404), and the National Natural Science Foundation of China (grant number 41101379). The funders had no role in study design, data collection and analysis, decision to publish, or preparation of the manuscript.

Competing Interests: The authors have declared that no competing interests exist.

* E-mail: lhli@ms.xjb.ac.cn (LHL); tianxin@caf.ac.cn (XT)

Introduction

Belowground biomass (BGB) is an important component of global terrestrial ecosystem carbon stocks and plays a critical role in global carbon cycling. Belowground biomass is more difficult and costly to measure and is a major source of uncertainties in large-scale biomass estimation and global carbon cycles [1–3]. The partitioning between aboveground and belowground biomass influences many of the functions performed by diverse terrestrial communities as well as the functions performed by individual plants (e.g., [4–7]). Root to shoot ratio (RS) is an effective parameter to describe the allocation between aboveground biomass (AGB) and BGB, and thus provides a practical tool to estimate BGB by relatively easily measured AGB. The RS reflects a plant's specific adaptive responses to its environment and has been widely used as a key descriptor for terrestrial ecosystem carbon modeling [8].

Biomass partitioning between belowground and aboveground parts can be predicted by plant allometric relationships [6,9]. At the level of individual plants, AGB scales nearly isometrically with respect to BGB, and this relationship has been validated across a broad spectrum of ecologically diverse vascular plants spanning several orders of magnitude in total body mass [10]. The allometric theory also estimates an isometric relationship between AGB and BGB at the community level [11], which has not been examined by adequate field measurements, especially its generality across diverse vegetation groups.

Recent studies have indicated that RS changes with environmental factors (e.g., climate, CO_2, soil texture, soil moisture, and nitrogen), biotic factors (e.g., plant type, stand age, and leaf traits), and forest origins [1,12–15]. Some improvements have been made in determining biomass allocation; however, many of the past studies focused on a particular ecosystem, and there have been few comprehensive investigations of RS across multiple ecosystems. Investigating biomass allocation patterns and their relationships with environmental factors crossing multiple ecosystems at regional scale may shed new light on this important topic.

China has rich vegetation resources and diverse vegetation types. China has the fifth highest total forest area and second highest total grassland area among all countries in the world, and thus the country plays an important role in global carbon stocks and cycling. The numerous vegetation types, diverse soil textures, complicated land use/cover patterns, and varying climatic zones offer a unique chance to examine the variation of RS and its influencing factors. Validating RS data from diverse biomass also provides a good opportunity to test the allometric relationships for ecosystems partitioning between AGB and BGB. Considerable work has been conducted to investigate the performance of allometric theory for a specific ecosystem and at sub-regional scales. For example, Wang et al. [16] examined RS relationships for major forest types in northeast China and pointed out that they differed significantly between natural and planted forests, and between broadleaf and coniferous forests. Yang et al. [17] focused their investigation on the Tibetan grasslands and suggested that the isometric relationships between AGB and BGB did not differ significantly between alpine steppe and alpine meadow. Further, Yang et al. [17] investigated relationships between AGB and BGB of grassland ecosystems across northern China and indicated that AGB was nearly proportional to BGB with a scaling exponent across various grassland types at the community level with no significant difference between temperate and alpine grasslands or between steppe and meadow. Comprehensive research on biomass allocation and its response to both climate and soil types for multiple ecosystems are essential to improving our understanding of the general rules of biomass allocation for biomes.

In this study, we compiled a comprehensive database of biomass and RS for the major terrestrial biomes across China. We used statistical approaches to address: 1) how RS varies across China's terrestrial biomes, 2) how RS responds to biotic and abiotic variables, and 3) how well the allometric theory performs in comparison with estimates of the empirical relationship between RS and environmental variables.

Materials and Methods

Datasets

We collected 2088 measurements of root and shoot biomass across China from hundreds of published papers (Fig. 1, Appendix S1). In the database, 276 records were extracted from Wang et al. [16], 377 records were extracted from Yang et al. [8], and 1139 records were extracted from Luo et al. [18]. The rest of the records in the dataset were compiled by the authors from related studies.

For each data set, we recorded the biomass data and whatever supporting information was available, including (1) shoot, root, and total biomass (dry weight of biomass per unit area); (2) site descriptions (including location, latitude, longitude, elevation, mean annual temperature, mean annual precipitation, mean annual potential evapotranspiration (PET), and soil texture); (3) vegetation descriptions (including vegetation type, dominate species, forest origin (primary, secondary, planted forest), age, height, density of stems, and mean diameter at breast height (DBH)); (4) descriptions of shoot and root biomass sampling methods (including sample size, dimensions of soil cores or soil pits, depth of sampling, whether root crowns were sampled, and whether existing allometric theories were applied).

The climatic variables mean annual precipitation (MAP) and temperature (MAT) at each biomass-sampling site were determined from the nearest surface meteorological stations across China (Fig. 1). Historically mean values of MAP and MAT during the period from 1961 to 2010 were used for analysis. Among 2088 sites, 77 sites have attached MAP and 88 sites have attached MAT. We assessed the accuracy of extracted MAP and MAT by comparing with site-attached values and found that extracted MAP was significantly correlated with the recorded measurements ($r^2 = 0.92$, $n = 77$, $p < 0.01$), and extracted MAT was in good agreement with the measurements ($r^2 = 0.9$, $n = 88$, $p < 0.01$) as well. We computed mean annual potential evapotranspiration (PET) using the FAO Penman–Monteith equation [19] from the meteorological variables recorded by the 664 surface stations, and then we computed potential water deficit index (PWDI) by the ratio of the mean annual precipitation to annual potential evapotranspiration (MAP/PET) [20].

Vegetation Types

The categorization of data into vegetation types relied primarily upon a soil texture classification scheme [17,18]. Different vegetation types had obviously different RS and soil texture may alter the value of RS for same vegetation types. Based on a soil texture map (Institute of Soil Science, Chinese Academy of Sciences, 1986), the dataset was divided into eight types: loam forest, sandy forest, clay forest, loam grassland, sandy grassland, clay grassland, loam shrubland, and sandy shrubland (Table 1). Recent studies have indicated that RS shows different physiolog-

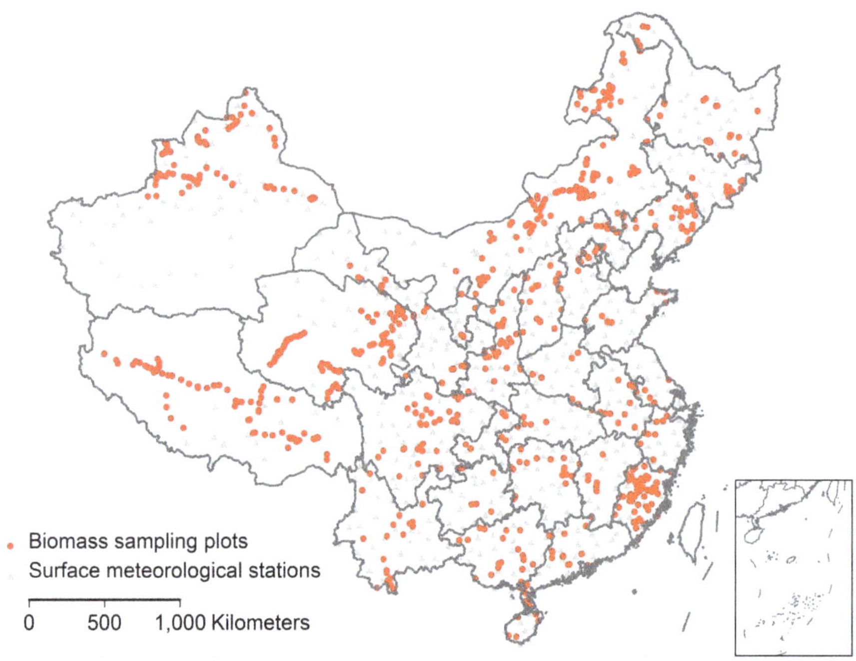

Figure 1. The geographical distribution of biomass sampling plots collected in this study. These data including forests, shrublands, and grasslands distributed across China.

Table 1. Median of above-ground biomass (AGB), below-ground biomass (BGB), root to shoot ratio (RS) and allometric relationships between AGB and BGB ($p<0.05$) across China's biomes.

Vegetation category	AGB(Mg ha^{-1})	BGB(Mg ha^{-1})	RS	Range	n	LogAGB–logBGB models		
						Slope	Intercept	r^2
Vegetation type								
Primary forest	99.1	24.0	0.24	0.03–1.20	454	0.87	−0.37	0.85
Secondary forest	93.7	22.6	0.23	0.08–0.58	97	0.95	−0.52	0.78
Planted forest	68.4	15.5	0.22	0.05–0.90	1002	0.81	−0.30	0.76
Temperate grassland	1.5	7.7	5.73	0.43–29.2	243	0.77	0.74	0.58
Alpine grassland	0.7	4.1	6.27	0.81–45.6	262	1.00	0.79	0.73
Soil texture								
Loam forest	76.5	16.7	0.22	0.02–1.20	967	0.90	−0.45	0.82
Sand forest	62.2	15.6	0.25	0.03–0.70	242	0.82	−0.30	0.79
Clay forest	91.1	19.3	0.22	0.05–0.50	345	0.87	−0.42	0.87
Loam grassland	1.0	5.5	6.10	0.43–29.2	336	0.78	0.74	0.66
Sand grassland	0.7	5.0	5.61	0.56–45.6	153	0.89	0.71	0.59
Clay grassland	1.1	7.2	7.60	2.0–12.72	10	0.96	0.87	0.74
Loam shrubland	9.5	11.3	0.71	0.26–1.56	12	1.00	−0.31	0.80
Sand shrubland	10.6	8.6	0.77	0.47–1.96	9	1.16	−0.44	0.95
Region								
China's forests	76.7	17.0	0.23	0.03–1.20	1551	0.86	−0.36	0.81
Global forests	81.0	22.5	0.25	0.09–1.16	207	0.88	−0.32	0.90
China's grasslands	1.0	5.2	6.00	0.43–45.6	502	0.82	0.74	0.63
Global grasslands	3.5	12.7	3.26	0.38–26.0	61	0.40	0.84	0.20
China's shrublands	10.6	10.6	0.73	0.26–1.96	22	1.05	−0.16	0.89
Global shrublands	16.6	14.0	1.84	0.34–4.25	8	–	–	–

n, number of samples; Slope, the scaling exponent of the log–log linear functions; Intercept, the allometric constant of log–log linear functions; r, the coefficient of determination.

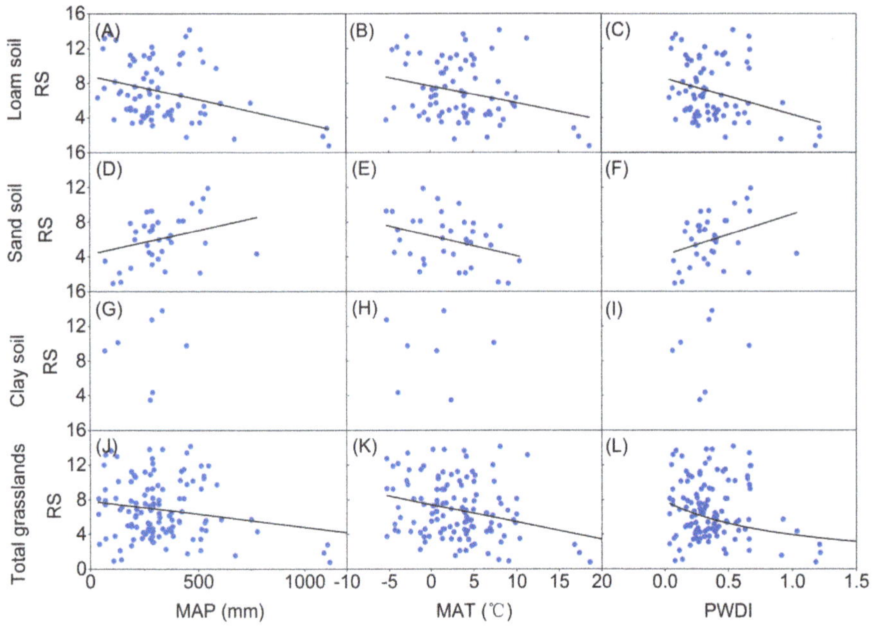

Figure 2. Root to shoot ratio (RS) for grasslands as a function of mean annual precipitation (MAP), temperature (MAT) and potential water deficit index (PWDI). Regression lines are given for loam, sandy, and claygrasslands and for all data together, if the relationships are significant at $p<0.05$.

ical and ecological responses to varied surroundings for a range of forest stand age and climate zones [8,12,16]. Summarizing these studies, the forests and grasslands were further investigated based on five groups divided by the forest origin and climatic zone: primary forest, secondary forest, planted forest, temperate grassland, and alpine grassland (Table 1).

Statistical Analysis

Descriptive analysis was used to determine the statistical characteristics (e.g., median value, range of data) of categorical data. Linear and nonlinear analyses were performed to develop regressions for RS dependent on the vegetation type, soil texture, and climatic variables. The significance of differences between slopes (scaling exponent) and y-intercepts (allometric constant) of the log-log transformed linear functions was evaluated by analyses of variance [21]. The standard error of the estimation (SEE) was used to measure the difference between predicted and actual values of root biomass.

Results

Variations of RS across China's Biomes

The biomass and RS values shown in Table 1 are referred to as the median rather than the mean values for the skewed distribution of data for many of the vegetation types. RS values ranged from 0.03 to 1.2, 0.43 to 45.6, and 0.26 to 1.96 for forests, grasslands, and shrublands, respectively (Table 1). The median RS tended to decrease from grasslands to shrublands to forests across China (Table 1).

The median RS of forests tended to increase from planted to secondary to primary stands. In addition, forests growing in sandy soil had a larger median RS than those in loam and clay soils. Further, loam soil forests had approximately the same median RS value as the clay soil forests (Table 1).

For grasslands, RS changed significantly with different grassland types. The median RS of alpine grassland was much higher than that of temperate grassland (Table 1). There was a general trend for RS to vary with soil texture changes across grasslands. The

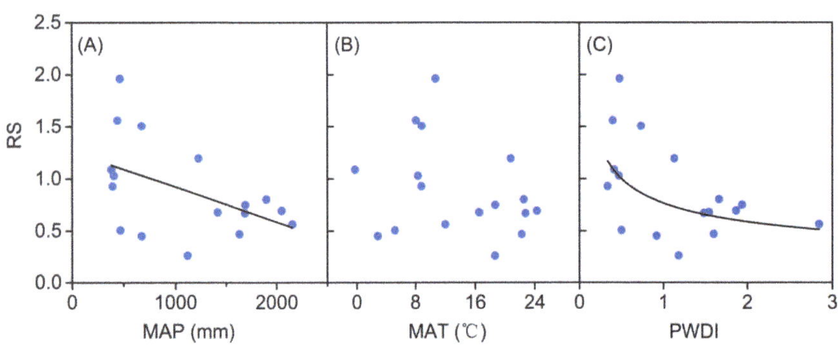

Figure 3. Root to shoot ratio (RS) for shrublands as a function of mean annual precipitation (MAP), temperature (MAT) and potential water deficit index (PWDI). Regression lines are given for all data together if the relationships are significant at $p<0.05$.

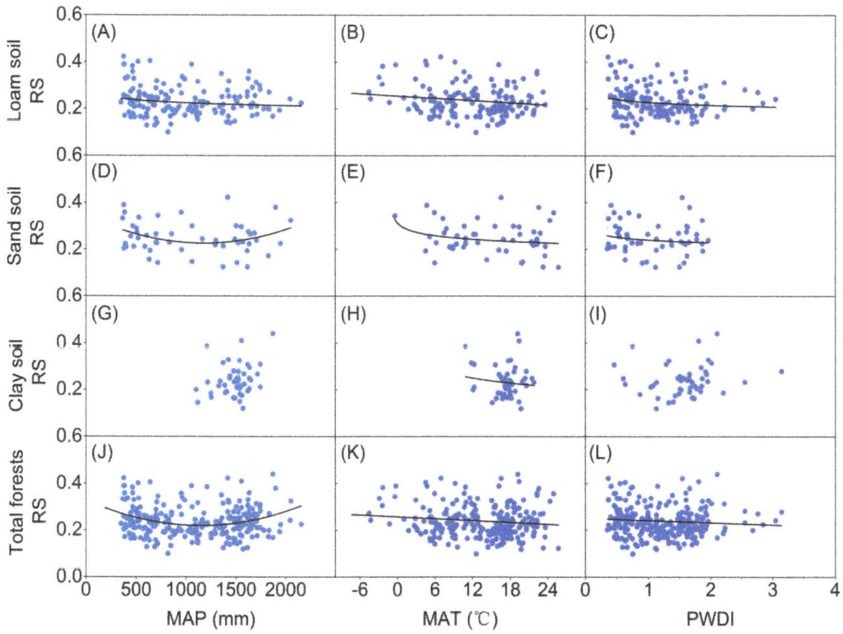

Figure 4. Root to shoot ratio (RS) for forests as a function of mean annual precipitation (MAP), temperature (MAT) and potential water deficit index (PWDI). Regression lines are given for loam, sandy, and clay forests and for all data together, if the relationships are significant at $p<0.05$.

median RS decreased from clay soil grassland to loam soil grassland to sandy soil grassland.

Responses of RS to Climatic Variables

China's forests, grasslands and shrublands responded differently to climatic variables. For both grasslands (Fig. 2J) and shrublands (Fig. 3A), RS decreased significantly ($p<0.01$) with increased MAP. A similar trend was found in RS for forests where MAP was less than 1250 $mm{\cdot}yr^{-1}$; however, RS values tended to increase significantly ($p<0.01$) where MAP was over 1250 $mm{\cdot}yr^{-1}$ (Fig. 4J). Generally, soil nutrient availability dominated RS where precipitation was high (>1250 mm) [26] and more biomass was allocated into roots for uptaking nutrient and RS tend to increase with MAP. Although RS decreased significantly with increasing MAT for both forests ($p<0.05$, Fig. 4K) and grasslands ($p<0.01$, Fig. 2K), there was no significant relationship between MAT and

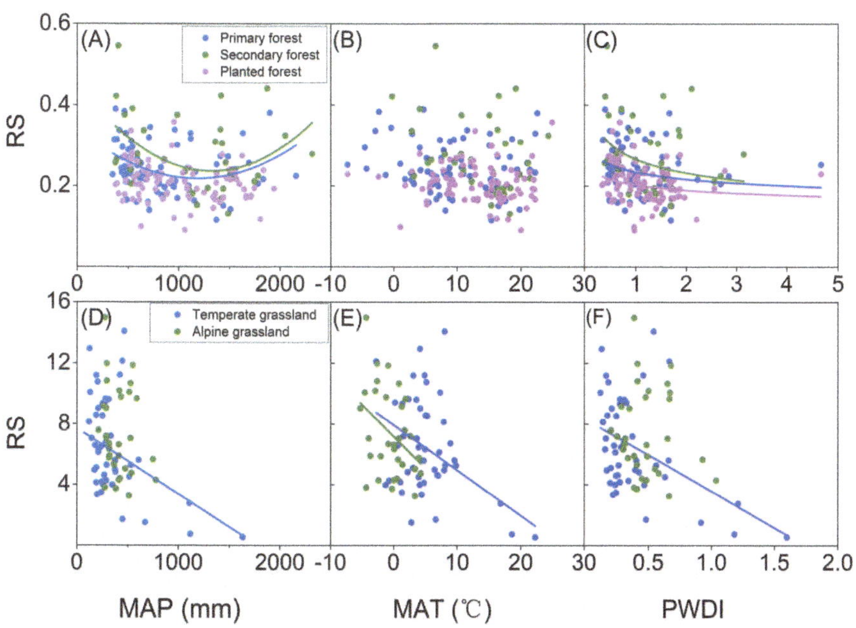

Figure 5. Root to shoot ratio (RS) as a function of mean annual precipitation (MAP), temperature (MAT) and potential water deficit index (PWDI): Forests (A–C) and Grasslands (D–F). Regression lines are given for every vegetation group if the relationships are significant at $p<0.05$.

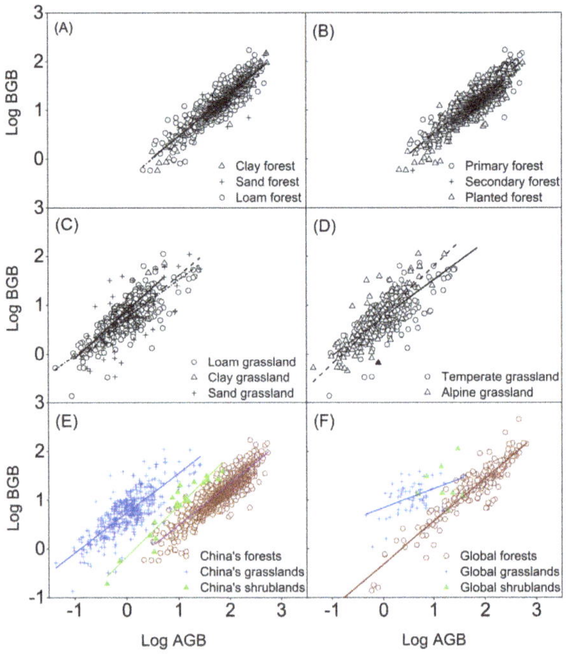

Figure 6. Allometric relationships between aboveground biomass (AGB) and belowground biomass (BGB) for different vegetation groups. (A) Comparison in between loam, sand and clay forest. (B) Comparison in between primary, secondary and planted forest. (C) Comparison in between loam, sand and clay grassland. (D) Comparison in between temperate and alpine grassland. (E) Comparison in between China's forest, grassland and shrubland. (F) Comparison in between global forest, grassland and shrubland. Regression lines and equations are given for every vegetation group if the relationships are significant at $p<0.05$. Global biomass data are from Mokany et al. [12].

RS for shrublands (Fig. 3B). RS values were negatively related to PWDI ($p<0.01$) for all forests (Fig. 4L), grasslands (Fig. 2L), and shrublands (Fig. 3C).

Soil texture affected RS's responses to climate. RS from both loam and sandy soils showed significant responses to climate variables for forests (Figs. 4A–F) and grasslands (Figs. 2A–F). However, no general trend was found between climate and RS from clay soil for either forests or grasslands (Figs. 2G–I). In addition to soil texture, the responses of RS to climate were influenced by forest origins. RS for natural forests showed a significant U-shaped trend ($p<0.05$) as MAP increased; however no trend existed between MAP and RS for planted forests (Fig. 5A). Neither natural nor planted forests showed a significant trend for RS as a function of MAT (Fig. 5B); however, RS decreased significantly ($p<0.01$) as PWDI increased for both natural and planted forests (Fig. 5C). Climatic zonation was an important factor influencing responses of grasslands' RS to climate. RS for temperate grasslands was negatively related to MAP, MAT, and PWDI (Fig. 5D–F, $p<0.01$). In contrast, no significant trend existed between RS and MAP (Fig. 5D) or PWDI (Fig. 5F) for alpine grassland, although RS for alpine grassland decreased significantly ($p<0.01$) with increasing MAT (Fig. 5E).

Allometric Relationships between Root and Shoot Biomass

The relationships between root biomass (y) and shoot biomass (x) across China's biomes fit well with an allometric power function, $y = ax^b$ (Fig. 6). The coefficients for a were 0.3, 8.02, and 2.35 and for b were 0.94, 0.72, and 0.66 for forests, grasslands, and shrublands, with corresponding r^2 of 0.75, 0.5, and 0.85, respectively. In general, the allometric relationships between root and shoot biomass were significantly ($p<0.01$) different from each other between China's major terrestrial ecosystems (forests, grasslands, and shrublands) and all slopes were significant ($p<0.01$) (Fig. 6; Table 1).

Across different forest types, none of the regressions between root and shoot biomass were significantly ($p<0.05$) different from each other and all slopes were significant ($p<0.01$) (Figs. 6A, B; Table 1). However, soil texture and climatic zonation altered grasslands' allometric relationships. The slope of the regression for clay grasslands was significantly ($p<0.01$) greater than that of both loam and sandy grasslands. The allometric relationship for loam soil is not significantly different from that of sandy soil ($p<0.01$) (Fig. 6C; Table 1). The slope for alpine grasslands was significantly greater ($p<0.01$) than that of temperate grasslands. The allometric relationship for alpine grasslands was significantly different from that of temperate grasslands ($p<0.01$) (Fig. 6D; Table 1). Meanwhile, the allometric relationship for China's grasslands was significantly different from the global average; however, the slope of the allometric equation for China's forests was similar to the global value (Figs. 6E, F).

Discussion

Variations in RS across China's Biomes

We reported that the overall median values of AGB were 76.7, 1.0 and 10.6, and the median of BGB were 17.0, 5.2 and 9.4 for China's forests, grasslands and shrublands, respectively (Table 2). As a consequence, the resulted median RS were 0.23, 0.60 and 0.74 for China's forests, grasslands and shrublands, respectively. Our estimated AGB, BGB and RS for grasslands were quite comparable with previous research using Northern China's dataset [8]. However, a remarkable difference existed in terrestrial root and shoot biomass and RS between China and global averages. Both root and shoot biomass in China were lower than global values for each vegetation type (Table 3). The median RS values for China's forests (0.23) and shrublands (0.73) were lower than global estimates (0.29 and 2.45, respectively [12]); however, the median RS for China's grasslands (6.0) was much higher than global estimates (3.3 [12] and 3.7 [22]). Such difference may come from several aspects. Firstly, sampling methods and the number of dataset used may be one of the main reasons. Makony et al. [24] pointed that data omitting was critical to the analysis of root and shoot ratio. Limited by axillary information for the dataset used in the current research, we ignored the selection and keep all data points we collected in analysis. Such treatment may avoid distorting the possible nature of biomass allocation pattern across wide biome, climate and soil conditions. Secondly, accuracy of climatic variables (mean annual precipitation and temperature) and soil properties (soil texture) may contribute the uncertainties of estimated RS values. Particularly for those data sites located in different altitude or slope of mountain areas, different climatic conditions should be expected but same climatic variables were used for analysis due to the lack of available meteorological observations. To overcome this drawback, we averaged all data points where shared meteorological measurements. However, the detailed response of RS to climatic variables may be also masked.

Factors Influencing Biomass Allocation

Biomass allocation is one of the most successful theories in modern ecology [36]. Although plant growth and biomass allocation are highly influenced by environmental and

Table 2. Statistical analysis for predicted root biomass (BGB) using median vegetation-specific root to shoot ratio (RS) and allometric equations.

Vegetation category		**Predicted root biomass by median vegetation-specific RS**			**Predicted root biomass by allometric equations**		
		Slope	r^2	**SEE**	**Slope**	r^2	**SEE**
Vegetation type							
Primary forest	443	0.95	0.77	12.1	1.09	0.76	12.1
Secondary forest	97	1.00	0.68	11.1	1.10	0.68	11.0
Planted forest	997	0.72	0.70	7.24	0.76	0.71	7.22
Temperate grassland	217	1.42	0.79	30.2	2.66	0.66	38.1
Alpine grassland	258	1.36	0.76	12.7	2.22	0.68	14.7
Soil texture							
Loam forest	951	0.97	0.72	9.68	1.01	0.72	9.66
Sand forest	242	0.82	0.78	8.50	0.97	0.78	8.63
Clay forest	342	0.89	0.81	7.85	0.93	0.80	7.92
Loam grassland	312	0.54	0.42	12.5	0.96	0.45	12.1
Sand grassland	152	0.73	0.56	11.8	1.18	0.60	11.3
Clay grassland	12	0.32	0.008	5.18	0.51	0.08	4.99
Loam shrubland	11	0.73	0.84	5.30	1.03	0.89	4.50
Sand shrubland	9	1.35	0.66	4.85	1.26	0.65	4.95
Region							
China's forests	1535	0.90	0.75	9.15	0.98	0.74	9.16
Global forests	207	1.02	0.56	30.1	0.98	0.56	30.1
China's grasslands	475	0.60	0.46	12.4	1.05	0.50	12.0
Global grasslands	57	0.13	0.04	9.66	0.71	0.04	9.66
China' s shrublands	21	0.76	0.79	5.26	1.04	0.85	4.53
Global shrublands	8	–	–	–	–	–	–

n is number of samples, Slope is the scaling exponent of the regressions, r is coefficient of determination and SEE is standard error of the estimation.

Table 3. Above-ground biomass (AGB), below-ground biomass (BGB) and root to shoot ratio (RS) for forests, grasslands and shrublands from different studies.

Vegetation category	AGB (Mg ha^{-1})	BGB (Mg ha^{-1})	RS	Reference
Forest				
China's forests	76.7	17.0	0.23	This study
China's forests	–	–	0.23	Luo et al. (2012)
Global forests	81.0	22.5	0.25	Mokany et al. (2006)
Grassland				
China grasslands	1.0	5.2	6.0	This study
China grasslands	1.0	5.7	5.7	Yang et al. (2009)
Global grasslands	3.5	12.7	3.3	Mokany et al. (2006)
Global grasslands	3.8	14.0	3.7	Jackson et al. (1996)
Shrubland				
China's shrublands	10.6	9.7	0.73	This study
Global shrublands	16.6	14.0	1.84	Mokany et al. (2006)

Arithmetic mean or median values are presented to compare with other studies.

non-environmental factors, a mass number of previous research have demonstrated that biomass allocation is impacted by climatic variables (particularly precipitation and temperature), soil type, nutrient availability and plant species [17,23–26,32,33].

Majority of previous studies reported that RS can be either positively or negatively and linearly correlated and even non-linearly related with the increase in annual precipitation at relatively larger regional or national scales [8,16,37,38,18,28] but the value of RS overall decreased with precipitation at global scale [3]. Similarly, the effect of temperature on RS can be different from region to region, vegetation type to type [18,38,27]. Various possible combinations of hydrothermal conditions (precipitation and temperature) may explain the variation of the diverse response of RS to individual factor of precipitation and temperature. Further, soil texture, vegetation type, composition and age, nutrient availability also induced variations in the RS across ecosystems [3]. Generally, there was an increasing trend for RS as soil texture changed from clay (0.22) to sand (0.25) for forests (Table 1). Lower water and nutrient availability from sandy soil may be responsible for the larger RS. In contrast, RS values of grasslands showed trend of decreasing when soil texture changed from clay to sand (Table 1). Our results also found that RS is more sensitive to MAT than MAP in grasslands. Lower temperatures in clay soil caused roots to gain more biomass to maintain energy and acquire water and nutrients [21,31], which may be responsible for the higher RS for clay grasslands.

One of key contributions of this study is that we comprehensively examined the responses of RS to multiply abiotic and biotic factors, including climatic variables (MAP, MAT, and PWDI), soil texture and vegetation type and age. The integrated analysis provided comprehensive understanding on the variations of RS and their response of RS to environmental factors. We found that RS could be significantly impacted by any of considered abiotic or biotic factors. Compared with any single factor from climate, a comprehensive index PWDI could be better to describe the dependency of RS on environmental factors. Biomass partition is commonly viewed as a result impacted by multiple environmental factors. Variation in RS and difference in their responses to individual or combined environmental factors supported the optimal partitioning hypothesis [29,30], suggesting an adaptation for plants in varied environments. It is widely recognized that RS varies with various biotic (e.g., stand age, vegetation type, and growing period) and abiotic factors (e.g., soil texture, soil nutrients, climate, and plant origin), as demonstrated in our analysis, although the mechanism is not well understood and inconsistencies between findings from existing studies are still quite large.

Differing from the analysis between RS and environmental factors, allometric equations have been also widely accepted as an effective method for investigating biomass partitioning between aboveground and belowground components (e.g., [1,34,35]). As an alternative, we attempt to examine whether the vegetation-specific RS presented in this study would provide a more accurate and easier means for estimating root biomass than allometric equations. We used the root and shoot biomass data collected in this study to compare the predictive accuracy of the two methods. We found that the application of median RS predicted root biomass with approximately the same accuracy as the allometric equations for 7 of the 13 vegetation groups (Table 2). Among the other 6 groups, we had a better prediction of root biomass using median RS for temperate and alpine grasslands, but lower predictive accuracy in loam, sandy, and clay grasslands and loam shrublands (Table 2). Moreover, the application of vegetation-specific RS for forests across soil textures had better predictive accuracy for root biomass than the single median RS for total forests in China. Similarly, the vegetation-specific RS for temperate and alpine grasslands also had better predictive accuracy for root biomass than the single median RS for total grasslands in China. This indicated that the accuracy of root biomass prediction varies significantly across vegetation types, particularly for grasslands, emphasizing that it is essential to advance our understanding of RS dynamics across vegetation types for improving the accuracy of root biomass predictions [12].

Conclusions

This paper systematically examined the variation in RS across China's terrestrial biomes and its influencing factors including vegetation type, soil properties, climatic variables, and stand age. We found that median RS tended to decrease from grasslands (6.0) to shrublands (0.73) to forests (0.23) in China. Although primarily decided by the plants' inherent isometric relationships, RS were significantly ($p<0.05$) influenced by vegetation type, soil texture, climatic variables, and stand age.

RS showed general trends with changing mean annual temperature, mean annual precipitation, and potential water deficit index. RS values were negatively related to mean annual precipitation and positively related to potential water deficit for forests, grasslands, and shrublands. For both forests and grasslands, RS were negatively related to mean annual temperature. Soil texture, forest origin, and climatic conditions affected the size of the RS and its relationships with climatic factors. Allometric equations performed better than empirical vegetation-specific RS equations in 6 out of 13 vegetation types.

Supporting Information

Appendix S1 Dataset on the root to shoot ratio, soil and vegetation types and climatic variables across China's terrestrial biomes.

Author Contributions

Conceived and designed the experiments: LHL. Performed the experiments: LMW XT. Analyzed the data: LMW. Contributed reagents/materials/analysis tools: XKW XC GPL. Wrote the paper: LHL XC GPL LMW. Compiled and formatted the data: LHL LMW.

References

1. Cairns MA, Brown S, Helmer EH, Baumgardner GA (1997) Root biomass allocation in the world's upland forests. Oecologia 111: 1–11.
2. Reich PB (2002) Root–shoot relations: optimality in acclimation and adaptation or the "Emperor's New Clothes"? In: Waisel Y, Eshel A, Kafkafi U, editors. Plant roots: The hidden half. Basel, Switzerland: Marcel Dekker. 205–220.
3. Mokany K, Raison RJ, Prokushkin AS (2005) Critical analysis of root:shoot ratios in terrestrial biomes. Glob Chang Biol 11: 1–13.
4. Poorter H, Nagel O (2000) The role of biomass allocation in the growth response of plants to different levels of light, CO_2, nutrients and water: A quantitative review. Aust J Plant Physiol 27: 595–607.
5. Binkley D, Stape JL, Ryan M (2004) Thinking about resource use efficiency in forests. For Ecol Manage 193: 5–16.
6. Niklas KJ (2005) Modeling below- and above-ground biomass for non-woody and woody plants. Ann Bot 95: 315–321.
7. Hui DF, Jackson RB (2006) Geographical and interannual variability in biomass partitioning in grassland ecosystems: a synthesis of field data. New Phytol 169: 85–93.
8. Yang YH, Fang JY, Ma WH, Guo DL, Mohammat A (2010) Large-scale pattern of biomass partitioning across China's grasslands. Global Ecol Biogeogr 19: 268–277.
9. Enquist BJ, Niklas KJ (2002) Global allocation rules for patterns of biomass partitioning across seed plants. Science 295: 1517–1520.
10. Niklas KJ (2006) A phyletic perspective on the allometry of plant biomass-partitioning patterns and functionally equivalent organ-categories. New Phytol 171: 27–40.
11. Cheng DL, Niklas KJ (2007) Above- and below-ground biomass relationships across 1534 forested communities. Ann Bot 99: 95–102.
12. Mokany K, Raison RJ, Prokushkin AS (2006) Critical analysis of root:shoot ratios in terrestrial biomes. Glob Chang Biol 12: 84–96.
13. Li LH, Yu Q, Zheng YF, Wang J, Fang QX (2006) Simulating the response of photosynthate partitioning during vegetative growth of winter wheat to environmental factors. Field Crops Res 96: 133–141.
14. Camila AC, Henrik S, Linda G, Annika N, Ulrika G, et al. (2011) Patterns of plant biomass partitioning depend on nitrogen source. PLoS One 6: e19211.
15. Yu Q, Wu H, He N, Lü X, Wang Z, et al. (2012) Testing the growth rate hypothesis in vascular plants with above- and below-ground biomass. PLoS One 7: e32162.
16. Wang XP, Fang JY, Zhu B (2008) Forest biomass and root–shoot allocation in northeast China. For Ecol Manage 255: 4007–4020.
17. Yang YH, Fang JY, Ji CJ, Han WX (2009) Above- and belowground biomass allocation in Tibetan grasslands. Journal of Vegetation Science 20: 177–184.
18. Luo YJ, Wang XK, Zhang XQ, Booth TH, Lu F (2012) Root:shoot ratios across China's forests: Forest type and climatic effects. For Ecol Manage 269: 19–25.
19. FAO (1998) Crop evapotranspiration–Guidelines for computing crop water requirements. Rome, Italy: Food and Agriculture Organization of the United Nations.
20. Brown S, Lugo AE (1982) The storage and production of organic matter in tropical forests and their role in the global carbon cycle. Biotropica 14: 161–187.
21. Crawley MJ (2002) Statistical computing–An introduction to data analysis using S-Plus. New York: John Wiley & Sons.
22. Jackson RB, Canadell J, Ehleringer JR, Mooney HA, Sala OE, et al. (1996) A global analysis of root distributions for terrestrial biomes. Oecologia 108: 389–411.
23. Ericsson T, Rytter L, Vapaavuori E (1996) Physiology of carbon allocation in trees. Biomass Bioenerg 11: 115–127.
24. Matsui Y, Fukuda Y, Inoue T, Matsushita T (2003) Effect of natural organic matter on powdered activated carbon adsorption of trace contaminants characteristics and mechanism of competitive adsorption. Water Res 37: 4413–4424.
25. Gholz HL, Vogel SA, Cropper WP Jr, McKelvey K, Ewel KC, et al. (1991) Dynamics of canopy structure and light interception in *Pinus elliottii* stands, north Florida. Ecol Monogr 61: 35–51.
26. Agren GI, Franklin O (2003) Root: shoot ratios, optimization and nitrogen productivity. Ann Bot 92: 795–800.
27. Read JJ, Morgan JA (1996) Growth and partitioning in *Pascopyrum smithii* (C3) and *Bouteloua gracilis* (C4) as influenced by carbon dioxide and temperature. Ann Bot 77: 487–496.
28. Li LH, Wang YP, Yu Q, Pak B, Eamus D, et al. (2012) Improving the responses of the Australian community land surface model (CABLE) to seasonal drought. J Geophys Res 117: G04002.
29. Friedlingstein P, Joel G, Field CB, Fung IY (1999) Toward an allocation scheme for global terrestrial carbon models. Glob Chang Biol 5: 755–770.
30. McConnaughay KDM, Coleman JS (1999) Biomass allocation in plants: ontogeny or optimality? A test along three resource gradients. Ecology 80: 2581–2593.
31. Gower ST, Vogt KA, Grier CC (1992) Carbon dynamics of Rocky Mountain Douglas scheme-fir: Influence of water and nutrient availability. Ecol Monogr 62: 43–65.
32. Schenk HJ, Jackson RB (2002) Rooting depths, lateral root spreads and below-ground/above-ground allometries of plants in water-limited ecosystems. J Ecol 90: 480–494.
33. Kellomäki S, Wang KY (2001) Photosynthetic responses to needle water potentials in Scots pine after a four-year exposure to elevated CO_2 and temperature. Tree Physiol 16: 765–772.
34. Vogt KA, Vogt DJ, Bloomfield J (1998) Analysis of some direct and indirect methods for estimating root biomass and production of forests at an ecosystem level. Plant Soil 200: 71–89.
35. Snowdon P, Eamus D, Gibbons P, Khanna PK, Keith H, et al. (2000) Synthesis of allometrics, review of root biomass and design of future woody biomass sampling strategies. NCAS Technical Report 17. Canberra: Australian Greenhouse Office.
36. Stearns SC (1992) The evolution of life histories. Oxford: Oxford University Press.
37. Kang M, Dai C, Ji W, Jiang Y, Yuan Z, et al. (2013) Biomass and its allocation in relation to temperature, precipitation, and soil nutrients in Inner Mongolia grasslands, China. PloS One, 8(7): e69561. doi:10.1371/journal.pone.0069561.
38. Luo YJ, Wang XK, Zhang XQ, Ren Y, Poorter H (2013) Variation in biomass expansion factors to forest type, climate and stand development. Annals of Forest Sciences, 70: 589–599.

13

Habitat Selection of a Large Carnivore along Human-Wildlife Boundaries in a Highly Modified Landscape

Chihiro Takahata[1]*, Scott Eric Nielsen[2], Akiko Takii[3], Shigeyuki Izumiyama[3]

1 Interdisciplinary Graduate School of Science and Technology, Shinshu University, Nagano, Japan, **2** Department of Renewable Resources, University of Alberta, Edmonton, Alberta, Canada, **3** Education and Research Center of Alpine Field Science, Faculty of Agriculture, Shinshu University, Nagano, Japan

Abstract

When large carnivores occupy peripheral human lands conflict with humans becomes inevitable, and the reduction of human-carnivore interactions must be the first consideration for those concerned with conflict mitigation. Studies designed to identify areas of high human-bear interaction are crucial for prioritizing management actions. Due to a surge in conflicts, against a background of social intolerance to wildlife and the prevalent use of lethal control throughout Japan, Asiatic black bears (*Ursus thibetanus*) are now threatened by high rates of mortality. There is an urgent need to reduce the frequency of human-bear encounters if bear populations are to be conserved. To this end, we estimated the habitats that relate to human-bear interactions by sex and season using resource selection functions (RSF). Significant seasonal differences in selection for and avoidance of areas by bears were estimated by distance-effect models with interaction terms of land cover and sex. Human-bear boundaries were delineated on the basis of defined bear-habitat edges in order to identify areas that are in most need of proactive management strategies. Asiatic black bears selected habitats in close proximity to forest edges, forest roads, rivers, and red pine and riparian forests during the peak conflict season and this was correctly predicted in our human-bear boundary maps. Our findings demonstrated that bears selected abandoned forests and agricultural lands, indicating that it should be possible to reduce animal use near human lands by restoring season-specific habitat in relatively remote areas. Habitat-based conflict mitigation may therefore provide a practical means of creating adequate separation between humans and these large carnivores.

Editor: Danilo Russo, Università degli Studi di Napoli Federico II, Italy

Funding: This research was funded in part by Grants-in-Aid for Scientific Research (B; 23310030) https://kaken.nii.ac.jp/d/p/23310030.en.html. The funders had no role in study design, data collection and analysis, decision to publish, or preparation of the manuscript. And no additional external funding received for this study.

Competing Interests: The authors have declared that no competing interests exist.

* E-mail: takahatacat@nifty.com

Introduction

As human land domination has expanded, many species have disappeared from their primary range. Nevertheless, some carnivores still survive within or near anthropogenic landscapes [1]. There are several reports of large carnivores periodically frequenting peri-farmlands: for instance, grey wolves (*Canis lupus*) [2]; bobcats (*Lynx rufus*) [3]; and American black bears (*Ursus americanus*) [4]. The inevitable consequence of the proximity of wildlife habitats to human-dominated lands is an increase in undesirable human-wildlife interactions, which in many cases involving a large carnivore can be fatal for both. The conservation of large carnivores newly adapted to human landscapes is one of the greatest challenges facing local wildlife managers because of the difficulty of reconciling the ecological requirements of animals with the need to preserve human life and property. The reduction of human-carnivore interactions is critical to the sharing of finite land, and this only seems possible through a better understanding of the processes and patterns involved in the use of human landscapes by wildlife.

Attention must first be directed to areas which animals find particularly attractive. Because only limited areas remain undisturbed in anthropogenic landscapes, animals need to find ways to derive some benefit from their habitat while simultaneously keeping their distance from the risks posed by humans. For instance, American black bears avoided frequent contact with people by shifting their core active time from day to night [4]. An intermediate level of housing density a short distance from a large forest edge was the main factor in human-bear interaction as it offered a combination of foraging opportunity and defensive refuge [5]. The effects which distance from human disturbance has on wildlife have been assessed worldwide [6–8]. Most previous research has been conducted on a broad scale, and there have been few studies done at a fine scale and in highly populated areas in which there is a substantial overlap with wildlife habitats.

Areas of overlap in human-wildlife habitat result in linear-shaped boundaries where both habitats are separated by distinct dichotomic geographical and ecological features (e.g. forest cover and open, flat and rugged terrain, etc.). The extent and structure of human-wildlife boundaries are an important factor in human-wildlife interaction. For example, sufficient space and environmental gradients within a boundary may give both people and animals a chance to avoid sudden and frequent contacts. In contrast, a boundary without enough space or tonal structure becomes a potential source of human-wildlife conflict. In some developing regions, such sharpening of human-wildlife boundaries has occurred due to massive land use expansion. It also happens in other regions where, conversely, land use changes by de-populated and aged rural societies cause wildlife habitat use to shift back toward the fringes of agriculture or urban areas. Increased tension is the inevitable consequence of sharpened human-wildlife boundaries. Large carnivores in particular suffer from a high risk of mortality due to the fears of local communities who have experienced fatal encounters, even if only on rare occasions. Although conflict mitigation is critical to the conservation of large carnivores in peripheral human lands, investigations of the habitats potentially associated with human-carnivore encounters have seldom been conducted [5]. Delineating naturally occurring human-wildlife boundaries also has the potential to help land use managers prioritize areas for management.

In Japan, current changes in human land use may have effects on the distribution and structure of human-wildlife boundaries. One of the more distinct land use changes has been a drastic reduction in the traditional use of coppice forests on the fringes of agricultural fields or settlements (see detail in [9]). Such secondary forests have become densely covered after the cessation of logging, and the area of unmanaged privately owned forests has reached about 30% of total forested lands [10]. Constant human presence and open patches in the secondary forests formally played an important role in preventing direct and frequent contact with wildlife. Moreover, about 11% of farmland has been abandoned since the 1960s due to a nationwide decline and the aging of the rural population [11]. For the last decade, Asiatic black bears (*Ursus thibetanus*) have experienced high mortality due to a surge in conflicts and contacts with people. In 2006, unusually large numbers of bears were sighted within and around rural and suburban lands, and as a result about 40% of the estimated total Japanese black bear population was killed [12]. In Nagano prefecture, the number of destroyed bears was 558, well in excess of the 150 estimated to be necessary to maintain viable populations [13]. This was partly due to intolerance of wildlife on the part of local communities [14], but also to an unreliable population estimate. Even though Huygens [15] reported no association between damage costs and prior-year bear kills, lethal control is still the major management method in Japan. A precautionary principle should be applied to Asiatic black bears in Japan because carnivore populations are seriously impacted by social intolerance and the prevalence of lethal control [16,2]. Under such circumstances, a reduction in the frequency of bear use near human-dominated lands is an urgent need if the number of bears killed is to be reduced.

We focused on habitat selection by Asiatic black bears near human landscapes to identify the key factors potentially associated with frequent contact and conflicts with humans. First, using GPS bear location data by sex-season groups, we examined how bears responded differently depending on the season to distance from roads, forest edges and rivers on the assumption that such linear landscape features influenced the shape of human-bear boundaries. We hypothesized that strongly selected variables during summer compared with autumn would be the key factors relevant to conflicts because summer is the peak season of human-bear conflict. Second, we estimated resource selection functions (RSF) to predict the distribution of relative probability of bear selection for each sex-season group. Finally, we delineated the human-bear boundaries by defining the edges of both bear habitat and human-dominated land. It was expected that this study would provide local wildlife managers with spatial models that help prioritize the location of management needs including the re-establishment of buffer functions and the development of adequate pre-avoidance schemes. This research aimed to explore a habitat-based approach to conflict mitigation to achieve co-existence with these elusive large carnivores occupying the fringes of human-dominated land.

Materials and Methods

Ethics Statement

Asiatic black bears were captured and fitted with a collar equipped with a global positioning system (GPS) (Televilt, Lindesberg, Sweden). All animals were handled in accordance with the "The Mammal Society of Japan Guidelines for the Treatment of Animal Samples (2009)" drafted with reference to the guidelines issued by the Institutional Animal Care and Use Committee (IACUC). For all locations used in the capture, release and tracking of bears, permission was granted by the Nagano Prefectural Office, the Ministry of the Environment and the District Forest Office.

Study Site and Landscape Covariates

Our research area (Figure 1) consists of two major landscapes with distinct configurations. One is a range of rugged mountain covered by various forest types including native subalpine coniferous forest, temperate broad-leaved deciduous forest and monocultural plantations converted from native deciduous forests. The other is valley basin dominated by farmlands and urban infrastructure with a highly dense road network (7.74 km/km^2, elevation<900 m). In the foothills, there extend unmanaged secondary-growth forests that were once coppice forests. Additionally, several riparian forests running across the human matrix connect the foothills and urban areas.

Landscape features relevant to Asiatic black bear habitat and anthropogenic lands were reclassified into 9 land cover types from a satellite image taken on 22nd June, 2007 (ALOS-AVNIR2) by a supervised mapping technique using Multispec v 3.2.1© [17]. The land cover categories consisted of 5 types of forest cover, 1 open-forest, 2 human-dominated landscapes and others (Table 1). Generally, a forest has diverse structures depending on its successional stage, and can provide different foraging opportunities for bears. For this reason, we generated a forest age map from the polygon layers on the Basic Planning maps produced by the National and Regional Forestry departments, and we reclassified 10 indices according to the age of the forests by referring to the historical records of forest management in the Basic Planning maps.

To explore how Asiatic black bears are affected by abiotic conditions, we created terrain raster predictors using various GIS algorithms applied to a 10 m Digital Cartographic Data Standards digital elevation model (DEM). Given the high ratio of mountains in our study area, terrain roughness was generated by calculating a square root of standard deviation of the DEM using the roughness tool in the Geomorphometric and Gradient Metrics Tool [18]. According to the general principles of ecology, conditions of sunlight and water are the major factors determining a plant community's potential to affect local wildlife food distribution. To include a variable of potential water content of soils, a compound topographic index (CTI) was developed by combining slope and flow accumulations calculated from the DEM. The degree of incoming solar energy for each pixel was computed by employing the solar radiation analysis tool in a spatial analyst extension (ArcInfo9.3©) capable of taking into account the effects of latitudinal and local terrain uniqueness on insolation. We set two days as representative for each season, 1st August for summer and 1st October for autumn. Four distance variables were

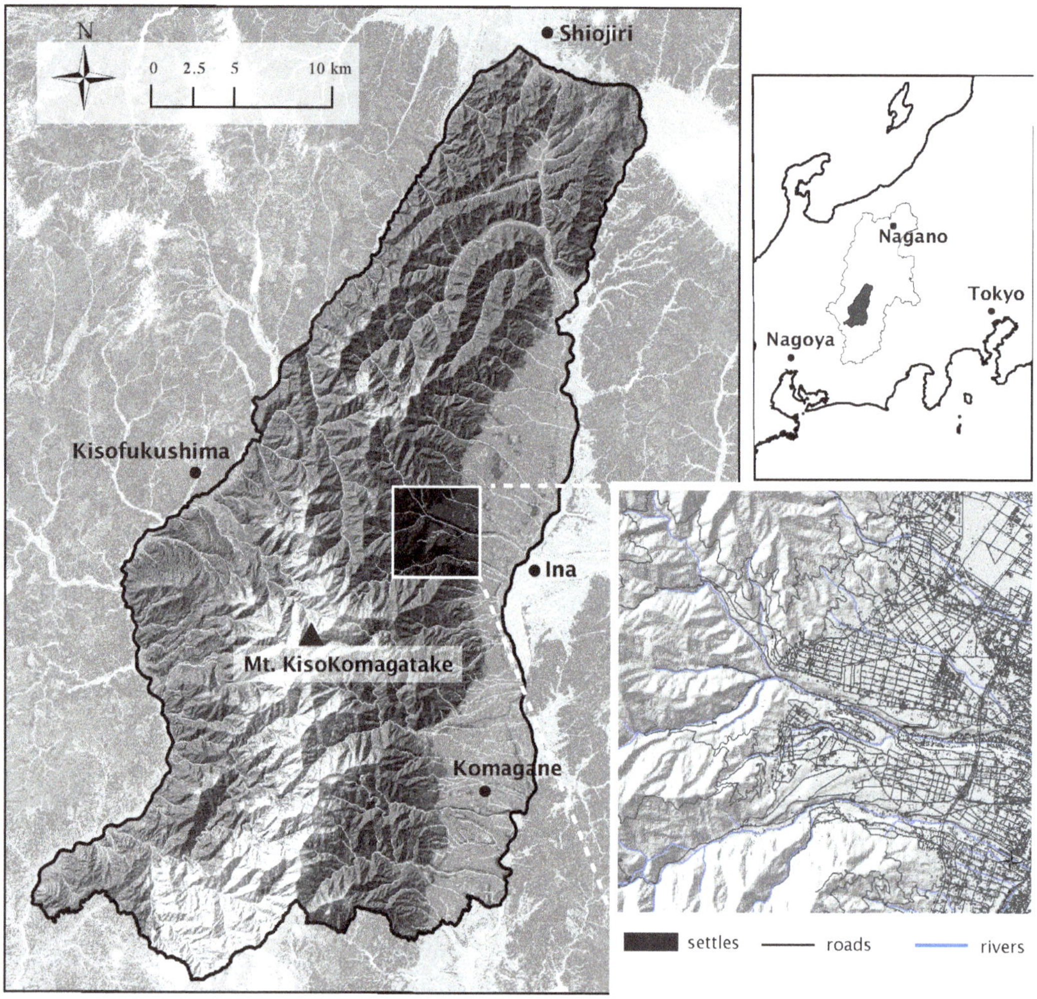

Figure 1. The study area of 1,023 km^2 (35°48′27N, 137°49′47E) including the central Japan Alps located between Ina and Kiso valley, the southern part of Nagano prefecture in the Honshu island of Japan. The example section (bottom right) has a typical distribution of the landscape features that proved most important in the production of human-bear boundary maps.

estimated: roads in open land (open roads), roads in forest cover (forest roads), forest edges and rivers. We then measured the nearest distances to each bear location. To create the forest-edge map, we reclassified the 9 land cover types into 2 major land cover types, open and forested lands. Using both the edge-enhancing and smoothing filters in the Neighborhood tool in the spatial analyst extension, we delineated the border lines between open and forest lands as forest edges. All maps were prepared as rasters at a 10 m resolution to facilitate the raster calculations necessary to develop and apply an RSF value to each pixel (Table 1).

Animal location data and sampling design

We used hourly GPS locations collected from 24 bears (14 females and 10 males) during both summer and autumn in 2008–2011. We defined summer as the period from 1st July to 10th September, and autumn as being from 11th September to denning according to dietary pattern correlated with seasonal shift. Resource selection functions (RSF) were estimated following a designIII(1) as suggested by Manley et al [19] where individual animals are identified with used resource units defined by telemetry data and available resource units randomly chosen from within the home ranges of individual animals. We delineated MCP (the minimum convex polygon) as individual home ranges derived from the GPS relocation data for each bear, and sampled 1000 points randomly from each MCP to yield the available resource units on the assumption that all resources within the MCP would be available to the bear during a season. The landscape attributes of the used units were contrasted with those of the available resource units using logistic regression to model habitat selection by Asiatic black bears.

Investigation of spatial distance effects

On the assumption that summer-specific habitat selection would be relevant to the frequency of human-bear interactions, we focused on seasonal differences in probability of bear selection in terms of distance from roads, rivers and forest edges using logistic regression based on used-available locations as the binary response variable. Starting with a model for the single explanatory variable of continuous distance (D) as the base line, we manually structured a distance-effect model by stepwise addition of quadratic terms of distance (D^2), season (SS), sex (SX), land cover (LC) and three interactions with season (D:SS, SS:SX, SS:LC). Log-likelihood, AIC (Akaike's information criteria) and Δ_i values were used as measures for the selection of the final distance-effect model for

Table 1. Landscape covariates considered to influence habitat selection by Asiatic black bears in the central Japan Alps. 9 land cover types were reclassified (6 types of vegetation, 2 human-landscape classes and others) from a satellite image with 10 m resolution.

Landscape Covariate	Description (Original data source, data type (unit) and range; % = a proportion of total study area)	Variable code
Land cover	Categorical	land cover
Larch plantation	Plantation of Larches *Larix leptolepis* (0 or 1), 26.2%	larch_pt
Coniferous plantation	Plantation of Japanese Cedars *Cryptomeria japonica* or Cypress *Chamaecyparis obtusa* (0 or 1), 7.3%	conifer_pt
Broadleaved forest	Forest of Deciduous broadleaved trees *Fagaceae, Rosaceae, Cornaceae, Juglans, Castanea* etc. (0 or 1), 13.7%	deciduous_f
Red pine forest	Forest of Japanese red pine *Pinus densiflora* (0 or 1), 7.4%	redpine_f
Open regenerating	Canopy openings with various regeneration stages consist of shrub and herbaceous plants (0 or 1), 5.3%	open_g
Subalpine forest	Forest mainly consists of evergreen conifers in subalpine zone (0 or 1), 6.2%	alpine_f
Farmlands	Multiple crop lands and livestock sheds (0 or 1), 14.6%	farmland
Near town	Paddy fields, villages and towns (0 or 1), 6.7%	neartown
Others	Alpine meadow, permanent snow, rocks, water surface and clouds (0 or 1), 10.6%	others
Forest Age	10 indices (1–10) from 0 to over 110 years old reclassified from the Basic planning maps of National and Regional Forestry	frstAge
Topographic features	Linear	
Elevation	Measured from 10 m DEM (m), mean 1268, range 456–2956	elev
CTI	Compound Topographic Index, range 0.4–21.9	cti
Roughness	Terrain Ruggedness Index, range 0–9.0	roughness
Solar Radiation		
Summer	Represented at 1st August, (Wh/m2), range 1087.8–7298.8	solar
Autumn	Represented at 1st October, (Wh/m2), range 115.6–5022.4	
Distance variables	Linear	
Forest edges	Distances from the nearest edge of forest cover (m), range 0–1000	frstEdg
Rivers	Distances from the nearest river (m), range 0–2686.8	rivers
Roads in open lands	Distances from the nearest road in open lands (m), range 0–6967.1	road_o
Roads in forests	Distances from the nearest road in forests (m), range 0–3284.9	road_f

Table 2. Comparison between distance-effect models using logistic regression with distance variables for each linear landscape features partly including interaction terms for seasons according to log-likelihood (LL), AIC, and AIC score as changes in AIC from the lowest model (Δi).

Model	Road_open_dist			Road_forest_dist			Forest_edge_dist			River_dist		
	LL	AIC	Δ_i	LL	AIC	Δ_i	LL	AIC	Δ_i	LL	AIC	Δ_i
D+D^2+D:SS+SS:LC+SS:SX	−58865	117777	0	−58943	117931	0	−59296	118638	0	−59780	119607	0
D+D^2+D:SS+SS:LC	−58878	117799	22	−58967	117976	45	−59316	118675	37	−59794	119631	24
D+D^2+D:SS+LC	−60092	120193	2416	−59889	119791	1859	−60415	120842	2204	−60929	121871	2264
D+D^2+D:SS+SS:SX	−60150	120315	2538	−59925	119863	1932	−60452	120919	2281	−60977	121968	2361
D+D^2+D:SS+SX	−60152	120197	2420	−59928	119869	1938	−60457	120925	2287	−60982	121977	2370
D+D^2+D:SS	−60155	120320	2544	−59939	119888	1957	−60462	120934	2296	−60985	121980	2373
D+D^2+SS	−60465	120938	3161	−60054	120117	2186	−61198	122405	3767	−61168	122344	2737
D+D^2	−60709	121423	3647	−60216	120439	2508	−61340	122687	4049	−61306	122617	3010
D	−60721	121447	3670	−60627	121259	3327	−61344	122692	4054	−61322	122648	3041

The variable names: continuous distance (D), quadratic terms of distance (D^2), season (SS), sex (SX), land cover (LC) and three interactions with season (D:SS, SS:SX, SS:LC).

each linear landscape. The final model was used to investigate which of the covariates had substantial effects on the differences between summer and autumn by predicting the odds ratio of selection by Asiatic black bears (Table 2).

The logit-odds of the final distance-effect model was calculated by using the following equation:

$$\log it(p) = \beta_0 + \beta_1 D + \beta_2 D^2 + \beta_3 SS + \beta_4 D : SS + \beta_5 SS : SX + \beta_6 SX + \beta_7 SS : LC + \beta_8 LC \quad (1)$$

where p was probability of bear selection, and β was a coefficient. And the logit-odds for bear selection during summer (season = 1) relative to autumn (season = 0) were given by the equation,

$$\log it(p, season = 1) - \log it(p, season = 0) = \beta_3 SS + \beta_4 D : SS + \beta_5 SS : SX + \beta_7 SS : LC \quad (2)$$

The aim of the distance-effect analysis was to determine which land cover types were significantly responsible for the large differences between the two seasons with respect to bear selection. In order to yield the odds-ratio between seasons, we ran models for each sex-land cover combination by controlling the interaction terms of the categorical variables SS:SX and SS:LC in equation (2) applying the coefficients β_5 and β_7. A total of 64 models (8 land cover: 2 sex for each linear landscape) were developed to estimate the odds ratio by adjusting the mean value of the continuous distance variable. We selected models containing land cover types that influenced the first and second largest differences between summer and autumn according to the odds ratio, their absolute differences being calculated by subtraction from 1 (= no selection) with elimination of the sign, and the significant level measured by the Wald test ($p<0.001$). Finally, we set 6 specific distances from 0 to 2000 m and predicted a mean probability of bear use at each distance to assess how much variation in seasonal response by bears would be explained by land cover variables as distances changed.

Habitat model structure and validation

Since effects of distance decline rapidly as bear locations occur beyond the linear landscape feature, we transformed distance variables into exponential decay functions, $e^{-\alpha d}$ where $-\alpha$ was the decay constant and d was distance (m) from the linear landscape. [20]. Quadratic terms in continuous predictors, except the non-linear exponential decay variables, were included if necessary, and interaction terms potentially relevant to foraging habitat and anthropogenic factors were added. Prior to variable selection, co-linearity among linear predictors was checked, and variables correlated with more than three other variables were removed. For model selection, we employed a mixed effect logistic regression with random intercept to balance the disproportional number of observations among individual animals [21]. To develop a plausible model in terms of bear biology, we conducted univariate logistic regression and a manual step-forward procedure [22]. After determining the full variable sets suitable to each sex-season group model, we constructed a global model incorporating all the variables in order to make comparisons among sex-season groups to determine how the predictive human-bear (HB) boundaries differed.

We built an individual RSF model separately for each bear, and averaged the coefficients within the sex-season group. We expected that disproportional error rates would be caused by the variety of sample sizes in some variables among individual animals, and that this would have substantial effects on the averaged coefficients. For example, the reliability of coefficients for individual bears were not equal when sample sizes in a land cover type differed significantly among the bears. To deal with these imbalances, we used an inverse variance weighted method to obtain appropriate averages that incorporated the differences in standard error for each parameter estimate. [20]. Next, we applied the mean coefficients to each predictor of GIS layers to develop RSF habitat maps. A reclassifying tool with quantile breakpoint was used to rank the RSF values into 10 classes to represent the spatial distribution of relative probability of habitat selection by Asiatic black bears across the target research area. To assess the credibility of the predictive performance of our RSF models, we prepared 2490 GPS observations for testing as samples independent of the training data used for model building. We first calculated the utilization function (Ux_i) for each predicted RSF class. The mid-points of the RSF values were multiplied by the area of each RSF class, and divided by the total value to obtain the Ux_i for each class [23]. The total number of the test data was multiplied by the Ux_i to determine the expected frequencies fell within each RSF class. And then, using a linear regression, we contrasted the expected frequency and observed frequency to assess the significant level of the slope by R^2 and χ^2 goodness of fit test [23].

Human-Bear (HB) boundary delineation

We identified HB boundaries by employing an edge-detection technique based on a focal statistical tool involving a moving window GIS operation. First, we combined farmlands, paddy fields, human-settled areas and roads into one landscape layer that represented "human lands". Second, we created bear-habitat edges by transforming the RSF maps with a circle-shaped moving window that determined the value of each pixel from the sum of its surrounding pixels. The generated bear-habitat maps were classified into three types (0 = others, 1 = sharp edge, 2 = moderate edge) with sharp edge defined by RSF classes 9 and 10, and moderate edge by RSF classes 7 and 8, respectively. In the same manner, the edges of human lands were detected and classified into two types (0 = others, 1 = edge). The edges of both bear habitat and human lands were determined by using a moving window circle with a 6-pixel radius (60 m) on the basis of animal movement. Here, we used the median step length (sequential distance between each location), which was about 70–100 m for the 4 groups of Asiatic black bears in our study. Finally, we multiplied the edges of bear habitat and human lands to generate 2 ranked boundary areas. The boundary was classified into two types (1 = sharp HB boundary, 2 = moderate HB boundary) by performing a raster calculation with the two edge layers as follows: (habitat edge; 0, 1, 2)×(human land edge; 0, 1). The non-overlapping edges were then zeroed out (0×0, 0×1 and 0×2) to leave only pixels of human-edge equal to 1 (1×1 and 1×2). All our GIS work was performed on ArcGIS (ESRI v9.3©) and statistical work was done in Stata (SE v12.0, College Station, Texas).

Results

We collected a total of 44,652 bear locations across sex-season groups (F-summer, N = 10297, mean 792±278 SD; F-autumn, 14357, 1044±411; M-summer, 9788, 979±430; M-autumn, 10210, 1021±456). On average, female MCP home ranges (summer, mean 21.96±40.83 km^2 SD; autumn, 24.45±25.72 km^2) were smaller than those of males (summer, 38.52±44.44 km^2; autumn, 99.55±107.64 km^2; t-test on the paired two samples of females and males, $p = 0.008$, $df = 19$), and the MCP ranges of most bears included human-dominated lands (F-summer, 9/13 = 9 of 13 bears; F-autumn, 10/13; M-summer, 10/10; M-autumn 9/10).

Distance effects of linear landscapes on bear responses

Asiatic black bears showed significant seasonal differences in their responses to distances from open roads, forest roads, rivers and forest edges (e.g. $\chi^2 = 1107$, $df = 2$, $p<0.001$ in a comparison of Road_open_dist models between $D+D^2$ and $D+D^2+D:SS$). The manually constructed model (stepwise variable entry) revealed that the land cover covariate had a substantial influence on model fit for all distance variables, and the inclusion of season and land cover interaction (SS:LC) dramatically improved models indicated by Δ_i which is the difference in AIC between models (e.g. $\chi^2 = 2197$, $df = 15$, $p<0.001$ in comparison of Forest_edge_dist models between $D+D^2+D:SS+LC$ and $D+D^2+D:SS+SS:LC$) (Table 2). As a result, for all linear landscapes we selected the final distance-effect model with the lowest AIC that contained quadratic terms of distance and season inclusive of all interactions. We then used the models to investigate the seasonally varied response of Asiatic black bears to distance from the linear landscapes.

Table 3 illustrates the odds ratio yield by equation (2) using coefficients of the selected distance-effect models that resulted in an effect size for each of the land cover covariates on a seasonal difference in the probabilities of selection by Asiatic black bears. As shown in the distance-effect models for open roads, the absolute difference in the odds ratio between the two seasons was the greatest in the model with alpine forest as a land cover covariate, followed by the model with open regenerating lands and deciduous forest ($p<0.001$). The other three distance-effect models revealed significant seasonal changes associated with deciduous forest and red pine forest (Table 3). In addition, we observed a significantly positive red pine forest selection in summer relative to autumn and, conversely, a significantly positive selection of deciduous forests in autumn relative to summer regardless of sex difference.

Note that the odds ratio for summer versus autumn in the distance-effect models referred to above was estimated only by holding the continuous distance variables at their mean. Therefore, we predicted the changes of probability of bear use as changes in distance from the linear landscapes by estimation of the mean probability at the specific distance (Figure 2 a–d). To develop graphs of each distance-effect prediction, we used models incorporating the top two most influential land cover covariates for seasonal difference without discriminating between sexes because sex differences had a less marked effect on seasonal changes than land cover covariates (Table 2). It was assumed that predictions produced using the open-road model with alpine forest were less relevant to bear selection because of the very long distances between geographical locations, so alpine forest was excluded as a predictor.

According to the predicted forest-edge distance-effect model (Figure 2a), the probability of use by Asiatic black bears increased dramatically as distances from forest edges shortened during summer. Selection within both deciduous and red pine forests exhibited similar trends in summer, and the difference between the mean probabilities for the two forests was much smaller than the difference during autumn (1/7 difference: mean of three distances; 200 m, 500 m, 1000 m). Probability of bear use increased with increasing proximity to forest road in both seasons (Figure 2b), but the pattern varied slightly. For example, there was a dramatic decline in probability of use at a distance of 200 m to 500 m in red pine forest during summer whereas bear use within a 500 m distance was relatively stable during autumn in both deciduous and red pine forests. In the open-road model, the probability of bear use in open regenerating lands near open roads was higher during summer than in autumn, while the probability in deciduous forests near roads was lower in summer than in autumn (Figure 2c). A high probability of bear use close to rivers occurred only in summer. The difference between the two forest types was very small in summer compared with the difference in autumn, and this was similar to patterns observed in forest-edge and forest-road models (Figure 2a,b). Overall, the changes in probabilities as distance increased were more gradual in the open-road and river models (Figure 2c,d). Therefore, it seems that open roads and rivers had more extensive distance effects on selection by Asiatic black bears than forest edges or forest roads.

Habitat Selection by Asiatic black bears

Elevation was removed from the final variable because of a close correlation with open roads ($R^2 = 0.75$), forest roads ($R^2 = 0.59$) and roughness of terrain ($R^2 = 0.53$). CTI was excluded due to its lower contribution. According to the results from the distance-effect models, the effects of open roads and rivers seemed more extensive than those of forest roads and edges (Figure 2c,d). Therefore, we determined the decay constant α for the exponential form of distance variables to be -0.005 for open roads and rivers, and -0.01 for forest roads and edges since a smaller value for the constant yields a more gradual decay. Those values, however, were roughly set as we had observed that the range of distance effects varied depending on the land cover types (Figure 2 a–d). In the final variables selected by univariate logistic regression (Table 4), there was a significant seasonal difference among predictors. In summer, both female and male bears strongly selected areas in close proximity to forest-edge. Although forest roads, rivers and solar radiation had a relatively weak influence as single predictors, once forest edges were associated with these covariates, it provided a positive or negative leverage to remain ranked in the top 5.

Our habitat selection models indicated that bears constantly selected deciduous forest year round (Table S1). In particular, the selection during autumn was highly significant ($p<0.001$). We found a remarkable selection of areas near forest edges during summer for both sexes. For instance, the probability of bear selection at a 20 m distance from forest edges was about two times higher than at a 100 m distance for female bears (odds 3.74/1.81), and about three times higher for male bears (odds 7.00/2.40). Distance to rivers was a strong predictor in the multivariate summer-habitat model, especially associated with forest edges (Table S1, S2). The difference between the odds ratio at 20 m and 100 m distances from rivers (Female: 3.03, Male: 7.36) increased where forest edges (20 m distance) were associated with rivers (Female: 21.45, Male: 18.01). Overall, forest-edge effects on bear selection nearly disappeared during autumn. Female bears generally avoided areas near forest roads in summer (odds 0.67 at 20 m), but selected in autumn (odds 2.26 at 20 m). In contrast, male bears selected areas near forest roads during summer (odds 1.40 at 20 m), but avoided such areas during autumn (odds 0.44 at 20 m).

Habitat predictions and Human-Bear (HB) boundary maps

Overall, the averaged coefficients from individual models were consistent with population models (Table S2). There were large error rates in the estimates of coefficients for some bears because their home ranges were exceedingly isolated from open roads. However, we confirmed that the inverse variance weights worked reasonably well to offset these unbalancing effects on the averaged

Table 3. Differences between summer and autumn depending on changes of land cover types and sex in response to distances from open roads, forest roads, forest edges and rivers by Asiatic black bears in the central Japan Alps.

		females			males		
model	**land cover**	**Odds Ratio**	**P>|z|[a]**	**Diff.Season[b]**	**Odds Ratio**	**P>|z|[a]**	**Diff. Season[b]**
Road_open_dist adjusted distance at mean = 1395.42 m LRT = 5022.88, *df* = 22, $p<0.001$	larch_pt	0.750	0.000	0.250	1.126	0.000	0.126
	alpine_f	1.867	**0.000**	**0.867**	1.920	**0.000**	**0.920**
	farmland	0.838	0.003	0.162	0.861	0.012	0.139
	conifer_pt	0.961	0.454	0.039	0.988	0.826	0.012
	deciduous_f	0.511	**0.000**	**0.489**	0.526	**0.000**	**0.474**
	open_g	1.607	**0.000**	**0.607**	1.653	**0.000**	**0.653**
	neartown	0.749	0.038	0.251	0.770	0.060	0.230
	others	1.140	0.596	0.140	1.172	0.520	0.172
	redpine_f	1.238	0.000	0.238	1.273	0.000	0.273
Road_forest_dist adjusted distance at mean = 304.78 m LRT = 4868.30, *df* = 22, $p<0.001$	larch_pt	0.842	0.000	0.158	0.880	0.000	0.120
	alpine_f	1.075	0.545	0.075	1.124	0.328	0.124
	farmland	1.065	0.279	0.065	1.113	0.062	0.113
	conifer_pt	0.874	0.011	0.126	0.913	0.089	0.087
	deciduous_f	0.536	**0.000**	**0.464**	0.560	**0.000**	**0.440**
	open_g	1.190	0.017	0.190	1.244	0.003	0.244
	neartown	0.984	0.906	0.016	1.028	0.840	0.028
	others	1.587	0.064	0.587	1.658	0.042	0.658
	redpine_f	1.443	**0.000**	**0.443**	1.508	**0.000**	**0.508**
Forest_edge_dist adjusted distance at mean = 193.52 m. LRT = 4161.50, *df* = 22, $p<0.001$	larch_pt	0.867	0.000	0.133	0.933	0.010	0.067
	alpine_f	0.774	0.027	0.226	0.834	0.113	0.166
	farmland	0.763	0.000	0.237	0.822	0.001	0.178
	conifer_pt	0.885	0.019	0.115	0.952	0.355	0.048
	deciduous_f	0.583	**0.000**	**0.417**	0.628	**0.000**	**0.372**
	open_g	0.824	0.008	0.176	0.887	0.101	0.113
	neartown	0.734	0.025	0.266	0.790	0.087	0.210
	others	0.905	0.682	0.095	0.974	0.915	0.026
	redpine_f	1.262	**0.000**	**0.262**	1.358	**0.000**	**0.358**
River_dist adjusted distance at mean = 367.10 m LRT = 3192.60, *df* = 22, $p<0.001$	larch_pt	0.890	0.000	0.110	0.950	0.057	0.050
	alpine_f	1.084	0.484	0.084	1.158	0.202	0.158
	farmland	1.007	0.910	0.007	1.075	0.209	0.075
	conifer_pt	0.893	0.029	0.107	0.953	0.362	0.047
	deciduous_f	0.533	**0.000**	**0.467**	0.570	**0.000**	**0.430**
	open_g	1.287	0.000	0.287	1.374	0.000	0.374
	neartown	0.889	0.395	0.111	0.949	0.706	0.051
	others	1.201	0.451	0.201	1.283	0.306	0.283
	redpine_f	1.518	**0.000**	**0.518**	1.621	**0.000**	**0.621**

Odds ratios for bear selection were calculated on the basis of logistic regression by controlling season inclusive interaction terms with sex and land cover. LRT denotes Log-likelihood Ratio Test in comparison with the constant only model.
[a]denotes the level of significance by Wald statistics at the point estimate and
[b]was the absolute difference from no selection (odds ratio = 1).

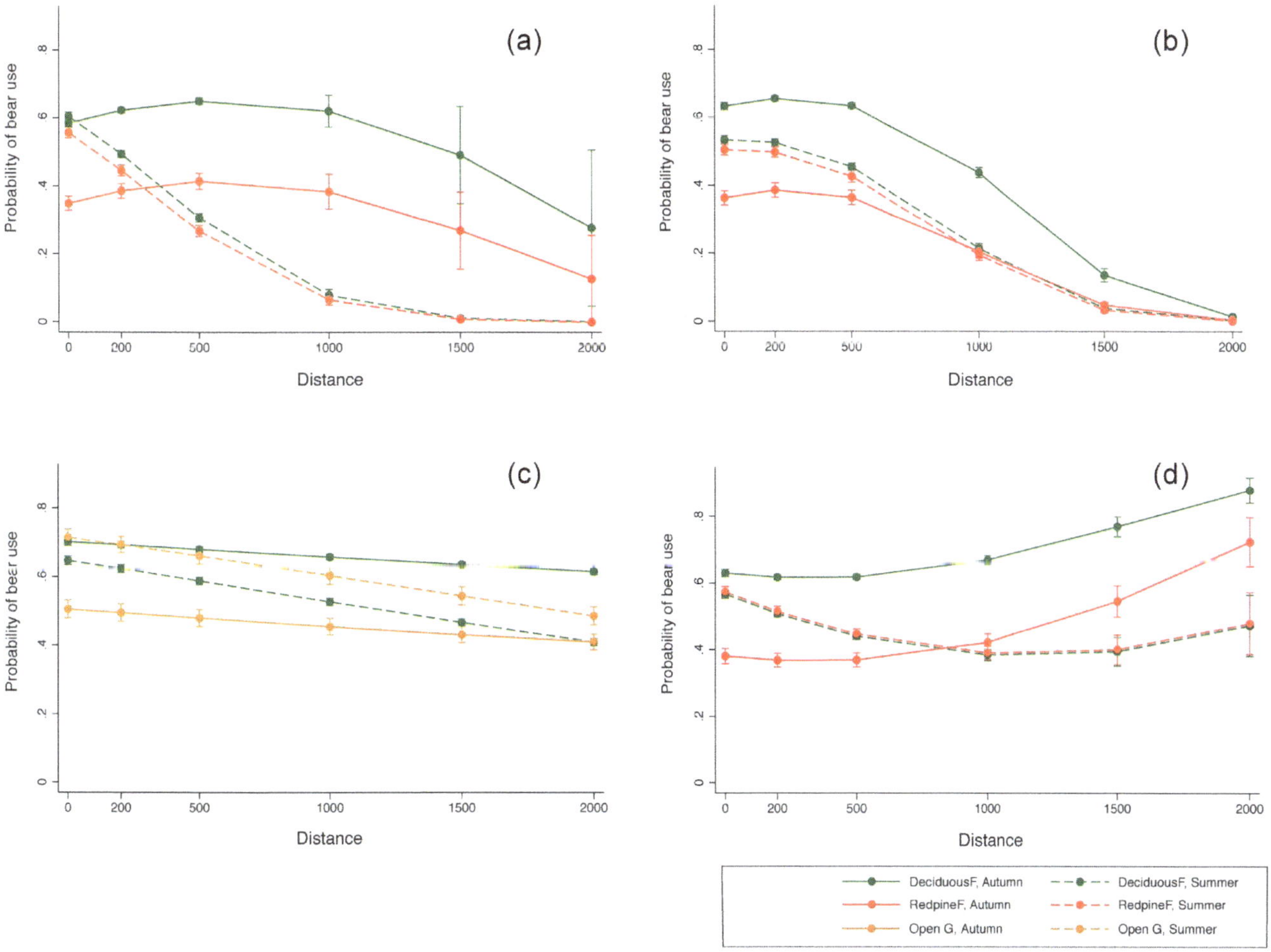

Figure 2. Changes in probability of use by Asiatic black bears along with increasing distances from linear landscape features; forest edges (a), forest roads (b), open roads (c) and rivers (d) in distance effect models. Dots represent the mean of probability at a distance, and error bars represent confidence intervals predicted in logistic regression as a function of the distance variables with interactions of land cover and season; summer (dashed line), autumn (continuous line), deciduous forest (green), red pine forest (red), and open regenerating lands (khaki).

coefficients for the population estimates. Our final RSF maps had acceptable predictive performances for the independent test data sets (F-summer, $R^2 = 0.84$, $df = 9$, $p < 0.01$; F-autumn, $R^2 = 0.93$, $df = 9$, $p < 0.01$; M-summer, $R^2 = 0.80$, $df = 9$, $p < 0.01$; M-autumn, $R^2 = 0.93$, $df = 9$, $p < 0.01$). The HB boundary zones were estimated for each sex-season group (Figure 3). Boundaries stretching along rivers and the fringes of foothills were a common characteristic among the 4 group models, and the extent was greater in male than female HB boundaries. The range of female-summer HB boundaries was about 28.6% of the area of high RSF classes, and it decreased to 4.3% in autumn as male HB boundaries decreased from 49.9% in summer to 10.5% in autumn. Notably, male bears exhibited a high probability of use in the HB boundary zones during summer, as was evidenced by the fact that 9.4% of test GPS locations fell within this boundary zone (Table 5).

Discussion

Key habitat components relevant to human-bear interactions

Despite the common perception of bears as mature forest-interior dwellers, the frequency with which Asiatic black bears selected sites near forest edges was greater than random. Forest-edge selection has been reported in several studies on Grizzly bears [24–25], Scandinavian brown bears [26] and American black bears [27]. To our knowledge, this is the first report of forest-edge selection by Asiatic black bears. The importance of phenology in bear habitat selection is well attested [28–30]. Therefore, the great seasonal change in red pine forest and deciduous forest indicates that the availability of food may be the main reason for the selection of forest edges, which provide herbaceous and fruiting food during the most food-scarce season [31]. On the other hand, bears may use edge space as a refuge for hiding. Grizzly bears selected forest-edge areas in open regenerating lands [24], and brown bears used habitat edges for a day bed to avoid diurnal human activities [32]. In our case, Asiatic black bears preferring to stay on the edge of a red pine forest during summer run the risk of encountering people because forests are commonly located adjacent to agricultural and residential areas in this landscape. Bears are known to have complex responses to roads, and our segregation of road types successfully revealed different responses to open roads and forest roads. There was no clear avoidance of open roads, which accords with previous findings of a neutral response to high-use roads by bears [33–34]. However, our distance-effect model clearly demonstrated that bear response to

Table 4. The final set of variables selected through univariate analysis in mixed effect logistic regression for the global RSF model across season-sex groups of Asiatic black bears; numbers indicate the rank of 15 variables ordered in accordance with Wald statistics.

	Summer				Autumn			
Variables	**females**		**males**		**females**		**males**	
frstEdg ×solar	**1**	*	**2**	*	6		8	
deciduousF ×frstEdg	**2**		**3**		**4**		**3**	*
frstEdg ×rivers	**3**		**1**	*	**5**		7	
road_f×frstEdg	**4**	*	**4**		8		**5**	
frstEdg	**5**	*	**5**	*	12		9	
road_o	6	*	6	*	15		10	
rivers ×solar	7		7		10	*	12	
land cover	8		12		**1**		**1**	
deciduousF ×rivers	9	*	8		**3**		**2**	*
rivers	10	*	9	*	13		14	
frstAge+frstAge^2	11		14		7		11	*
roughness	12	*	10	*	11	*	15	
deciduousF ×frstAge	13		13		**2**	*	**4**	
road_f	14	*	11	*	14		6	
solar+solar^2	15		15	*	9		13	

Significant mark * = $p<0.05$, all parameters were included the constant.
Bold letters = the top 5 ranks.

roads differs according to the land cover near the road. Despite the differences, Asiatic black bears selected red pine forests and deciduous forests near forest roads during both summer and autumn. This might be explained by the fact that roads in forests provide surrogates for the kind of natural openings which have recently decreased in our region. Our habitat selection models support the findings of sex-season specific responses to the forest roads in earlier studies [35–36]. For example, subordinate individual bears such as females with cubs or young males made use of human-influenced areas [37], and adult male bears displayed a greater tolerance of using roads as a movement corridor [34]. In the case of Asiatic black bears, the greater seasonal difference found for males can be explained by the fact that their larger home range contains a relatively large portion of deciduous forest, rich in autumn food resources, at a moderate distance from forest roads.

Riverside environment strongly affected the summer habitat selection of Asiatic black bears, functioning as an ecotonal belt providing various edible plants with a combination of light and moisture (e.g. *Cardiocrinum* sp, *Petasites* sp) [31]. The selection of red pine forest can be explained by the abundant understory of fruiting shrub species, attractive to bears during the season of food shortages. The significant reduction in the selection of red pine forest during autumn may reflect the fact that there is an abundance of food such as insects, particularly ants [38], in the summer only. The forests extend continuously from the foothills to the riparian forests in the lower plains that draw bears to areas encompassed by agriculture and urban landscapes. Remnant riparian forests offer bears a linear habitat with sufficient cover for foraging and bedding [27] and a dispersal corridor for a large range of movement [39–40]. Despite this potential, the green corridor can undermine populations as it leads to frequent contacts with people [41]. In our study landscape, the riparian forest that ended up facing the urban fringe does not function as a corridor. Young adult males are both the largest dispersers and the group of bears with the highest mortality [10]. Therefore, the riparian forests in our region have an adverse effect on the survival of these bears by playing the role of "false dispersal corridors".

In autumn, the land-cover variable was the best predictor in our univariate logistic regression (Table 2), and a preference for deciduous forest was prominent during the hyperphagia season due to a high correlation with the availability of fruited oak trees [38–39]. The substantial difference between deciduous and red pine forest was revealed by our distance-effect models in autumn. In contrast, the smaller difference between the two forests in summer implied the importance of bear food occurring in edges or understory regardless of forest type. Over all, the great seasonal differences in our habitat model may reflect the amplified temperature gradient caused by the combination of temperate climate and montane slope exerting a significant influence on phenology. On the one hand, it should be noted that our habitat models did not include variables of time of day. The importance of temporal aspects in habitat selection has been stressed by several recent studies [42], and adding temporal variables may bring an important new dimension to our understanding of how bears use peripheral human lands.

Predicted human-bear boundaries

Our human-bear boundary maps gave a reasonably accurate picture of remarkable seasonal differences, and the patterns of bear response to linear landscape features were successfully shown. Consequently, we are confident that the largest extent of human-bear boundaries during summer corresponds to the peak season of human-bear conflicts and incidents [43]. The boundaries of male bears were similarly distributed in both summer and autumn against significant seasonal differences in the females' map. We observed that some male bears were attracted by specific croplands and stayed for longer periods after the date chosen to

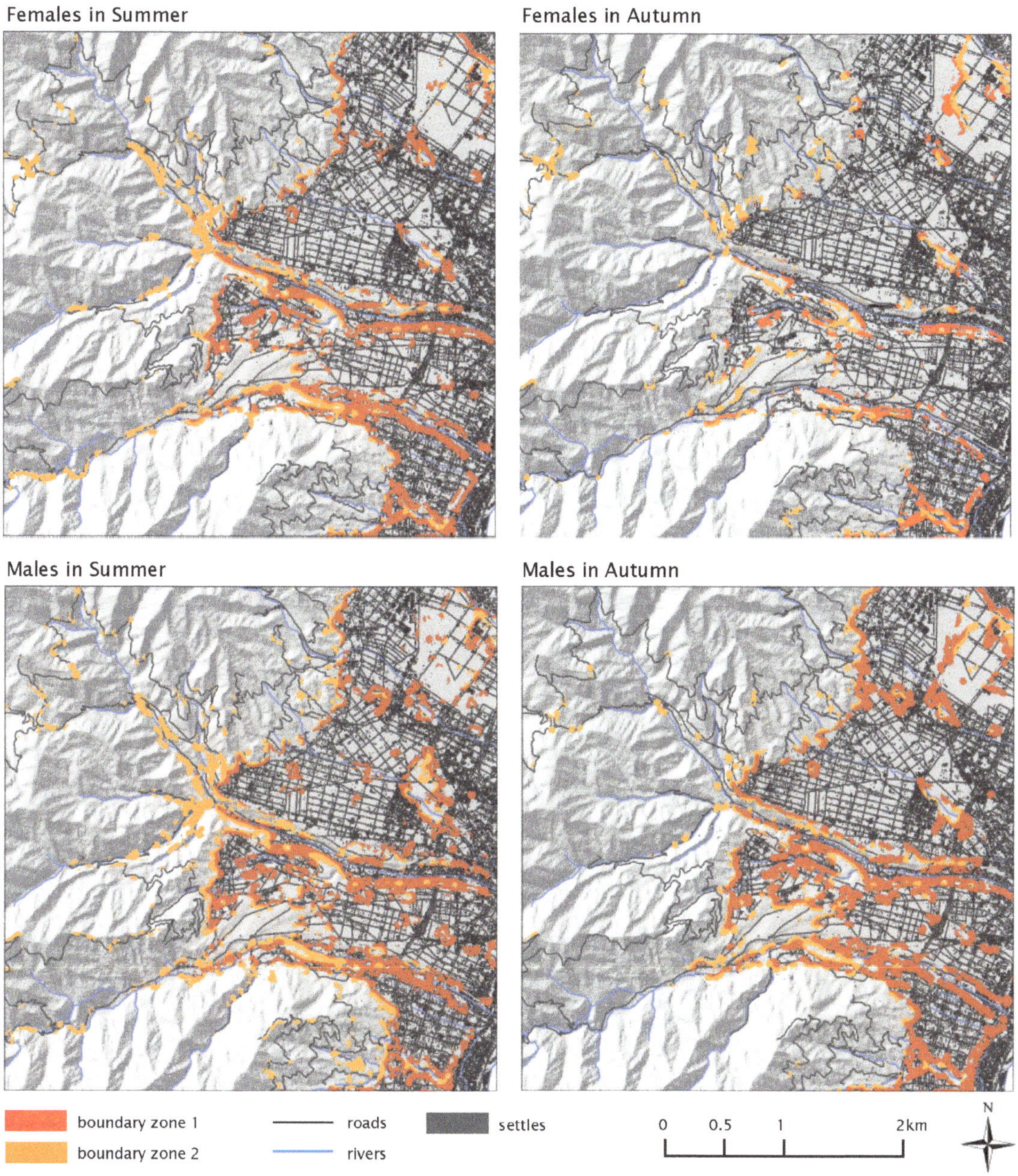

Figure 3. Distribution of the boundaries between humans and Asiatic black bears in the central Japan Alps (10×10 m resolution). The colors indicate sharp boundary: the overlapped edges of RSF class 9–10 and human lands (red) and moderate boundary: the overlapped edges of RSF class 7–8 and human lands (orange). The different panels show HB boundaries for females in summer (top left), females in autumn (top right), males in summer (bottom left) and males in autumn (bottom right).

Table 5. Summary of human-bear (HB) boundaries for each sex-season group of Asiatic black bears in the central Japan Alps.

		Boundary zone				bear locations
		Area of sharp boundary zone 1 (km^2)	**Area of moderate boundary zone 2 (km^2)**	**Total area of boundary zone 1 and 2 (km^2)**	**Boundary zone overlapped with 9–10 RSF class (%)**	**Test bear locations within boundary zone (%)**
females	summer	30.2	32.8	63.1	28.6	8.4
	autumn	22.6	12.7	35.3	4.3	1.2
males	summer	65.3	36.8	102.1	49.9	9.4
	autumn	22.4	44.2	66.6	10.5	5.8

partition the seasons. This is why the pattern of autumn boundaries remained similar to that in the summer. During the inter-crop season, croplands producing food attractive to bears (e.g. corn, apples and other fruits) have potentially negative impacts on the seasonal migration of bears by detaining bears longer, changing their intrinsic behavior and so causing chronic conflicts [44].

Influence of human land use changes on human-wildlife interactions

Although the negative impacts of large-scale deforestation on primary bear habitat is obvious [45–46], it has also been recognized that open regenerating lands after timber harvests are beneficial to American black bears [47–48] and Grizzly bears [24]. Brodeur [49] reported that open regenerating shrubs offer a high density of fruit plants. Likewise, patchily distributed early serial fruits were found to be an important resource for Asiatic black bears during the summer food shortage [50–51]. The increasing number of abandoned villages and farmlands might offer Asiatic black bears an attractive alternative to open shrub land. We were not able to distinguish between actually used and abandoned farmlands from the satellite image, and we recognize this may be the reason for the positive association of farmlands and bear habitat in our models. Additionally, red pine forest is a typical semi-natural growth after cessation of logging. Given that they are commonly located near or within human landscape, abandoned farmlands and red pine forests are key areas with respect to the need to strike a balance between the conservation of summer bear habitat and the reduction of human-bear contacts.

Bears occurring in forest edges adjacent to human settlements risk being sighted by people, and in many cases this results in bear mortality. Mountain roads in this region are used for both forestry and recreation, such as hiking, fishing and picking wild edible plants. The unexpected human presence resulting from these irregular activities can threaten bears in nearby forest edges or on forest roads and lead to tragic encounters. It is equally important to exercise caution in riparian forest areas. And there is also a high risk that croplands located near the linear green belt will change the natural behavior of bears. Given the inevitable bear use of areas near human-dominated lands during summer, pre-avoidance schemes to regulate access to the HB boundary zones should be established to minimize encounters with bears.

Management implications and future directions

Creating spatial separation between humans and large carnivores is a challenging task for wildlife managers worldwide. One simple and efficient method is to construct physical barriers, and electric fencing around cropland has become widely used to prevent crop damage by wildlife. Another option is to create buffer zones. For example, cover can be removed by clear-cutting forest-edge shrubs to increase permeability and so deter wildlife from lingering near human lands. Such indirect means of preventing encounters by changing wildlife behavior would be effective only if implemented in the key areas predicted on the basis of reliable habitat estimates. Given the complexity of the problem, it is necessary to endeavour to change not only wildlife behavior, but also human behavior. Human security is a genuine concern for people living close to occupied carnivore habitats. There are specific areas and periods in which human-wildlife interaction becomes more likely. For example, our RSF model indicated that there is an about 80% probability that farmlands producing bear attractants within 100 m of a forest edge will suffer damage. Such quantifying of risk on a fine scale in this way can be useful for wildlife managers seeking to persuade local farmers to change crops. Knowledge sharing with local communities is critically important [52], particularly in the case of large carnivore conservation where excessive fear exists, and our visualized maps are likely to be helpful in attempts to change local attitudes.

Our findings indicate that red pine forest is the main reason for bear occurrence in peripheral human lands in our region. It is, however, not practical and against conservation practices to eradicate all red pine forests in order to drive bears from the foothills. Traditional coppice forest (called *Satoyama* in Japanese) management has been shown to be beneficial to species diversity [9], insofar as it entails only intermediate levels of human disturbance. For this reason, restoration of *Satoyama* management of red pine forest is probably the most suitable way of creating buffer structures without the removal of the critical summer habitat of Asiatic black bears. At a broader scale, large areas of plantation are left unmanaged, resulting in increased canopy closure and a dramatic reduction of open regenerating lands and edge habitat in mountain areas. We expect that the creation of open lands in unmanaged plantation, by providing a suitable summer habitat for bears in areas isolated from human landscape, would provide a new opportunity to reduce human-bear interaction.

The rapid increase in human-wildlife interaction in Japan against a background of an aged and de-populated rural society, and the abandonment of forestry and agricultural practices, may be an exceptional case in global terms. Although the demographic process is the opposite of that found in nations of rapid human population growth, the problems caused by the sharpening of the human-wildlife boundaries are similar. As many researchers have recognized, fencing is not a panacea [53], and the creation of distance between large carnivores and humans would be a more sophisticated approach to long-term mitigation. One example of this in our case would be the restoration of season-specific habitats in remote areas. We conclude that a habitat-based approach has enormous potential as a means of creating adequate separation between humans and wildlife. The results of our study imply that human land use has an indirect but fundamental influence on the frequency of human-wildlife interaction. However, this linkage has rarely been investigated. Thus, we recommend that further research be undertaken to understand the mechanisms of increased human-wildlife interaction in relation to the influence of anthropogenic land modification and management processes. We believe this would provide a new direction in the search for ways to achieve co-existence with large carnivores by resolving the complex issue of conflict.

Supporting Information

Table S1 Logistic regression coefficients, odds ratio and 95% Confidence Interval for the odds ratio estimated by the final habitat selection model on the basis of mixed effect logistic regression for each season-sex group of Asiatic black bears. β was a coefficient and LRT indicated Log-likelihood Ratio Test in comparison with the constant only model. n denoted were the number of observations. The odds ratios and their 95% CI were given as 10^5 (denoted [a]) and 10^2 (denoted [b]) times the original value of their coefficients.

Table S2 Estimated coefficients for RSF models of habitat selection by Asiatic black bears; averaged coefficients (mean β) estimated by individual level, and coefficients (β), standard errors (S.E.), and significance (* = p<0.01) estimated by population level (each sex-season group) in the mixed effect logistic regressions.

Acknowledgments

For collecting field data, special thanks are due to A Hiyoshi, Y Hosokawa, K Kido, K Nakane, A Kawai and other staff in the wildlife research lab in Shinshu University. We thank T Shimizu and the staff of the District Forest Office for providing forest record data and logistical support, and local hunters and government officers for help with capturing animals. CT appreciates the support and encouragement of the following: Dr. H Hayashi, colleagues in the Shinshu Bear Research Group and Associate Prof. M Uehara. We also thank Prof. K Hoshikawa and M Naruse for GIS technical advice, and we are grateful to Dr. D Ruzicka for proofreading and Dr. M Shaheed for valuable comments on the manuscript. The constructive suggestions made by Dr. Damiano G. Preatoni, one anonymous referee and the editor Dr. D Russo substantially improved our manuscript.

Author Contributions

Conceived and designed the experiments: CT SEN. Performed the experiments: AT SI. Analyzed the data: CT SEN. Contributed reagents/ materials/analysis tools: SI. Wrote the paper: CT. Obtained permission for use of forestry records and setting animals traps: SI.

References

1. Bateman PW, Fleming PA (2012) Big city life: carnivores in urban environments. J Zool 287: 1–23.
2. Treves A, Karanth KU (2003) Human-Carnivore Conflict and Perspectives on Carnivore Management Worldwide. Conserv Biol 17: 1491–1499.
3. Tigas LA, Van Vuren DH, Sauvajot RM (2002) Behavioral responses of bobcats and coyotes to habitat fragmentation and corridors in an urban environment. Biol Conserv 108: 299–306.
4. Beckmann JP, Berger J (2003) Rapid ecological and behavioural changes in carnivores: the responses of black bears (Ursus americanus) to altered food. J Zool 261: 207–212.
5. Merkle JA, Krausman PR, Decesare NJ, Jonkel JJ (2011) Predicting spatial distribution of human–black bear interactions in urban areas. J Wildl Manage 75: 1121–1127.
6. Apps CD, McLellan BN, Woods JG, Proctor MF (2004) Estimating grizzly bear distribution and abundance relative to habitat and human influence. J Wildl Manage 68:138–152.
7. Waller JS, Servheen C (2005) Effects of transportation infrastructure on grizzly bears in northwestern Montana. J Wildl Manage 69: 985–1000.
8. Reynolds-Hogland MJ, Mitchell MS (2007) Effects of roads on habitat quality for bears in the southern Appalachians: a long-term study. J Mamm 88: 1050–1061.
9. Katoh K, Sakai S, Takahashi T (2009) Factors maintaining species diversity in *satoyama*, a traditional agricultural landscape of Japan. Biol Conserv 142: 1930–1936.
10. Forestry Agency (2011) Annual Report on Forest and Forestry in Japan. Available: http://www.rinya.maff.go.jp/j/kikaku/hakusyo/23hakusyo/. Accessed 20 December 2012
11. Ministry of Agriculture, Forestry and Fisheries (2012). FY2011 Annual Report on Food, Agriculture and Rural Areas in Japan. Available: http://www.maff.go.jp/j/wpaper/w_maff/h23/. Accessed 20 December 2012.
12. Ministry of the Environment (2008). The manual for responding to bear appearance. Available: http://www.env.go.jp/nature/choju/docs/docs5-4a/ (in Japanese). Accessed 14 January 2011.
13. Kishimoto R (2009) Status of the 2006 drastic rise of Asiatic black bear (*Ursus thibetanus*) intrusions into residential areas in Nagano Prefecture. Biology of Bear Intrusions. 35.
14. Sakurai R, Jacobson SK (2011) Public perceptions of bears and management interventions in Japan. Human–Wildl Interactions 5:123–134.
15. Huygens OC, van Manen FT, Martorello DA, Hayashi H, Ishida J (2004) Relationships between Asiatic black bear kills and depredation costs in Nagano Prefecture, Japan. Ursus 15: 197–202.
16. Ferguson SH, Larivière S (2002) Can comparing life histories help conserve carnivores? Anim Conserv 5: 1–12.
17. Multispec v 3.2.1©. Purdue University, Available: https://engineering.purdue.edu/biehl/MultiSpec/index.html. Accessed 5 November 2011.
18. The Geomorphometry and Gradient Metrics Toolbox, Available: http://www.conservationgateway.org/ConservationPlanning/ToolsData/Pages/tools-and-data.aspx. Accessed 10 July 2012.
19. Manly BFJ, McDonald LL, Thomas DL, McDonald TL, and Erickson WP (2002) Resource Selection by Animals: Statistical Analysis and Design for Field Studies. Second edition. The Netherlands: Kluwer Academic Publishers.
20. Nielsen SE, Cranston J, Stenhouse GB (2009) Identification of Priority Areas for Grizzly Bear Conservation and Recovery in Alberta. J Conserv Plan 5: 38–60.
21. Gillies CS, Hebblewhite M, Nielsen SE, Krawchuk MA, Aldridge CL, Frair JL, Saher DJ, Stevens CE, Saher CL (2006) Application of random effects to the study of resource selection by animals. J Anim Ecol 75: 887–898.
22. Hosmer DW, Lemeshow S (2000) Applied Logistic Regression. Second Edition. Wileys Series in Probability and Statistics.
23. Johnson CJ, Nielsen SE, Merrill EH, McDonald TL, Boyce MS (2006) Resource selection functions based on use-availability data: theoretical motivation and evaluation methods. J Wildl Manage 70: 347–357.
24. Nielsen SE, Boyce MS, Stenhouse GB (2004a) Grizzly bears and forestry: I. Selection of clearcuts by grizzly bears in west-central Alberta, Canada. For Ecol Manage 199: 51–65.
25. Stewart BP, Nelson TA, Laberee K, Nielsen SE, Wulder MA, Stenhouse GB (2013) Quantifying grizzly bear selection of natural and anthropogenic edges. The J Wildl Manage.
26. May R, Van Dijk J, Wabakken P, Swenson JE, Linnell JD, Zimmermann B (2008) Habitat differentiation within the large-carnivore community of Norway's multiple-use landscapes. J Appl Ecol 45:1382–1391.
27. Lyons AL, Gaines WL, Servheen C (2003) Black bear resource selection in the northeast Cascades, Washington. Biol Conserv 113: 55–62.
28. Munro RHM, Nielsen SE, Price MH, Stenhouse GB, Boyce MS (2006) Seasonal and diet patterns of grizzly bear diet and activity in west-central Alberta. J Mamm 87: 1112–1121.
29. Nielsen SE, Boyce MS, Stenhouse GB, Munro RH (2003) Development and testing of phenologically driven grizzly bear habitat models. Ecoscience 10: 1–10.
30. Davis H, Weir RD, Hamilton AN, Deal JA (2006) Influence of phenology on site selection by female American black bears in coastal British Columbia. Ursus 17: 41–51.
31. Huygens OC, Miyashita T, Dahle Carr M, Izumiyama S, Sugawara T, Hayashi H (2003) Diet and feeding habits of Asiatic black bears in the Northern Japanese Alps. Ursus 14: 236–245.
32. Ordiz A, Støen OG, Delibes M, Swenson JE (2011) Predators or prey? Spatio-temporal discrimination of human-derived risk by brown bears. Oecologia 166: 59–67.
33. Van Manen FT, Mccollister MF, Nicholson JM, Thompson LM, Kindall JL, Jones MD (2012) Short-Term Impacts of a 4-Lane Highway on American Black Bears in Eastern North Carolina. Wildl Monographs 181: 1–35.
34. Roever CL, Boyce MS, Stenhouse GB (2010) Grizzly bear movements relative to roads: application of step selection functions. Ecography 33: 1113–1122.
35. Gaines WL, Lyons AL, Lehmkuhl JF, Raedeke KJ (2005) Landscape evaluation of female black bear habitat effectiveness and capability in the North Cascades, Washington. Biol Conserv 125: 411–425.
36. Gibeau ML, Clevenger AP, Herrero S, Wierzchowski J (2002) Grizzly bear response to human development and activities in the Bow River Watershed, Alberta, Canada. Biol Conserv 103: 227–236.
37. Elfström M, Zedrosser A, Støen OG, Swenson JE (2012) Ultimate and proximate mechanisms underlying the occurrence of bears close to human settlements: review and management implications. Mamm Rev 43: 167–255.
38. Fujiwara S, Koike S, Yamazaki K, Kozakai C, Kaji K (2012) Direct observation of bear myrmecophagy: Relationship between bears' feeding habits and ant phenology. Mamm 78: 34–40.
39. Pereira M, Rodríguez A (2010) Conservation value of linear woody remnants for two forest carnivores in a Mediterranean agricultural landscape. J Appl Ecol 47: 611–620.
40. Singleton PH, Gaines WL, Lehmkuhl JF (2004) Landscape permeability for grizzly bear movements in Washington and southwestern British Columbia. Ursus 15: 90–103.
41. Wilson SM, Model MJ, Mattson DJ, Graham JM, Merrill T (2006) Landscape conditions predisposing grizzly bears to conflicts on private agricultural lands in the western USA. Biol Conserv 130: 47–59.
42. Ordiz A, Støen OG, Sæbø S, Sahlén V, Pedersen BE, Kindberg J, Swenson JE (2013) Lasting behavioural responses of brown bears to experimental encounters with humans. J Appl Ecol 50: 306–314.
43. Takahata C, Nishino S, Kido K, Izumiyama S (2013) An evaluation of habitat selection of Asiatic black bears in a season of prevalent conflicts. Ursus 24: 16–26.
44. Spencer RD, Beausoleil RA, Martorello DA (2007) How agencies respond to human-black bear conflicts: a survey of wildlife agencies in North America. Ursus 18: 217–229.
45. Mattson DJ (1990) Human impacts on bear habitat use. Ursus 8:33–56.
46. Schoen JW (1990) Bear habitat management: a review and future perspective. Ursus 8: 143–154.
47. Cunningham SC, Ballard WB, Monroe LM, Rabe MJ, Bristow KD (2003) Black bear habitat use in burned and unburned areas, central Arizona. Wildl Soc Bull: 786–792.
48. Jones MD, Pelton MR (2003) Female American black bear use of managed forest and agricultural lands in coastal North Carolina. Ursus 14: 188–197.
49. Brodeur V, Ouellet JP, Courtois R, Fortin D (2008) Habitat selection by black bears in an intensively logged boreal forest Can J Zool 86: 1307–1316.
50. Koike S (2010) Long-term trends in food habits of Asiatic black bears in the Misaka Mountains on the Pacific coast of central Japan. Mamm Biol 75:17–28.

51. Hashimoto Y, Kaji M, Sawada H, Takatsuki S (2003) Five-year study on the autumn food habits of the Asiatic black bear in relation to nut production. Ecol Res 18: 485–492.
52. Glikman JA, Vaske JJ, Bath AJ, Ciucci P, Boitani L (2012) Residents' support for wolf and bear conservation: the moderating influence of knowledge. Eur J Wildl Res 58: 295–302.
53. Hayward MW, Kerley GI (2009) Fencing for conservation: restriction of evolutionary potential or a riposte to threatening processes? Biol Conserv 142: 1–13.

Spatial Variation in Soil Properties among North American Ecosystems and Guidelines for Sampling Designs

Henry Loescher[1,2]*, Edward Ayres[1,2,3], Paul Duffy[4], Hongyan Luo[1,2], Max Brunke[1]

1 Science Team, National Ecological Observatory Network (NEON), Boulder, Colorado, United States of America, **2** Institute of Arctic and Alpine Research (INSTAAR), University of Colorado, Boulder, Colorado, United States of America, **3** Natural Resource Ecology Laboratory, Colorado State University, Fort Collins, Colorado, United States of America, **4** Neptune and Company Inc., Lakewood, Colorado, United States of America

Abstract

Soils are highly variable at many spatial scales, which makes designing studies to accurately estimate the mean value of soil properties across space challenging. The spatial correlation structure is critical to develop robust sampling strategies (*e.g.*, sample size and sample spacing). Current guidelines for designing studies recommend conducting preliminary investigation(s) to characterize this structure, but are rarely followed and sampling designs are often defined by logistics rather than quantitative considerations. The spatial variability of soils was assessed across ~1 ha at 60 sites. Sites were chosen to represent key US ecosystems as part of a scaling strategy deployed by the National Ecological Observatory Network. We measured soil temperature (T_s) and water content (SWC) because these properties mediate biological/biogeochemical processes below- and above-ground, and quantified spatial variability using semivariograms to estimate spatial correlation. We developed quantitative guidelines to inform sample size and sample spacing for future soil studies, *e.g.*, 20 samples were sufficient to measure T_s to within 10% of the mean with 90% confidence at every temperate and sub-tropical site during the growing season, whereas an order of magnitude more samples were needed to meet this accuracy at some high-latitude sites. SWC was significantly more variable than T_s at most sites, resulting in at least 10× more SWC samples needed to meet the same accuracy requirement. Previous studies investigated the relationship between the mean and variability (*i.e.*, sill) of SWC across space at individual sites across time and have often (but not always) observed the variance or standard deviation peaking at intermediate values of SWC and decreasing at low and high SWC. Finally, we quantified how far apart samples must be spaced to be statistically independent. Semivariance structures from 10 of the 12-dominant soil orders across the US were estimated, advancing our continental-scale understanding of soil behavior.

Editor: Ben Bond-Lamberty, DOE Pacific Northwest National Laboratory, United States of America

Funding: The National Ecological Observatory Network (NEON) is a project sponsored by the National Science Foundation (NSF) and managed under cooperative agreement by NEON, Inc. Any opinions, findings, and conclusions or recommendations expressed in this material are those of the author(s) and do not necessarily reflect the views of the NSF.' be formally acknowledged. The funders had no role in study design, data collection and analysis, decision to publish, or preparation of the manuscript.

Competing Interests: The National Ecological Observatory Network (NEON) is solely supported by the National Science Foundation (NSF). NEON is incorporated as a 501(c)3 non-profit entity, and needs to do this to engage in a cooperative agreement with the NSF. Loescher, Ayres, and Luo are employed by NEON. Brunke was employed by NEON at the time of data collection. The Neptune Company Inc. provides superlative statistical and experimental design services on contract with NEON. Duffy is employed by Neptune as a professional statistical consultant, and acted in that capacity here. Both NEON and Neptune adhere to the highest ethical standards and scientific integrity, and consider this a scholarly work with no competing interests as defined by PLOS ONE, i.e, there are no competing interests can be financial or non-financial, professional, or personal that would limit the scientific creativity and scholarly activities presented in this manuscript, nor limit any capability to provide free open-source of our data. These affiliations do (does) not alter the authors' adherence to all the PLOS ONE policies on sharing data and materials.

* E-mail: hloescher@neoninc.org

Introduction

Researchers interested in measuring the mean value for a soil property across space must make decisions about the number of samples required and how far apart they should be spaced. These decisions are inherently dependent on the variance structure and spatial independence of the data, but quantitative guidelines to inform these decisions are unavailable at most sites. For instance, intuitively researchers know that a large number of widely-spaced samples would be more likely to better measure the spatial mean of a soil property than a few samples all located close to one another. But it is typically not clear what "a large number" or "widely-spaced" means. Moreover, since increasing the sample size consumes resources (*e.g.*, time, money, and analytical capacity), it is important that sufficient samples are collected to meet the accuracy requirement to answer a specific question without wasteful over-sampling. Likewise, choosing an appropriate distance between samples is important for two reasons; 1) samples should ideally be spaced sufficiently far apart to avoid correlation and pseudoreplication in measurements at a given scale [1], thereby maximizing the amount of information provided by a limited number of samples, and 2) spacing between samples should be minimized to constrain costs, which typically increase with distance.

Here, we focus on sampling strategies designed to estimate the spatial mean value of a soil property with the smallest possible

Table 1. Site characteristics, where Lat = Latitude, Long = Longitude, El = Elevation, Temp = Mean Annual Air Temperature, Precip = Mean annual Bulk Precipitation, CH = Canopy Height, CS = Canopy Structure, CL = closed, SS = short stature, O = open, SO = semi-open.

Site	Lat (°)	Long (°)	El (m.a.s.l)	Temp (°C)	Precip (mm)	CH (m)	Ecosystem type	CS	Soil order	Owner
Harvard Forest	42.54	−72.17	348	7	1066	26.0	Temperate deciduous forest	CL	Spodosol	Harvard U
Burlington	42.52	−71.18	38	9	1095	24.0	Temperate deciduous forest	CL	Inceptisol	Town of Burlington MA
Bartlett Exp. Forest	44.06	−71.29	273	7	1219	23.0	Temperate deciduous forest	CL	Spodosol	USDA FS
Blandy Farm	39.06	−78.07	182	12	972	1.0	Shrubland	SS	Ultisol-Alfisol	U Virginia
Smithsonian CRC (SCBI)	38.89	−78.14	355	13	1057	35.0	Temperate deciduous forest	CL	Alfisol	Smithsonian Institution
Smithsonian SERC	38.89	−76.56	10	13	1097	38.0	Temperate deciduous forest	CL	Ultisol	Smithsonian Institution
Ordway-Swisher BS	29.69	−81.99	48	20	750	23.0	Temperate coniferous forest	O	Entisol	U Florida Foundation
J. Jones Ecological Research station	31.19	−84.47	47	19	1344	27.0	Temperate coniferous forest	CL	Ultisol	Robert W. Woodruff Foundation
Disney Wilderness Preserve	28.13	−81.44	20	22	1219	1.5	Grassland	SS	Spodosol	Nature Conservancy
Guanica State Forest	17.97	−66.87	126	23	840	10.0	Tropical deciduous forest	CL	Aridisol	State Forest, Dept. Natural and Environmental Resources, Puerto Rico
Lajas Ag. Exp. St.	18.02	−67.08	16	25	830	0.4	Agricultural	SS	Vertisol	U Puerto Rico
Ponce	17.99	−66.64	1	27	901	5.0	Savanna	O	Mollisol	Municipality of Ponce, PR
Steigerwaldt	45.51	−89.58	477	5	809	5.5	Temperate deciduous forest	SO	Spodosol	Steigerwaldt Land Services
Tree Haven	45.49	−89.59	461	5	809	23.0	Temperate deciduous forest	CL	Spodosol	University of Wisconsin Stevens Point
UNDERC	46.23	−89.54	520	5	751	24.0	Temperate deciduous forest	CL	Spodosol	U Notre Dame
Konza - Core	39.10	−96.56	415	13	835	1.5	Grassland	SS	Mollisol	Nature Conservancy/Kansas State U
Konza - Relocatable	39.11	−96.61	323	13	835	1.5	Agricultural	SS	Mollisol	Nature Conservancy/Kansas State U
U Kansas Bio Station	39.04	−95.19	321	14	940	19.0	Temperate deciduous forest	CL	Inceptisol	U Kansas
Oak Ridge Nat'l Lab	35.96	−84.28	342	15	1352	28.0	Temperate deciduous forest	CL	Ultisol	US DOE
Great Smokey Mtns NP	35.69	−83.50	661	13	1453	30.0	Temperate deciduous forest	CL	Ultisol	*US NPS
Mountain Lake Res. Stn.	37.38	−80.52	1170	8	1131	18.0	Temperate deciduous forest	CL	Entisol	U Virginia
Talladega Nat'l Forest	32.95	−87.39	164	15	1384	25.0	Temperate coniferous forest	SO	Ultisol	USDA FS
Choctaw NWA	31.85	−88.17	11	18	1524	35.0	Temperate deciduous forest	CL	Inceptisol	DOI NWA
Dead Lake WMA	32.54	−87.80	24	17	1415	30.0	Temperate deciduous forest	CL	Ultisol	US Army Corps of Engineers
Dakota-Coteau	47.16	−99.11	575	4	478	0.4	Grassland	SS	Mollisol	State of North Dakota Land Trust
Northern Great Plains Res. Lab	46.77	−100.92	589	5	434	0.4	Agricultural	SS	Mollisol	USGS/DOI FWS
Woodworth	47.13	−99.24	590	4	478	1.0	Agricultural	SS	Mollisol	USGS/DOI FWS
Central Plains Exp. Range	40.82	−104.75	1653	9	322	0.4	Grassland	SS	Mollisol	USDA ARS
Sterling	40.46	−103.03	1365	10	422	1.0	Agricultural	SS	Mollisol	Private,
Rocky Mtn. NP (Castnet)	40.28	−105.55	2742	7	998	19.0	Temperate coniferous forest	SO	-	DOI NPS
LBJ Nat'l grasslands	33.40	−97.57	272	18	864	13.0	Temperate deciduous forest	SO	Alfisol	USDA FS
M. Klemme Range Res. Station	35.41	−99.06	520	15	780	1.0	Agricultural	SS	Inceptisol	Oklahoma State U, Oklahoma Agricultural Experimental Station

Table 1. Cont.

Site	Lat (°)	Long (°)	El (m.a.s.l)	Temp (°C)	Precip (mm)	CH (m)	Ecosystem type	CS	Soil order	Owner
Northcutt	33.89	−96.84	212	17	970	4.0	Savanna	O	Inceptisol	Private,
Bozeman	45.66	−111.05	1503	7	491	1.0	Agricultural	SS	Mollisol	Montana State U
Loch Leven	45.46	−110.62	1448	6	432	0.4	Grassland	SS	Mollisol	Montana Dept of Fish, Wildlife and Parks
Yellowstone NP	44.95	−110.54	2129	7	249	14.0	Temperate coniferous forest	SO	-	*DOI NPS
Niwot Ridge	40.05	−105.58	3478	−4	930	0.2	Tundra	SS	Inceptisol	USDA FS
Fraser Exp. Forest	39.86	−105.86	3526	1	584	14.0	Boreal/montane coniferous forest	SO	Inceptisol	USDA FS
Moab	38.25	−109.39	1800	14	239	0.2	Grassland	SS	Aridisol	BLM
Santa Rita Exp. Range	31.91	−110.84	999	22	303	2.0	Shrubland	SS	Entisol	U Arizona
Jornada Exp. Range	32.59	−106.84	1321	18	241	0.4	Grassland	SS	Aridisol	USDA ARS
Onaqui-Benmore	40.18	−112.45	1654	9	274	1.2	Shrubland	SS	Aridisol	BLM/USDA FS
Murray	40.65	−111.92	1302	12	450	1.0	Grassland	SS	- (anthropodisol)	City of Murray, UT
Red Butte	40.78	−111.80	1676	12	450	10.0	Temperate deciduous forest	CL	Mollisol	USDA FS/U Utah
Wind River RNA	45.82	−121.95	368	9	2223	50.0	Temperate coniferous forest	CL	Andisol	USDA FS
Abby Road	45.76	−122.33	367	9	2249	1.0	Temperate coniferous forest	SS	Andisol	State of Washington Dept. Natural Resources
Thayer	45.71	−122.34	557	9	2249	12.0	Temperate coniferous forest	CL	Inceptisol	State of Washington Dept. Natural Resources
Thyme	45.71	−122.38	301	9	2249	1.0	Temperate coniferous forest	SS	Ultisol	State of Washington Dept. Natural Resources
Good Seed	45.78	−122.30	559	9	2249	1.0	Temperate coniferous forest	SS	Ultisol	State of Washington Dept. Natural Resources
Soaproot Saddle	37.03	−119.26	1210	13	957	32.0	Temperate coniferous forest	SO	Alfisol	USDA FS
Lower Teakettle	37.01	−119.01	2149	8	957	35.0	Temperate coniferous forest	SO	Entisol-Inceptisol	USDA FS
San Joaquin Exp. Range	37.11	−119.73	397	17	375	21.0	Savanna	O	Alfisol-Inceptisol	USDA FS
Toolik Lake LTER	68.66	−149.37	827	−9	316	0.3	Tundra	SS	Gelisol	BLM
Pump Station 2	69.60	−148.67	142	−9	316	0.3	Tundra	v	Gelisol	BLM
Barrow Exp. Observatory	71.28	−156.62	7	−12	105	0.3	Tundra	SS	Gelisol	*Ukpeaġvik Iñupiat Corporation*
Caribou-Poker	65.15	−147.50	239	−3	262	8.0	Boreal/montane coniferous forest	SC	Inceptisol	UAF
Poker Flats	65.11	−147.42	472	−3	262	8.0	Boreal/montane coniferous forest	SS	Inceptisol	Alaska Dept Natural Resource/UAF
Eight Mile Lake	63.88	−149.22	684	−2	366	0.3	Tundra	SS	Gelisol	Alaska Dept Natural Resource
Delta Junction	63.88	−145.75	504	−3	305	10.0	Boreal/montane coniferous forest	SC	Inceptisol	BLM
Kenai NWA	60.53	−150.58	110	1	483	12.0	Boreal/montane coniferous forest	CL	Spodosol	DOI NWA

All sites were given access to measure soil properties by the owner, no soil samples were taken, and (*) denotes that a research permit was needed to gain access to the site.
Note: - = soil order unknown, USDA = US Department of Agriculture, FS = Forest Service, ARS = Agricultural Research Service, DOI = US Department of Interior, BLM = US Bureau of Land Management, DOE = US Department of Energy, UAF = University of Alaska, Fairbanks, FWS = US Fish and Wildlife Service, USGA = US Geological Survey, NPS = National Park Service, LTER = long term ecological research site NWA = National Wildlife Area, RNA = Research Natural Area, WMA = Wildlife Management Area, NP = National Park, BS = Biological Station.

sample size and sample spacing. However, it is also important to note that other goals also exist that require different sampling designs, *e.g.*, redundant sampling (*i.e.*, field duplicates) involving two or more measurements at the same location to aid error detection (*e.g.*, sensor malfunction) and increase confidence in the data, while varying the distance between adjacent samples can be useful when characterizing the spatial variance structures.

The number of samples and sample spacing is rarely justified in research papers or presentations. In many, if not most studies, decisions relating to the sampling strategy appear to be based on a combination of subjective opinion, repeating previously used sampling designs (often from a different ecosystem or in relation to a different soil property), and available resources. There are only a few methodological articles that discuss sampling designs recognize the importance of choosing an appropriate sample size and sample spacing, and provide only vague- or no recommendations at all [2–4]. The advice often given is to conduct a preliminary spatial variability study (of soil properties) at the research site to determine the number of samples and sample size required to achieve a given accuracy [4,5]. While this is good advice, it is rarely followed, presumably due to a lack of time, money, and/or an appreciation of its importance.

Soils are highly variable and embody systematic (*e.g.*, soil forming processes) and random sources of variability across space [6,7]. Some studies have used the coefficient of variation to assess this variability at a given site, and used in sample size analyses [8–14]. Geostatistical techniques, in particular semivariograms, have increasingly been used over the last few decades to characterize spatial variation in soil properties [15,16]. Semivariograms provide two useful components for designing a robust sampling strategy: 1) an estimate of the variance (*i.e.*, the semivariogram sill), which can be used to inform sample size in future studies; and 2) an estimate of the minimum distance required for samples to be considered spatially independent (*i.e.*, the semivariogram range), which can be used to inform sample spacing (Figure S1). We emphasize that the use of sample size analyses in this paper is to inform decisions relating to future studies, not to assess the power of a sampling design retrospectively, which may be fundamentally flawed [17].

Site-specific studies that have assessed spatial variation in soil properties have shown that spatially structured variability is ubiquitous. For instance, Robertson et al. [11] studied variation in soil physical, chemical, and biological properties across an agricultural field in Michigan, USA. Almost every soil property exhibited spatially structured variation, which in most cases accounted for the majority of the variability that was observed Robertson et al. [11], and indicated that measurements of soil properties at one location could be used to predict values for those properties up to ~60 m away (*i.e.*, the range). Similar findings have been reported elsewhere, across a range of ecosystems and soils [9,12,18–24], although there are a few exceptions, with some properties exhibiting little or no spatial dependence at the scale studied [9,11,19,20]. The identification of spatial patterns in soil properties among sites, in particular at larger continental scales has been hampered, because most studies have investigated these quantities at a single site.

The few cross-site comparisons studies have focused on soil moisture variability located only within a small region with relatively narrow ecoclimatic variability [13,14,25] or, if they spanned larger scales, involved only a few sites [26] or combined data from different studies with widely differing designs *e.g.*, spatial extent, physical quantities, sampling depth, seasonal timing, measurement technique, etc. [12,27–30]. Moreover, previously published data are often unavailable for additional analyses. A single study that employs a uniform sampling and analysis approach, *i.e.*, the above mentioned quantities, as well as being representative of site characteristics (*e.g.*, ecosystem structure, soil type, etc), made under similar environmental conditions for a wide range of ecosystems is needed for meaningful cross-site comparisons and to develop quantitative sampling guidelines that can inform sampling strategies in future soil studies.

Here, we studied the spatial variation in soil properties using a consistent approach at 60 sites throughout the USA and developed quantitative relationships to provide: 1) an ecological understanding of spatial variability in soils, and 2) guidance on sampling designs for future soil studies. We studied local scale (~1 ha) variation in soil temperature (T_s) and soil water content (SWC) at sites that included representatives from all major terrestrial ecosystem types, soil types, and climates found in the USA. Soil temperature and moisture were used as proxies for other soil properties because they can be measured quickly and accurately in

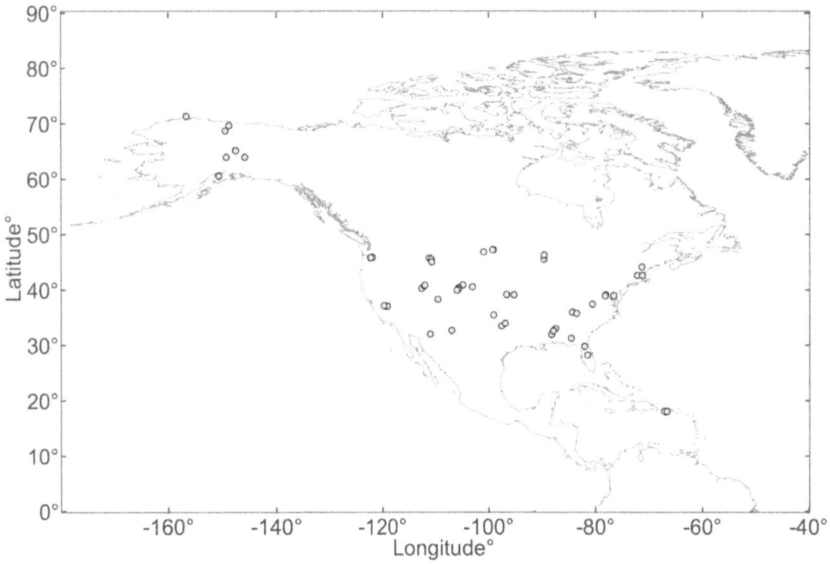

Figure 1. The sampling sites (open circles) were widely distributed throughout the US.

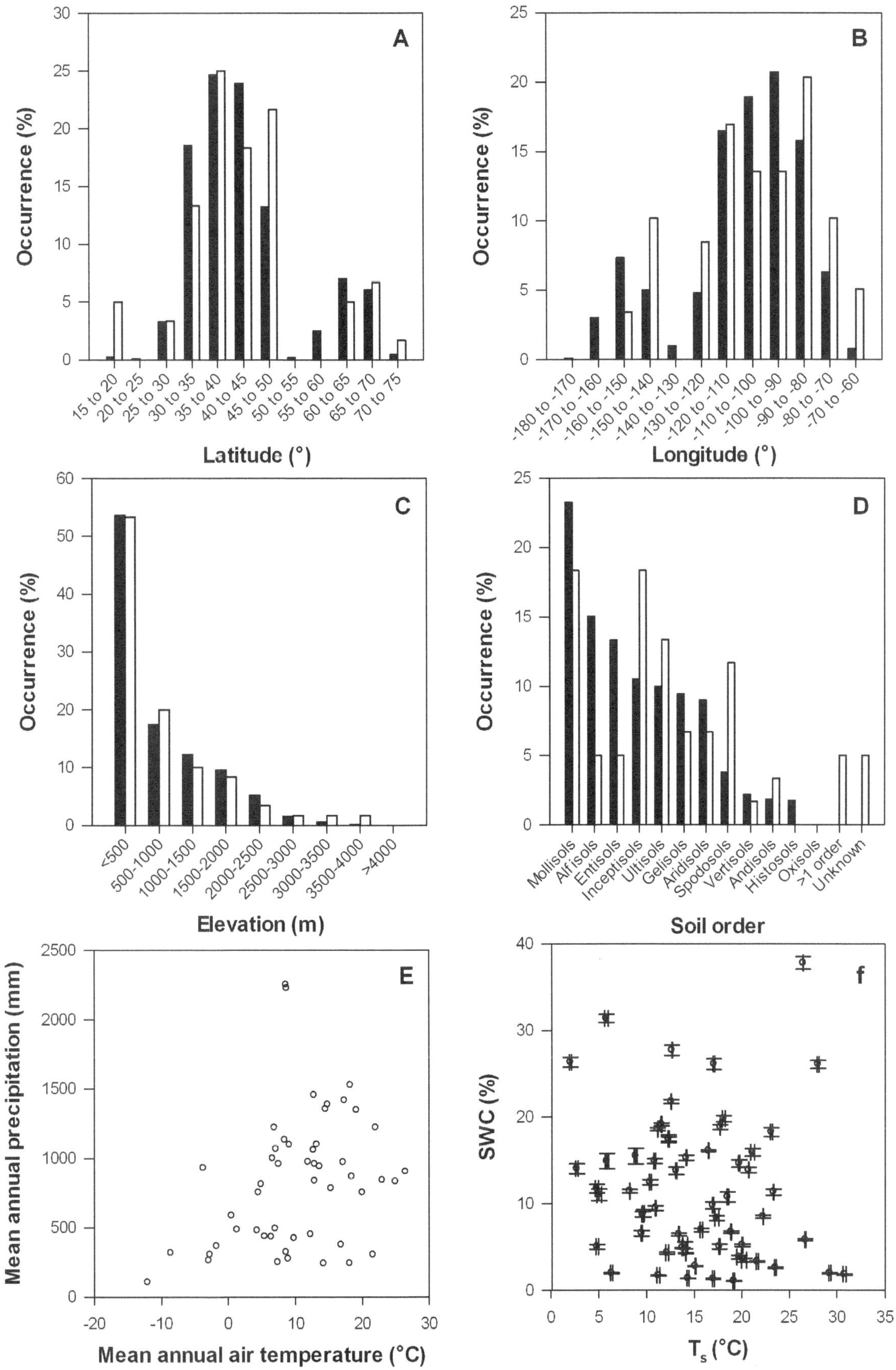
A
Occurrence (%)
0
5
10
15
20
25
30
15 to 20
20 to 25
25 to 30
30 to 35
35 to 40
40 to 45
45 to 50
50 to 55
55 to 60
60 to 65
65 to 70
70 to 75
Latitude (°)
B
Occurrence (%)
0
5
10
15
20
25
-180 to -170
-170 to -160
-160 to -150
-150 to -140
-140 to -130
-130 to -120
-120 to -110
-110 to -100
-100 to -90
-90 to -80
-80 to -70
-70 to -60
Longitude (°)
C
Occurrence (%)
0
10
20
30
40
50
60
<500
500-1000
1000-1500
1500-2000
2000-2500
2500-3000
3000-3500
3500-4000
>4000
Elevation (m)
D
Occurrence (%)
0
5
10
15
20
25
Mollisols
Alfisols
Entisols
Inceptisols
Ultisols
Gelisols
Aridisols
Spodosols
Vertisols
Andisols
Histosols
Oxisols
>1 order
Unknown
Soil order
E
Mean annual precipitation (mm)
0
500
1000
1500
2000
2500
-20
-10
0
10
20
30
Mean annual air temperature (°C)
f
SWC (%)
0
10
20
30
40
0
5
10
15
20
25
30
35
T_s (°C)

Figure 2. The sites were broadly representative of the US. Percent occurrence of land in latitudinal (a), longitudinal (b), elevational (c), and soil order (d) categories in the US (black bars), and our study sites (white bars). The study sites also occupied a wide range of climates (e) and T_s and SWC conditions at the time of samplings (f). Error bars in (f) represent ±1 SE.

the field, they are major drivers of soil biotic and biogeochemical properties and processes, their spatial pattern integrates the spatial pattern of ecosystem canopy structure and has been shown to correlate with spatial patterning of other soil properties at the local scale, and they are increasingly recognized as important variables in understanding land-atmosphere interactions *e.g.*, soil moisture is recognized as a Global Climate Observing System Essential Climate Variable [9,11,31–36]. The sampling scale (~1 ha) corresponds to many ecosystem-level soil studies and is the basis of many large emergent observatories (National Ecological Observatory Network, NEON, www.neoninc.org; Integrated Carbon Observing System, www.icos-infrastructure.eu). We used these data to further examine the relationship between the spatial structure of variability in soil properties and site characteristics at broader landscape- to continental scales (*e.g.*, latitude, climate, ecosystem type, soil type, and vegetation structure). Lastly, we used analyses to estimate the sample size required to accurately estimate the spatial mean T_s and SWC at each site.

This study was made to inform the sensor-based soil plot designs at NEON sites. In addition, several hypotheses were tested. We hypothesized that;

1) the spatial variability in soil properties would be associated with variability in ecosystem structure, *e.g.*, we expect open canopy ecosystems, which exhibit large spatial variability in vegetation height (*e.g.*, savannas) to have larger variability in soil properties than closed canopy ecosystems,
2) agricultural ecosystems would exhibit lower spatial variability in soil properties than other ecosystems, since management activities (*e.g.*, tillage and irrigation) aim to homogenize properties,
3) younger soils (*e.g.*, inceptisols) would exhibit both lower spatial variability and have smaller range values than older soils (*e.g.*, ultisols). We expected that some soil properties (*e.g.*, soil texture) that influence the spatial structure of variability are controlled by processes that operate over long temporal scales (*e.g.*, weathering),
4) the spatially structured variability (*e.g.*, semivariogram range, nugget, and sill), as well as the sample size required to accurately estimate the mean, would be positively correlated for soil temperature (T_s) and soil water content (SWC), becuase some underlying site characteristics (*e.g.*, canopy gaps, etc.) are likely to influence both variables,
5) some forms of semivariograms may be more common than other forms, *e.g.*, , that small semivariogram range values would be more commonly associated with small semivariogram sills than large sills, and
6) variability would be larger for SWC than T_s since water is more mobile than heat and sensitive to microtopography, as well as being actively transported by plants.

We also aimed to develop empirically-based rules that could be used by other researchers to guide sampling designs to accurately estimate the spatial mean of soil properties for future studies.

Methods

Site descriptions and data collection

The study was conducted at 60 NEON sites (Table 1). These sites were chosen for landscape level representativeness using the approaches of Hargrove and Hoffman [37,38], which divided the USA into 20 eco-climatic domains. Within each domain, 3 representative site were chosen; 1 wildland (*i.e.*, relatively undisturbed) site and typically 2 sites that address key ecological issues for that domain (*e.g.*, urbanization, land management, advance of invasive species, or climate change) [39,40]. Measurements at these sites included ecosystem scale (*i.e.*, 10^3 to 10^6 m^2) tower-based estimates of eddy covariance [41–44], and the long term sensor-based soil measurements are required to be made in the flux footprint of the tower, *i.e.*, at the flux scale [45–47] and occur on locally (co-) dominant soil series. No sampling occurred requiring any sampling permits (re soil sampling), no human sampling occurred, nor the use or impact to any endangered or protected animals were involved in this study. The sampling area for this study corresponded to the expected location of NEON's long term sensor-based soil measurements at each site. We received permission to access and to measure soil properties by the owner at all sites, and no soil samples were taken (hence not requiring a sampling permit of any kind). Two National Park Service sites required a research permit to gain access to the site (see Table 1).

Soil taxonomic information at a site was gathered from the USDA Natural Resource Conservation Service's Web Soil Survey, except for Wind River [48], Bartlett Experimental Forest (www.fs.fed.us/ne/durham/4155/bartlett.htm#SOI), and Eight Mile Lake [49]. Soil taxonomic information was not available for Yellowstone National Park, WY, or Rocky Mountain National Park, CO, while the soil at Murray, UT, was classified by the Web Soil Survey as *dumps* (*i.e.*, anthropogenosol), which cannot be assigned to a specific soil order. Mean annual temperature and mean annual total precipitation data were gathered from local site sources when available, and from climate data for nearby towns (www.usclimatedata.com) when site-specific information was unavailable.

Field-based soil temperature and moisture data were collected over 1 to 2 days at each site between August 2009 and April 2011 (with most sites visited between March and October 2010) and were timed to coincide with peak growing season. The only exception was sampling at Mountain Lake, VA, which occurred prior to budburst due to logistical constraints. Our measurements corresponded to the growing season because this is the time of year when most biological activity occurs. It should be noted that spatial variability of soil properties can change seasonally [44], *i.e.*, there is often intra-annual nonstationarity in both functional linkages and spatial correlation), although this does not always occur [12,21,23,26,50]. As such, the spatial variability that we observed may not reflect spatial variation among other seasons or years.

Soil temperature, T_s (°C), was measured at 0–12 cm depth with platinum resistance temperature sensors (RTD 810, Omega Engineering Inc., Stamford CT) and soil water content, SWC (vol H_2O vol^{-1} soil expressed as a percentage) was measured at a 0–15 cm depth with time domain dielectric sensors (CS616, Campbell Scientific Inc., Logan UT). Due to the large number of sites and soil types it was not possible to calibrate the soil moisture sensor to the soil type at each site; instead a manufacturer recommended equation was used to convert the sensor measurement of the dielectric constant to SWC [51,52]. The sensors were inserted into the soil and allowed to stabilize prior to data

Table 2. Summary statistics of T_s and SWC measurements, as well as range, sill, and nugget estimates based on the semivariogram model, at each site.

	T_s										SWC										Sampling date‡
Site	Mean (°C)	SD	$CV_{Traditional}$	Maximum lag	Range (m)*	Nugget (% of sill)*	Sill*	CV_{Sill}	Sample size†	n	Mean (%)	SD	$CV_{Traditional}$	Maximum lag	Range (m)*	Nugget (% of sill)*	Sill*	CV_{Sill}	Sample size†	n	
Harvard Forest	17.7	0.6	0.04	100	74	55	0.46	0.05	0.8	144	5.0	3.5	0.69	100	17	62	9.60	0.87	203.8	144	7/8/2010
Burlington	19.2	0.8	0.04	100	9	56	0.67	0.06	1.0	119	1.1	0.6	0.52	100	10	31	0.44	0.88	209.5	120	7/7/2010
Bartlett Exp. Forest	17.0	0.9	0.05	100	40	67	0.98	0.08	1.8	126	9.8	4.6	0.47	100	68	27	29.41	0.78	166.1	126	7/6/2010
Blandy Farm	22.2	0.4	0.02	100	-	-	-	-	-	129	8.5	1.8	0.21	100	47	53	3.56	0.31	26.8	129	8/18/2010
Smithsonian CRC (SCBI)	20.7	0.4	0.02	80	5	16	0.05	0.02	0.1	126	13.9	3.6	0.26	80	-	-	-	-	-	126	8/19/2010
Smithsonian ERC	23.3	0.4	0.02	80	34	15	0.16	0.02	0.2	123	11.3	3.9	0.34	80	32	41	13.40	0.46	56.7	123	8/17/2010
Ordway-Swisher BS	14.3	1.2	0.08	100	6	5	0.58	0.08	1.5	144	1.4	0.3	0.22	100	23	64	0.09	0.31	25.7	144	3/24/2010
J. Jones Ecological Research station	14.1	0.3	0.02	100	35	31	0.08	0.03	0.2	129	4.3	0.7	0.17	100	8	96	0.54	0.24	15.8	129	3/21/2010
Disney Wilderness Preserve	16.5	0.6	0.04	100	4	25	0.31	0.05	0.6	144	16.1	1.0	0.06	95	78	77	1.00	0.09	2.1	144	3/23/2010
Guanica State Forest	26.7	0.7	0.03	84	-	-	-	-	-	129	5.8	1.4	0.24	84	15	32	2.35	0.37	37.3	129	3/18–19/2010
Lajas Ag. Exp. St.	28.1	1.2	0.04	65	12	22	0.63	0.04	0.4	120	26.0	5.3	0.20	65	24	57	27.17	0.28	21.7	120	3/17–18/2010
Ponce	26.5	0.4	0.02	100	25	18	0.21	0.02	0.2	126	37.8	7.9	0.21	100	25	9	73.66	0.32	28.0	121	10/12/2010
Steigerwaldt	19.8	0.8	0.04	100	-	-	-	-	-	144	14.6	4.9	0.33	100	-	-	-	-	-	144	8/12/2010
Tree Haven	18.6	0.9	0.05	80	-	-	-	-	-	129	10.8	6.5	0.60	80	12	40	15.87	0.52	74.2	129	8/11/2010
UNDERC	18.9	0.4	0.02	100	70	68	0.13	0.03	0.2	141	6.7	1.6	0.23	100	20	75	2.70	0.35	32.4	141	8/10/2010
Konza - Core	14.2	0.2	0.02	100	18	58	0.04	0.02	0.1	129	15.3	3.6	0.23	100	52	75	13.70	0.34	31.7	129	5/12/2010
Konza - Relocatable	12.6	0.2	0.02	100	35	34	0.04	0.02	0.1	123	21.8	2.6	0.12	100	67	38	9.02	0.20	10.3	123	5/13/2010
U Kansas Bio Station	12.7	0.5	0.04	84	7	18	0.06	0.03	0.2	129	27.7	6.7	0.24	84	-	-	-	-	-	129	5/11/2010
Oak Ridge Nat'l Lab	13.4	0.4	0.03	84	30	16	0.18	0.04	0.5	129	6.4	1.7	0.26	84	-	-	-	-	-	129	4/19/2010
Great Smokey Mtns NP	11.2	0.3	0.02	80	70	54	0.05	0.03	0.2	129	18.6	1.9	0.10	80	-	-	-	-	-	129	4/21/2010
Mountain Lake Res. Stn.	8.2	0.4	0.05	100	79	42	0.11	0.06	0.9	129	11.4	2.2	0.19	100	70	70	5.55	0.29	23.1	129	4/22/2010

Table 2. Cont.

	T_s										SWC										Sampling date‡
Site	Mean (°C)	SD	$CV_{Traditional}$	Maximum lag	Range (m)*	Nugget (% of sill)*	Sill*	CV_{Sill}	Sample size†	n	Mean (%)	SD	$CV_{Traditional}$	Maximum lag	Range (m)*	Nugget (% of sill)*	Sill*	CV_{Sill}	Sample size†	n	
Talladega Nat'l Forest	20.2	3.0	0.15	100	94	36	4.78	0.15	6.3	129	3.5	2.5	0.72	100	20	20	5.63	0.97	252.9	129	4/16/2010
Choctaw NWA	17.1	1.1	0.06	84	16	8	0.63	0.07	1.2	129	26.1	7.4	0.28	84	45	42	51.93	0.39	41.3	129	4/14/2010
Dead Lake WMA	17.8	0.3	0.02	80	-	-	-	-	-	132	18.8	3.1	0.17	80	-	-	-	-		132	4/19/2011
Dakota-Coteau	12.3	0.8	0.07	100	10	19	0.43	0.08	1.5	144	17.4	3.2	0.19	100	19	60	8.06	0.23	14.4	144	5/26/2010
Northern Great Plains Res. Lab	11.6	0.5	0.04	100	8	23	0.12	0.04	0.5	144	19.1	2.0	0.11	100	4	63	4.37	0.15	6.5	144	5/25/2010
Woodworth	15.7	1.3	0.08	100	14	65	0.89	0.09	2.0	144	6.9	2.4	0.35	100	8	53	2.89	0.35	32.5	144	5/27/2010
Central Plains Exp. Range	23.5	1.0	0.04	100	95	32	0.54	0.04	0.5	144	2.6	0.8	0.32	100	29	40	0.72	0.46	57.8	144	8/13/2009
Sterling	23.1	1.3	0.06	100	50	57	0.60	0.05	0.6	156	18.2	6.5	0.36	100	2	0	49.25	0.54	80.1	156	8/10/2009
Rocky Mtn. NP (Castnet)	11.2	1.7	0.15	84	-	-	-	-	-	132	1.7	0.5	0.32	84	12	42	0.31	0.46	57.3	132	8/18–19/2009
LBJ Nat'l grasslands	20.1	0.6	0.03	100	9	2	0.48	0.05	0.6	129	5.2	1.5	0.29	100	3	17	2.49	0.43	50.0	129	9/28/2010
M. Klemme Range Res. Station	18.1	1.6	0.09	100	22	11	2.31	0.12	3.8	129	19.8	4.1	0.21	100	72	23	21.98	0.33	30.4	128	4/30/2010
Northcutt	19.7	2.8	0.14	100	11	27	2.49	0.11	3.5	141	3.8	2.3	0.59	100	78	15	7.59	1.02	280.8	141	4/28/2010
Bozeman	21.1	2.3	0.11	100	25	7	3.47	0.12	4.2	123	15.9	4.8	0.30	100	23	49	25.99	0.45	55.6	123	7/20/2010
Loch Leven	21.6	1.4	0.06	100	15	44	0.82	0.06	0.9	138	3.3	0.6	0.19	100	70	36	0.57	0.32	27.9	138	7/21/2010
Yellowstone NP	12.1	1.7	0.14	100	31	13	3.02	0.20	11.1	132	4.3	1.4	0.33	100	15	28	2.36	0.51	69.0	132	7/22/2010
Niwot Ridge	8.8	1.7	0.20	80	5	3	2.60	0.26	18.0	123	15.5	10.0	0.65	84	27	9	83.27	0.83	187.9	123	7/30/2010
Fraser Exp. Forest	6.3	1.2	0.19	84	21	23	1.00	0.22	13.5	132	2.0	0.7	0.33	80	15	94	0.44	0.47	60.3	132	9/15/2010
Moab	15.1	0.5	0.04	100	16	64	0.12	0.03	0.3	144	2.8	0.4	0.16	100	37	58	0.22	0.24	15.6	144	10/6/2010
Santa Rita Exp. Range	30.8	1.9	0.06	100	2	0	0.79	0.04	0.5	141	1.8	0.5	0.30	100	17	38	0.32	0.44	52.7	141	8/25/2010
Jornada Exp. Range	29.3	1.2	0.04	100	46	73	0.57	0.04	0.4	141	2.0	0.5	0.26	100	-	-	-	-	-	141	8/23/2010
Onaqui-Benmore	13.8	0.8	0.06	84	14	12	0.61	0.08	1.7	135	4.9	0.8	0.16	84	-	-	-	-	-	135	5/18/2010
Murray	17.4	2.6	0.15	84	-	-	-	-	-	135	8.3	3.1	0.37	84	2	0	6.86	0.45	54.0	135	5/19/2010

Table 2. Cont.

	T_s										SWC										Sampling date[‡]
Site	Mean (°C)	SD	$CV_{Traditional}$	Maximum lag	Range (m)*	Nugget (% of sill)*	Sill*	CV_{Sill}	Sample size[†]	n	Mean (%)	SD	$CV_{Traditional}$	Maximum lag	Range (m)*	Nugget (% of sill)*	Sill*	CV_{Sill}	Sample size[†]	n	
Red Butte	10.8	1.0	0.09	65	22	42	0.81	0.12	3.7	132	14.9	3.2	0.21	65	15	79	10.44	0.31	25.4	132	5/20/2010
Wind River RNA	10.9	0.4	0.03	100	4	31	0.09	0.04	0.4	138	9.5	3.1	0.33	100	3	63	10.27	0.48	62.1	138	6/8/2010
Abby Road	9.6	1.0	0.11	100	3	30	1.02	0.15	6.0	141	8.8	3.7	0.42	100	21	44	16.15	0.65	114.1	141	10/20/2010
Thayer	9.5	0.6	0.06	70	40	58	0.42	0.10	2.5	111	6.6	3.4	0.51	70	-	-	-	-	-	111	10/21/2010
Thyme	12.3	0.6	0.05	100	17	32	0.40	0.07	1.4	129	17.5	5.1	0.29	100	27	53	27.31	0.43	49.3	129	6/10/2010
Good Seed	13.1	0.8	0.06	100	20	13	0.79	0.10	2.5	126	13.8	5.0	0.36	100	43	46	25.47	0.52	72.7	126	6/9/2010
Soaproot Saddle	17.0	1.1	0.07	84	9	13	1.18	0.09	2.2	129	1.3	0.5	0.36	84	3	18	0.24	0.52	73.5	129	7/15/2010
Lower Teakettle	14.0	2.9	0.20	80	11	2	6.75	0.26	18.6	129	5.2	4.4	0.86	80	13	2	20.06	1.22	404.6	129	7/15/2010
San Joaquin Exp. Range	9.7	1.2	0.12	100	6	0	0.96	0.14	5.6	144	8.8	5.0	0.57	100	8	6	19.91	0.71	137.8	144	2/22/2011
Toolik Lake LTER	2.7	1.6	0.59	100	3	40	2.65	0.85	194.6	144	14.0	6.9	0.49	100	5	58	52.77	0.73	145.1	144	6/19/2010
Pump Station 2	2.1	1.2	0.59	100	31	42	1.63	0.88	207.6	128	26.3	7.1	0.27	100	17	47	55.17	0.40	43.1	144	6/20/2010
Barrow Exp. Observatory	5.8	2.2	0.39	100	14	15	4.31	0.51	69.1	141	31.4	5.4	0.17	100	-	-	-	-	-	141	7/26/2010
Caribou-Poker	5.0	3.1	0.61	100	6	44	9.90	0.88	211.2	144	11.0	8.1	0.73	100	7	49	67.34	1.05	300.9	144	6/22/2010
Poker Flats	10.4	1.4	0.14	100	3	3	1.58	0.17	8.0	144	12.5	3.6	0.29	100	8	59	13.29	0.41	46.4	144	6/28/2010
Eight Mile Lake	5.9	2.1	0.36	80	72	81	4.65	0.52	73.5	138	14.9	10.6	0.71	80	14	74	94.74	0.92	230.4	138	6/29/2010
Delta Junction	4.8	1.9	0.40	100	44	31	4.28	0.61	100.4	138	11.7	5.8	0.49	100	6	34	33.23	0.70	131.2	138	6/23/2010
Kenai NWA	4.8	1.5	0.32	100	84	80	2.66	0.48	61.9	144	5.0	3.2	0.63	100	44	55	11.47	0.95	245.9	144	7/6/2010

The standard deviation (SD) is reported here because it was required to calculate $CV_{Traditional}$.
*- indicates that the semivariogram model did not reach its sill, which prevented sill, range, and nugget values from being accurately estimated.
[†]Sample size required to estimate to with 10% of the spatial mean with 90% confidence.
[‡]Month/day/year format.

Table 3. Correlation coefficients indicating covariation among many site characteristics.

	Latitude	Longitude	Mean annual air temperature	Mean annual precipitation	Elevation	Canopy height	Mean T_s	Mean SWC
Latitude	1	−0.808	−0.921	−0.288	−0.001	−0.211	−0.700	0.084
Longitude	−0.808	1	0.693	0.277	−0.173	0.309	0.654	−0.018
Mean annual air temperature	−0.921	0.693	1	0.284	−0.252	0.196	0.701	−0.061
Mean annual precipitation	−0.288	0.277	0.284	1	−0.271	0.412	0.005	0.025
Elevation	−0.001	−0.173	−0.252	−0.271	1	−0.154	−0.129	−0.325
Canopy height	−0.211	0.309	0.196	0.412	−0.154	1	0.012	−0.199
Mean T_s	−0.700	0.654	0.701	0.005	−0.129	0.012	1	−0.115
Mean SWC	0.084	−0.018	−0.061	0.025	−0.325	−0.199	−0.115	1

acquisition. Data at each measurement point were acquired at 1 s (execution interval), and descriptive statistics were calculated over 30 s averaging periods with a datalogger (CR3000, Campbell Scientific Inc.). The sampling design was similar among the sites, but it was not identical due to site-specific space constraints (*e.g.*, property boundaries) and obstacles (*e.g.*, rock outcrops, roads, streams, tree trunks, and stone walls). The measurements were taken along 2 to 4 intersecting transects at each site, with the number and length of transects depending on site topography and property size. The intersection point of the transects was typically at the center of each transect. The length of transects ranged from 42 m to 210 m. The starting, center and ending points of each transect were recorded by a GPS unit, which was used to calculate distances during semivariogram analysis.

The determination of sampling locations followed the cyclic sampling logic used by Bond-Lamberty *et al.* [23], with measurements taken at 0, 2, 8, 28, 38, and 42 m (noted as transect points) and with this spacing scheme repeated until the end of the transect was reached (*i.e.*, the next sampling points would occur at 44, 50, 70, 80, 84, 86, 92 … m). In addition, two more T_s and SWC measurements were made at −0.3 m and +0.3 m from each transect point along the axis of the transect, thus resulting in a total of 3 T_s and 3 SWC measurements made at each transect point (Figure S1). The minimum and maximum distance between measurements ranged from 0.3 m to at least 84 m at each site, and up to 210 m at some sites, as a result, we were able to capture both small spatial scale (~0.3 m) and the plot-to-footprint scale (~100's m^{-2}) variability. Measurements were taken at 134±1 (mean ±1 standard error, convention used throughout the text, unless otherwise noted) locations per site (maximum = 156, minimum = 111). The number of measurement points at each site exceeded the recommendation

Table 4. Statistical significance (p values) of site characteristics on mean T_s and SWC, as well as semivariogram properties and derived values.

	T_s					SWC				
	Mean	Range	Sill	Nugget	Sample size	Mean	Range	Sill	Nugget	Sample size
Latitude	<0.001	NS	NS	NS	NS	NS	NS	NS	NS	NS
Longitude	NS	NS	0.013	NS	0.003	NS	0.008	0.023	0.002	0.004
Elevation	NS	0.091	NS	NS	NS	0.032	NS	NS	NS	NS
Mean annual air temperature	NS	NS	0.003	NS	0.004	NS	NS	0.004	0.082	0.009
Mean annual precipitation	NS	NS	NS	NS	NS	0.007	NS	0.050	NS	0.002
Canopy height	NS	0.065	NS	0.024	0.087	NS	0.037	NS	NS	NS
Ecosystem type	<0.001	NS	NS	NS	NS	<0.001	NS	NS	0.015	0.071
Canopy structure	NS	0.019	<0.001	0.006	<0.001	NS	NS	0.053	<0.001	<0.001
Soil type	NS	0.009	NS	0.046	0.021	NS	NS	NS	0.026	NS
Mean T_s	NA	NS	0.012	NS	NS	0.032	NS	0.046	0.001	0.004
Mean SWC	NS	NS	NS	0.005	NS	NA	NS	<0.001	0.072	0.037
Model r^2	0.73	0.40	0.57	0.48	0.83	0.51	0.18	0.70	0.62	0.66

The r^2 of the statistical model (observed versus predicted) is also shown.
NS: not significant (*i.e.*, $p>0.1$ and factor was removed from the statistical model), NA: not applicable.

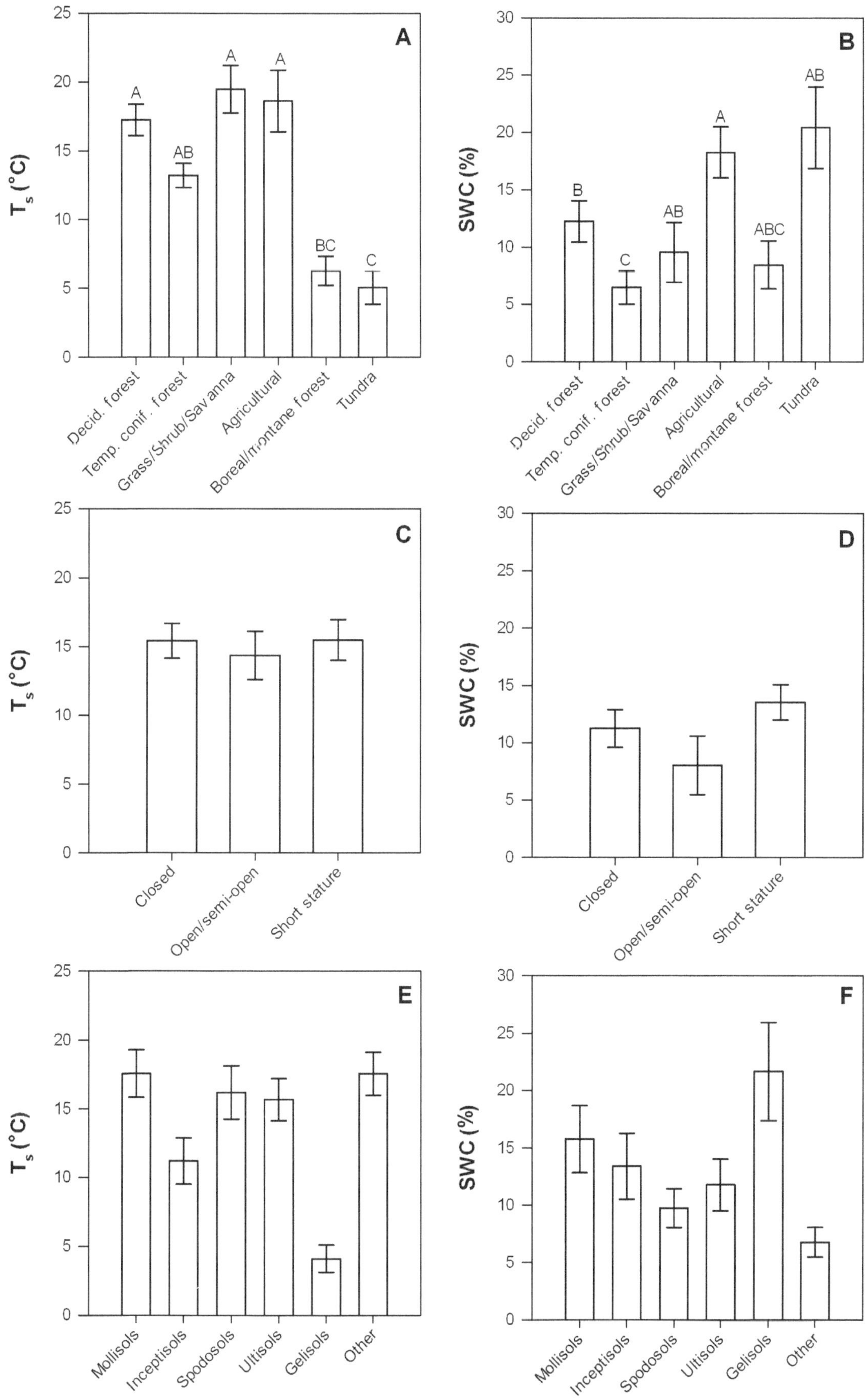
A
B
C
D
E
F
T_s (°C)
SWC (%)
0
5
10
15
20
25
30
Decid. forest
Temp. conif. forest
Grass/Shrub/Savanna
Agricultural
Boreal/montane forest
Tundra
Closed
Open/semi-open
Short stature
Mollisols
Inceptisols
Spodosols
Ultisols
Gelisols
Other
AB
BC
ABC

Figure 3. Mean T_s (left column) and SWC (right column) at each site in relation to ecosystem type (a, b), canopy structure (c, d), and soil type (e, f). Error bars represent ±1 SE. Bars with different letter were significantly different ($p<0.05$) based on Tukey HSD tests. The letter "A" corresponds to the largest least square mean(s) and subsequent letters (B, C, …) correspond to progressively smaller least square means.

of at least 100 points to construct representative isotropic (*i.e.*, non-directional) semivariograms with soil data [53,54].

It typically took ~2.5 to 5 hours to collect the data at each site, during which time natural changes to T_s and SWC could occur, *e.g.*, solar heating of the soil, evaporation and/or transpiration. Thus, we also continuously collected T_s and SWC with a second, stationary suite of 3 pairs of identical T_s and SWC sensors (spaced 0.3 m apart) adjacent to the point where all the transects intersected to estimate diurnal and other weather-related influences on the measurements. Data at were acquired at 1 s (execution interval), and descriptive statistics were calculated over 30 sec averaging periods with a datalogger (Models CR1000, Campbell Scientific Inc.). T_s and SWC data from the transects and the stationary system, as well as measurement coordinates, that were used to create the semivariograms are available available by completing the request form at the bottom of the webpage: http://neoninc.org/pds/FIU007.php [55].

Semivariograms and model fitting

Two primary approaches were used to de-trend temporal changes from the data collected along the transects, such that the remaining variability in the data can be attributed to its spatial sources and quantified with a semivariogram. First, mean T_s and SWC values from the stationary location were subtracted from transect data at each corresponding time of day to produce *corrected* data. Second, a linear regression based on the relationship between time of day and the *corrected* T_s and SWC data was fitted, and the residuals calculated. These residuals were then used to construct the semivariogram. In the context of geostatistics, de-trending is driven by the need to satisfy the second-order stationarity requirement, which needs to be met in order to ensure that the resulting covariance function is valid, *e.g.*, [24]. At 5 sites (Sterling, Rocky Mountain NP, Central Plains Exp. Range, Caribou Poker, and Northcutt) the stationary system was either unavailable or malfunctioned. Therefore, de-trending was done using the linear regression approach only. If additional patterns (*i.e.*, gradients in T_s or SWC across the sampling site or non-normal distribution) were still visible in the data after removing temporal trends, we used a third method to de-trend the data based on topographic relief (elevation, aspect, and slope calculated from a digital elevation map) to meet the assumptions of the semivariogram model. This third method consisted of adding elevation, aspect, and slope in all combinations (*i.e.*, elevation×aspect×slope, elevation×aspect, elevation×slope, etc.) to the time of day linear regression described above. Linear regressions that were not significant ($p>0.05$) were excluded, and the residuals were recalculated. De-trending using elevation, slope, and/or aspect was necessary at 18 sites (Barrow Environmental Observatory, Poker Flats, Eight Mile Lake, Woodworth, Northern Great Plains Research Lab, Tree Haven, Steigerwaldt, Upper Teakettle, Konza – Core, UK Biological Station, Harvard Forest, Niwot Ridge, Smithsonian Conservation Biology Institute, Ordway-Swisher BS, Great Smoky NP, Talladega Nat'l Forest, Choctaw WMA, and Murray).

Data collected were used for geospatial analyses in the R statistical computing language with the geoR package [56,57]. At each site the empirical semivariogram, $\hat{\gamma}(h)$, which is half the average squared difference between data pairs, was calculated using the following equation,

$$\hat{\gamma}(h)=\frac{1}{2N(h)}\sum_{a=1}^{N(h)}[z(u_a)-z(u_a+h)]^2 \quad (1)$$

where, $z(u_\alpha)$, $\alpha=1, 2, \ldots, n$ denotes the set of T_s or SWC data, u_α is the vector of spatial coordinates of the α_{th} observation, h represents a distance separating pairs of data, and $N(h)$ is the number of data pairs separated by a given distance. In general, the covariance function for any pair of data that is h units apart can be represented as,

$$C(h)=s\gamma(h) \quad (2)$$

where, s is a variance parameter and $\gamma(h)$ is any positive definite correlation function. The correlation function used in this work was a spherical model with the following form,

$$\hat{\gamma}(h)=\begin{cases}(1-1.5\left(\frac{h}{a}\right)+0.5\left(\frac{h}{a}\right)^3, & if \quad h<a\\ 0 & , otherwise\end{cases} \quad (3)$$

where, a is the semivariogram range. In addition, we allowed for a nugget term, which is simply added to the covariance function. Approximate values for the range, sill, and nugget were estimated from the experimental semivariogram to seed the model and best-fit values were subsequently estimated using weighted least squares. It is important to note that a key goal of this study is to provide meaningful comparisons of the results across a number of sites, and this comparability was weighed as more important than a potentially negligible difference in the fit of a model for varying functional forms of the variogram, i.e., same sources of uncertainty in the model fit for all sites. Based on these two constraints, we selected the functional form of the variograms (spherical) that provided the best fit at most of the sites and applied it to all sites for the sake of consistency.

Lag spacing was set to 1 m for every empirical semivariogram. Preliminary tests with T_s and SWC data from Eight Mile Lake, AK, and Klemme, OK, showed that changing the lag spacing to 1, 2, or 5 m did not alter the estimated range value in any systematic manner, and the change was ±9% on average, which corresponded to ±4 m. Similarly, the coefficient of variation estimated from the semivariogram model was strongly positively correlated with the coefficient of variation calculated via the traditional method (see *Coefficient of variation and sample size* subsection in the *Results* section), indicating that 1 m lag spacing allowed the model to accurately describe the variability in the dataset.

Calculating CVs and sample size

The coefficient of variation for T_s and SWC was calculated for each site in two ways: 1) the traditional method of dividing the standard deviation by the mean (referred to as $CV_{Traditional}$); and 2) using the sill to estimate standard deviation and dividing that value by the mean (referred to as CV_{Sill}) as follows,

$$CV_{sill}=\frac{\sqrt{2\times\hat{S}}}{\bar{x}} \quad (4)$$

Table 5. Site characteristics where the range value was larger than or less than the maximum lag distance covered by the semivariogram model, (mean ±1 SE). Range>maximum lag indicates that the semivariogram did not reach an asymptot.

	Latitude (°)	Longitude (°)	Elevation (m.a.s.l.)	Mean annual temperature (°C)	Mean annual precipitation (mm)	Canopy height (m)	Mean T_s (°C)	Mean SWC (%)	Maximum lag of semivariogram (m)	n
T_s										
Range<maximum lag	43±2	−106±3	741±111	9±1	861±80	12±2	15±1	12±1	94±1	53
Range>maximum lag	37±4	−90±6	759±367	12±3	899±109	13±4	19±2	10±2	87±3	7
P value	NS	NS	NS	NS	NS	NS	0.098	NS	NS	
SWC										
Range<maximum lag	42±2	−105±3	778±122	9±1	841±76	11±2	15±1	11±1	95±1	50
Range>maximum lag	42±4	−102±8	572±168	10±3	990±211	16±4	15±2	14±3	86±3	10
P value	NS	NS	NS	NS	NS	NS	NS	NS	0.024	

NS: not significant (*i.e.*, p>0.1 and factor was removed from the statistical model).

where, $\bar{x}$ represents the sample mean, and $\hat{S}$ represent the semivariogram sill.

We calculated the sample size (n) needed to estimate the mean T_s and SWC to within 10% of the spatial mean and with 90% confidence using [4],

$$n = \frac{(1.645)^2\left(\frac{\sqrt{2\times\hat{S}}}{\overline{X}}\right)}{(0.1)^2} \quad (5)$$

where, $\bar{x}$ represents the sample mean, and $\hat{S}$ represents the semivariogram sill. The constants 0.1 and 1.645 represent the accuracy requirement (*i.e.*, 10%) expressed as a proportion and the value of t statistic that corresponds to the 90% confidence interval, respectively. These accuracy and confidence thresholds were chosen because they are conservative enough to meet the goals of most studies. An assumption of the hypothesis test underlying this analysis is that the data are normally distributed, which we ensured prior to constructing the semivariograms.

Statistical analyses. Several of the hypotheses that we tested involved relating site characteristics to dependent variables. Some site characteristics (*i.e.*, factors in the statistical tests) were continuous variables (latitude, longitude, mean annual air temperature, mean annual precipitation, elevation, canopy height, mean T_s, and mean SWC), while others were categorical variables (ecosystem type, canopy structure, and soil type). Some levels of the categorical site characteristics contained too few sites to allow statistical analyses to be conducted. For example, Andisols were only found at 2 sites (Wind River and Abby Road, WA) and a tropical deciduous forest ecosystem only occurred at 1 site (Guanica, PR). As a result, levels were combined to increase the number of sites within each level prior to conducting the statistical tests. Ecosystem types were grouped into 6 levels: deciduous forests (both tropical and temperate), temperate coniferous forests, grasslands (incl. shrublands, and savannas), agricultural, boreal (including montane forests), and tundra. While we recognize that short stature ecosystems (*e.g.*, grasslands) can have closed canopies with only small gaps between neighboring plants or open canopies with substantial amounts of bare ground between plants, they clearly differ in structure from taller stature ecosystems. The ecosystem and canopy structure classifications were somewhat subjective and some sites could have been reasonably placed in two or more different groups, but the chosen category represents our best judgment. Canopy structure was grouped into 3 levels: closed canopy ecosystems, open and semi-open canopy ecosystems, and short stature ecosystems. Vegetation canopy heights were measured in the field and reflect the average height of the

Table 6. Correlation coefficients and statistical significance for T_s and SWC semivariogram properties and derived values.

	Spearman correlation coefficient	p value
T_s range – SWC range	0.26	0.089
T_s sill – SWC sill	0.36	0.016
T_s nugget – SWC nugget	0.34	0.021
T_s CV_{Sill} – SWC CV_{Sill}	0.59	<0.001
T_s sample size – SWC sample size	0.59	<0.001

tallest component of the plant community (*e.g.*, in a savanna ecosystem the canopy height reflects the average height of the trees, rather than the grasses). Soils were grouped according to soil orders: Inceptisols, Spodosols, Ultisols, Mollisols, Gelisols, and other (incl. Vertisols, Andisols, Alfisols, Entisols, Aridisols, as well as sites with >1 soil order and sites where the soil order was unknown).

In all statistical models described below, factors in the model that had a p value of >0.1 were removed in a stepwise manner and the model was re-run until all remaining factors had a p value of <0.1. Statistical significance was assumed at $p<0.05$ for all tests. Post-hoc Tukey HSD tests were performed when ANOVAs identified a significant effect of the categorical factor(s).

We expected that many of the site characteristics would covary (*e.g.*, latitude and mean annual air temperature) and it is necessary to understand this covariation when interpreting the statistical tests that we performed. To assess covariation in the explanatory variables used in the statistical tests, we calculated correlation coefficients for each combination of site characteristics (latitude, longitude, mean annual air temperature, mean annual precipitation, elevation, canopy height, mean T_s, and mean SWC). Categorical site characteristic factors (*i.e.*, ecosystem type, canopy structure, and soil type) were excluded from this analysis. Stepwise removal of non-significant factors from statistical tests tends to remove factors that covary with one-another, therefore, interpretation of a significant effect caused by a site characteristic should also consider other covarying site characteristics.

At some sites it was not possible to estimate the range or sill because the semivariogram did not reach an asymptote (see *Semivariograms* subsection in the *Results* section); therefore these sites were excluded from further statistical analysis. We tested whether these sites had different characteristics than sites where the semivariograms did reach an asymptote to determine whether this could influence the interpretation of our results. To achieve this, nominal logistic models were performed in JMP (SAS Institute Inc., Cary, NC) to determine whether significant differences in latitude, longitude, elevation, mean annual air temperature, mean annual precipitation, canopy height, mean T_s, mean SWC, and maximum semivariogram lag distance existed between sites where semivariograms did and did not reach an asymptote. Categorical site characteristic (*i.e.*, ecosystem type, canopy structure, and soil type) were excluded from this analysis due to the limited number of sites within each level of the categories.

ANOVAs were performed in JMP to determine the relationship between site characteristics and T_s and SWC. The factors in the ANOVAs were latitude, longitude, elevation, mean annual air temperature, mean annual precipitation, canopy height, ecosystem type, canopy structure, and soil type and the dependent variables were mean T_s and mean SWC. Both mean T_s and mean SWC were $\log_{10}$ transformed to meet assumptions of normality and homogeneity.

ANOVAs were also used to test hypotheses 1–4. ANOVAs were performed in JMP to assess the relationship between site characteristics and semivariogram range, nugget, and sill, as well as estimated sample size required to meet the accuracy requirement for T_s and SWC. Semivariogram nugget data for T_s and SWC met the assumptions of normality and homogeneity; however, range, sill, and sample size data required $\log_{10}$ transformations to meet these assumptions.

Because some factors often explained the majority of variation in an ANOVA, relatively weak relationships that were statistically significant were sometimes masked by the dominant factors when plotted as graphs. This is because the ANOVA accounts for variability in the data caused by others factors, whereas the graphs,

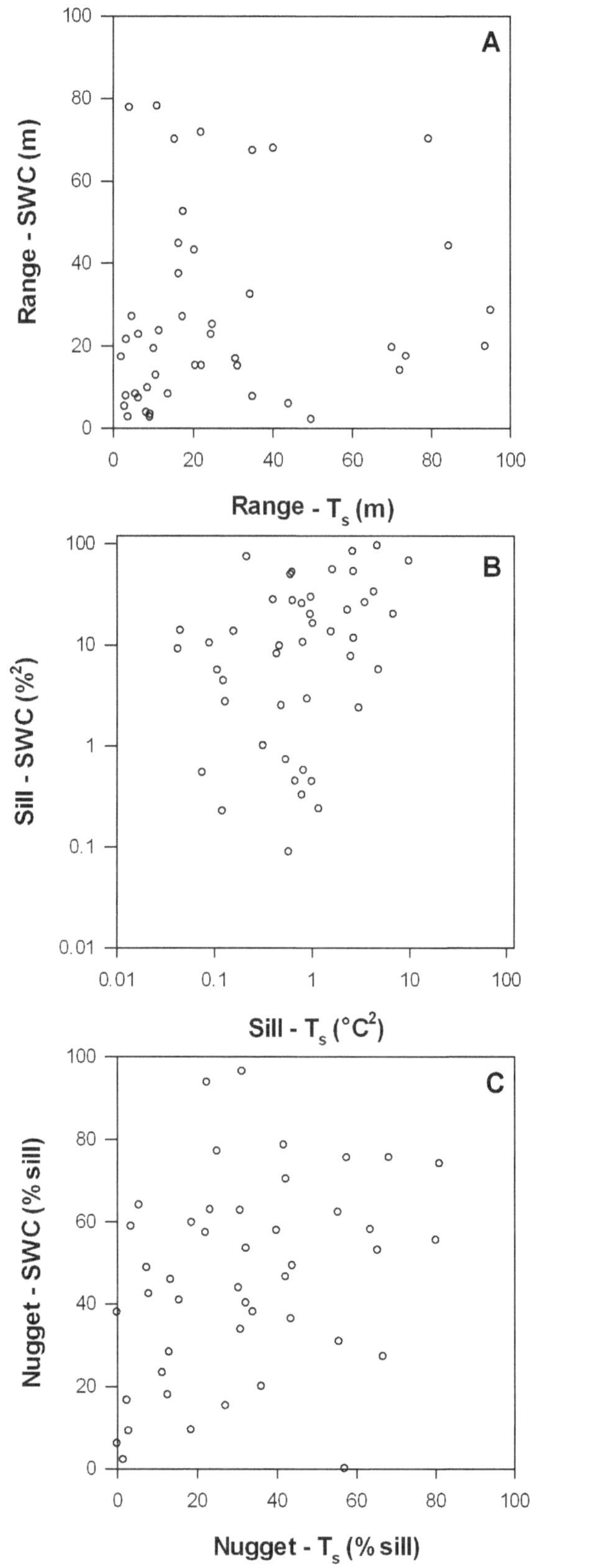

Figure 4. Semivariogram range (a), sill (b), and nugget (c) for T_s versus SWC at the sites. Sill values are plotted on a log scale due to large differences in magnitude across the sites.

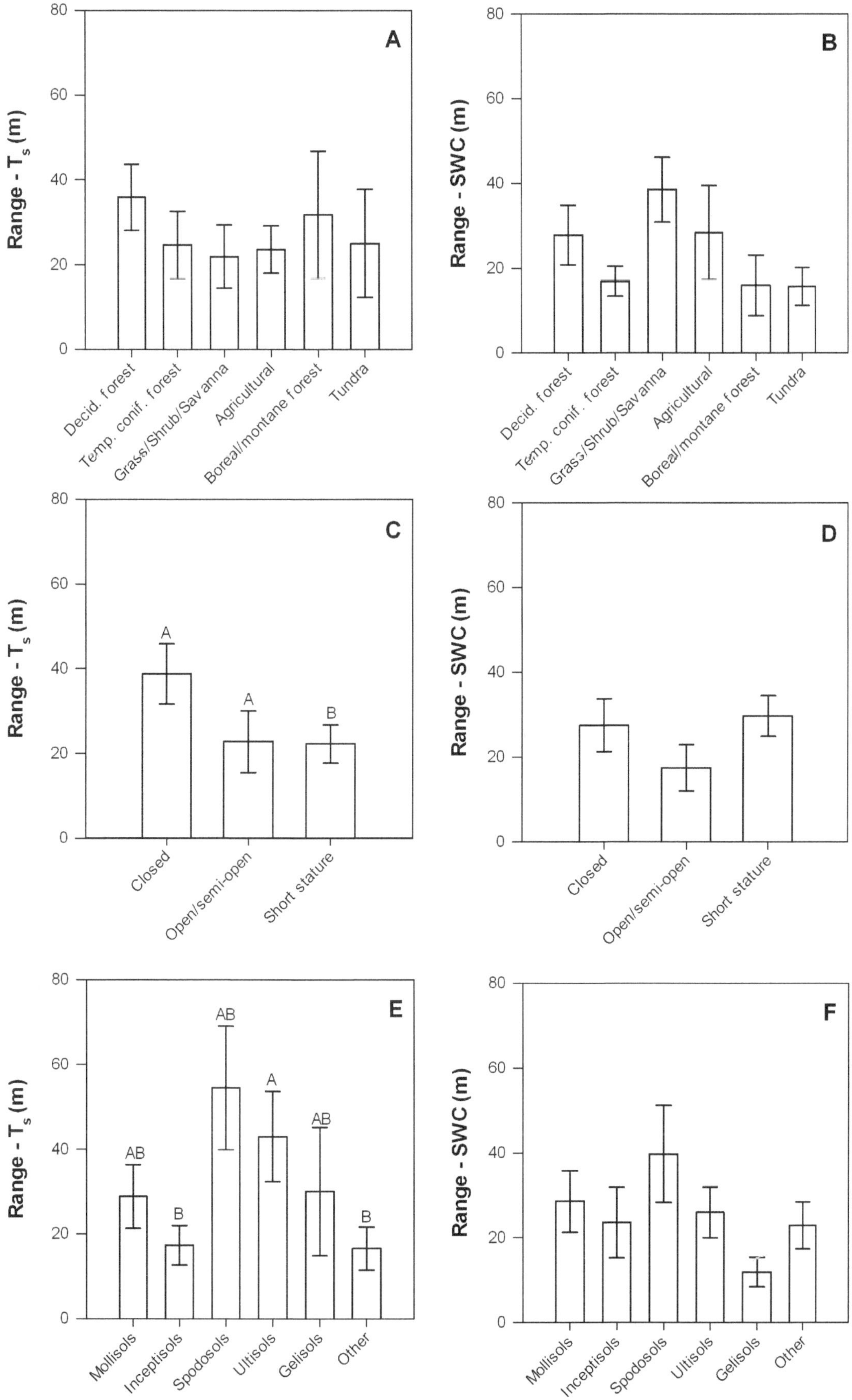
A
Range - T_s (m)
0
20
40
60
80
Decid. forest
Temp. conif. forest
Grass/Shrub/Savanna
Agricultural
Boreal/montane forest
Tundra
B
Range - SWC (m)
C
A
A
B
Closed
Open/semi-open
Short stature
D
E
AB
B
AB
A
AB
B
Mollisols
Inceptisols
Spodosols
Ultisols
Gelisols
Other
F

Figure 5. Semivariogram range for T_s (left column) and SWC (right column) at each site in relation to ecosystem type (a, b), canopy structure (c, d), and soil type (e, f). Error bars represent ±1 SE. Bars with different letter were significantly different ($p<0.05$) based on Tukey HSD tests. The letter "A" corresponds to the largest least square mean(s) and subsequent letters (B, C, …) correspond to progressively smaller least square means.

which are based on raw data, do not. This should be taken into account when comparing the statistical results to the graphs.

A Spearman's rank correlation was performed in JMP to assess the relationships between mean T_s and mean SWC to determine whether they were significantly correlated. Additional Spearman's rank correlations were performed in JMP to assess the relationships among T_s range and SWC range, T_s nugget and SWC nugget, T_s sill and SWC sill, and T_s sample size and SWC sample size (Hypothesis 5); and T_s range, nugget, and sill, as well as SWC range, nugget, and sill (Hypothesis 6).

Paired one sample *t*-tests (JMP) were used to assess whether range, nugget, sill, and sample size values differed between T_s and SWC among the sites to test Hypothesis 6. In addition, to test whether the semivariograms accurately described the variability in T_s and SWC at each site, linear regressions (JMP) were used to assess the relationship between $CV_{Traditional}$ and CV_{Sill} for both T_s and SWC.

The range and sample size for T_s and SWC were plotted against the cumulative proportion of sites to investigate the frequency distribution. These relationships were modeled with a 3-parameter logistic curve using SigmaPlot (Systat Software Inc., Chicago, Illinois).

In the interest of brevity, we only present graphs relating to a subset of site characteristics in this paper, namely ecosystem type, canopy structure, and soil order. The relationship between dependent variables and other site characteristics, including latitude, longitude, elevation, mean annual air temperature, mean annual precipitation, mean T_s, and mean SWC, are presented in the Appendices.

Results

Site characteristics

The sites encompassed a continental-scale range of eco-climatic properties (Table 1, Figure 1), and were broadly representative of the proportion of US land in different latitudinal, longitudinal, and elevational categories, as well as spanning a wide range of soil types, climate space, and soil temperature and moisture conditions at the time of sampling (Figure 2). T_s varied from 2.1±0.1°C (mean ±1SE) at Pump Station 2, AK, to 30.8±0.2°C at Santa Rita Exp. Range, AZ, while SWC varied from 1.1±0.1% at Burlington, MA, to 37.8±0.7% at Ponce, PR (Table 2). T_s and SWC estimates were not correlated with each other among the sites ($p=0.143$), resulting from several sites with contrasting T_s and SWC combinations (*i.e.*, warm and wet, warm and dry, cool and wet, and cool and dry, Figure 2f).

As expected, correlations demonstrated a substantial amount of covariation in the site characteristics (Table 3). For example, mean annual air temperature, mean T_s, and longitude were all strongly negatively correlated with latitude. Weaker correlations were observed between mean annual precipitation and canopy height, and elevation and mean SWC. These correlations should be considered when interpreting the results, since the stepwise removal of non-significant factors from the ANOVA models tends to remove factors that covary from the model. For example, a significant relationship between longitude and a dependent variable would likely also be significant for latitude, mean annual air temperature, or mean T_s if longitude had not been included as a factor.

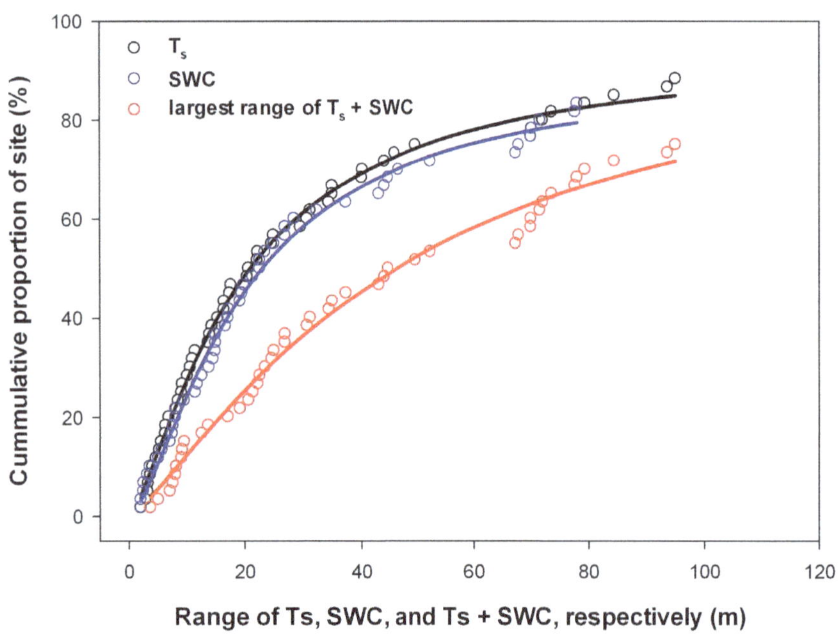

Figure 6. Semivariogram range for T_s, SWC, and the largest range for T_s or SWC at each site versus the cumulative proportion of sites. A logistic curve was fitted to the data with R^2 of 1.00, 0.99, and 0.99 for T_s, SWC, and T_s and SWC, respectively. Equations for the curves can be found in the results section (Eq. 6a–c).

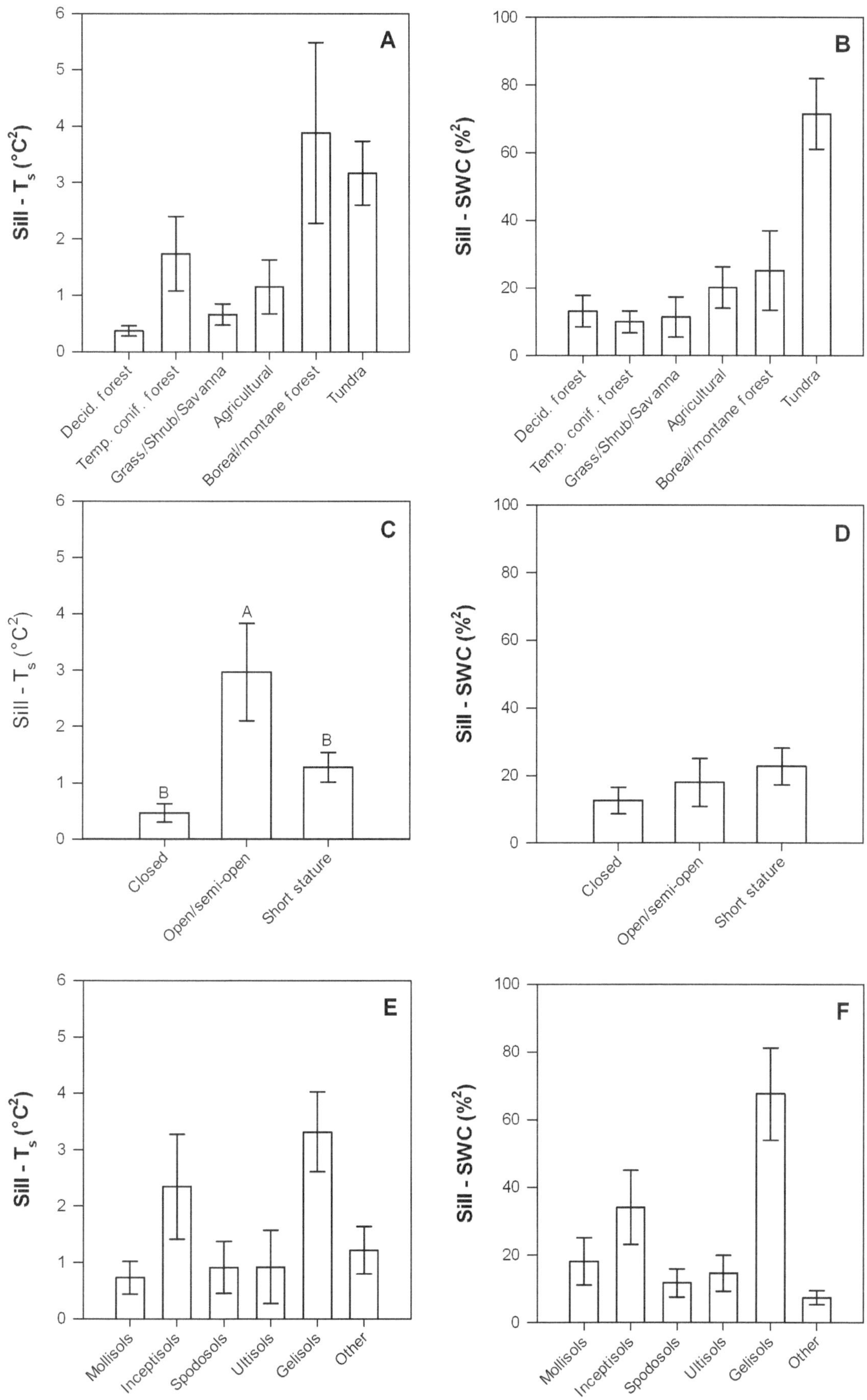
A
Sill - Ts (°C²)
0
1
2
3
4
5
6
Decid. forest
Temp. conif. forest
Grass/Shrub/Savanna
Agricultural
Boreal/montane forest
Tundra
B
Sill - SWC (%²)
0
20
40
60
80
100
C
A
B
B
Closed
Open/semi-open
Short stature
D
E
Mollisols
Inceptisols
Spodosols
Ultisols
Gelisols
Other
F

Figure 7. Semivariogram sill for T_s (left column) and SWC (right column) at each site in relation to ecosystem type (a, b), canopy structure (c, d), and soil type (e, f). Error bars represent ±1 SE. Bars with different letter were significantly different ($p<0.05$) based on Tukey HSD tests. The letter "A" corresponds to the largest least square mean(s) and subsequent letters (B, C, ...) correspond to progressively smaller least square means.

Mean T_s among sites was negatively related to latitude and differed among ecosystems, *i.e.*, not surprisingly being higher in temperate ecosystems, and lower in montane and high-latitude ecosystems (Table 4, Figure 3). Mean T_s among sites followed expected relationships to mean annual air temperature, elevation, and soil type (*e.g.*, being lower in Gelisols, which are associated with cold climates; Figure 3), but the effects were not significant due to covariation among these factors and the factors that were statistically significant in the model (*i.e.*, latitude and ecosystem type). Mean SWC was positively related to mean total annual precipitation, decreased with increasing elevation and mean T_s, and was significantly related to ecosystem type, being largest in tundra ecosystems, which have low evapotranspiration rates, and in agricultural ecosystems, several of which were irrigated, and the lowest in temperate coniferous forests (Table 4, Figure 3).

Semivariograms

In 12% and 17% of cases the semivariograms for T_s and SWC, respectively, had not reached an asymptote at the maximum lag distance used to construct the modeled semivariogram (Table 2). As a result, sill, range, and nugget (expressed as a percent of sill) values for these semivariograms were based on extrapolation, which may not accurately represent the *true* values. At 75% of sites both the T_s and SWC semivariogram reached an asymptote, at 13% of sites only the T_s semivariogram reached an asymptote, at 8% of sites only the SWC semivariogram reached an asymptote, and at the remaining 3% of sites no asymptote was reached for T_s or SWC. Sites with semivariograms that did not reach an asymptote were not different from other sites in terms of their environmental characteristics (*i.e.*, latitude, longitude, elevation, mean annual air temperature, mean annual precipitation, canopy height, mean soil temperature, or mean soil moisture). However, the SWC semivariograms that did not reach an asymptote were associated with shorter maximum lag distances (Table 5). Unless stated otherwise, all subsequent results are based solely on estimated sill, range, and nugget values from semivariograms that reached an asymptote. Since there were no significant differences in environmental characteristics between sites where the semivariogram did and did not reach an asymptote, there is no evidence that excluding sites where the semivariogram did not reach an asymptote introduced a systematic bias into the interpretation of our results.

Semivariogram range

The mean range value was 27±4 m and 26±3 m for T_s and SWC, respectively. A paired *t*-test demonstrated that T_s range values did not significantly differ from SWC range values ($p = 0.773$). Despite the similarity in the mean range, range values for T_s were not correlated with range values for SWC (Table 6; Figure 4a). At some sites, the T_s and SWC range values were similar, but at many sites one range value was substantially larger than the other.

Range values for T_s were significantly influenced by canopy structure and soil type, while those for SWC were significantly related to longitude and canopy height (Table 4, Figure 5). T_s range values were lower in short stature ecosystems than those in closed canopy or open and semi-open canopy ecosystems, and also lower in Inceptisols and "other" soils than in Ultisols (Figure 5). SWC range values increased with longitude (*i.e.*, increased from west to east) and decreased with canopy height (Table 4).

The relationship between the T_s range and the cumulative proportion of sites exhibited a logistic pattern (Figure 6). For example, 88%, 75%, and 55% of sites had a range value for T_s of less than 100, 50, and 25 m, respectively. Similar patterns were observed for SWC range values, as well as the largest range value for T_s or SWC (*i.e.*, whichever was largest at a site; Figure 6). The equations that described these relationships were,

$$\text{Soil temperature}: \quad y = \frac{94.3705}{\left[1 + (x/19.1852)^{-1.3735}\right]} \qquad (6a)$$

$$\text{Soil moisture}: \quad y = \frac{90.0096}{\left[1 + (x/19.801)^{-1.4759}\right]} \qquad (6b)$$

$$\text{Soil temperature/moisture}: \quad y = \frac{100.4066}{\left[1 + (x/46.6724)^{-1.2824}\right]} \qquad (6c)$$

where, x is the range distance expressed in meters, and y is the percent of sites that had smaller ranges than x.

Semivariogram sill

Sill values for T_s and SWC were positively related to one another, suggesting that underlying site properties, in part, mediate the spatial variability in both variables (Table 6, Figure 4b). The sill values for T_s were negatively related to longitude (*i.e.*, decreasing from west to east), and mean annual air temperature, positively related to mean T_s, and higher for open and semi-open ecosystems than for closed canopy or short stature ecosystems (Table 4, Figure 7). The sill values for SWC were also negatively related to longitude and mean annual air temperature, and positively related to mean T_s as well as mean SWC (Table 4). Notably, neither ecosystem type nor soil type were significantly related to sill values for T_s or SWC (Table 4).

Semivariogram nugget

The nugget was expressed as a percent of the sill and indicates the proportion of variation that occurred at scales of <1 m. The nugget ranged from 0–81% for T_s and 0–96% for SWC, and a paired *t*-test demonstrated that the mean T_s nugget (32±3%) was significantly different from the mean SWC nugget (44±3%, $p< 0.001$). Similar to the sill values, T_s nugget values were positively correlated with SWC nugget values, indicating that underlying site characteristics mediate sub-meter variability of both parameters (Table 6, Figure 4c).

Nugget values for T_s were negatively related to canopy height and mean SWC, and were higher in closed canopy ecosystems than short stature or open/semi-open ecosystems (Table 4, Figure 8). There was also a significant effect of soil type on T_s nugget values, however, a subsequent post-hoc test did not reveal significant differences among soil types (Figure 8). Nugget values for SWC were positively related to longitude (*i.e.*, increasing from

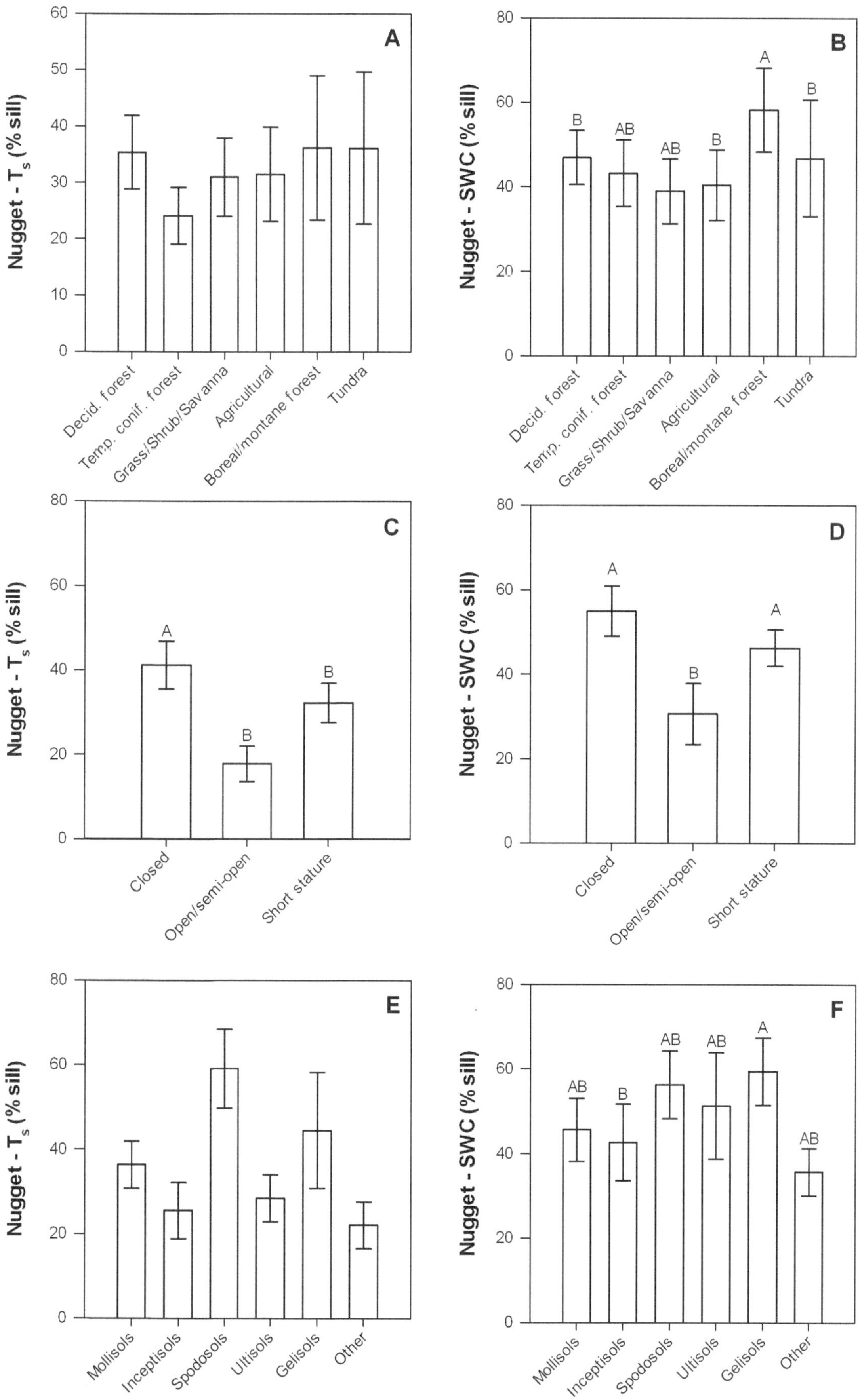
A
Nugget - T_s (% sill)
0
10
20
30
40
50
60
Decid. forest
Temp. conif. forest
Grass/Shrub/Savanna
Agricultural
Boreal/montane forest
Tundra
B
Nugget - SWC (% sill)
0
20
40
60
80
B
AB
AB
B
A
B
Decid. forest
Temp. conif. forest
Grass/Shrub/Savanna
Agricultural
Boreal/montane forest
Tundra
C
Nugget - T_s (% sill)
0
20
40
60
80
A
B
B
Closed
Open/semi-open
Short stature
D
Nugget - SWC (% sill)
0
20
40
60
80
A
B
A
Closed
Open/semi-open
Short stature
E
Nugget - T_s (% sill)
0
20
40
60
80
Mollisols
Inceptisols
Spodosols
Ultisols
Gelisols
Other
F
Nugget - SWC (% sill)
0
20
40
60
80
AB
B
AB
AB
A
AB
Mollisols
Inceptisols
Spodosols
Ultisols
Gelisols
Other

Figure 8. Semivariogram nugget expressed as a percent of the sill for T_s (left column) and SWC (right column) at each site in relation to ecosystem type (a, b), canopy structure (c, d), and soil type (e, f). Error bars represent ±1 SE. Bars with different letter were significantly different (p<0.05) based on Tukey HSD tests. The letter "A" corresponds to the largest least square mean(s) and subsequent letters (B, C, ...) correspond to progressively smaller least square means.

west to east), negatively related to mean T_s, as well as being significantly related to ecosystem type, canopy structure, and soil type (Table 4, Figure 8). SWC nugget values were larger in forest ecosystems (particularly boreal/montane forests), and lower in tundra ecosystems, as well as being larger in closed canopy and short stature ecosystems and lower in open/semi-open canopies ecosystems (Figure 8). SWC nugget values were larger for Gelisols than for Inceptisols (Figure 8).

Relationships among the semivariogram sill, range, and nugget

Range values were positively related to nugget values (expressed as a percent of the sill) for the T_s semivariograms (Table 7, Figure 9a). However, there was no relationship between range and sill values or nugget and sill values for T_s (Table 7). Likewise, range, sill, and nugget values were all unrelated to one another for the SWC semivariograms (Table 7, Figure 9b).

The relative abundance of different semivariogram shapes was assessed by assigning them to four categories: small nugget (<50% sill) and small range (<50 m); large nugget (>50% sill) and small range (<50 m); small nugget (<50% sill) and large range (>50 m); and large nugget (>50% sill) and large range (>50 m). For T_s the most common combination was a small nugget and small range, which accounted for 37 of the 60 sites (62%; Figure 9c). Small nugget and range values were also the most common combination for SWC, accounting for 24 of the 60 sites (40%). In addition, 18 of the sites (30%) had SWC semivariograms with a large nugget combined with a small range value. For both T_s and SWC, it was relatively rare to observe a large range value with either a small or large nugget value (Figure 9c).

Coefficient of variation and sample size

The two estimates of the coefficient of variation ($CV_{Traditional}$ and CV_{Sill}) were strongly and positively correlated with one another (T_s, $r^2=0.96$, $p<0.001$; SWC, $r^2=0.95$, $p<0.001$; Figure 10), demonstrating that the semivariograms accurately described the pattern of spatial variability among the sites. However, $CV_{Tradtional}$ values were typically smaller than corresponding CV_{Sill} values for both T_s and SWC ($p<0.001$). This occurred because the traditional method of calculating CV does not account for spatial correlation in data, yet spatial correlation was ubiquitous at our sampling sites (data not shown). Across all sites, CV_{Sill} values were 24±3% and 44±2% larger than $CV_{Tradtional}$ values for T_s and SWC, respectively.

The CV_{Sill} values ranged from 0.02 to 0.88, with a mean of 0.16±0.03, and 0.09 to 1.22 with a mean of 0.51±0.04 for T_s and SWC, respectively (Table 2). A paired *t*-test demonstrated that T_s CV_{Sill} values significantly differed from SWC CV_{range} values ($p<0.001$), demonstrating that SWC was more variable across space. Similar to the sill values, there was a positive correlation between CV_{Sill} values for T_s and SWC (Table 6, Figure 10), again suggesting that variability in these parameters may be controlled by the same site characteristics.

We used a sample size analysis based on CV_{Sill} (Eq. 5) to calculate the number of samples required to estimate T_s and SWC to within 10% of the mean with 90% confidence (Table 2). We used CV_{Sill} rather than $CV_{Traditional}$ to calculate sample size because $CV_{Traditional}$ underestimated variability (Figure 10), which in turn would cause the sample size to be underestimated. For example, at Harvard Forest the sample size required to estimate SWC to within 10% of the mean with 90% confidence was 127 when calculated with $CV_{Traditional}$ (0.69), but was 204 when calculated with CV_{Sill} (0.87). Across all the sites, the sample size calculated using $CV_{Traditional}$ underestimated the sample size calculated with CV_{Sill} by a factor of 1.6±0.1 (maximum: 2.6) and 2.1±0.1 (maximum: 3.0) for T_s and SWC, respectively. Hereafter, we focus on sample sizes calculated using CV_{Sill}.

Since we used the sill to represent the variance in the sample size analysis, the number of samples required assumes that the sample spacing will be larger than or equal to the range value from the semivariogram (*i.e.*, the distance at which the sill is reached). In addition, because the sample size was based on CV_{Sill}, both parameters (CV_{Sill} and sample size) exhibited similar patterns. For example, the sample size analyses indicated that the number of samples necessary to meet the accuracy requirement varied from 1 to 211 for T_s (mean of 20±7) and from 2 to 405 for SWC (90±13), and a paired *t*-test demonstrated that sample size was significantly larger for T_s than SWC ($p<0.001$), which is similar to findings for CV_{Sill}. In addition, the sample size required for T_s was positively correlated to the sample size required for SWC (Table 6, Figure 11).

The number of samples required to estimate T_s to within 10% of the mean with 90% confidence was negatively related to longitude (*i.e.*, decreasing from west to east) and mean annual air temperature, and was also influenced by canopy structure and soil types (Table 4, Figure 12). More samples were required in open/semi-open canopy ecosystems than other ecosystems, and in Gelisols than in Mollisols (Figure 12). At 87% of sites, 20 or fewer samples were sufficient to meet the accuracy requirement for T_s. The remaining 13% of sites that required more samples were all Alaskan sites that shared a number of similarities, including high latitudes, westerly longitudes, low mean annual air temperature (< 2°C), low mean T_s at the time of sampling (<6°C), low mean annual precipitation (<500 mm), tundra or boreal forest ecosystem types, primarily Gelisols or Inceptisols, low to mid elevations (<1000 m.a.s.l.), and low to mid-vegetation canopy heights (< 15 m; Figure 12).

Table 7. Spearman correlation coefficients and statistical significance for T_s and SWC semivariogram properties.

	Correlation coefficient	P value
T_s		
Range - nugget	0.53	<0.001
Range - sill	−0.07	0.628
Nugget - sill	−0.15	0.278
SWC		
Range - nugget	0.10	0.492
Range - sill	0.03	0.863
Nugget - sill	−0.17	0.237

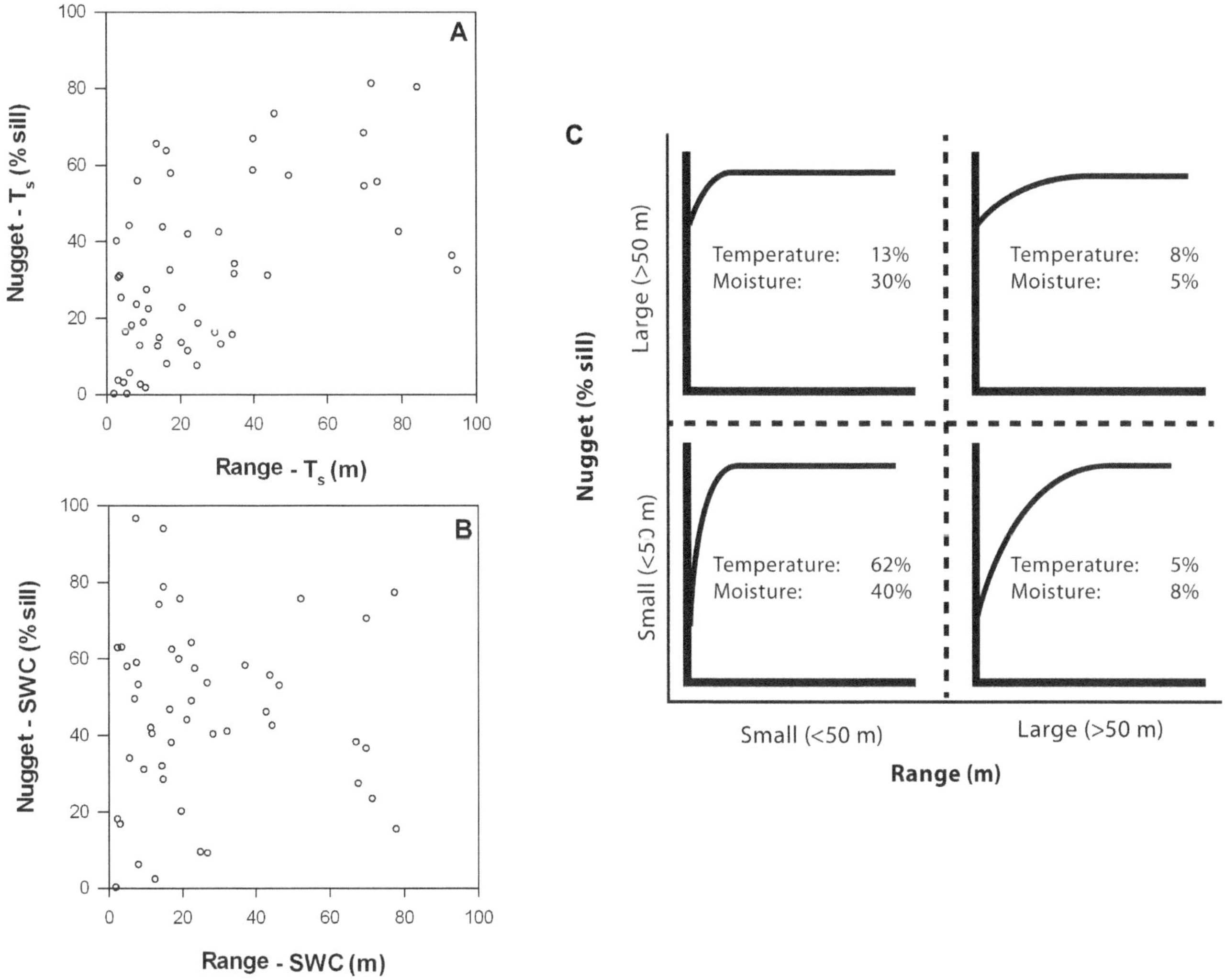

Figure 9. Semivariogram range versus nugget for T_s (a) and SWC (b), and the relative occurrence of semivariograms with small (>50 m) or large (>50 m) range values and small (<50% sill) or large (>50% sill) nugget values. The percentages in (c) are based on all 60 sites, including the sites where range values exceeded the maximum distance of the semivariogram model (*i.e.*, 7 sites (12%) for T_s and 10 sites (17%) for SWC). As a result, the percentages do not add up to 100%. Since the semivariograms that did not reach an asymptote all had range values that exceeded 50 m, the percentages in the right two quadrants are underestimated by a total of 12 and 17 percentage points for T_s and SWC, respectively.

The number of samples required to estimate SWC to within 10% of the mean with 90% confidence was positively related to mean annual precipitation and mean T_s, negatively related to longitude, mean annual air temperature, mean SWC, and was also influenced by canopy structure (Table 4, Figure 12). More samples were required to meet the accuracy requirement in closed and open/semi-open canopy ecosystems than short stature ecosystems (Figure 12). Unlike T_s, sites that required a large number of samples (*i.e.*, >100) to accurately measure SWC did not share many similar characteristics, although none had a mean soil moisture of >16% (data not shown).

When the required sample size for T_s was plotted against the cumulative proportion of sites the relationship could be accurately described by a logistic equation (Figure 13). For example, based on the sample size analyses, one sample was sufficient to estimate T_s to within 10% of the spatial mean with 90% confidence at 45% of sites, while 10 samples was sufficient to meet this requirement at 79% of sites. A similar pattern emerged for SWC sample size plotted against the cumulative proportion of sites, although at least an order of magnitude more samples were required than for T_s (Figure 13). In addition, the pattern was similar for the number of samples required for T_s or SWC (whichever was largest at a site; Figure 13). The equations that described these relationships were,

$$\text{Soil temperature}: \quad y=\frac{94.3653}{\left[1+(x/1.2306)^{-0.9435}\right]} \tag{7a}$$

$$\text{Soil moisture}: \quad y=\frac{95.9175}{\left[1+(x/49.799)^{-1.8399}\right]} \tag{7a}$$

$$\text{Soil temperature/moisture}: \quad y=\frac{99.5316}{\left[1+(x/56.6125)^{-1.4619}\right]} \tag{7c}$$

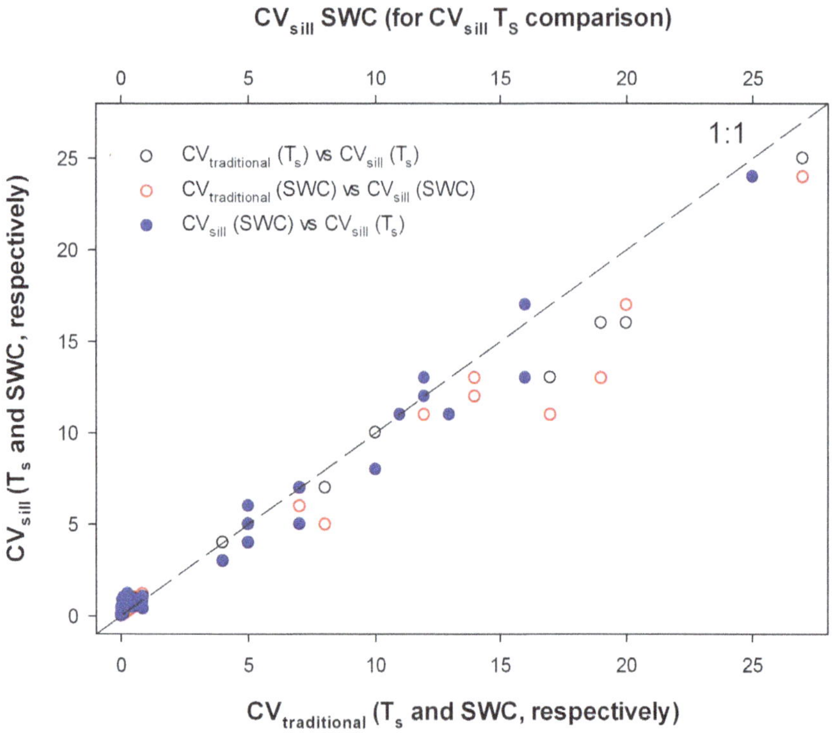

Figure 10. Coefficient of variation calculated by dividing the standard deviation by the mean ($CV_{Traditional}$) versus coefficient of variation calculated using the semivariogram sill (CV_{Sill}) for T_s, SWC, and CV_{Sill} for T_s versus SWC. Dotted line is 1:1.

where, x represents a number of samples and y is the percent of sites that require x (or fewer) samples to estimate the spatial mean to within 10% of the mean with 90% confidence.

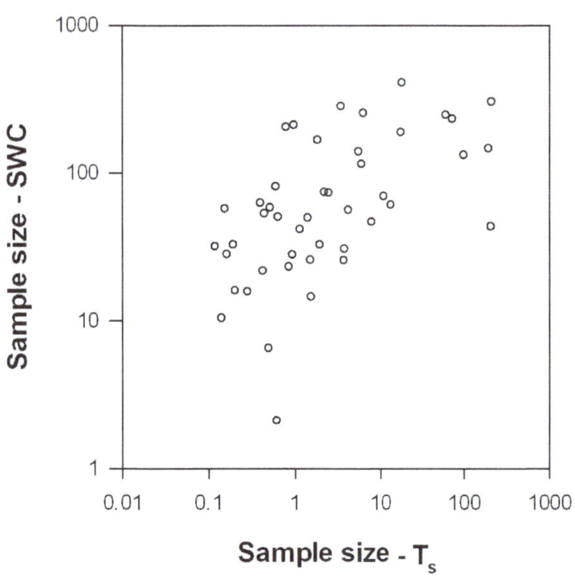

Figure 11. Number of samples required to measure T_s to within 10% of the spatial mean with 90% confidence versus the number of samples required to meet the same accuracy requirement for SWC. Values are plotted on a semi-log scale due to large differences in magnitude across the sites.

Efficacy of sampling strategies

To estimate the mean T_s across the area we sampled at a particular site with the fewest samples, sample spacing should equal or exceed the range value estimated from the T_s semiovariogram at that site. This is because samples spaced this distance apart are effectively independent, thereby providing the maximum amount of information from each individual sample. In contrast, if the sample spacing was less than the range value then adjacent samples would be correlated (*i.e.*, each sample would provide less than the maximum amount of information), which meant that more samples would be needed to accurately estimate the spatial mean of T_s. We emphasize that using the semivariogram range to inform sample spacing is appropriate here because we are only constraining the sampling to accurately estimate the mean. However, this may not be appropriate for studies with different goals.

Since Eq. 6a describes the proportion of sites with a soil temperature range less than a given value and Equation 7a describes the proportion of sites where a given sample size would be required to estimate soil temperature to within 10% of the mean with 90% confidence, we created a matrix that outlines the efficacy of different sampling strategies based on these two equations (Figures 14–16). For example in Figure 14, two temperature measurements spaced 5 m apart would only be sufficient to estimate mean T_s to within 10% of the mean with 90% confidence over ~1 ha (the approximate size of the area we sampled) at 7% of our sites. Interestingly, increasing the number of samples while keeping sample spacing at 5 m barely improved the efficacy of the sampling strategy and this occurs for two reasons, i) because spatial variation in T_s is often low (Figure 10), so few samples are needed at most sites (Figure 13), and ii) because only 12% of sites had a range value of 5 m or less (Figure 6). However, if the sample spacing were increased to 100 m, just two samples

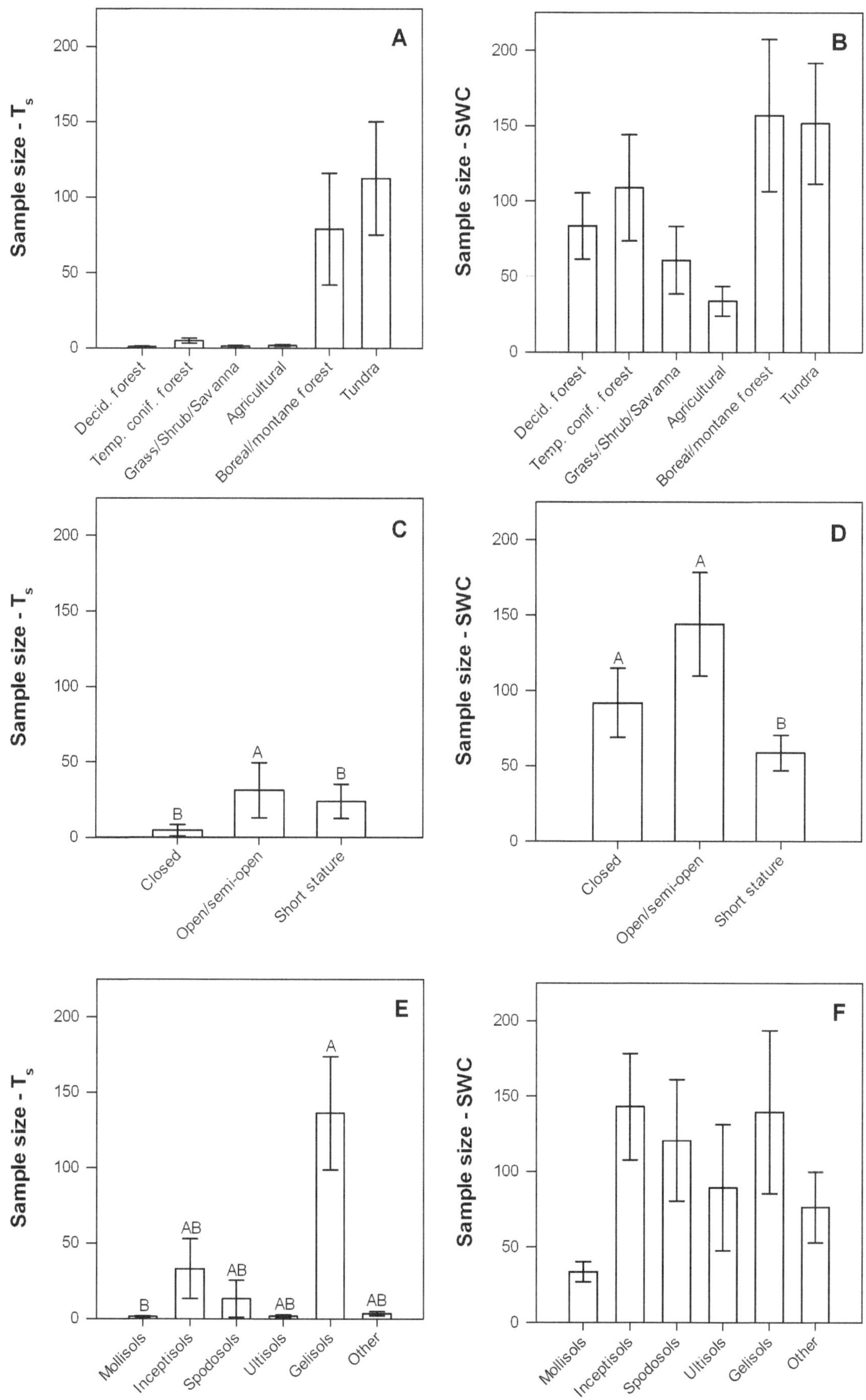
A
Sample size - T_s
200
150
100
50
0
Decid. forest
Temp. conif. forest
Grass/Shrub/Savanna
Agricultural
Boreal/montane forest
Tundra
B
Sample size - SWC
200
150
100
50
0
Decid. forest
Temp. conif. forest
Grass/Shrub/Savanna
Agricultural
Boreal/montane forest
Tundra
C
Sample size - T_s
200
150
100
50
0
B
A
B
Closed
Open/semi-open
Short stature
D
Sample size - SWC
200
150
100
50
0
A
A
B
Closed
Open/semi-open
Short stature
E
Sample size - T_s
200
150
100
50
0
B
AB
AB
AB
A
AB
Mollisols
Inceptisols
Spodosols
Ultisols
Gelisols
Other
F
Sample size - SWC
200
150
100
50
0
Mollisols
Inceptisols
Spodosols
Ultisols
Gelisols
Other

Figure 12. Number of samples to measure T_s (left column) and SWC (right column) to within 10% of the spatial mean with 90% confidence at each site in relation to ecosystem type (a, b), canopy structure (c, d), and soil type (e, f). Error bars represent ±1 SE. Bars with different letter were significantly different ($p<0.05$) based on Tukey HSD tests. The letter "A" corresponds to the largest least square mean(s) and subsequent letters (B, C, …) correspond to progressively smaller least square means.

would be needed to meet the requirement of measuring T_s to within 10% of the spatial mean with 90% confidence at half the sites, although common sense dictates that more samples would be preferable (*e.g.*, 5 samples). This matrix can be used as a guide to quantitatively evaluate the efficacy of increasing sample size and/or sample spacing to measure the spatial mean of T_s at scales of ~1 ha at our sampling sites.

Using Equation 6b and 7b, we created a similar matrix to evaluate the efficacy of different sampling strategies for measuring SWC at spatial scales of ~1 ha (Figure 15). As with T_s, 2 samples spaced 5 m apart would not be an effective sampling strategy to measure SWC to within 10% of the mean with 90% confidence. Increasing the sample size or sample spacing only slightly improves the efficacy, however, by increasing both the sample size and sample spacing the sampling strategy can become substantially more effective at meeting the accuracy requirement. Figure 15 also demonstrates how a sampling strategy that is effective at a given proportion of sites can be achieved in many different ways. For instance, 150 samples spaced 10 m apart would meet the accuracy requirement for SWC at the same proportion of sites as 60 samples 15 m apart or 30 samples 55 m apart.

Lastly, using Equation 6c and 7c we created a matrix to evaluate the efficacy of measuring both T_s and SWC to within 10% of the mean with 90% confidence at spatial scales of ~1 ha (Figure 16). I n this case, more samples spaced further apart were always required to meet the accuracy requirement at a given proportion of sites than for T_s or SWC alone.

Discussion

We characterized the spatial structure of variability in soil properties over ~1 ha at 60 sites throughout the US. Our study included nearly an order of magnitude more sites than any previous study that examined the spatial variation in soil properties, and as a result, allowed us to derive some empirical rules to describe the observed variability in soils. Moreover, we were able to evaluate the efficacy of different sampling strategies (*i.e.*, sample size and sample spacing) at measuring the spatial mean of T_s and SWC, which can guide future soil studies. We emphasize that our guidelines for sampling strategies are designed to accurately estimate the spatial mean of a soil property, and as a result, our guidelines may not be suitable for studies with different goals. Because our sites were broadly representative of US geography and ecology (*i.e.*, including every major terrestrial ecosystem type and soil type), our findings are broadly applicable to US soils, and to a lesser extent other regions of the world.

Variability

Generally, variability was larger for SWC than T_s, which resulted in a larger sample size required to measure SWC to within 10% of the spatial mean with 90% confidence than for T_s. This is consistent with other studies that have also observed larger CV values for SWC than T_s [23,24,58–60]. We did not specifically set out to determine the sources of variability, or why variability was larger for SWC than T_s, but a likely explanation is that water is more mobile than heat, and as a result more sensitive to

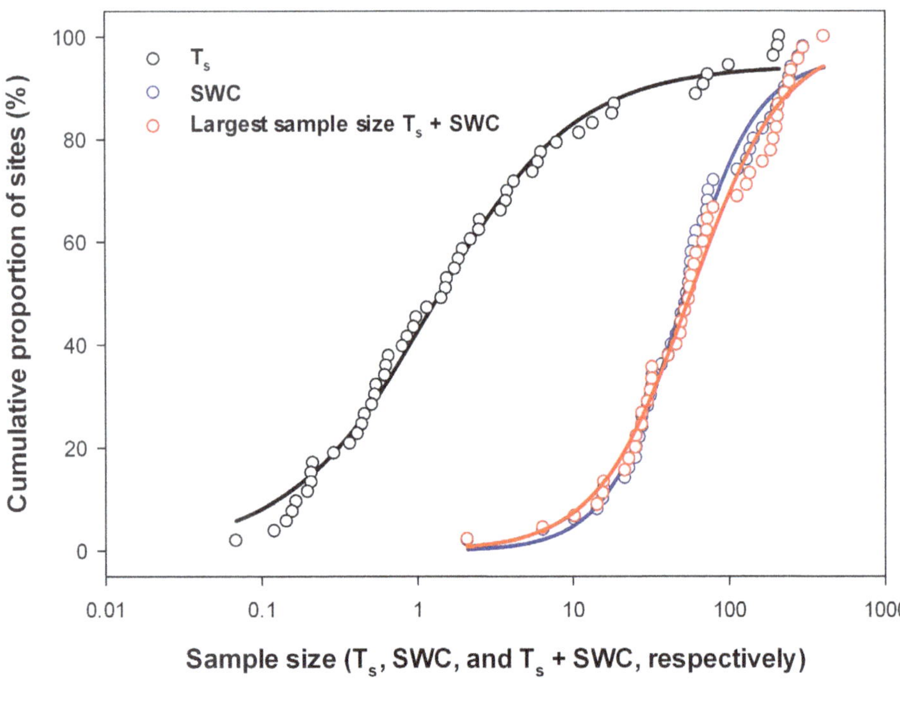

Figure 13. Number of samples required to measure T_s, SWC, or T_s and SWC to within 10% of the spatial mean with 90% confidence versus the cumulative proportion of sites. A logistic curve was fitted to the data with $R^2 = 0.99$ for each equation (Eqs. 7a–c). Values are plotted in a semi-log scale due to large differences in magnitude across the sites.

Legend

0	10	20	30	40	50	60	70	80

Sample spacing (m)

Sample size	5	10	15	20	25	30	35	40	45	50	55	60	65	70	75	80	85	90	95	100
2	7	16	23	28	32	35	38	40	42	43	44	45	46	47	47	48	48	49	49	49
3	8	18	26	32	37	40	43	46	47	49	50	51	52	53	54	55	55	56	56	56
4	9	19	28	34	40	43	47	49	51	53	54	55	56	57	58	59	59	60	60	61
5	10	20	29	36	41	46	49	52	54	55	57	58	59	60	61	62	62	63	63	64
6	10	21	30	37	43	47	51	53	56	57	59	60	61	62	63	64	64	65	65	66
8	10	22	32	39	45	49	53	56	58	60	62	63	64	65	66	67	67	68	68	69
10	11	23	33	40	46	51	54	57	60	62	63	65	66	67	68	69	69	70	70	71
15	11	24	34	42	48	53	57	60	62	64	66	67	69	70	71	71	72	73	73	74
20	11	24	35	43	49	54	58	61	63	65	67	69	70	71	72	73	74	74	75	75
25	11	24	35	43	50	55	59	62	64	66	68	70	71	72	73	74	74	75	76	76
30	12	25	35	44	50	55	59	62	65	67	69	70	72	73	74	74	75	76	76	77
35	12	25	36	44	50	55	59	63	65	67	69	71	72	73	74	75	76	76	77	77
40	12	25	36	44	51	56	60	63	66	68	69	71	72	73	74	75	76	77	77	78
45	12	25	36	44	51	56	60	63	66	68	70	71	73	74	75	76	76	77	78	78
50	12	25	36	44	51	56	60	63	66	68	70	71	73	74	75	76	77	77	78	78
60	12	25	36	45	51	56	60	64	66	68	70	72	73	74	75	76	77	78	78	79
70	12	25	36	45	51	57	61	64	67	69	71	72	73	75	76	76	77	78	78	79
80	12	25	36	45	52	57	61	64	67	69	71	72	74	75	76	77	77	78	79	79
90	12	25	36	45	52	57	61	64	67	69	71	72	74	75	76	77	77	78	79	79
100	12	25	36	45	52	57	61	64	67	69	71	73	74	75	76	77	78	78	79	79
125	12	26	37	45	52	57	61	64	67	69	71	73	74	75	76	77	78	79	79	80
150	12	26	37	45	52	57	61	65	67	69	71	73	74	75	76	77	78	79	79	80
175	12	26	37	45	52	57	61	65	67	70	71	73	74	75	76	77	78	79	79	80
200	12	26	37	45	52	57	61	65	67	70	71	73	74	76	77	77	78	79	79	80
250	12	26	37	45	52	57	62	65	68	70	72	73	75	76	77	78	78	79	80	80

Figure 14. Efficacy of different sampling strategies at measuring T_s to within 10% of the spatial mean with 90% confidence at scales of ~1 ha. Values represent the percent of sites where the corresponding sampling strategy would achieve the accuracy requirement. The x-axis is based on Equation 6a and the y-axis is based on Equation 7a.

microtopography (*i.e.*, water drains from hummocks to depressions, whereas heat does not), and is actively transported by plants (*i.e.*, transpiration induced uptake and hydrologic lift), resulting in larger variability at this spatial scale.

Despite generally larger spatial variability in SWC than T_s, (*i.e.*, sill and CV_{Sill}), T_s and SWC were positively correlated suggesting that they were likely controlled by similar site properties at the scale we sampled. Indeed, ANOVAs also indicated that sill values for both measures decreased with increasing longitude (*i.e.*, from west to east) and mean annual air temperature, while they increased with mean T_s. In addition, among the sites sill values for temperature and moisture increased with mean T_s and SWC, respectively, exhibiting the commonly observed pattern of increased variability with increasing mean values, *i.e.*, heteroscedastic [61]. Similarly, data presented by Brocca et al. [12] shows that the standard deviation for SWC increased from the driest site (MON, mean ± SD: 22.8±2.0) to the wettest site (LEC, 38.5±3.0) for seven Italian sites. In contrast, Western et al. [26] found no clear relationship between SWC and variance at locations in Australia and New Zealand, while data reported by Famiglietti et al. [25] show a negative relationship among six sites in Oklahoma. These studies however, differ in their design (*e.g.*, spatial scale, sampling frequency, geographic region), making it difficult to determine why different patterns were observed among the studies. But one explanation could be that the driest sites in the other studies had a mean SWC of between 10% and 23%, whereas half of our sites had a mean SWC of <10% (all of which had relatively low semivariogram sill values) and previous studies that have monitored SWC over time at individual sites often observed that low mean SWC coincided with relatively low variability, *e.g.*, [13,14]. We studied the relationship between the mean and variability (*i.e.*, sill) across space, whereas most previous studies have investigated this relationship at individual sites across time and have often (but not always) observed the variance or standard deviation increasing with intermediate SWC and decreasing at low and high SWC [12–14,20,26].

Ecosystem type did not significantly influence sill values for either T_s or SWC, despite the common assumption that agricultural soils were less variable than soils in other ecosystems [4,11]. Similarly, Robertson et al. [11] were surprised by the large variability in a wide range of soil properties from a soybean field that visibly appeared homogeneous, supporting the notion that agricultural ecosystems are typically more variable than many researchers appreciate.

The ANOVA models left 43% and 30% of the variability in T_s and SWC sill values unexplained for, respectively, indicating that other site properties (*i.e.*, besides the ones we measured) also are sources of variability. In part this may relate to vegetation canopy structure at the sites. For example, soil beneath canopy gaps would receive both more radiation and precipitation inputs than soil

Sample spacing (m)

Sample size	5	10	15	20	25	30	35	40	45	50	55	60	65	70	75	80	85	90	95	100
2	0	0	0	0	0	0	0	0	0	0	0	0	0	0	0	0	0	0	0	0
3	0	0	0	0	0	0	0	0	0	0	0	0	0	0	0	0	0	0	0	0
4	0	0	0	0	0	1	1	1	1	1	1	1	1	1	1	1	1	1	1	1
5	0	0	0	1	1	1	1	1	1	1	1	1	1	1	1	1	1	1	1	1
6	0	0	1	1	1	1	1	1	1	1	1	1	1	1	2	2	2	2	2	2
8	0	1	1	1	2	2	2	2	2	2	2	2	2	2	3	3	3	3	3	3
10	0	1	2	2	3	3	3	3	3	3	4	4	4	4	4	4	4	4	4	4
15	1	2	3	4	5	6	6	6	7	7	7	7	7	7	8	8	8	8	8	8
20	2	4	5	7	8	9	9	10	10	11	11	11	12	12	12	12	12	12	12	12
25	2	5	8	10	11	12	13	14	15	15	16	16	16	16	17	17	17	17	17	17
30	3	7	10	12	14	16	17	18	19	19	20	20	21	21	21	22	22	22	22	22
35	3	8	12	15	17	19	21	22	23	24	24	25	25	26	26	26	27	27	27	27
40	4	9	14	17	20	22	24	26	27	28	28	29	29	30	30	31	31	31	31	32
45	5	10	16	20	23	25	27	29	30	31	32	33	33	34	34	35	35	35	36	36
50	5	12	17	22	25	28	30	32	33	35	35	36	37	38	38	38	39	39	39	40
60	6	13	20	25	30	33	35	37	39	40	41	42	43	44	44	45	45	46	46	46
70	7	15	22	28	33	36	39	42	43	45	46	47	48	49	49	50	50	51	51	52
80	7	16	24	31	36	39	43	45	47	49	50	51	52	53	53	54	55	55	55	56
90	7	17	26	33	38	42	45	48	50	51	53	54	55	56	57	57	58	58	59	59
100	8	18	27	34	40	44	47	50	52	54	55	57	58	59	59	60	61	61	62	62
125	8	19	29	37	43	47	51	54	56	58	60	61	62	63	64	65	65	66	66	67
150	9	20	30	38	45	49	53	56	59	61	62	64	65	66	67	68	68	69	69	70
175	9	21	31	40	46	51	55	58	61	63	64	66	67	68	69	70	70	71	71	72
200	9	21	32	40	47	52	56	59	62	64	66	67	68	69	70	71	72	72	73	73
250	10	22	33	41	48	53	57	61	63	65	67	69	70	71	72	73	74	74	75	75

Figure 15. Efficacy of different sampling strategies at measuring SWC to within 10% of the spatial mean with 90% confidence at scales of ~1 ha. Values represent the percent of sites where the corresponding sampling strategy would achieve the accuracy requirement. The x-axis is based on Equation 6b and the y-axis is based on Equation 7b. Legend in Figure 14.

beneath a closed-canopy, which would increase spatial variability in both measures at a site [62]. Indeed, T_s sill values were significantly larger in open/semi-open ecosystems (*i.e.*, sites with high variability in vegetation cover) than in closed-canopy or short stature ecosystems, but this pattern did not occur for SWC. In a rangeland ecosystem, diurnal maximum T_s at 5 cm were found to be much lower beneath sagebrush plants (10°C) than in interspaces between plants (17°C), indicating that canopy structure strongly influenced spatial variability in T_s [63]. Similar results have been found with SWC [64,65]. In addition, previous studies have shown that topography influences the spatial variability of SWC [12,66,67], which may also account for some of the unexplained variation found here.

Sample size

We estimated the number of samples required to measure T_s and/or SWC to within 10% of the spatial mean with 90% confidence at each site for prospective purposes (*i.e.*, to inform future study designs). This represents a conservative number of samples for most studies, as this level of accuracy is often deemed more than sufficient and some have supported even lower accuracy requirements [4,5]. Of course each study should evaluate their required accuracy, and if these criteria were relaxed, design elements like sample size could be reduced.

Previous estimates of the sample size required to measure soil properties to a given accuracy have often assumed that their data was not spatially correlated [12,13,14,24]. However, our data, and data from many other studies [9,18,23,24,26,59,68–72], show that spatial correlation is ubiquitous for T_s and SWC. Since $CV_{Traditional}$ did not account for spatial correlation (*i.e.*, it assumes that all measurements are independent regardless of how closely they are spaced) it underestimated the true variability in the data, causing the required sample size to be underestimated [73]. In contrast, CV_{Sill} did account for spatial correlation, therefore, it reliably estimated the variability in the data and is appropriate for calculating the required sample size. Had we used $CV_{Traditional}$ rather than CV_{Sill} to calculate sample size, we would have substantially underestimated the required sample size (*i.e.*, by 60% for T_s and 109% for SWC when averaged across all sites). As a result, caution should be used when relying on sample size estimates that assume no spatial correlation in T_s and SWC data.

Other methods can be used to estimate the number of samples required to accurately estimate the spatial mean of a soil property. For example, previous research has shown that fewer samples are sufficient to meet an accuracy requirement if the spatial mean is calculated by kriging, rather than by averaging the data [74–77]. However, the kriging method requires knowledge of the semivariogram at the sampling site, which is frequently unknown. Moreover, our data shows that it is not possible to reliably estimate the key features of T_s and SWC semivariograms (*i.e.*, sill, nugget, and range) based on the site characteristics we studied. Similarly, the temporal stability (or rank stability) approach and the random combination method can also be used to reduce the required sample size, but these approaches require detailed knowledge of spatial variation of soil properties at the sampling site, which is also often unknown [14,24,77,78]. It should be noted that these approaches address a slightly different goal (*i.e.*, estimating the number of samples required to estimate mean soil

Sample size \ Sample spacing (m)	5	10	15	20	25	30	35	40	45	50	55	60	65	70	75	80	85	90	95	100
2	0	0	0	0	0	0	0	0	0	0	0	0	0	0	0	0	1	1	1	1
3	0	0	0	0	0	0	1	1	1	1	1	1	1	1	1	1	1	1	1	1
4	0	0	0	1	1	1	1	1	1	1	1	1	1	1	1	1	1	1	1	1
5	0	0	1	1	1	1	1	1	1	1	2	2	2	2	2	2	2	2	2	2
6	0	0	1	1	1	1	1	2	2	2	2	2	2	2	2	2	2	3	3	3
8	0	1	1	1	2	2	2	2	3	3	3	3	3	3	4	4	4	4	4	4
10	0	1	1	2	2	3	3	3	4	4	4	4	4	5	5	5	5	5	5	5
15	1	2	2	3	4	5	5	6	6	7	7	7	8	8	8	8	9	9	9	9
20	1	2	3	5	6	6	7	8	9	9	10	10	11	11	12	12	12	13	13	13
25	1	3	4	6	7	8	9	10	11	12	13	13	14	15	15	15	16	16	17	17
30	2	3	5	7	9	10	12	13	14	15	16	16	17	18	18	19	19	20	20	21
35	2	4	6	8	10	12	14	15	16	17	18	19	20	21	21	22	23	23	24	24
40	2	5	7	9	12	14	15	17	18	20	21	22	23	24	24	25	26	26	27	27
45	2	5	8	11	13	15	17	19	20	22	23	24	25	26	27	28	28	29	30	30
50	2	6	9	11	14	16	19	20	22	24	25	26	27	28	29	30	31	32	32	33
60	3	6	10	13	16	19	21	23	25	27	29	30	31	33	34	35	36	36	37	38
70	3	7	11	15	18	21	24	26	28	30	32	33	35	36	37	38	39	40	41	42
80	3	8	12	16	19	23	25	28	30	33	34	36	38	39	40	42	43	44	44	45
90	4	8	13	17	21	24	27	30	32	35	37	38	40	42	43	44	45	46	47	48
100	4	8	13	18	22	25	28	31	34	36	38	40	42	44	45	46	48	49	50	51
125	4	9	14	19	24	28	31	34	37	40	42	44	46	48	49	51	52	53	54	55
150	4	10	15	20	25	29	33	36	39	42	45	47	49	51	52	54	55	56	57	59
175	5	10	16	21	26	30	34	38	41	44	46	49	51	53	54	56	57	59	60	61
200	5	11	16	22	27	31	35	39	42	45	48	50	52	54	56	57	59	60	62	63
250	5	11	17	23	28	32	37	40	44	47	50	52	54	56	58	60	61	63	64	65

Figure 16. Efficacy of different sampling strategies at measuring T_s and SWC to within 10% of the spatial mean with 90% confidence at scales of ~1 ha. Values represent the percent of sites where the corresponding sampling strategy would achieve the accuracy requirement. The x-axis is based on Equation 6c and the y-axis is based on Equation 7c. Legend in Figure 14.

moisture over a number of time steps) than the statistical method to calculate sample size, which focuses on a single time step. Because of these limitations, we calculated the required sample size using the classical statistical approach (Eq. 5) to allow our findings to be more easily applied to sampling designs at sites where the spatial structure of variation may be unknown.

Spatial variability in T_s was so low at many sites that the sample size analyses indicated that even just one sample was sufficient to meet our accuracy requirement (although common sense dictates that more samples would be preferable), while 10 samples were sufficient at 79% of sites, and 100 samples were sufficient at all but 3 sites. In contrast, SWC often required at least an order of magnitude more samples to meet the accuracy requirement at a similar proportion of sites (*e.g.*, 100 samples were sufficient at 72% of sites). Because more samples were required to meet our accuracy requirement for SWC than T_s at almost every site, the sample size needed when examining both T_s and SWC was very similar to that found for SWC alone.

The large sample sizes needed to accurately measure the spatial mean SWC based on our findings differ substantially from previous studies that assumed SWC was not spatially correlated [12–14]. For example, Brocca et al. [12] reported that 15 or fewer samples would be needed at 4 out of 8 sites to measure SWC to within ±2% volumetric with 95% confidence, and at most 40 samples would be needed to meet this accuracy requirement at any of the sites. We largely attribute this difference to accounting for spatial correlation in our estimates of sample size, but other factors may also play a role including that many previous studies have focused on agricultural and grassland sites, which tend to require smaller sample sizes (Figure 13). In addition, at site with a mean SWC of <20%, our relative error threshold (10% of the mean) is more stringent than the 2% absolute error that has often been used in previous SWC studies. We are not aware of previous studies that have estimated the sample size required to measure T_s to a given accuracy, but our data suggests these estimates will only be reliable if they account for spatial correlation.

Based on our finding, 20 samples would be more than sufficient to meet the accuracy requirement for T_s at most sites, although Alaskan sites typically required many more samples to meet the same requirement. As a result, measuring T_s across space in a temperate or sub-tropical ecosystem one might reasonably assume that 20 samples would be sufficient (assuming appropriate sample spacing), while working at high latitude (>50°N) one would likely need an order of magnitude more samples to have confidence in meeting the same accuracy requirement. Sites where a large number of samples were required were more closely associated with high latitudes, westerly longitudes, and cold mean T_s at the time of sampling (<6°C).

There are plausible reasons why sites with both low mean T_s and high latitude would result in the need for large sample sizes. First, sites with low mean T_s statistically result in high CV values because the equation divides by the mean T_s (see Eq. 4), which in turn results in large sample sizes (see Eq. 5). Second, the low angle of the sun at high latitude sites, which results in large differences in shading (*i.e.*, radiation inputs) of locations north or south of an obstacle (*e.g.*, tree trunk, tussock, or rock) will spend a large amount of time shaded by the obstacle and when unshaded, the sun will be at a low angle (*i.e.*, low radiation inputs), and collectively cause large spatial variability in T_s. The converse is true at lower latitude sites where shading by an obstacle will be much less than at high latitude sites creating less spatial variation in T_s and in turn results in fewer samples necessary to meet the accuracy requirement. Indeed, the positive relationship that we observed between T_s sill values and latitude lends additional support to the first hypothesis, suggesting that high latitudes, as well as low mean T_s values, both result in large sample sizes. In contrast, we are not aware of a plausible reason why westerly longitudes should be expected to increase the number of samples required, instead it seems likely that relationship between longitude and the number of samples is an artifact of US geography, since all high latitude sites in the US are located in Alaska, which is located further west than most of the temperate and sub-tropical US. Thus, it appears that the low mean T_s and high latitudes, not westerly longitudes, of the Alaskan sites resulted in the large sample sizes for T_s required at those sites.

Unlike T_s, the sample sizes required to accurately measure spatial SWC were only weakly related to site characteristics making it difficult to estimate a sample size based solely these variables. However, it is notable that a sample size of 100 was sufficient at most sites (72%), while sites that needed >100 samples all had mean SWC of <16%, and sites that needed >250 samples had mean SWC of ≤11%. Similar to T_s, the reduction in the required sample size at high mean SWC is presumably partly due to the fact that low mean SWC results in large CVs because this is calculated by dividing by the mean SWC (see Eq. 4), which in turn causes large sample sizes to be needed to accurately estimate the spatial mean (see Eq. 5).

The large sample size that was required to accurately estimate the spatial mean SWC across ~1 ha at most sites is beyond the scope of many research projects. However, the recent development of sensors that aim to measure SWC at scales of hundreds and thousands of m^2 may hold promise for estimating mean SWC at this scale with just 1 or a handful of sensors [79,80] and may represent a more cost-effective way of meeting the accuracy requirement than deploying tens or hundreds of sensors that each measure SWC in a small volume of soil.

Sample spacing

The number of samples required to meet our accuracy requirement assumes that the samples are spaced at least at the distance of the range in order to avoid correlation and pseudoreplication [1]. The mean range for both T_s and SWC was ~27 m. Similar range values have been observed for T_s and SWC in other studies that sampled at a similar scale [8,18,20,59,71,81], demonstrating that range values of tens of meters are the norm for temperature and moisture at this scale.

Unlike the sill values, the range values for T_s and SWC were not correlated, indicating that they are controlled by different site properties. The ANOVAs supported this conclusion, showing that range values for T_s were related to canopy structure and soil type, whereas for SWC they were related to longitude and canopy height. However, the ANOVA models left 60% and 82% of the variability in range values unexplained for T_s and SWC, respectively, indicating that we did not measure the most important factor(s) that control the spatial structure of variability. As a result it is difficult to determine what controls the spatial structure of variability at our sites, but potential factors may include microtopography, canopy gap spacing, and variation in soil texture.

Since we cannot reliably predict the semivariogram range for T_s or SWC based on the site characteristics that we measured, a researcher wishing to measure T_s and/or SWC at scales of ~1 ha could use Equations 6a–c to estimate if a given sampling spacing would likely be sufficient (*i.e.*, larger than the range value). For instance, a sample spacing of 5 m would likely be deemed insufficient since only ~12% of sites had a smaller T_s or SWC range, whereas a sample spacing of 40 m may be considered more appropriate since ~2/3 of sites had a smaller T_s or SWC range than this. This sample spacing could then be coupled with the estimated number of samples required based on latitude and mean T_s (for spatial measurements of temperature) and/or mean SWC (for spatial measurements of moisture) to develop a site-specific sampling strategy.

Sampling strategies

Figures 14–16 can guide a sampling design. These figures clearly demonstrate the potential benefit of increasing sample number and/or sample spacing at the spatial scale we sampled. In particular, there are many different ways of achieving a sampling strategy that is likely to meet our accuracy requirement at a given proportion of sites (Figures 14, 15, 16). This demonstrates a particularly pertinent finding that increasing sample spacing is as important as increasing sample number to achieve a given sampling efficacy. This is significant because doubling the samples size often creates much more work and/or expense than simply doubling sample spacing.

Several existing networks that monitor T_s and SWC (*e.g.*, USDA Soil Climate Analysis Network (SCAN) and the US NOAA Climate Reference Network (USCRN)) use sampling designs that would likely require supplemental sampling effort in order to accurately estimate the spatial (~1 ha) mean of these properties at most sites (rf. Figures 14, 15, 16). To illustrate this point, USCRN deploys 3 T_s and SWC sensors 5.2 m apart in a range of depths at >100 sites throughout the US (M Palecki, pers. comm.), while SCAN deploys 1 T_s and SWC sensor at several depths at sites throughout the US [33]. Both the number of sensors and their spatial distribution would be inadequate to estimate these quantities within 10% of the mean with 90% confidence. To highlight the utility of our data, an effective way to increase the local scale (~1 ha) spatial accuracy in this case would be to increase sample spacing and add more SWC sensors (rather than T_s sensors).

Using the results found here, we have constructed a simplified matrix to inform researchers on the efficacy of different sampling designs (Figures 14, 15, 16), given the caveat that they do not account for information provided by samples spaced at distances less than the range; 2) recognize that one T_s sample (*i.e.*, 0 m sample spacing) is sufficient at several sites; and 3) recognize that in practice a sampling design operates at multiple spatial scales, *e.g.*, 9 samples spaced 5 m apart and arranged in a 3×3 grid also includes 4 samples spaced 10 m apart (Figure 14). Similarly, 9 samples spaced 5 m apart along a transect also includes 5 samples spaced 10 m apart, 3 samples spaced 20 m apart, and 2 samples spaced 40 m apart (Figure 17). Nonetheless, these figures provide an empirically-derived quantitative starting point for developing robust soil sampling strategies.

In addition to the limitations associated with Figures 14, 15, 16, researchers interested in using our findings to guide their sampling strategy should also bear in mind the conditions under which our data were generated. For example: 1) the sites were typically dominated by a single vegetation type, therefore our findings may not adequately capture variability at locations that cross major vegetation ecotones; 2) data were collected during growing season, so may not represent other seasons; 3) since our sampling typically occurred on a single day, which may have represented anomalous conditions (*e.g.*, usually hot, cold, wet, or dry), we expect that the general patterns we observed across multiple sites are more

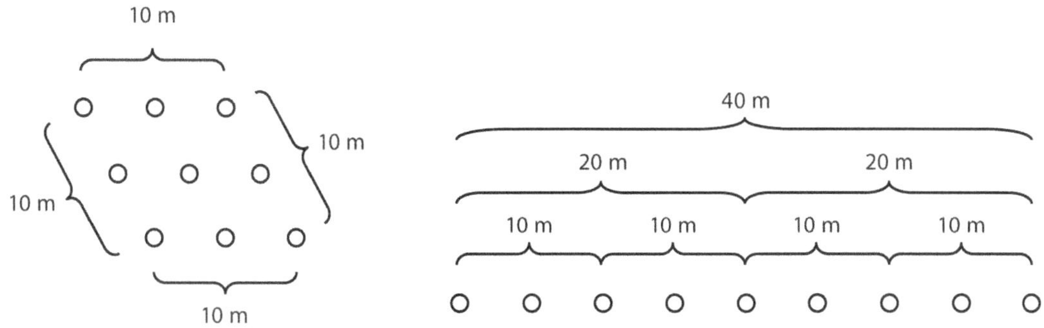

Figure 17. Nine sampling points (open circles) spaced 5 m apart and arranged (left panel) in a 3×3 grid and (right panel) along a transect. Besides 9 samples spaced 5 m apart, the 3×3 grid also includes 4 samples spaced 10 m apart, while the transect also includes 5 samples spaced 10 m apart, 3 samples spaced 20 m apart, and 2 samples spaced 40 m apart.

reliable than data from an individual site; 4) our data are representative of the scale that we sampled (~1 ha) and may not be applicable at substantially smaller or larger scales; and 5) our data are based on surface soil measurements, which typically exhibits larger spatial variability than deeper soils [63,82].

Emergent continental scale properties

Soil forming processes, parental material, time, climate, weathering type and rates, and biological activity have been recognized in soil classification schemes [83,84]. But because soil genesis is difficult to measure directly, the current US soil taxonomy, rf. [84] relies on soil properties (pedology) that only implies genesis. This classification system provides an underlying framework of physical mechanisms that contribute towards the spatial correlation that we attempt to quantify. Here, we have shown new evidence of emergent continental scale soil properties, *i.e.*, how ecosystem scale spatial variability is patterned across the US. Granted, this variability can be either interpreted directly as functions of soil temperature and moisture, or by proxy for other soil properties. We also fully recognize that there are a variety of local scale controls on this variability, *i.e.*, canopy structure, canopy height, etc. (that mostly control the temperature and/or water microclimate), but this does not preclude the existence of continental scale patterns that we identified, as these influences manifest themselves at smaller spatial resolutions.

Continental scale patterns included: 1) SWC range values increased with longitude (*i.e.*, increased from west to east), 2) sill values for T_s and SWC were negatively related to longitude (*i.e.*, decreasing from west to east), 3) nugget values for SWC were positively related to longitude (*i.e.*, increasing from west to east), and 4) range values for T_s increased from inceptisols (younger soils) to spodosols (intermediate age) to ultisols (older soils), suggesting that the spatial structure of variability is in part related to soil development. Such patterns could not have been identified from previous studies since they involved too few sites (≤6), and a meta-analysis approach was not possible due to large methodological inconsistencies among the studies (*i.e.*, different sampling scales, measurement methods, sampling depths, etc).

Conclusions

We characterized the spatial variability of soil properties at 60 US sites to inform NEON's sensor-based soil sampling strategy. In addition, we developed quantitative guidelines that can be used to inform soil sampling decisions throughout the US and elsewhere. While our data can be used as a guide, the most reliable way to generate a robust sampling design would be to conduct a site specific assessment of variability in soil properties on several occasions during the relevant season and at the relevant scale, as others have previously suggested [4,5]. Indeed, as Klironomos *et al.* [5] highlighted, although a preliminary study of spatial variation may seem like a large amount of extra work, it significantly contributes towards avoiding a sampling strategy that does not provide data that are both necessary and sufficient to meet the design constraints. However, since soil researchers have only rarely followed these recommendations, we also present our findings as a useful tool to help guide future soil sampling designs.

Supporting Information

Figure S1 Data collected across (a) space can be used to construct a (b) semivariogram, which describes the relationship between semivariance (*i.e.*, half the variance) and distance. A typical sampling layout used in this study is shown in (a). In addition to collecting data at each point shown on the graph, data were also collected at −0.3 m and −0.3 m from each point along the axis of each transect. The three components that describe the shape of the (b) semivariogram are the sill, nugget, and range. The sill represents the maximum semivariance that is encountered at a site and is equivalent to half the variance in the data set used to create the semivariogram. The nugget represents the variance that exists at spatial scales smaller than the minimum sampling distance, as well as sampling error. The range represents the distance beyond which samples are effectively independent at the scale sampled.

Acknowledgments

Staff associated with each site provided invaluable assistance and logistical support. Authors wish to thank M. Slater and L. Goldman for graphical assistance. We thank Don Julio for inspiring many late night scientific discussions.

Author Contributions

Conceived and designed the experiments: HWL EA PD. Performed the experiments: HWL EA HL MB. Analyzed the data: HWL EA PD HL MB. Contributed reagents/materials/analysis tools: HWL EA PD. Wrote the paper: HWL EA PD MB. Secured funding and grantsmanship: HWL.

References

1. Hurlbert SH (1984) Pseudoreplication and the design of ecological field experiments. Ecological Monographs 54: 187–211.
2. Peck TR, Melsted SW (1973) Field sampling for soil testing. In: Walsh LM, Beaton JD, editors. Soil Testing and Plant Analysis. Madison, Wisconsin: Soil Science Society of America, Inc., pp 67–75.
3. Petersen RG, Calvin LD (1996) Sampling. In: Sparks DL, Page AL, Helmke PA, Loeppert RH, Soltanpour PN, Tabatabai MA, Johnston CT, Summer ME, editors. Methods of Soil Analysis Part 3: Chemical Methods. Madison, Wisconsin: Soil Science Society of America, Inc., pp 1–18.
4. Boone RD, Grigal DF, Sollins P, Ahrens RJ, Armstrong DE (1999) Soil sampling, preparation, archiving, and quality control. In: Robertson GP, Coleman DC, Bledsoe CS, Sollins P, editors. Standard Soil Methods for Long-Term Ecological Research. Oxford: Oxford University Press. pp 3–28.
5. Klironomos JN, Rillig MC, Allen MF (1999) Designing belowground field experiments with the help of semi-variance and power analyses. Applied Soil Ecology 12: 227–238.
6. Jenny H (1941). *Factors of Soil Formation*. New York: McGraw-Hill Press. 281 p.
7. Burgess TM, Webster R (1980) Optimal interpolation and isarithmic mapping of soil properties. I. The variogram and punctual kriging. European Journal of Soil Science 31: 315–331.
8. Beckett PHT, Webster R (1971) Soil variability: A review. Soils and Fertilizers 34: 1–15.
9. Robertson GP, Huston MA, Evans FC, J M Tiedje JM (1988) Spatial variability in a successional plant community - Patterns of nitrogen availability. Ecology 69: 1517–1524.
10. Grigal DF, McRoberts RE, Ohmann LF (1991) Spatial variation in chemical-properties of forest floor and surface mineral soil in the North Central United-States. Soil Science 151: 282–290.
11. Robertson GP, Klingensmith KM, Klug MJ, Paul EA, Crum JR, et al. (1997) Soil resources, microbial activity, and primary production across an agricultural ecosystem. Ecological Applications 7: 158–170.
12. Brocca L, Morbidelli R, Melone F, Moramarco T (2007) Soil moisture spatial variability in experimental areas of central Italy. Journal of Hydrology 333: 356–373.
13. Famiglietti JS, Ryu DR, Berg AA, Rodell M, Jackson TJ (2008) Field observations of soil moisture variability across scales. Water Resources Research 44: W01423.
14. Brocca L, Melone F, Moramarco T, Morbidelli R (2010) Spatial-temporal variability of soil moisture and its estimation across scales. Water Resources Research 46: W02516.
15. Burgess TM, R. Webster R (1980) Optimal interpolation and isarithmic mapping of soil properties. I. The variogram and punctual kriging. European Journal of Soil Science 31: 315–331.
16. Goovaerts P (1998) Geostatistical tools for characterizing the spatial variability of microbiological and physico-chemical soil properties. Biology and Fertility of Soils 27: 315–334.
17. Hoenig J, Heisey DM (2001) The abuse of power: The pervasive fallacy of power calculations for data analysis. The American Statistician 55: 1–6.
18. Nyberg L (1996) Spatial variability of soil water content in the covered catchment at Gardsjon, Sweden. Hydrological Processes 10: 89–103.
19. Dobermann A, Goovaerts P, Neue HU (1997) Scale-dependent correlations among soil properties in two tropical lowland rice fields. Soil Science Society of America Journal 61: 1483–1496.
20. Ettema CH, Coleman DC, Vellidis G, Lowrance R, Rathbun SL (1998) Spatiotemporal distributions of bacterivorous nematodes and soil resources in a restored riparian wetland. Ecology 79: 2721–2734.
21. Western AW, Bloschl G, Grayson RB (1998) Geostatistical characterisation of soil moisture patterns in the Tarrawarra a catchment. Journal of Hydrology 205: 20–37.
22. Castrignano A, Giugliarini L, Risaliti R, Martinelli N (2000) Study of spatial relationships among some soil physico-chemical properties of a field in central Italy using multivariate geostatistics. Geoderma 97: 39–60.
23. Bond-Lamberty B, Brown KM, Goranson C, Gower ST (2006) Spatial dynamics of soil moisture and temperature in a black spruce boreal chronosequence. Canadian Journal of Forest Research-Revue Canadienne De Recherche Forestiere 36: 2794–2802.
24. Herbst M, Prolingheuer N, Graf A, Huisman JA, Weihermuller L, et al. (2009) Characterization and understanding of bare soil respiration spatial variability at plot scale. Vadose Zone Journal 8: 762–771.
25. Famiglietti JS, Devereaux JA, Laymon CA, Tsegaye T, Houser PR, et al. (1999) Ground-based investigation of soil moisture variability within remote sensing footprints during the Southern Great Plains 1997 (SGP97) Hydrology Experiment. Water Resources Research 35: 1839–1851.
26. Western AW, Zhou SL, Grayson RB, McMahon TA, Bloschl G, et al. (2004) Spatial correlation of soil moisture in small catchments and its relationship to dominant spatial hydrological processes. Journal of Hydrology 286: 113–134.
27. Choi M, Jacobs JM, Cosh MH (2007) Scaled spatial variability of soil moisture fields. Geophysical Research Letters 34: L01401.
28. Brocca L, Tullo T, Melone F, Moramarco T, Morbidelli R (2012). Catchment scale soil moisture spatial-temporal variability. Journal of Hydrology 422–423: 63–75.
29. Li B, Rodell M (2012) Spatial variability and its scale dependency of observed and modeled soil moisture under different climate conditions. Hydrology and Earth System Sciences 17:1177–1188.
30. Mittelbach H, Seneviratne SI (2012) A new perspective on the spatio-temporal variability of soil moisture: Temporal dynamics versus time invariant contributions. Hydrology and Earth System Sciences 16: 2169–2179.
31. Kaspar TC, Bland WL (1992) Soil temperature and root growth. Soil Science 154: 290–299.
32. Sitaula BK, Bakken LR (1993) Nitrous-oxide release from spruce forest soil - Relationships with nitrification, methane uptake, temperature, moisture and fertilization. Soil Biology and Biochemistry 25: 1415–1421.
33. Kirschbaum MUF (1995) The temperature-dependence of soil organic-matter decomposition, and the effect of global warming on soil organic-C storage. Soil Biology and Biochemistry 27: 753–760.
34. Rey A, Petsikos C, Jarvis PG, Grace J (2005) Effect of temperature and moisture on rates of carbon mineralization in a Mediterranean oak forest soil under controlled and field conditions. European Journal of Soil Science 56: 589–599.
35. Global Carbon Observing System GCOS (2007) Global climate observing system: Ensuring the availability of global observations for climate. Available: http://www.wmo.int/pages/prog/gcos/Publications/gcos-138.pdf. Accessed: 1 July 2013
36. Schaefer GL, Cosh MH, Jackson TJ (2007) The USDA Natural Resources Conservation Service Soil Climate Analysis Network (SCAN). Journal of Atmospheric and Oceanic Technology 24: 2073–2077.
37. Hargrove WW, Hoffman FM (1999) Using multivariate clustering to characterize ecoregion borders. Computing in Science and Engineering 1: 18–25.
38. Hargrove WW, Hoffman FM (2004) Potential of multivariate quantitative methods for delineation and visualization of ecoregions. Environmental Management 34: S39–S60.
39. Keller M, Schimel DS, Hargrove WW, Hoffman FM (2008) A continental strategy for the National Ecological Observatory Network. Frontiers in Ecology and the Environment 6: 282–284.
40. Schimel D, Keller M, Berukoff S, Kao R, Loescher HW, et al. (2011) NEON 2011 Science Strategy; Enabling continental-scale ecological forecasting. Boulder CO: NEON Inc. 55 p.
41. Lemon ER (1960) Photosynthesis under field conditions 2. An aerodynamic method for determining the turbulent carbon dioxide exchange between the atmosphere and a corn field. Agronomy Journal 52: 697–703.
42. Monteith JL, Szeicz G (1960) The carbon-dioxide flux over a field of sugar beet. Quarterly Journal of the Royal Meteorological Society 86: 205–214.
43. Kaimal JC, Wyngaard JC (1990) The Kansas and Minnesota experiments. Boundary-Layer Meteorology 50: 31–47.
44. Loescher HW, Law BE, Mahrt L, Hollinger DY, Campbell J, et al. (2006) Uncertainties in, and interpretation of, carbon flux estimates using the eddy covariance technique. Journal of Geophysical Research-Atmospheres 111. D21S90, doi:10.1029/2005JD006932.
45. Schmid HP (1994) Source areas for scalars and scalar fluxes. Boundary-Layer Meteorology 67: 293–318.
46. Schmid HP (1997) Experimental design for flux measurements: matching scales of observations and fluxes. Agricultural and Forest Meteorology 87: 179–200.
47. Horst TW, Weil JC (1995) How far is far enough - the fetch requirements for micrometeorological measurement of surface fluxes. Journal of Atmospheric and Oceanic Technology 12: 447–447.
48. Shaw DC, Franklin JF, Bible K, Klopatek J, Freeman E, et al. (2004) Ecological setting of the wind river old-growth forest. Ecosystems 7: 427–439.
49. Lee H, Schuur EAG, Vogel JG (2010) Soil CO2 production in upland tundra where permafrost is thawing. Journal of Geophysical Research-Biogeosciences 115: G01009.
50. Gorres JH, Dichiaro MJ, Lyons JB, Amador JA (1998) Spatial and temporal patterns of soil biological activity in a forest and an old field. Soil Biology & Biochemistry 30: 219–230.
51. Giese K, Tiemann R (1975) Determination of the complex permittivity from thin-sample time domain reflectometry, Improved analysis of the step response waveform. Adv Molec Relax Processes 7: 45–49.
52. Topp GC, Davis JL, Annan AP (1980) Electromagnetic determination of soil water content: measurements in coaxial transmission lines. Water Resources Research, 16: 574–582.
53. McBratney AB, Webster R (1986) Choosing functions for semi-variograms of soil properties and fitting them to sampling estimates. Journal of Soil Science 37: 617–639.
54. Webster R, Oliver MA (1992) Sample adequately to estimate variograms of soil properties. Journal of Soil Science 43: 177–192.
55. Ayres E, Loescher HW, Brunke M, Luo H (2011) Soil temperature and moisture data from NEON candidate sites. National Ecological Observatory Network pub., Boulder, Colorado. http://neoninc.org/pds/FIU007.php.
56. Trangmar BB, Yost RS, Uehara G (1986) Application of geostatistics to spatial studies of soil properties. Advances in Agronomy: 45–94.
57. Riberiro JR, Diggle PJ (2001) geoR: A package for geostastical analysis. R-NEWS 1: 15–18. ISSN 1609-3631.

58. Kosugi Y, Mitani T, Ltoh M, Noguchi S, Tan M, et al. (2007) Spatial and temporal variation in soil respiration in a Southeast Asian tropical rainforest. Agricultural and Forest Meteorology 147: 35–47.
59. Huang XD, Grace P, Mengersen K, Weier K (2011) Spatio-temporal variation in soil derived nitrous oxide emissions under sugarcane. Science of the Total Environment 409: 4572–4578.
60. Panosso AR, Marques J, Pereira GT, La Scala N (2009) Spatial and temporal variability of soil CO2 emission in a sugarcane area under green and slash-and-burn managements. Soil and Tillage Research 105: 275–282.
61. Sokal RR, Rohlf FJ (1995) Biometry. 3rd edition. New YorkNY: Freeman Press, 937 p.
62. Loescher HW, Powers JS, Oberbauer SF (2002) Spatial variation of throughfall volume in an old growth tropical rain forest. Journal of Tropical Ecology 18: 397–407.
63. Pierson FB, Wight JR (1991) Variability of near-surface soil-temperature on sagebrush rangeland. Journal of Range Management 44: 491–497.
64. Elkins NZ, Sabol GV, Ward TJ, Whitford WG (1986) The influence of subterranean termites on the hydrological characteristics of a Chihuahuan desert ecosystem. Oecologia 68: 521–528.
65. Schlesinger WH, Reynolds JF, Cunningham GL, Huenneke LF, Jarrell WM, et al. (1990) Biological feedbacks in global desertification. Science 247: 1043–1048.
66. Famiglietti JS, Rudnicki JW, Rodell M (1998) Variability in surface moisture content along a hillslope transect: Rattlesnake Hill, Texas. Journal of Hydrology 210: 259–281.
67. Chaplot V, Walter C (2003) Subsurface topography to enhance the prediction of the spatial distribution of soil wetness. Hydrological Processes 17: 2567–2580.
68. Mohanty BP, Klittich WM, Horton R, Vangenuchten MT (1995) Spatio-temporal variability of soil temperature within three land areas exposed to different tillage systems. Soil Science Society of America Journal 59: 752–759.
69. Fitzjohn C, Ternan JL, Williams AG (1998) Soil moisture variability in a semi-arid gully catchment: implications for runoff and erosion control. Catena 32: 55–70.
70. Anctil F, Mathieu R, Parent LE, Viau AA, Sbih M, et al. (2002) Geostatistics of near-surface moisture in bare cultivated organic soils. Journal of Hydrology 260: 30–37.
71. Gallardo A (2003) Spatial variability of soil properties in a floodplain forest in northwest Spain. Ecosystems 6: 564–576.
72. Ryu D, Famiglietti JS (2006) Multi-scale spatial correlation and scaling behavior of surface soil moisture. Geophysical Research Letters 33: L08408, doi: 10.1029/2006GL025831.
73. Legendre P (1993) Spatial autocorrelation - Trouble or new paradigm? Ecology 74: 1659–1673.
74. McBratney AB, Webster R (1983) How many observations are needed for regional estimation of soil properties. Soil Science 135: 177–183.
75. Di HJ, Trangmar BB, Kemp RA (1989) Use of geostatistics in designing sampling strategies for soil survey. Soil Science Society of America Journal 53: 1163–1167.
76. Chung CK, Chong SK, Varsa EC (1995) Sampling strategies for fertility on a stony silt loam soil. Communications in Soil Science and Plant Analysis 26: 741–763.
77. Wang CM, Zuo Q, Zhang RD (2008) Estimating the necessary sampling size of surface soil moisture at different scales using a random combination method. Journal of Hydrology 352: 309–321.
78. Teuling AJ, Uijlenhoet R, Hupet F, van Loon EE, Troch PA (2006) Estimating spatial mean root-zone soil moisture from point-scale observations. Hydrology and Earth System Sciences 10: 755–767.
79. Larson KM, Small EE, Gutmann ED, Bilich AL, Braun JJ, et al. (2008) Use of GPS receivers as a soil moisture network for water cycle studies. Geophysical Research Letters 35: L2445, doi:10.1029/2008GL03613.
80. Zreda M, Desilets D, Ferre TPA, Scott RL (2008) Measuring soil moisture content non-invasively at intermediate spatial scale using cosmic-ray neutrons. Geophysical Research Letters 35: L21402, doi: 10.1029/2008GL035655.
81. Loague K (1992) Soil-water content at R-5.1. Spatial and temporal variability. Journal of Hydrology 139: 233–251.
82. Scharringa M (1976) Representativeness of soil temperature measurements. Agricultural Meteorology 16: 263–276.
83. Dokuchaev VV (1883) Russian Chernozems (Russkii Chernozems). Israel Prog. Sci. Trans., Jerusalem, 1967. Transl. from Russian by N. Kaner. Springfield, VA: U.S. Dept. of Commerce.
84. Coffey GN (1912) A Study of the Soils of the United States. U. S. Dept. of Agriculture Bur., Soils Bull. 85, US Govt. Printing Office, Washington, D.C. 114 p.

15

Interannual Variation of the Bowen Ratio in a Subtropical Coniferous Plantation in Southeast China, 2003-2012

Yakun Tang[1,2], Xuefa Wen[1]*, Xiaomin Sun[1]*, Huimin Wang[1]

1 Key Laboratory of Ecosystem Network Observation and Modeling, Institute of Geographic Sciences and Natural Resources Research, Chinese Academy of Sciences, Beijing, China, **2** University of Chinese Academy of Sciences, Beijing, China

Abstract

The interannual variation of the Bowen ratio, through its effect on the warming extent of available energy to the ecosystem land surface air, heavily influences the ecosystem microclimate and affects the hydrological cycle at both regional and global scales. Although the precipitation amount in southeast China is not expected to change greatly as a result of climate change, the precipitation frequency may be altered in the future. We explored the interannual variation of the Bowen ratio and its affecting mechanisms based on eddy covariance measurements in a subtropical plantation in southeast China during 2003–2012. The results indicated that the annual mean Bowen ratio was 0.35 ± 0.06, with a range of 0.29–0.45. The Bowen ratio during the dry season (July-October) positively correlated with the annual Bowen ratio ($R^2 = 0.85$, $p < 0.001$). The effective precipitation frequency during the dry season, through its positive effect on shallow soil water content, indirectly and negatively affected the annual Bowen ratio. Between 2003 and 2012, the annual Bowen ratio exhibited a marginally significant decreasing trend ($p = 0.061$), meanwhile the effective precipitation frequency and shallow soil water content during the dry season increased significantly ($p < 0.001$). The annual Bowen ratio may decrease further if the effective precipitation frequency and shallow soil water content during the dry season follow similar trends in the future. The warming effect of available energy to the surface air of our studied plantation may decline with the decreasing annual Bowen ratio.

Editor: Gil Bohrer, The Ohio State University, United States of America

Funding: This study was supported by the Strategic Priority Research Program-Climate Change: Carbon Budget and Relevant Issues of the Chinese Academy of Sciences (grant XDA05050601), the National Natural Science Foundation of China (grant 31070408), and the Strategic Program of Knowledge Innovation of Chinese Academy of Sciences (grant KZCX2-EW-QN305). The funders had no role in study design, data collection and analysis, decision to publish, or preparation of the manuscript.

Competing Interests: The authors have declared that no competing interests exist.

* E-mail: wenxf@igsnrr.ac.cn (XW); sunxm@igsnrr.ac.cn (XS)

Introduction

The consumption of available energy as sensible heat flux warms the land surface air [1]. While energy consumption as latent heat flux, which is equivalent to evapotranspiration and responsible for a big part of precipitation in many parts of the world, mitigates this warming effect [1,2]. The Bowen ratio, which is the ratio of the sensible heat flux to latent heat flux, through its effect on the warming extent of available energy to the ecosystem land surface air, heavily influences the ecosystem microclimate and the hydrological cycle at both regional and global scales [3–5].

Forests cover approximately 31% of the total global land area [6] and are vital components of energy cycles between the terrestrial surface and atmosphere [4,7]. Numerous models have suggested that the Bowen ratio will decline in the future because the increased energy due to carbon dioxide (CO_2) is primarily consumed as latent heat flux rather than as sensible heat flux [8,9]. However, diverse long-term trends of the annual Bowen ratio that measured directly by the eddy covariance method have been observed among forest ecosystems [10–13]. Keenan et al. [10] demonstrated that the Bowen ratio shows a global increasing trend due to the adjustment of leaf stomata in response to increasing atmospheric CO_2, based on the energy fluxes of 21 forest sites ranging from 7 to 18 years. While the annual Bowen ratio shows a decreasing trend in a boreal Scots pine forest [11]. And no significant trend is observed in a broadleaved forest [3], in a temperate spruce forest [12], or in a Japanese cypress forest [13]. There is a lack of consensus between these models and the observations because these models do not fully considering the interactive relationships between the atmosphere-biosphere energy fluxes and vegetation in response to climate change, especially for forests [8,14].

Because the forest Bowen ratio is influenced by the interactive effects of environmental and physiological factors, the dominant factors affecting the variation of the Bowen ratio are also diverse [2,4,7]. Wilson et al. [7] demonstrated that physiological factor (vegetation surface resistance) dominate the discrepancy of the Bowen ratio among different forest types in temperate regions. While, environmental factors are shown to primarily affect the discrepancy of the Bowen ratio among forest types in East Asia [2]. For a specific forest, the interannual variation of the Bowen ratio in a boreal Scots pine forest [11] is primarily attributed to net radiation, and the precipitation amount dominates the interannual variation of the Bowen ratio in a temperate spruce forest [12]. In addition, seasonal precipitation affects forest ecosystem energy fluxes in a nonlinear fashion [15–17]. There is a need to recognize the influence of the seasonal distribution of precipitation on the interannual variation of the Bowen ratio [17].

Coniferous forests account for approximately 36% of all global forested areas and are predominantly distributed in subtropical, temperate, and boreal regions [6]. Southern China has the largest global evergreen subtropical forest (covering 53 million ha), and coniferous plantations account for 41% of the total forested area in south China [18]. Although the precipitation amount in southeast China is not expected to change significantly because of climate change, the precipitation frequency is expected to change in the future [19]. This region is characterized by a subtropical humid monsoon climate with abundant water and energy resources [20], while drought stress occurs during July-October as a result of different distributions of temperature and precipitation [21]. In southeast China, previous studies in a subtropical coniferous plantation at the Qianyanzhou site mainly focus on the effect of drought stress on the interannual variation of the carbon exchange process [20,22]. However, the interannual variation of the Bowen ratio in response to drought stress has yet to be investigated. In this study, we examined the energy fluxes during 2003–2012 at the Qianyanzhou site. The objectives of this study were (1) to investigate the interannual variation of the Bowen ratio and its affecting mechanism, (2) to identify the seasonal drought effect on the interannual variation of the Bowen ratio, (3) to study and analyze the long-term trend of annual Bowen ratio.

Materials and Methods

Ethics Statement

The study site is maintained by the Institute of Geographic Sciences and Natural Resources Research, Chinese Academy of Sciences. This area is a practice base for researchers at the Chinese Academy of Sciences. All necessary permits were obtained for this field study. The field study did not involve endangered or protected species.

Site Description

The Qianyanzhou site (26°44′52″N, 115°03′47″E, and elevation 102 m) is a member of ChinaFLUX and is located at the Qianyanzhou station of the Chinese Ecosystem Research Network (CERN) in southeast China. This area is influenced by a subtropical monsoon climate, where the annual mean temperature and precipitation are 17.9 °C and 1472.8 mm, respectively, according to meteorological records of 1985–2012. The soil is red earth, largely weathered from red sandstone, and is classified as Typic Dystrudepts Udepts Inceptisols according to the United States soil taxonomy [21]. The soil texture is divided into the following particle grades: 2.0–0.05 mm (17%), 0.05–0.002 mm (68%), and <0.002 mm (15%) [20]. The soil bulk density at the surface (0–40 cm) is 1.57 g cm^{-3}.

The evergreen coniferous plantation was planted in 1985, and the dominant tree species are Masson pine (*Pinus massoniana L.*), Slash pine (*Pinus elliottii E.*), and Chinese fir (*Cunninghamia lanceolata L.*). According to a survey conducted in 2008, the stem densities of Masson pine, Slash pine, and Chinese fir were 700, 545, and 93 stems ha^{-1}, respectively; the mean heights were 11.2, 14.3, and 11.8 m, respectively; and the mean diameters at the breast height were 13.6, 18.2, and 13.8 cm, respectively. Further characteristics of the site are provided in Wen et al. [20] and Zhang et al. [23].

Eddy Covariance and Meteorological Measurements

The eddy covariance instruments were mounted at 39.6 m on a tower. The concentrations of carbon dioxide (CO_2) and water vapor were measured using an LI-7500 open-path CO_2/H_2O analyzer (Model LI-7500, Licor Inc., USA), and the three-dimensional wind speed and virtual temperature were detected by a three-dimensional sonic anemometer (Model CSAT3, Campbell Scientific Inc., USA). All raw data were sampled at 10 Hz, and the 30 min mean fluxes were calculated and stored by a CR5000 datalogger (Campbell Scientific Inc., USA).

A suite of meteorological measurements was also measured at the site. A four-component net radiometer (Model CNR-1, Kipp & Zonen Inc., The Netherlands), a quantum sensor of photosynthetically active radiation (Model LI190SB, Licor Inc., USA), and a pyranometer (Model CM11, Kipp & Zonen Inc., The Netherlands) were used to measure radiation. Air temperature and relative humidity sensors (Model HMP45C, Campbell Scientific Inc., USA) and wind speed sensors (A100R, Vector Inc., USA) were mounted at heights of 1.6, 7.6, 11.6, 15.6, 23.6, 31.6, and 39.6 m above the ground. Soil temperatures were made at depths of 2, 5, 20, 50, and 100 cm with thermocouples (105T and 107-L, Campbell Scientific Inc., USA). Soil volumetric water contents were recorded at depths of 5, 20, and 50 cm using three TDR probes (Model CS615-L, Campbell Scientific Inc., USA). Two soil heat flux plates (Model HFT-3, Campbell Scientific Inc., USA) were placed at a depth of 5 cm below the ground surface. Precipitation was monitored with a rain gauge (Model 52203, RM Young Inc., USA). All data were collected by three CR10X dataloggers (Campbell Scientific Inc., USA) and a CR23X datalogger (Campbell Scientific Inc., USA) with a 25-channel solid-state multiplexer (Model AM25T, Campbell Scientific Inc., USA). Meteorological factors including solar radiation, net radiation, and precipitation have been observed at a meteorological station less than 1 km from the flux site since 1985. Additional details on data acquisition are described further in our previous studies [20,21].

Flux Correction and Gap Filling

Latent heat flux (LE) and sensible heat flux (H) (W m^{-2} s^{-1}) between the forest and atmosphere over 30 min intervals were calculated as follows:

$$LE=\lambda\overline{w'\rho_v'(z_r)}+\lambda\rho\sum_{i=1}^{n}((\Delta e_i/\gamma\Delta t)\Delta z_i) \quad (1)$$

$$H=\rho c_p\overline{w'T'(z_r)}+\rho c_p\sum_{i=1}^{n}((\Delta T_a/\Delta t)\Delta z_i) \quad (2)$$

where the first term on right-hand side is the eddy flux of latent heat flux or sensible heat flux, the second term is the storage of latent heat flux or sensible heat flux below the height of observation (z_r), and all advective terms in the mass conservation equation are ignored. The latent heat flux storage term was calculated based on the profile of air temperature and relative humidity, as suggested by McCaughey [24] and Oliphant et al. [25]. The sensible heat flux storage term was computed using the change in the air temperature profile according to Oliphant et al. [25].

Planar fit rotation was applied to remove the effect of instrument tilt or irregularity on the airflow at monthly intervals [20,26]. The Webb-Pearman-Leuning (WPL) correction was performed to adjust density changes resulting from fluctuations in heat and water vapor [27]. We screened and eliminated the eddy covariance flux data for anomalous or spurious values caused by precipitation, water condensation, and system failure. Any value that exceeded 5 standard deviations in a window of 10 values was deleted. To avoid the possible underestimation of the fluxes under stable conditions during the night, the effect of friction velocity was identified for each year according to the method described by Reichstein et al. [28]. The latent heat flux and sensible heat flux at night (solar elevation angle<0) were excluded when the friction velocity was less than 0.19 m s^{-1}, which was the maximum friction velocity threshold among 2003–2012.

Several strategies were adopted to fill the meteorological data and energy flux gaps. First, gaps of meteorological data that were less than 2 h were interpolated linearly from adjacent data points. Longer gaps of solar radiation and net radiation were filled based on the linear regression between these data observed at the flux site and at the meteorological station less than 1 km from the flux site. The regression was calculated at bi-monthly intervals. Longer gaps of air temperature and relative humidity were interpolated from adjacent levels in the measurement profile. Meanwhile, longer gaps of soil water content were filled using the mean diurnal variation method [29]. The annual short and longer data gaps coverage for all meteorological data were no more than 0.6% and 3.1% during 2003–2012, respectively. Then, we filled short (<2 h) latent heat flux and sensible heat flux data gaps using linear interpolation and longer gaps using a marginal distribution sampling method [28]. From 2003 to 2012, the annual mean short gaps coverage for latent and sensible heat fluxes were 1.6% and 1.4%, respectively. The annual mean daytime longer gaps coverage for latent and sensible heat fluxes were 19.1% and 19.2%, respectively. Meanwhile, the annual mean nighttime longer gaps coverage for latent (79.3%) and sensible heat (80.1%) fluxes were larger than that in daytime, respectively, due to the nighttime data filtration through the friction velocity.

The energy balance ratio (*EBR*) was used to assess the performance of the eddy covariance system. The ratio was calculated from 30 min mean available values for each year as [30]:

$$EBR = (LE + H)/(R_n - G - S) \quad (3)$$

where R_n is the net radiation, G is the soil heat flux, and S is the heat storage below the observation height.

The energy balance ratio in this study ranged from 0.59 (2005) to 0.76 (2009), which fell within the range of 0.34 to 1.69 reported at FLUXNET sites [30].

Bulk Canopy Parameters and Potential Evapotranspiration

Daily canopy conductance (g_c) was calculated by averaging 30 min mean values from 10:00 to 16:00 of the Chinese Standard Time (available data>80%). Data on rainy days were omitted to avoid wet canopy. Canopy conductance is sensitive to drought stress [5] and was computed by inverting the Penman-Monteith equation [31,32],

$$g_c = \frac{\gamma LE g_a}{s(R_n - G) + \rho c_p VPD - LE(s+\gamma)} \quad (4)$$

$$g_a = ((u/u^{*2}) + 6.2u^{*-0.67})^{-1} \quad (5)$$

where s is the slope of a function relating saturation vapor pressure to temperature (kPa K^{-1}), γ is the psychrometric constant (kPa K^{-1}), LE is the latent heat flux (W m^{-2}), C_p is the specific heat of air (J kg^{-1} K^{-1}), VPD is the air vapor pressure deficit (kPa), ρ is the air density (kg m^{-3}), g_a is the aerodynamic conductance (mm s^{-1}), u is the wind speed (m s^{-1}), and u^* is the friction velocity (m s^{-1}).

The Priestley-Taylor (α) is the ratio of the measured evapotranspiration (ET) to the equilibrium evapotranspiration (ET_{eq}). The equilibrium evapotranspiration is considered the climatologically expected evapotranspiration, which assumes a closed volume with constant energy input over a 'wet' surface [33]. A small value for Priestley-Taylor (α) is indicative of a lack of water supply for evapotranspiration [5,34]. Potential evapotranspiration was calculated as 1.26 × the equilibrium evapotranspiration [35].

$$a = ET/ET_{eq} \quad (6)$$

$$ET_{eq} = ((R_n - G)s)/(s+\gamma) \quad (7)$$

Budyko's Aridity Index

Budyko's aridity index is the ratio of the precipitation amount to potential evapotranspiration [36]. When Budyko's aridity index is smaller than one, the system is water limited. This criterion has been used to detect periods of drought stress in coniferous forest and grassland ecosystems [34,37]. Ryu et al. [34] also suggested that a seasonal timescale is appropriate for indicating the drought stress period, especially considering the influence of vegetation.

Remotely Sensed Enhanced Vegetation Index

The enhanced vegetation index (EVI) reflects both the leaf area and the canopy biophysical characteristics [38]. This index is resistant to the soil background and less susceptible to atmospheric disturbance [38,39], and has been considered a physiological affecting factor to the net ecosystem production for evergreen coniferous and broadleaved deciduous forests [25,40]. The 8-day enhanced vegetation index at a spatial resolution of 500 m was calculated according to Huete et al. [38]. The MODIS/Terra Surface Reflectance 8-Day L3 Global 500 m (MOD09A1) data were obtained from the data portal of the Earth Observation and Modeling Facility at the University of Oklahoma. The Savitzky-Golay method was adopted to detect and replace abnormal enhanced vegetation index values [23], and a window length parameter of 4 was used in this method.

Effective Precipitation Frequency

The effective precipitation frequency of the studied plantation was defined as the number of days during a specific period when the daily precipitation amount exceeded 1.5 mm, as suggested by Wang et al. [21]. This criterion was chosen because it is the minimum daily precipitation amount that can influence the shallow soil water content in this studied plantation.

Data Statistics Analysis Method

In this study, the monthly and annual latent and sensible heat fluxes were referred to the sum of these fluxes at each timescale. The monthly and annual Bowen ratio was calculated as a ratio of the sensible heat flux to the latent heat flux at each timescale. Baldocchi et al. [41] suggested that monthly timescale is a pronounced spectral gap in an ecosystem to capture seasonal change of energy fluxes. The independent factors (environmental or physiological) that affect the annual Bowen ratio can be used to explain the interannual variation of the Bowen ratio. In this

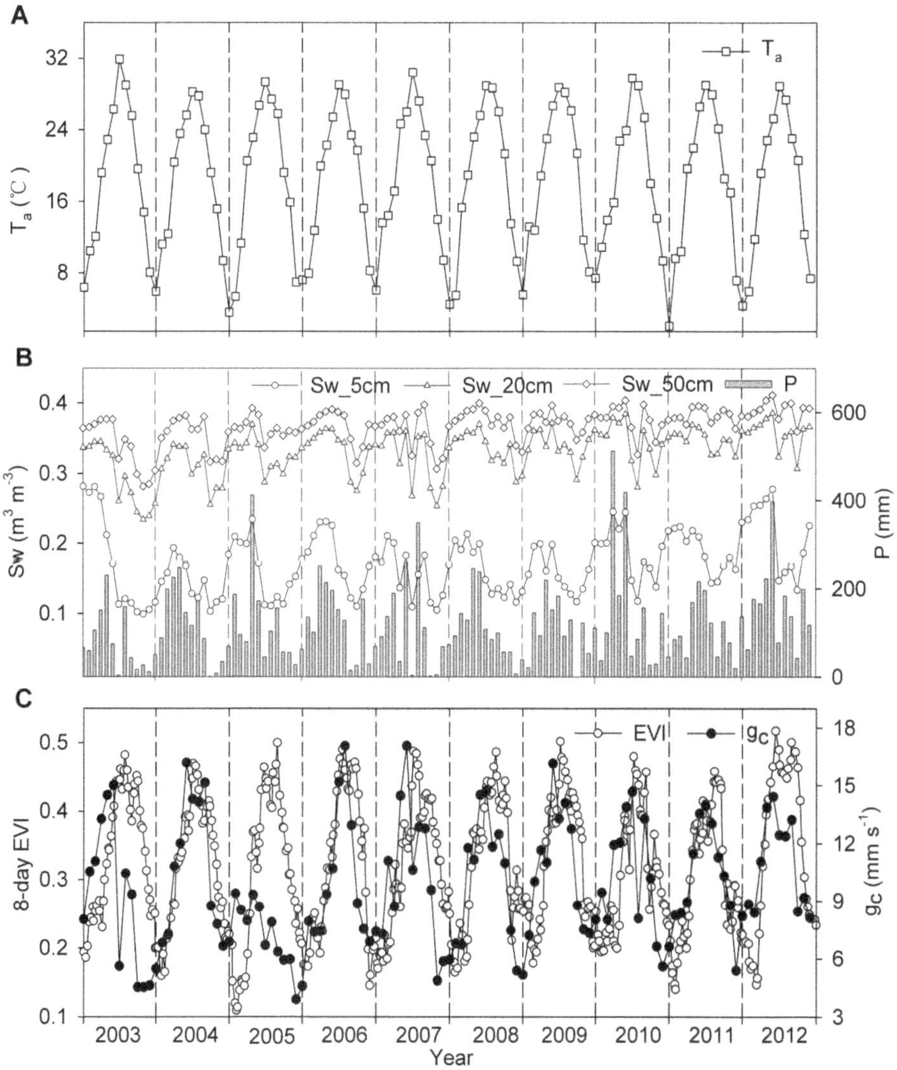

Figure 1. Seasonal and interannual variation of environmental and physiological factors. Factors include (A) monthly mean air temperature (T_a), (B) monthly mean soil water content at 5, 20, and 50 cm (Sw_5 cm, Sw_20 cm, and Sw_50 cm) and monthly precipitation amount (P), and (C) monthly mean canopy conductance (g_c) and 8-day timescale enhanced vegetation index (EVI).

study, through a single factor regression analysis between the independent factors and the annual Bowen ratio, the dominant affecting factor for the interannual variation of the Bowen ratio was detected. The environmental factors include net radiation, air temperature, vapor pressure deficit, soil water content at three depths (5, 20, and 50 cm) and precipitation amount. The physiological factor is referred to the enhanced vegetation index.

For the single factor regression analysis, the coefficient of determination (R^2) and *p*-value were reported. First, for each timescale, the dominant affecting factor for the Bowen ratio was selected based on the maximum coefficient of determination. The residual between the Bowen ratio and the fitted Bowen ratio, which is a function of the dominant factor, was also calculated for each timescale. Then, the regression analysis was performed between other independent factors and the residual. The dominant affecting factor for the residual was also selected based on the maximum coefficient of determination [42]. All statistical analyses were performed using SPSS 13.0.

Results

Interannual Variation of Environmental and Physiological Factors

Seasonal variation of monthly environmental factors, canopy conductance (g_c), and the 8-day enhanced vegetation index (EVI) during 2003–2012 are plotted in Fig. 1. The air temperature exhibited a single peak, with the maximum and minimum values occurring in summer (June-August) and winter (December-February), respectively. In general, precipitation experienced more frequently in the first half of the year than in the second half of the year, and precipitation deficiency often occurred in summer. This high temperature and a lack of sufficient precipitation in summer may result in seasonal drought stress in this studied plantation [20]. Seasonal variation of the soil water content at 5, 20, and 50 cm were closely related to the precipitation variability (Fig. 1B). The magnitude of the seasonal variation in soil water content decreased with the increasing soil depth. Generally, canopy conductance exhibited the maximum value in summer (Fig. 1C), while, canopy conductance declined

Table 1. Interannual variation of annual net radiation (Rn, MJ m^{-2} yr^{-1}), air temperature (T_a, °C), vapor pressure deficit (VPD, kPa), soil water content at 5 cm, 20 cm, and 50 cm (Sw_5 cm, Sw_20 cm, and Sw_50 cm (m^3 m^{-3})), enhanced vegetation index (EVI), total precipitation amount (P, mm), latent heat flux (LE, MJ m^{-2} yr^{-1}), sensible heat flux (H, MJ m^{-2} yr^{-1}), and the Bowen ratio (β) during 2003–2012 at the Qianyanzhou site.

Year	Rn	T_a	VPD	Sw_5 cm	Sw_20 cm	Sw_50 cm	EVI	P	LE	H	β
2003	2849.3	18.9	0.62	0.18	0.30	0.34	0.34	944.9	1864.8	845.5	0.45
2004	2871.2	18.6	0.54	0.14	0.30	0.35	0.32	1375.5	2061.7	704.4	0.34
2005	2620.8	17.9	0.46	0.16	0.33	0.36	0.30	1455.4	1253.1	564.7	0.45
2006	2658.1	18.4	0.46	0.18	0.33	0.37	0.32	1485.3	1732.5	568.8	0.33
2007	2776.6	18.9	0.52	0.16	0.32	0.36	0.33	1318.7	2125.5	631.8	0.30
2008	2861.1	18.4	0.53	0.16	0.34	0.38	0.33	1332.9	2068.9	768.9	0.37
2009	2939.0	18.7	0.55	0.16	0.34	0.37	0.33	1253.9	2175.3	721.8	0.33
2010	2670.1	18.4	0.51	0.19	0.35	0.38	0.31	1854.3	1911.2	581.9	0.30
2011	2776.5	17.9	0.51	0.19	0.35	0.38	0.30	1237.3	1882.4	547.7	0.29
2012	2725.9	17.4	0.40	0.21	0.36	0.39	0.34	1921.9	1745.0	610.3	0.35
Mean	2774.9	18.4	0.51	0.17	0.33	0.37	0.32	1418.0	1882.0	654.6	0.35
SD	105.38	0.48	0.06	0.02	0.02	0.02	0.01	289.16	268.80	100.63	0.06
CV	0.04	0.03	0.12	0.12	0.06	0.04	0.04	0.20	0.14	0.15	0.17

SD is standard deviation. CV is the coefficient of variation (ratio of standard deviation to mean value).

during the summers of 2003, 2007, and 2010. Except in 2005, the seasonal variation of the enhanced vegetation index followed that of canopy conductance. During the summers of 2003, 2007, and 2010, the decreasing magnitude of the enhanced vegetation index was smaller than that of the canopy conductance.

Table 1 lists the annual environmental and enhanced vegetation index values. Except in 2006 and 2010, the precipitation amounts in other years were below the long-term average value (1985–2012, 1472.8 mm). Meanwhile, only the air temperature in 2012 was below its long-term average value (1985–2012, 17.9°C). The maximum and minimum coefficients of variation (CV) of environmental factors were observed in the precipitation amount (20%) and air temperature (3%), respectively.

According to the seasonal variation of Budyko's aridity index during 2003–2012, the dry season was defined from July to October, and the wet season spanned the remainder of the year (Fig. 2). Wang et al. [21] defined the same period for the dry season in this studied plantation. The annual mean (mean ± one

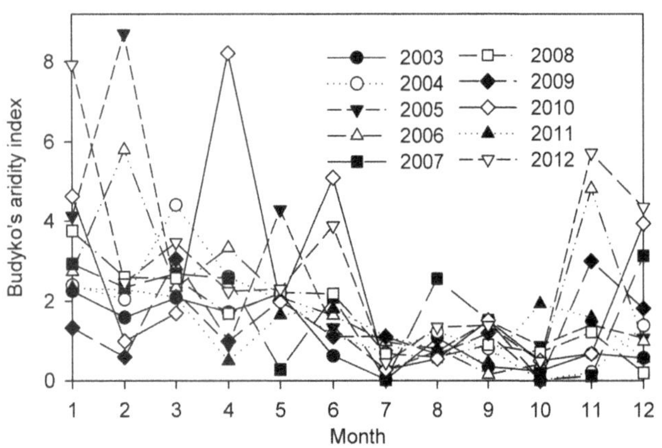

Figure 2. Seasonal and interannual variation of the monthly Budyko's aridity index.

standard deviation) Budyko's aridity index during the dry season was 0.75±0.17 and ranged from 0.43 (2003) to 0.99 (2007). During the wet season, the Budyko's aridity index ranged from 1.46 (2003) to 3.45 (2012), with a mean value of 2.22±0.76. The annual mean Priestley-Taylor (α) during the dry season (0.58±0.09) was lower than that during the wet season (0.66±0.09), which also indicated the impact of drought stress during the dry season.

Interannual Variation of the Bowen Ratio

Fig. 3 shows the monthly sum of latent heat flux, monthly sum of sensible heat flux and monthly Bowen ratio. The latent heat flux increased from a minimum during winter to a maximum during summer. The monthly averaged sensible heat flux also showed a similar seasonal variation as latent heat flux, although there was no obviously seasonal variation in 2004. The seasonal variation of the Bowen ratio was 'U' shaped, as the latent heat flux overwhelmed the sensible heat flux in almost all months, particularly in summer (Fig. 3C). The Bowen ratio began to decline in winter and the lowest value occurred in summer. However,the Bowen ratio obviously increased during the summers of 2003, 2007, and 2010 because of severe drought stress (Fig. 3C).

The annual means and interannual variation of the latent heat flux, sensible heat flux, and Bowen ratio are listed in Table 1. The annual means of the latent heat flux, sensible heat flux, and Bowen ratio were 1882±269 MJ m^{-2} yr^{-1}, 655±101 MJ m^{-2} yr^{-1}, and 0.35±0.06, respectively. The maximum values (0.45) of the Bowen ratio were observed in 2003 and 2005, which correspond to a maximum sensible heat flux in 2003 and a minimum latent heat flux in 2005, respectively. Meanwhile, the minimum Bowen ratio (0.29) was observed in 2011 and corresponds to a minimum sensible heat flux. The coefficient of variation of the annual Bowen ratio reached up to 17%, which was higher than those of the latent heat flux (14%) and sensible heat flux (15%).

The Bowen ratio decreased during the vigorous growth period (March-October), as the latent heat flux was higher than the sensible heat flux in these months (Fig. 3). From 2003 to 2012,

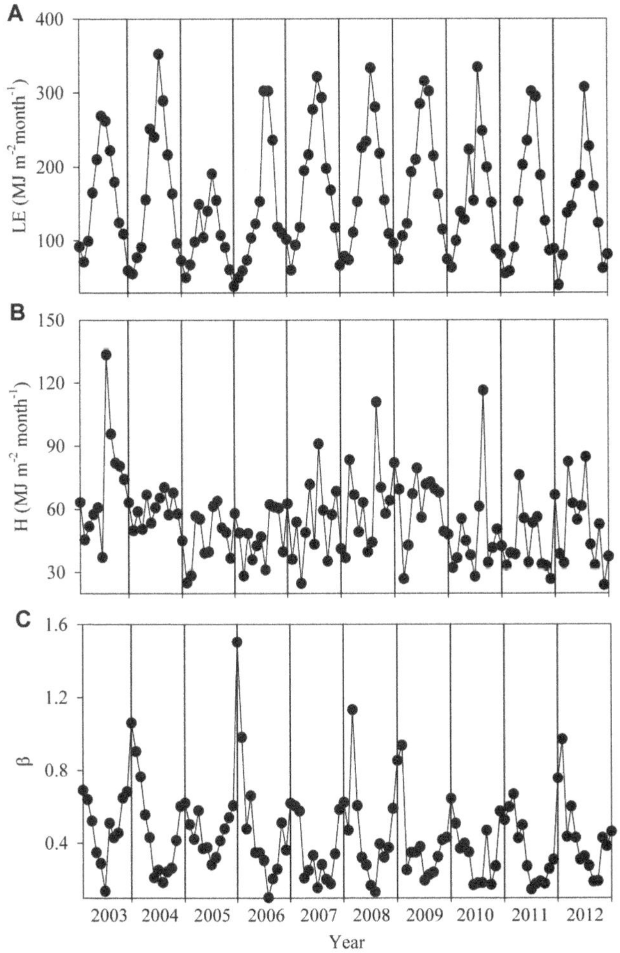

Figure 3. Seasonal and interannual variation of the monthly energy fluxes. Energy fluxes include (A) monthly sum of latent heat flux (LE), (B) monthly sum of sensible heat flux (H), and (C) monthly Bowen ratio (β).

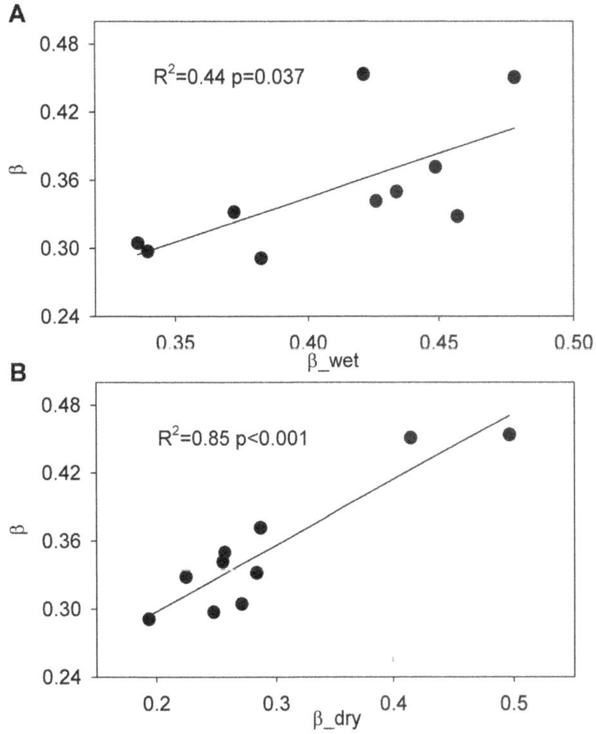

Figure 4. Relationships between the annual Bowen ratio (β) and Bowen ratio during wet and dry seasons. Bowen ratio during (A) the wet season (β_wet) was calculated as the ratio of sensible heat flux to latent heat flux from November to June at each year, and during (B) the dry season (β_dry) was calculated as the ratio of sensible heat flux to latent heat flux from July to October at each year, respectively.

because the dry season occurred during the vigorous growth period of the forest, the mean value of the Bowen ratio during the dry season (0.29±0.09) was lower than that during the wet season (0.41±0.05). The maximum (0.50) and minimum (0.11) Bowen ratio during the dry season were observed in 2003 and 2011, respectively, which correspond to the maximum (391 MJ m^{-2}) and minimum (176 MJ m^{-2}) sensible heat flux during the dry season. The Bowen ratio during the dry season and the wet season significantly correlated with the annual Bowen ratio, with coefficients of determination were 0.85 ($p<0.001$) and 0.44 ($p=0.037$), respectively (Fig. 4).

Factors Affecting the Interannual Variation of the Bowen Ratio

Fig. 5 presents the dominant affecting factors for the Bowen ratio on a monthly timescale. The Bowen ratio was primarily affected by air temperature ($R^2=0.50$, $p<0.001$) (Fig. 5A). The net radiation, vapor pressure deficit, soil water content at 20 and 50 cm, and enhanced vegetation index also significantly correlated with the monthly Bowen ratio, with coefficients of determination were 0.43, 0.29, 0.11, 0.22, and 0.36, respectively. While, neither the soil water content at 5 cm nor the precipitation amount significantly correlated with the monthly Bowen ratio. Removing the influence of air temperature on the monthly Bowen ratio, the residual was mainly affected by the soil water content at 20 cm ($R^2=0.24$, $p<0.001$) (Fig. 5B).

On an annual timescale, no environmental or physiological factors were observed that significantly affected the annual Bowen ratio. However, soil water content at 5 cm during the dry season played a key role in affecting the annual Bowen ratio ($R^2=0.70$, $p=0.003$), followed by the soil water content at 20 cm ($R^2=0.48$, $p=0.026$) and 50 cm ($R^2=0.45$, $p=0.035$) (Fig. 6). By removing the influence of soil water content at 5 cm on the annual Bowen ratio, there was no affecting factor for this residual. Further, the Bowen ratio during the dry season was also primarily affected by the soil water content at 5, 20, and 50 cm during the dry season, with coefficients of determination were 0.66 ($p=0.002$), 0.63 ($p=0.026$), and 0.63 ($p=0.036$), respectively. Note that, there were no environmental and physiological factors during the wet season that significantly affected the annual Bowen ratio and the Bowen ratio during the wet season.

Discussion

Precipitation Amount and Frequency Affecting the Interannual Variation of the Bowen Ratio

The affecting effect of the precipitation amount to the annual Bowen ratio was diverse at specific forest sites [11–13]. Grünwald and Bernhofer (2007) [12] indicated that the annual Bowen ratio in a temperate spruce forest was primarily affected by annual precipitation amount. However, in a Japanese cypress forest [13], in a boreal Scots pine forest [11], and in this studied plantation, no

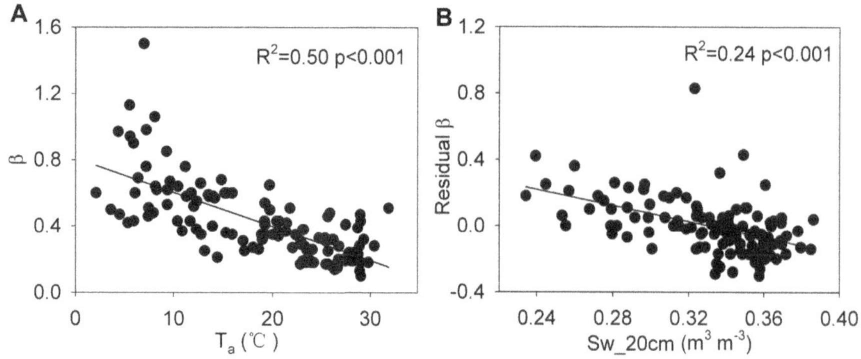

Figure 5. Factors affecting the Bowen ratio on a monthly timescale. The relationships (A) between the monthly Bowen ratio (β) and air temperature (T_a), (B) between residual of the monthly Bowen ratio (β) as a function of air temperature (T_a) and soil water content at 20 cm (Sw_20 cm).

affecting effect between the annual precipitation amount and annual Bowen ratio was found.

In this studied plantation, the annual Bowen ratio was mainly affected by the soil water content at 5 cm during the dry season (Fig. 6A). Unlike a broadleaved temperate plantation [43], a beech forest [44], and a mixed-conifer forest [45], the soil water content at 5 cm during the dry season was not significantly influenced by the precipitation amount ($R^2 = 0.38$, $p = 0.065$). Instead, due to the precipitation interception capacity of the forest canopy [21], the soil water content at 5 cm during the dry season was significantly affected by the effective precipitation frequency (no less than

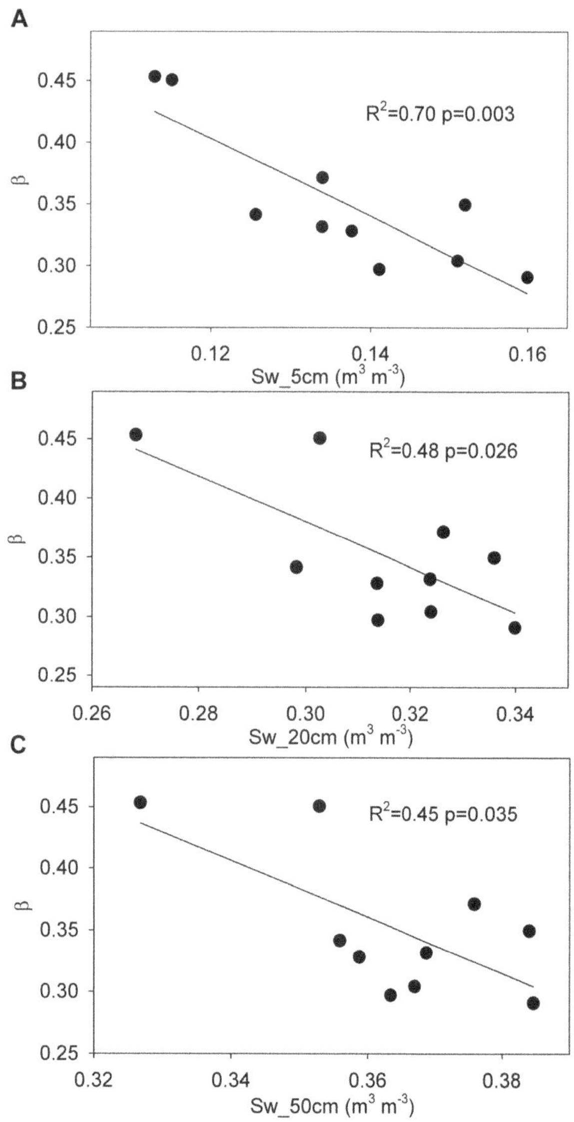

Figure 6. Relationships between the annual Bowen ratio (β) and dry season soil water content. Dry season (July-October) soil water content was measured at (A) 5 cm (Sw_5 cm), (B) 20 cm (Sw_20 cm), and (C) 50 cm (Sw_50 cm).

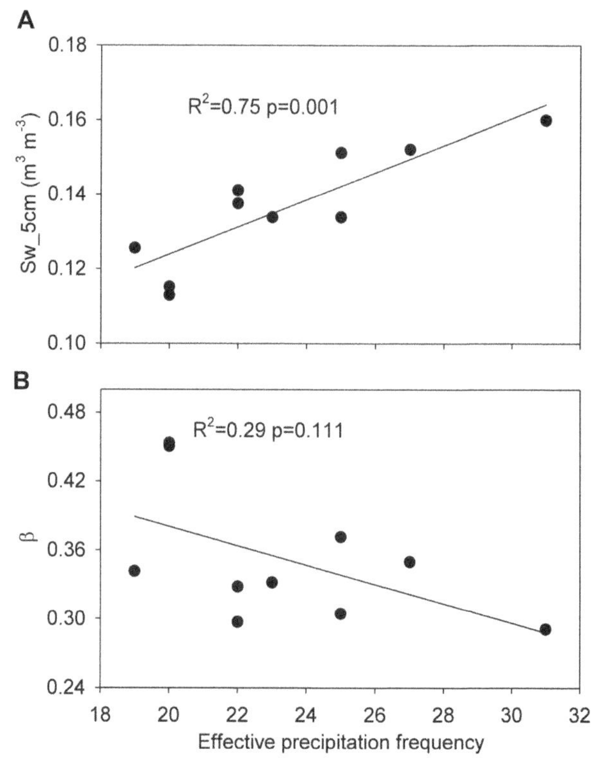

Figure 7. Impact of the effective precipitation frequency on soil water content and annual Bowen ratio (β). The effective precipitation frequency was measured as the number of days with a precipitation amount ≥ 1.5 mm during the dry season, and soil water content was measured at 5 cm (Sw_5 cm) during the dry season.

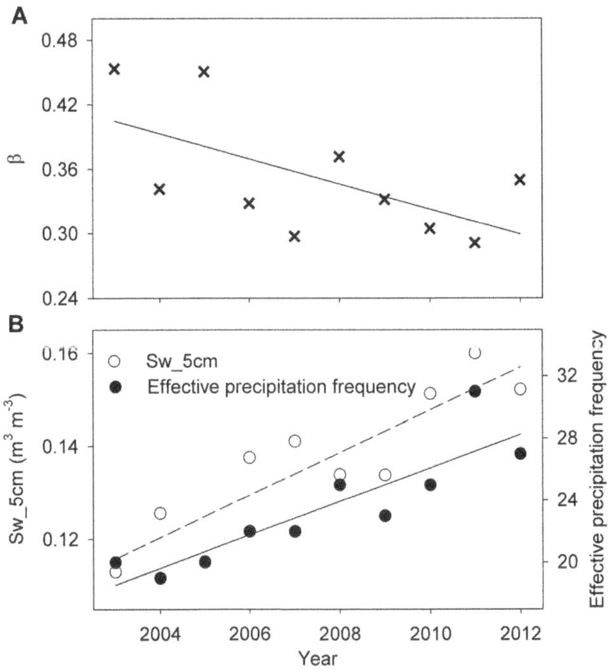

Figure 8. Long-term trends of the Bowen ratio (β), soil water content, and effective precipitation frequency. Long-term trends were marginally significant for (A) the annual Bowen ratio ($p = 0.06$), and significant for (B) the soil water content at 5 cm (Sw_5 cm; dashed line, $p<0.001$) and the effective precipitation frequency (solid line, $p<0.001$) during the dry season.

1.5 mm day^{-1}) in this studied plantation (Fig. 7A). Because the maximum leaf area index was 5.6 m^2 m^{-2} and the variation of the leaf area index was less than 1.5 m^2 m^{-2} in this studied plantation [20], the precipitation amount of less than 1.5 mm day^{-1} was mostly intercepted by the forest canopy and did not influence the soil water content at 5 cm [21]. The results also indicated that the effective precipitation frequency during the dry season did not directly affect the annual Bowen ratio ($R^2 = 0.29$, $p = 0.111$) (Fig. 7B). Therefore, we concluded that the effective precipitation frequency during the dry season, not the precipitation amount, indirectly affected the annual Bowen ratio through its effect on the soil water content at 5 cm during the dry season.

The annual Bowen ratio in this studied plantation exhibited a marginally significant decreasing trend ($p = 0.061$) during 2003–2012 (Fig. 8A). During the dry season, although the effective precipitation frequency did not exhibit a noticeable trend ($p = 0.772$) between 1985 and 2002, the effective precipitation frequency significantly increased ($p<0.001$) between 2003 and 2012 (Fig. 8B). The soil water content at 5 cm during the dry season also significantly increased ($p<0.001$) between 2003 and 2012 (Fig. 8B). Therefore, we predicted that the annual Bowen ratio will likely decrease in the future, if the effective precipitation frequency and shallow soil water content during the dry season continue to follow an increasing trend similar to that observed over the last 10 years.

Comparison of the Bowen Ratio with Other Forests

Generally, the seasonal variation of Bowen ratio was 'U' shaped with the minimum value occurred in summer, particularly in temperate forests [46,47]. While, the Bowen ratio increased during the summers of 2003, 2007, and 2010 due to severe drought stress in this studied plantation (Fig 3C). Similar to this studied plantation, the Bowen ratio increased in a ponderosa pine forest in the western United States [48] and in a deciduous broadleaved forest in the southern United States [3] due to severe drought stress. However, drought stress had no discernible impact on the Bowen ratio in a tropical rain forest in the Amazon [31]. This discrepancy could be attributed to differences in drought stress conditions and in water use strategies of forests to cope with drought stress [3,31,32,49].

The annual mean Bowen ratio in this studied plantation was 0.35±0.06 during 2003–2012. This value was lower than the reported values of 0.55 in a temperate Douglas-fir [50], 0.74 in a temperate mixed forest [46], 0.81 in a boreal scots pine forest [11], 0.89 in a loblolly pine plantation [47], and was similar to 0.39 in a tropical palm forest [51]. A tropical rainforest in eastern Amazonia experienced a lower annual Bowen ratio (0.17) [31] compared to the studied plantation.

Relationships between Latent Heat Flux and Sensible Heat Flux and Their Effect on the Bowen Ratio

Although the Bowen ratio is the relative change between the sensible heat flux and latent heat flux, the sensitivity of the Bowen ratio to the latent heat flux and sensible heat flux, and affecting factors of these fluxes are diverse [3,11–13]. For example, in a temperate spruce forest [12] and a boreal Scots pine forest [11], the annual Bowen ratio was significantly correlated with the latent heat flux and sensible heat flux. In the temperate spruce forest, both the annual sensible heat flux and Bowen ratio were significantly affected by the annual precipitation amount [12]. Meanwhile, both the annual sensible heat flux and Bowen ratio were influenced by annual net radiation in the boreal Scots pine forest [11]. However, there were no factors that significantly affected the annual latent heat flux in these two forests.

In this studied plantation, the annual Bowen ratio was neither significantly correlated with the latent heat flux ($R^2 = 0.32$, $p = 0.087$) nor correlated with sensible heat flux ($R^2 = 0.20$, $p = 0.193$). The annual net radiation played a key role in affecting both the annual latent heat flux ($R^2 = 0.61$, $p = 0.008$) and sensible heat flux ($R^2 = 0.59$, $p = 0.009$). However, on an annual timescale there were no factors that significantly influenced the annual Bowen ratio. On a monthly timescale, the Bowen ratio was significantly correlated with latent heat flux ($R^2 = 0.49$, $p<0.001$) and sensible heat flux ($R^2 = 0.03$, $p = 0.045$). The monthly latent heat flux and sensible heat flux were mainly affected by net radiation ($R^2 = 0.86$, $p<0.001$) and vapor pressure deficit ($R^2 = 0.36$, $p<0.001$), respectively. However, the monthly Bowen ratio was primary influenced by air temperature ($R^2 = 0.50$, $p<0.001$). Thus, we could not predict the factors affecting the interannual variation of the Bowen ratio by directly inferred from the affecting factors of the latent heat flux or sensible heat flux.

Conclusions

This study demonstrated that the effective precipitation frequency during the dry season, not the precipitation amount, through its positive effect on the soil water content at 5 cm, indirectly affected the annual Bowen ratio during 2003–2012. Meanwhile, the annual Bowen ratio exhibited a marginally significant decreasing trend ($p = 0.061$). Assuming that the effective precipitation frequency and soil water content at 5 cm during the dry season continue to increase ($p<0.001$) due to climate change, as was observed during 2003–2012, the annual Bowen ratio may decrease further. Thus, the warming effect of available energy on the surface air may also decline in this studied plantation. This consequence is consistent with the studies reported by Lee et al. [52] and Wickham et al. [53]. These studies demonstrated that the warming effect of available energy on the forest surface air is weaker in mid and low latitudes than in high latitudes, due to the forest biophysical adjustment mechanisms [52,53]. Further studies of long-term trend of Bowen ratio and its affecting mechanism are necessary to acknowledge the warming effect of available energy on the forest surface air in response to climate change.

Author Contributions

Conceived and designed the experiments: YT XW XS. Performed the experiments: XW XS HW. Analyzed the data: YT XW XS. Contributed reagents/materials/analysis tools: YT XW XS. Wrote the paper: YT XW XS.

References

1. Beringer J, Chapin FS, Thompson CC, McGuire AD (2005) Surface energy exchanges along a tundra-forest transition and feedbacks to climate. Agric For Meteorol 131: 143–161.
2. Matsumoto K, Ohta T, Nakai T, Kuwada T, Daikoku K, et al. (2008) Energy consumption and evapotranspiration at several boreal and temperate forests in the Far East. Agric For Meteorol 148: 1978–1989.
3. Wilson KB, Baldocchi DD (2000) Seasonal and interannual variability of energy fluxes over a broadleaved temperate deciduous forest in North America. Agric For Meteorol 100: 1–18.
4. Baldocchi DD, Kelliher HM, Black TA, Jarvis P (2000) Climate and vegetation controls on boreal zone energy exchange. Global Change Biol 6: 69–83.
5. Li ZH, Zhang YP, Wang SS, Yuan GF, Yang Y, et al. (2010) Evapotranspiration of a tropical rain forest in Xishuangbanna, southwest China. Hydrol Process 24: 2405–2416.
6. FAO (Food and Agriculture Organization) (2010) Global Forest Resources Assessment. FAO, Rome, Italy. 333 p.
7. Wilson KB, Baldocchi DD, Aubinet M, Berbigier P, Bernhofer C, et al. (2002) Energy partitioning between latent and sensible heat flux during the warm season at FLUXNET sites. Water Resour Res 38: 30-1-30-11.
8. Richardson AD, Keenan TF, Migliavacca M, Ryu Y, Sonnentag O, et al. (2013) Climate change, phenology, and phenological control of vegetation feedbacks to the climate system. Agric For Meteorol 169: 156–173.
9. Lamarque JF, Shindell DT, Josse B, Young PJ, Cionni I, et al. (2013) The Atmospheric Chemistry and Climate Model Intercomparison Project (AC-CMIP): overview and description of models, simulations and climate diagnostics. Geosci Model Dev 6: 179–206.
10. Keenan TF, Hollinger DY, Bohrer G, Dragoni D, Munger JW, et al. (2013) Increase in forest water-use efficiency as atmospheric carbon dioxide concentrations rise. Nature 499: 324–327.
11. Launiainen S (2010) Seasonal and inter-annual variability of energy exchange above a boreal Scots pine forest. Biogeosciences 7: 3921–3940.
12. Grünwald T, Bernhofer C (2007) A decade of carbon, water and energy flux measurements of an old spruce forest at the Anchor Station Tharandt. Tellus Ser B 59: 387–396.
13. Kosugi Y, Takanashi S, Tanaka H, Ohkubo S, Tani M, et al. (2007) Evapotranspiration over a Japanese cypress forest. I. Eddy covariance fluxes and surface conductance characteristics for 3 years. J Hydro 337: 269–283.
14. Black TA, Chen WJ, Barr AG, Arain MA, Hogg EH, et al. (2000) Increased carbon sequestration by a boreal deciduous forest in years with a warm spring. Geophys Res Lett 27: 1271–1274.
15. Law B, Williams M, Anthoni P, Baldocchi DD, Unsworth MH (2000) Measuring and modelling seasonal variation of carbon dioxide and water vapour exchange of a Pinus ponderosa forest subject to soil water deficit. Global Change Biol 6: 613–630.
16. Ciais P, Reichstein M, Viovy N, Granier A, Ogee J, et al. (2005) Europe-wide reduction in primary productivity caused by the heat and drought in 2003. Nature 437: 529–533.
17. Ross I, Misson L, Rambal S, Arneth A, Scott RL, et al. (2012) How do variations in the temporal distribution of rainfall events affect ecosystem fluxes in seasonally water-limited Northern Hemisphere shrublands and forests? Biogeosciences 9: 1007–1024.
18. Wang YD, Wang ZL, Wang HM, Guo CC, Bao WK (2012) Rainfall pulse primarily drives litter fall respiration and its contribution to soil respiration in a young exotic pine plantation in subtropical China. Can J For Res 42: 657–666.
19. IPCC (Intergovernmental Panel on Climate Change) (2007) Climate change 2007: The Physical Science Basis. Solomon, S., Qin, D., Manning, M, et al. eds. Contribution of Working Group I to the Fourth Assessment Report of the Intergovernmental Panel on Climate Change. Cambridge, UK and New York, USA: Cambridge University Press. 996 p.
20. Wen XF, Wang HM, Wang JL, Yu GR, Sun XM (2010) Ecosystem carbon exchanges of a subtropical evergreen coniferous plantation subjected to seasonal drought, 2003–2007. Biogeosciences 7: 357–369.
21. Wang YD, Li QK, Wang HM, Wen XF, Yang FT, et al. (2011) Precipitation frequency controls interannual variation of soil respiration by affecting soil moisture in a subtropical forest plantation. Can J For Res 41: 1897–1906.
22. Saigusa N, Ichii K, Murakami H, Hirata R, Asanuma J, et al. (2010) Impact of meteorological anomalies in the 2003 summer on Gross Primary Productivity in East Asia. Biogeosciences 7: 641–655.
23. Zhang WJ, Wang HM, Yang FT, Yi YH, Wen XF, et al. (2011) Underestimated effects of low temperature during early growing season on carbon sequestration of a subtropical coniferous plantation. Biogeosciences 8: 1667–1678.
24. Mccaughey JH (1985) Energy-balance storage terms in a mature mixed forest at Petawawa, Ontario - a Case-Study. Bound-Layer Meteorol 31: 89–101.
25. Oliphant AJ, Grimmond CSB, Zutter HN, Schmid HP, Su HB, et al. (2004) Heat storage and energy balance fluxes for a temperate deciduous forest. Agric For Meteorol 126: 185–201.
26. Wilczak JM, Oncley SP, Stage SA (2001) Sonic anemometer tilt correction algorithms. Bound-Layer Meteorol 99: 127–150.
27. Webb EK, Pearman GL, Leuning R (1980) Correction of flux measurements for density effects due to heat and water vapour transfer. Q J Roy Meteor Soc 106: 85–100.
28. Reichstein M, Falge E, Baldocchi DD, Papale D, Aubinet M, et al. (2005) On the separation of net ecosystem exchange into assimilation and ecosystem respiration: review and improved algorithm. Global Change Biol 11: 1424–1439.
29. Falge E, Baldocchi DD, Olson R, Anthoni P, Aubinet M, et al. (2001) Gap filling strategies for long term energy flux data sets. Agric For Meteorol 107: 71–77.
30. Wilson KB, Goldstein A, Falge E, Aubinet M, Baldocchi DD, et al. (2002) Energy balance closure at FLUXNET sites. Agric For Meteorol 113: 223–243.
31. da Rocha HR, Goulden ML, Miller SD, Menton MC, Pinto L, et al. (2004) Seasonality of water and heat fluxes over a tropical forest in eastern Amazonia. Ecological Applications 14: 22–32.
32. Tang YK, Wen XF, Sun XM, Zhang XY, Wang HM (2014) The limiting effect of deep soil water on evapotranspiration of a subtropical coniferous plantation subjected to seasonal drought. Adv Atmos Sci 31: In press.
33. McNaughton KG, Spriggs TW (1986) A mixed-layer model for regional evaporation. Bound-Layer Meteorol 34: 243–262.
34. Ryu Y, Baldocchi DD, Ma S, Hehn T (2008) Interannual variability of evapotranspiration and energy exchange over an annual grassland in California. J Geophys Res 113: doi: 10.1029/2007JD009263.
35. Priestley CHB, Taylor RJ (1972) On the assessment of surface heat flux and evaporation using large-scale parameters. Mon Weather Rev 100: 81–92.
36. Budyko MI (1974) Climate and Life. New York, USA: Academic Press. 508 p.
37. Jassal RS, Black TA, Spittlehouse DL, Brümmer C, Nesic Z (2009) Evapotranspiration and water use efficiency in different-aged Pacific Northwest Douglas-fir stands. Agric For Meteorol 149: 1168–1178.
38. Huete A, Didan K, Miura T, Rodriguez EP, Gao X, et al. (2002) Overview of the radiometric and biophysical performance of the MODIS vegetation indices. Remote Sens Environ 83: 195–213.
39. Wu CY, Chen JM, Huang N (2011) Predicting gross primary production from the enhanced vegetation index and photosynthetically active radiation: Evaluation and calibration. Remote Sens Environ 115: 3424–3435.
40. Churkina G, Schimel D, Braswell B, Xiao X (2005) Spatial analysis of growing season length control over net ecosystem exchange. Global Change Biol 11: 1777–1787.
41. Baldocchi DD, Falge E, Wilson KB (2001) A spectral analysis of biosphere-atmosphere trace gas flux densities and meteorological variables across hour to multi-year time scales. Agric For Meteorol 107: 1–27.
42. Fisher JB, Malhi Y, Bonal D, da Rocha H, de Araujo S, et al. (2009) The land–atmosphere water flux in the tropics. Global Change Biol 15: 2694–2714.
43. Borken W, Savage K, Davidson EA, Trumbore SE (2006) Effects of experimental drought on soil respiration and radiocarbon efflux from a temperate forest soil. Global Change Biol 12: 177–193.
44. Epron D, Ngao J, Granier A (2004) Interannual variation of soil respiration in a beech forest ecosystem over a six-year study. Ann For Sci 61: 499–505.
45. Concilio A, Chen J, Ma S, North M (2009) Precipitation drives interannual variation in summer soil respiration in a Mediterranean-climate, mixed-conifer forest. Climatic Change 92: 109–122.
46. Wu JB, Guan DX, Han SJ, Shi TT, Jin CJ, et al. (2007) Energy budget above a temperate mixed forest in northeastern China. Hydrol Process 21: 2425–2434.
47. Sun G, Noormets A, Gavazzi MJ, McNulty SG, Chen J, et al. (2010) Energy and water balance of two contrasting loblolly pine plantations on the lower coastal plain of North Carolina, USA. Forest Ecol and Manag 259: 1299–1310.
48. Goldstein AH, Hultman NE, Fracheboud JM (2000) Effects of climate variability on the carbon dioxide, water, and sensible heat fluxes above a ponderosa pine plantation in the Sierra Nevada (CA). Agr Forest Meteorol 101: 113–129.

49. Bracho R, Powell TL, Dore S, Li J, Hinkle CR, et al. (2008) Environmental and biological controls on water and energy exchange in Florida scrub oak and pine flatwoods ecosystems. J Geophys Res 113: doi: 10.1029/2007JG000469.
50. Humphreys ER, Black TA, Ethier GJ, Drewitt GB, Spittlehouse DL, et al. (2003) Annual and seasonal variability of sensible and latent heat fluxes above a coastal Douglas-fir forest, British Columbia, Canada. Agric For Meteorol 115: 109–125.
51. Roupsard O, Bonnefond JM, Irvine M, Berbigier P, Nouvellon Y, et al. (2006) Partitioning energy and evapo-transpiration above and below a tropical palm canopy. Agric For Meteorol 139: 252–268.
52. Lee XH, Goulden ML, Hollinger DY, Barr A, Black TA, et al. (2011) Observed increase in local cooling effect of deforestation at higher latitudes. Nature 479: 384–387.
53. Wickham JD, Wade TG, Riitters KH (2013) Empirical analysis of the influence of forest extent on annual and seasonal surface temperatures for the continental United States. Glob Ecol and Biogeogr 22: 620–629.

Rates of Litter Decomposition and Soil Respiration in Relation to Soil Temperature and Water in Different-Aged *Pinus massoniana* Forests in the Three Gorges Reservoir Area, China

Wenfa Xiao[1*⁹], Xiaogai Ge[1,2*⁹], Lixiong Zeng[1], Zhilin Huang[1], Jingpin Lei[3], Benzhi Zhou[2], Maihe Li[4,5]

1 State Forestry Administration Key Laboratory of Forest Ecology and Environment, Research Institute of Forest Ecology, Environment and Protection, Chinese Academy of Forestry, Beijing, PR China, **2** Research Institute of Subtropical Forestry, Chinese Academy of Forestry, Fuyang, Zhejiang, PR China, **3** State Forestry Administration Key Laboratory of Forest Silviculture, Research Institute of Forestry, Chinese Academy of Forestry, Beijing, PR China, **4** Swiss Federal Research Institute WSL, Birmensdorf, Switzerland, **5** State Key Laboratory of Forest and Soil Ecology, Institute of Applied Ecology, Chinese Academy of Sciences, Shenyang, PR China

Abstract

To better understand the soil carbon dynamics and cycling in terrestrial ecosystems in response to environmental changes, we studied soil respiration, litter decomposition, and their relations to soil temperature and soil water content for 18-months (Aug. 2010–Jan. 2012) in three different-aged *Pinus massoniana* forests in the Three Gorges Reservoir Area, China. Across the experimental period, the mean total soil respiration and litter respiration were 1.94 and 0.81, 2.00 and 0.60, 2.19 and 0.71 μmol CO_2 m^{-2} s^{-1}, and the litter dry mass remaining was 57.6%, 56.2% and 61.3% in the 20-, 30-, and 46-year-old forests, respectively. We found that the temporal variations of soil respiration and litter decomposition rates can be well explained by soil temperature at 5 cm depth. Both the total soil respiration and litter respiration were significantly positively correlated with the litter decomposition rates. The mean contribution of the litter respiration to the total soil respiration was 31.0%–45.9% for the three different-aged forests. The present study found that the total soil respiration was not significantly affected by forest age when *P. masonniana* stands exceed a certain age (e.g. >20 years old), but it increased significantly with increased soil temperature. Hence, forest management strategies need to protect the understory vegetation to limit soil warming, in order to reduce the CO_2 emission under the currently rapid global warming. The contribution of litter decomposition to the total soil respiration varies across spatial and temporal scales. This indicates the need for separate consideration of soil and litter respiration when assessing the climate impacts on forest carbon cycling.

Editor: Dafeng Hui, Tennessee State University, United States of America

Funding: This work was supported by the Special Research Programme In Forestry Industry, State Forestry Administration of China (Project No. 201104008), National 'Twelfth Five-Year" Plan for Science & Technology Project of China (2011BAD38B04), Zigui National Forest Ecological Research Station and CFERN&GENE Award Funds on Ecological Paper. The funders had no role in study design, data collection and analysis, decision to publish, or preparation of the manuscript.

Competing Interests: This study was partly supported by CFERN&GENE Award Funds. There are no patents, products in development or marketed products to declare. This does not alter the authors' adherence to all the PLOS ONE policies on sharing data and materials, as detailed online in the guide for authors.

* Email: xiaowenf@caf.ac.cn (WX); gexiaogai2006@163.com (XG)

⁹ These authors contributed equally to this work.

Introduction

Soil respiration is a major process controlling carbon (C) loss from terrestrial ecosystems [1]. Globally, soil respiration releases approximately 80 Pg C into the atmosphere per year [2], which is estimated to account for 20–38% of the total annual biogenic CO_2 emissions to the atmosphere [3]. Given the predicted increase in atmospheric CO_2 concentrations, litter as a main C source for the total soil CO_2 efflux will increase [4]. Litterfall represents a major flux of the vegetative C to soil, and hence, changes in litter inputs are likely to have wide-reaching consequences for soil C dynamics [5]. Decomposition plays an integral role in determining soil C entering atmosphere [6,7], and the fate of C contained in litter, therefore, plays an important role in the long-term C sequestration in forest soils [8,9]. The total respiration together with litter inputs data could be used to evaluate C dynamics in soils [10,11]. To better understand and predict the soil C dynamics and cycling in terrestrial ecosystems in a changing world, studies of the relationships between soil respiration and litter decomposition in response to environmental changes are still required.

The aboveground plant litter fluxes have been found to be strongly correlated with soil respiration, and litterfall manipulations have strong effects on soil CO_2 efflux [12]. Several studies have investigated the contribution of the leaf litter layer to soil respiration [4,13,14,15]. Raich [2] showed that soil C flux from respiration was 2.8–3.0 times higher than the C flux from the aboveground litter production. Buchmann [16] found that the annual C loss through soil respiration (710 g C m^{-2} yr^{-1}) was higher than the C input of the annual aboveground litterfall (240 g C m^{-2} yr^{-1}). Prévost-Bouré et al. [4] indicated that addition of fresh litter significantly increased the total soil CO_2 efflux. Sayer et al. [5] indicated that soil respiration was on average 20% lower in the litter removal and 43% higher in the litter addition treatment compared to the controls.

On a global scale, soil respiration is mainly controlled by temperature and precipitation [17]. Many studies have shown that soil temperature is the primary factor determining the rates of soil respiration [18] and litter decomposition, and thus, soil warming increased the decomposition rate of fine woody debris in temperate forests [19]. However, Yuste et al.[17] found that soil respiration was insensitive to temperature in temperate maritime forests during late spring and summer when soil water content was limited. Fierer et al. [20] showed that the temperature sensitivity of litter decomposition varied with litter type and the extent of decomposition. These results indicate that soil respiration and litter decomposition depend on climatic variables such as soil temperature and precipitation [21]. It is thus necessary to see whether the temperature sensitivity of soil respiration and litter decomposition keeps consistent in a changing world.

Pinus massoniana Lamb. is a native species with high adaptability and tolerance to drought and low fertility soil but with low productivity in the Chinese subtropical regions. *P. massoniana* forests cover an area of ~2 million square kilometers in China, and are one of the most important vegetation types with key ecological importance in the Three Gorges Reservoir Area [22]. However, the relationships between soil respiration and litter decomposition in *P. massoniana* forests have not yet been studied. Based on long-term litter decomposition studies, Aerts and de Caluwe [23] concluded that the initial litter respiration rates are reliable indicators for long-term litter decomposability. The present study aimed to investigate the linkage of soil respiration and litter decomposition in different-aged *P. massoniana* forests, to provide basic knowledge for assessing the forest carbon budget and dynamics in those forests. We hypothesized that 1) the dynamics of soil respiration are reliable indicators for litter decomposition rate, 2) the soil respiration and litter decomposition have similar sensitivity to temperature, and 3) the co-variant effects of temperature and soil water contents on both litter decomposition and soil respiration are significant in different-aged *P. massoniana* forests.

Materials and Methods

Ethics Statement

This work was approved by Zigui National Forest Ecological Research Station. Our study related to litter decomposition and soil respiration did not involve endangered or protected species, and did not damage or destroy the vegetation and animals, although the study sites were located within a protected area for wildlife in the Three Gorges Reservoir Area, China.

Study site and forests

The study sites were located in Zigui county (110°00′14″–111°18′41″E, 30°38′14″–31°11′31″N), Hubei province, China. The region has a subtropical monsoon climate with mean annual temperature of 17–19°C and mean annual precipitation of 1000–1250 mm mainly occurring between April and September. The forests studied are *P. massoniana* pure stands on Haplic Luvisol soil [24], with *Camellia oleifera, Loropetalum chinensis, Cotinus coggygria* and *Echinochloa crusgalli, Veronicastrum villosulum* in the understorey. The characteristics of the forests/sites were summarized in Table 1. Table 1 showed that the 30-year-old forests had lower mean values of soil organic matter and total nitrogen contents compared to the other two forests.

The experiment included 20-, 30-, and 46-year-old *P. massoniana* forests each with three spatially separated stands ($n = 3$) at similar elevations (±50 m). Each stand has an area of more than 1 ha. In the center of each stand, a 20 m×30 m plot was established in June 2010. The rates of litter decomposition and soil respiration were determined in each plot.

Soil respiration, soil temperature and soil water content

Six polyvinyl chloride (PVC) collars (19.6 cm in inner diameter and 10 cm in height) were randomly installed to a soil depth of 5 cm for soil respiration sampling within each plot in July 2010. The litters were kept intact in three of these six collars, and were removed in the other three randomly selected collars ($n = 3$). The first measurement was conducted in August 2010, i.e. one month after the collar establishment. Soil respiration was measured using Li-8100 portable soil CO_2 flux system (LI-COR Inc., Lincoln, NE, USA) [25]. All soil respiration measurements were made between 08:00 hours and 11:00 hours (local time), to avoid bias of diurnal changes, according to our pilot tests (data not shown). Each measurement usually takes about 12 min. All the PVC collars were installed permanently throughout the experimental period. The soil respiration was measured monthly from August 2010 to January 2012 (i.e. a total of 18 times of repeated measures).

Soil temperature (T_5) and soil moisture at 5 cm soil depth were recorded adjacent to each soil respiration collar. Soil temperature was measured with a soil temperature probe. Volumetric soil water content (SWC) was measured using a theta probe (EM50, Decagon, USA), which was calibrated to the soil type in the plots following the procedure described by Delta-T (Luan et al. 2011).

Litter decomposition experiment

Leaf litter decomposition was studied using the standard litterbag technique [26]. In each stand, freshly fallen leaf litters were collected by hand and air-dried in May 2010, and then well-mixed for the decomposition experiment. Sub-samples of air-dried leaf litters were oven-dried at 70°C to constant weight for calculating the conversion coefficient between air-dried and oven-dried leaf. The initial litter quality was analyzed and summarized in Table 2. Table 2 showed that the leaf litter of the 30-year-old stands had higher C/N and lignin/N ratios than the leaf litters of the other two stands.

Twenty grams of air-dried leaf litters were placed in a nylon-mesh bag with 20 cm×20 cm in size and 1-mm mesh. We placed 36 litter-filled bags in each plot in June 2010, and a total of 324 decomposition bags (3 forests ×3 stands ×36 bags/stand) were placed. Litterbags lying flat on the litter layer surface were fastened using steel screens. Freshly fallen litters above the litterbags were removed monthly. Six bags in each stand were retrieved at 90 d (2nd November, 2010), 180 d (2nd February, 2011), 270 d (2nd May, 2011), 360 d (2nd August, 2011), 450 d (2nd November, 2011) and 540 d (31st January, 2012) of field decomposition, respectively. Each litterbag was brushed to remove the external soils and litters, and then the remaining litter was placed in a paper bag and oven-dried at 70°C to constant weight. Litter water content (LWC) was also calculated for each sample.

Data analysis

The total soil respiration rate was defined as the sum of the litter respiration plus the litter-free soil respiration. According to Sayer et al. [5], soil and litter respiration was calculated using the equation (1)

$$R_{\mathrm{L}} = R_{\mathrm{S}} - R_{(\mathrm{L-})} \quad (1)$$

where R_L is the litter respiration, $R_{(L-)}$ is the litter-free soil respiration (i.e. litter in collar removed), and Rs is the total soil respiration. Each R was calculated using the equation (2) [3]:

Table 1. Summary of stand and site characteristics measured in summer 2010 (mean ±1 SD; $n=3$).

	Forests		
	20-year-old	**30-year-old**	**46-year-old**
Stand characteristics			
Elevation (m)	950	350	990
DBH (cm)	16.1±1.15	25.3±0.93	33.1±1.09
Height (m)	15.6–19.0	17.5–21.6	18.4–24.8
Stand density (ha^{-1})	800	710	575
Litter layer depth (cm)	4.03±1.59	5.77±0.85	6.34±1.99
Litter litterfall ($t \cdot ha^{-1} \cdot a^{-1}$)	3.38±0.07	4.69±0.20	5.60±0.23
Litter standing crop ($t \cdot ha^{-1}$)	9.35±5.14	9.26±2.97	14.05±6.40
Soil characteristics			
Soil depth (cm)	70-100	60-80	70-100
Soil organic matter ($g \cdot kg^{-1}$)	51.90±8.38	19.87±6.78	69.62±11.57
Total nitrogen ($g \cdot kg^{-1}$)	2.36±0.58	1.24±0.18	3.33±0.70
Soil bulk density ($g \cdot cm^{-3}$)	1.13±0.14	1.53±0.08	1.08±0.14
pH value	5.12±0.33	5.43±0.09	4.50±0.15

Note DBH = mean diameter at breast height.

$$R=\alpha\, e^{\beta T} \quad (2)$$

where R is the soil or litter respiration, T is the temperature at 5 cm soil depth (T_5), α is the respiration rate at $T=0°C$, and β is a fitted temperature-response coefficient, according to Rey [27].

The temperature sensitivity (Q_{10}) of soil respiration (equation 3) [28] and litter mass loss (equation 4) [29] was calculated using the following equations:

$$\text{Soilrespiration}\,Q_{10}=e^{10\beta} \quad (3)$$

$$\text{Litterdecomposition}\,Q_{10}=(k_2/k_1)^{[10/(T2-T1)]} \quad (4)$$

where β is a fitted parameter, k_1 and k_2 are the rate constants for a process of interest at two observed temperatures T_1 and T_2.

The co-variant effects of soil temperature and soil moisture on soil respiration were fitted using equation (5), according to Saiz et al. [5].

$$Rs=(ae^{bT})(cSWC+dSWC^2) \quad (5)$$

where Rs is the total soil respiration, T is the soil temperature at 5 cm depth, SWC is the soil water content at 5 cm soil depth, a, b, c, d are fitted parameters.

The decomposition rates were estimated using a single exponential decay model (6) [30]:

$$M_t=M_0\, e^{kt} \quad (6)$$

where M_t is the litter dry mass at time t, M_0 is the initial litter mass, t is the sampling time interval, and k is the annual decay constant.

Repeated measures ANOVAs were used to test the effects of forest age (between subject), time (within subject), and their interaction on soil respiration, litter respiration, litter-free soil respiration, and litter decomposition rate. One-way and Univariate ANOVAs were applied to compare the litter leaf loss rate followed by Tukey's test. Exponential regression analyses were used to examine the relationships between environmental factors (soil temperature and soil water content) and respirations or litter decomposition rates, as well as the relationships between soil respirations and litter decomposition rates. A log-transformed

Table 2. Initial substrate quality (mean ±1SD, $n=3$) of leaf litter in *Pinus massoniana* forests studied.

Initial litter quality	**20-yr-old forests**	**30-yr-old forests**	**46-yr-old forests**
C %	56.25±0.10a	56.54±0.20a	57.21±2.45a
N %	0.98±0.40a	0.74±0.09a	0.85±0.15a
C/N ratio	62.81±20.56b	76.94±9.56a	68.17±14.45ab
Lignin %	34.27±1.31a	34.31±0.42a	34.24±0.16a
Lignin/N ratio	34.89±3.19b	46.63±5.23a	33.49±9.96b

Note: different letters within a row indicate significant difference ($P<0.05$) among the stands.

Table 3. Effects of forest age and sampling time on the total soil respiration, litter-free soil respiration, and litter respiration in *Pinus massoniana* forests tested using repeated measures ANOVAs.

	Between subject			Within subject					
	Forests			Time			Forest × Time		
	df	F	*P*	df	F	*P*	df	F	*P*
Across the experimental period									
Total soil respiration	2	0.323	0.725	17	88.023	0.00	34	4.2	0.00
Litter-free soil respiration	2	0.782	0.463	17	10.401	0.00	34	2.985	0.00
Litter respiration	2	1.114	0.336	17	120.884	0.00	34	3.873	0.00
April - October									
Total soil respiration	2	1.215	0.312	9	28.301	0.00	18	2.563	0.001
Litter-free soil respiration	2	0.669	0.521	9	42.848	0.00	18	2.719	0.000
Litter respiration	2	5.986	0.007	9	3.826	0.00	18	1.842	0.022
November - March									
Total soil respiration	2	3.355	0.050	7	25.834	0.00	14	6.190	0.00
Litter-free soil respiration	2	7.855	0.003	7	12.238	0.00	14	6.043	0.00
Litter respiration	2	0.541	0.59	7	14.635	0.00	14	7.196	0.00

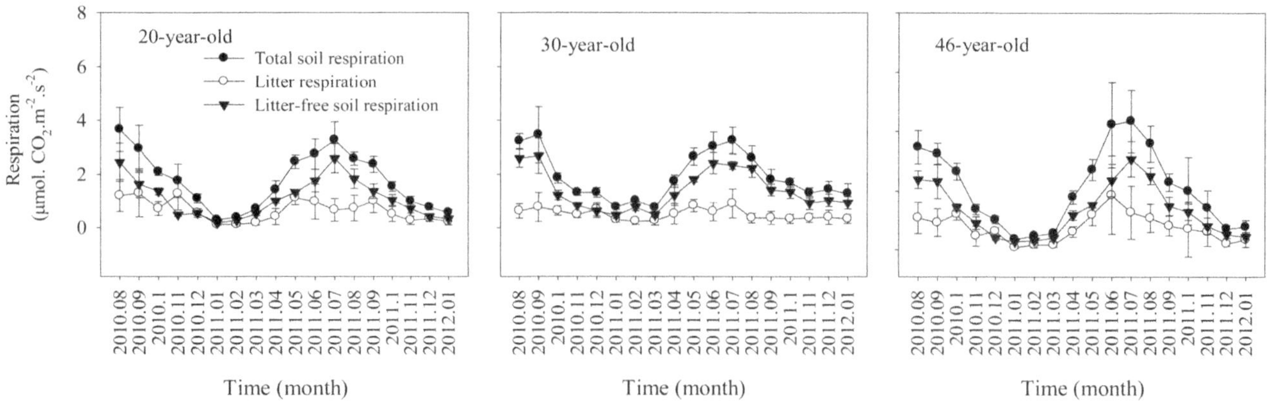

Figure 1. Seasonal patterns and monthly dynamics of the total soil respiration (•), litter respiration (○), and the litter-free soil respiration (▼) in different-aged *Pinus massoniana* forests (mean ±1SD, *n*=9).

model was applied to calculate the annual decay rate constants (*k*) from litter dry mass remaining. All statistical analyses were performed using SPSS 16.0 software package for windows (SPSS Inc., Chicago, USA).

Results

Dynamics of soil and litter respiration

The litter respiration, litter-free soil respiration, and the total soil respiration rates did not vary with forest age across the experimental period (Table 3). Values of all those parameters changed significantly with time, showing obvious seasonality. The forest age effects on litter and soil respiration changed with time, showing significant stand × time interactions (Table 3). The litter respiration had the lowest value in January, and peaked in May/June, whereas the total soil respiration and the litter-free soil respiration had lower rates in winter and higher rates in summer, with the highest rates in August and lowest values in January (Fig. 1). Thus, the highest value of litter respiration occurred nearly one month earlier than the peak values of the total soil respiration and the litter-free soil respiration (Fig. 1).

The mean annual litter-free soil respiration was 1.47±0.92 (mean±1SD), 1.58±0.83, and 1.68±0.99 μmol CO_2 m^{-2} s^{-1} in the 20-, 30 and 46-year-old stands, respectively (Fig. 1). The mean litter respiration was 0.81±0.66, 0.60±0.66, and 0.71±0.77 μmol CO_2 m^{-2} s^{-1} in the 20-, 30-, and 46-year-old stands, respectively (Fig. 1). The litter respiration rates were always lower than the litter-free soil respiration rates, except for some cases in November and December (Fig. 1).

The total soil respiration tended to be somewhat higher ($P>0.05$) in the 46-year-old stands than in the 20- and 30-year-old stands in summer, but it was significantly higher ($P=0.05$) in the 30-year-old stands than in the other two stands in winter (Table 3, Fig. 1). The total soil respiration rates increased from spring to summer and reached the maximum values of 3–4 μmol CO_2 m^{-2} s^{-1} in July/August (Fig. 1). The mean total soil respiration across the experimental period was 1.94±1.28, 2.00±1.0, and 2.19±1.51 μmol CO_2 m^{-2} s^{-1} in the 20-, 30-, and 46-year-old stands, respectively (Fig. 1).

Litter decomposition and its relationship to soil respiration

The leaf litter decomposition rates did not statistically differ among the different-aged stands (data not shown) and followed a similar pattern through the whole decomposition time (Fig. 2). After 540 days' decomposition in the field, the remaining litter ranged from 57.6% in the 20-year-old stands, to 56.1% in the 30-year-old stands, and 61.3% in the 46-year-old stands compared to their initial mass (Fig. 2). The litter decomposition rates were relatively slow and constant during the first six month, and then increased with the onset of the rainy season. The decomposition rates were lower during the dry period from March to June (180–270 d) than those during the hot season in summer.

An exponential model of decomposition was fitted to the experimental data (Fig. 2). The annual decomposition rate coefficients (*k*) were 0.29 (20-year-old stands), 0.33 (30-year-old stands), and 0.30 (46-year-old stands). The predicated remaining litter was 64.9%, 61.1%, and 64.5% in the 20-, 30-, and 46-year-old stands after 540 days of field decomposition, respectively. The predicted data of litter mass remaining slightly overestimated the measured data (Fig. 2). The unexpected higher values of litter mass remaining measured at 270 d in the 30- and 46-year-old stands may be caused by bias with respect to uncontrolled field experimental conditions and to lower decomposition rate during the dry period from March to May (e.g. 180–270 days) (Fig. 2).

The total soil respiration, litter respiration, and litter-free soil respiration in relation to litter decomposition had similar patterns for the three stands, showing a quadratic function ($R^2=0.37$, 0.45, 0.30, and $P=0.03$, 0.01, 0.07, respectively; see Fig. 3). The mean percentage contribution of litter respiration to the total soil respiration was 45.9% (ranging from 19.1 to 72.6% across the experimental period) in the 20-year-old stands, 31.0% (14.3–47.6%) in the 30-year-old stands, and 38.3% (15.6–61.0%) in the 46-year-old stands. The contribution of litter respiration to the total soil respiration was larger in winter than in summer.

Soil respiration and litter decomposition rate in relation to soil temperature and water content

The annual mean soil temperature (T_5) was 12.64°C (ranging from 2.40°C in January to 23.60°C in July) in the 20-year-old stands, 15.08°C (2.80 to 25.80°C) in the 30-year-old stands, and 12.83°C (2.70 to 25.53°C) in the 46-year-old stands (Fig. 4). The annual mean soil water content and litter water content (LWC) were 19.70% and 20.86%, 16.28% and 16.70%, 20.29% and 20.36% in the 20-, 30-, and 46-year-old stands, respectively (Fig. 4).

There were significant exponential relationships between soil temperature (T_5) and the total soil respiration ($R^2=0.80$, $P<0.001$; Fig. 5a), litter respiration ($R^2=0.53$, $P<0.001$; Fig. 5c), or litter-

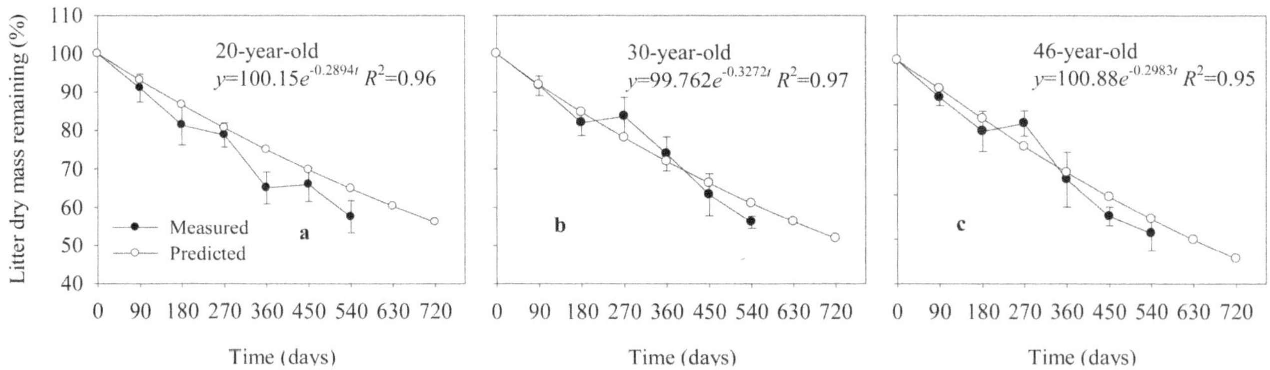

Figure 2. Observed (•) and predicted (○) values of litter dry mass remaining across the experimental period in different-aged *Pinus massoniana* forests (mean ±1SD, *n*=9).

free soil respiration ($R^2=0.83$, $P<0.001$; Fig. 5e). The corresponding Q_{10} was 2.25 for the total soil respiration, 2.10 for the litter respiration, and 2.36 for the litter-free soil respiration. No clear relationships between respirations and soil water contents were found (Fig.5b, d, f).

The co-variant effects of soil temperature and moisture on soil respiration were significant for the three stands ($R^2=0.70$, 0.42, 0.69, all $P<0.001$; Table 4). There were no combined effects of soil temperature and soil or litter water contents on litter respiration (Table 4).

Soil moisture was found to be marginally significantly linearly correlated with litter decomposition ($R^2=0.043$, $P=0.051$; Fig. 6b). The litter decomposition was significantly quadratically correlated with soil temperature ($R^2=0.58$, $P<0.001$; Fig. 6a). The temperature sensitivity (Q_{10} value) of litter mass loss was 1.5.

Discussion

Effects of forest age on soil and litter respiration

Ewel et al. [31] suggested that soil respiration decreased with forest age in temperate forests, while it increased with forest age in tropical and subtropical forests. The present study found that forest age affected the litter respiration during the growing season and influenced the litter-free soil respiration and the total soil respiration during the dormancy period (Table 3). Our results suggest that the effects of forest age on soil respiration may be mainly caused by stand roots in terms of root quantity (roots biomass) and quality (fine roots fraction) which increase with increasing forest age [32]. Previous studies indicated that forest age had a significantly positive influence on soil respiration in *Pinus taeda* plantations [33] and in warm-temperate oak forests [32]. Tang et al. [34] and Bolstad et al. [35] found that soil respiration first increased and then decreased with increasing forest age. Saiz et al. [36] found that soil respiration decreased with forest age during the younger stages of Sitka spruce forests. These results indicate that the effects of forest age on soil respiration differ with tree species and stand structure probably associated with the amount of fine roots and the quality of soil C pools [36].

Our results showed that the forest age effect on litter and soil respiration was nonlinear. For example, the mean litter respiration rate was 0.66, 0.54, and 0.74 μmol CO_2 m^{-2} s^{-1} for the 20-, 30-, and 46-year-old stands, respectively. This lower litter respiration in the 30-year-old stands seemed to be a result of the lower soil organic matter and nitrogen contents (Table 1), which may have led to decreases in soil microbial activities [37]. The lower litter respiration in the 30-year-old stands may also be resulted from the lower initial litter quality compared to that in the other two stands (Table 2) because litter quality directly affects litter decomposition rates [1]. Hättenschwiler and Gasser [38] showed that the decomposition rate was significantly positively correlated with litter N concentration ($R^2=0.75$, $P=0.025$) but negatively correlated with litter C/N ratio ($R^2=0.72$, $P=0.032$) and

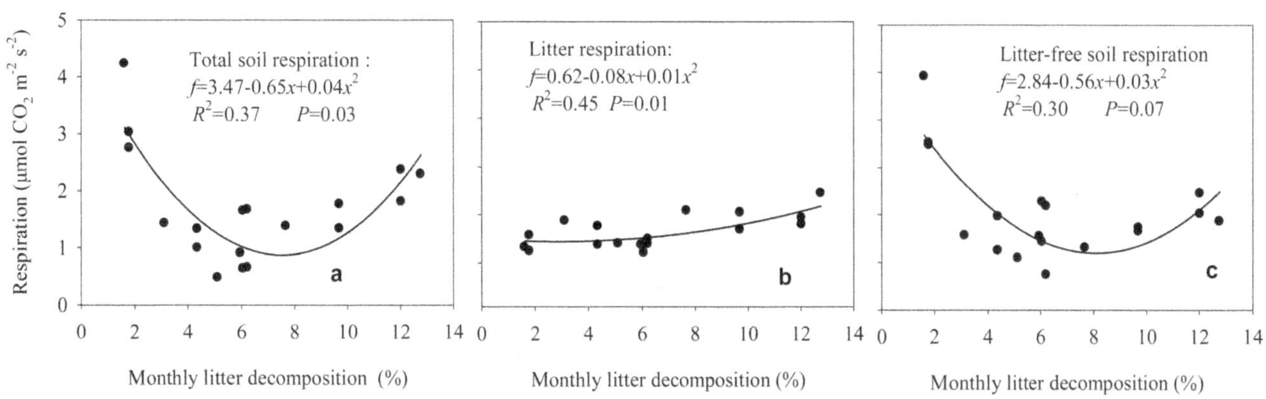

Figure 3. Relationships between mean monthly litter decomposition and the total soil respiration (a), litter respiration (b), and litter-free soil respiration rate (c) for different-aged *Pinus massoniana* forests. Note: respiration rate is the mean value across the litter decomposition period.

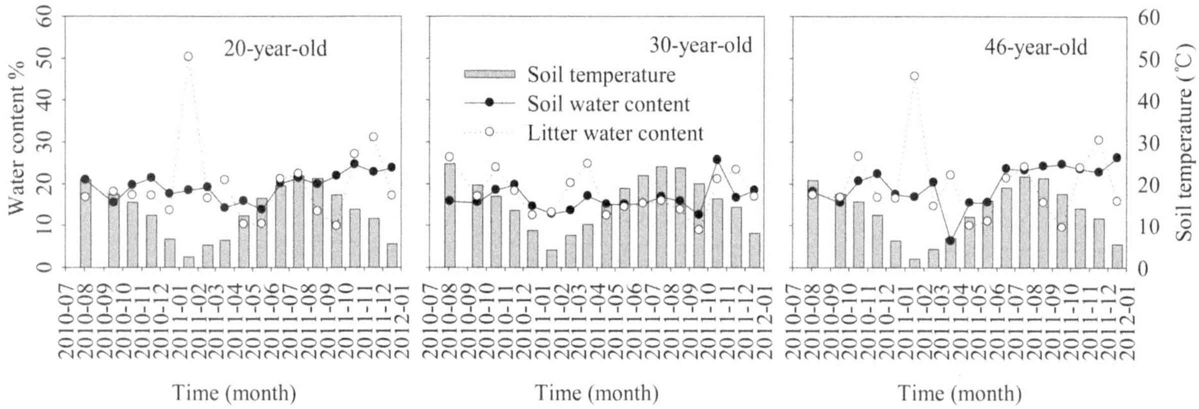

Figure 4. Soil temperature (5 cm soil depth), soil water content (5 cm soil depth) and litter water content in different-aged *Pinus massoniana* forests studied across the experiment period.

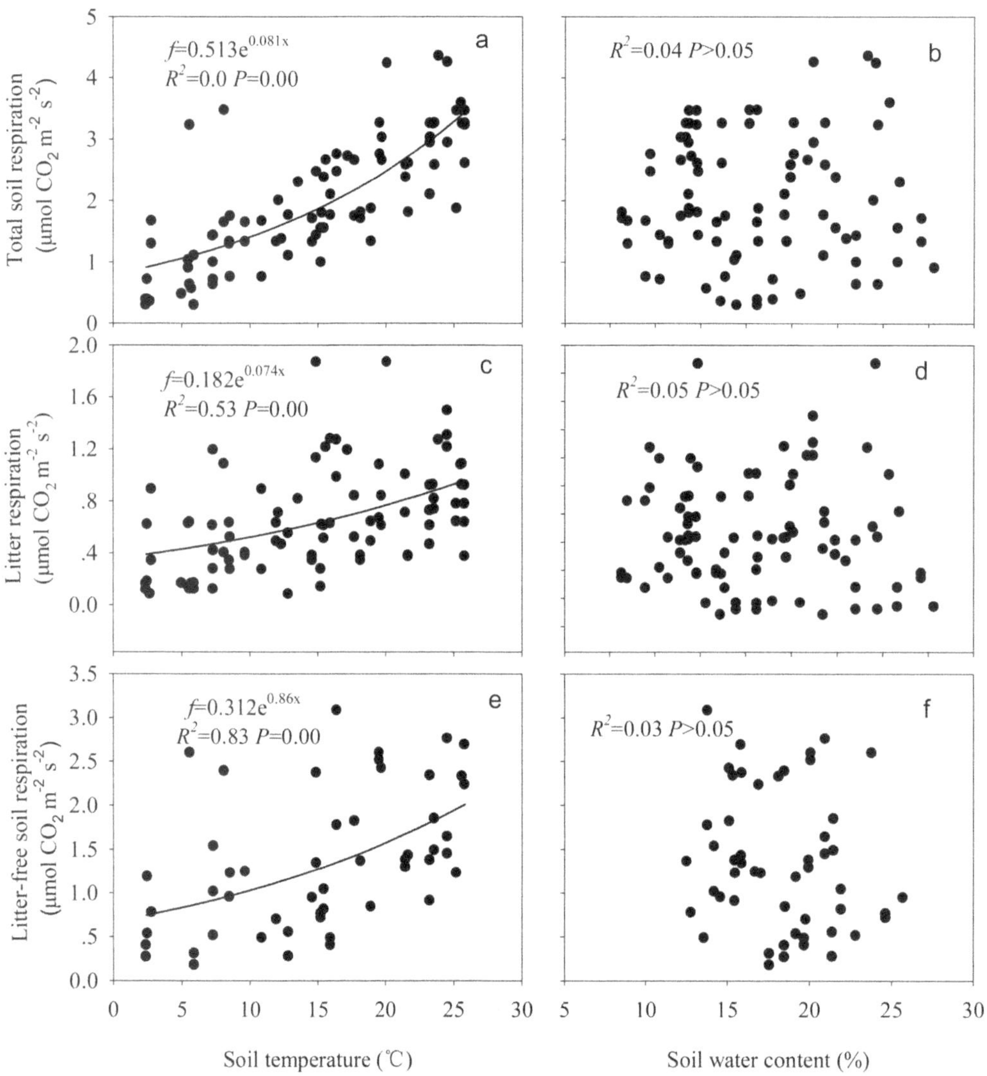

Figure 5. Relationships between the total soil respiration (a, b), litter respiration (c, d), litter-free soil respiration (e, f) and soil temperature at 5 cm depth or soil water content at 5 cm depth in different-aged *Pinus massoniana* forests during the 18-months field observations.

Table 4. The functions and model fit parameters (n, R^2 and P-values) between measured soil respiration with soil temperature (T), soil moisture content (SWC) and litter water content (LWC) in *Pinus massoniana* forests.

Stand	Equations	n	R^2	P
Total soil respiration in relation to T and SWC				
20-year-old stands	$Rs = 0.040e^{0.072T}$ (1.555SWC–0.037SWC^2)	162	0.695	<0.001
30-year-old stands	$Rs = 0.060e^{0.048T}$ (1.729SWC–0.045SWC^2)	162	0.420	<0.001
46-year-old stands	$Rs = 0.040e^{0.062T}$ (1.380SWC–0.017SWC^2)	162	0.693	<0.001
Litter respiration in relation to T and SWC				
20-year-old stands	$Rs = 0.038e^{0.051T}$ (1.158SWC–0.035SWC^2)	162	0.201	>0.05
30-year-old stands	$Rs = 0.042e^{0.002T}$ (1.618SWC–0.049SWC^2)	162	0.031	>0.05
46-year-old stands	$Rs = 0.020e^{0.047T}$ (1.106SWC–0.007SWC^2)	162	0.240	>0.05
Litter respiration in relation to T and LWC				
20-year-old stands	$Rs = 0.039e^{0.047T}$ (0.841LWC–0.019LWC^2)	162	0.163	>0.05
30-year-old stands	$Rs = 0.052e^{0.002T}$ (0.564LWC–0.017LWC^2)	162	0.157	>0.05
46-year-old stands	$Rs = 0.049e^{0.054T}$ (0.603LWC–0.011LWC^2)	162	0.256	>0.05

lignin/N ratio ($R^2 = 0.75$, $P = 0.026$). Bini et al. [39] found that plant litter with a C/N ratio of <25 degraded easily. Fanin et al. [40] reported that litter substrate quality (N, P content) was the most important factor explaining the observed spatial variations in soil respiration, and higher respiration rates were associated with high litter N and P contents.

Saiz et al. [36] indicated that the interactive effects of abiotic and biotic factors may modify the influences of forest age on soil respiration. Previous studies have reported both increased [41] and decreased [42] decomposition rates along a chronosequence of forest stands. Our present study found that litter decomposition rates did not significantly differ with forest age (Table 3), this may be caused by the difference in soil nutrients among the three stands (Table 1). Our previous study with *P. massoniana* stands found that the decomposition rates were faster in soil nutrient-poor than in nutrient-rich stands [26]. Similarly, Barlow et al. [43] found that the secondary species (*Vismia* sp. and *Bellucia* sp.) decomposed slower than the primary forest (*Bertholettia* sp.) along successional stages. Pandey et al. [44] showed that natural Oak (*Quercus* sp.) forests had lower soil nutrient level and greater litterfall with faster decomposition rates compared to Oak plantation forests. These studies suggest that not only forest age but also other factors such as site conditions (e.g. soil nutrients) and litter quality interact to affect litter decomposition [26,43].

Effects of soil temperature and water contents on respiration and litter decomposition

Variation in soil temperature is one of the most important factors determining the seasonal and diurnal variations in soil respiration [17], and soil respiration is highly sensitive to changes in surface temperature [45]. The present study found that the significant effects of soil temperature on respirations can be well-described with simple exponential regression models (Fig5a, c, e), which support the results of Shi et al. [46] and Zimmermann et al. [1]. Fang and Moncrieff [45] showed that the responses of soil respiration to temperature were commonly described using exponential equations, and thus the temperature sensitivity of soil respiration (Q_{10}) reduced at high temperature range. On the other hand, Davidson et al. [28] found that the temperature sensitivity of soil respiration (Q_{10}) decreased with drought or water deficit. Luan et al. [47] suggested that both soil chemical and physical characters contributed to the Q_{10} variations. Hence, our study

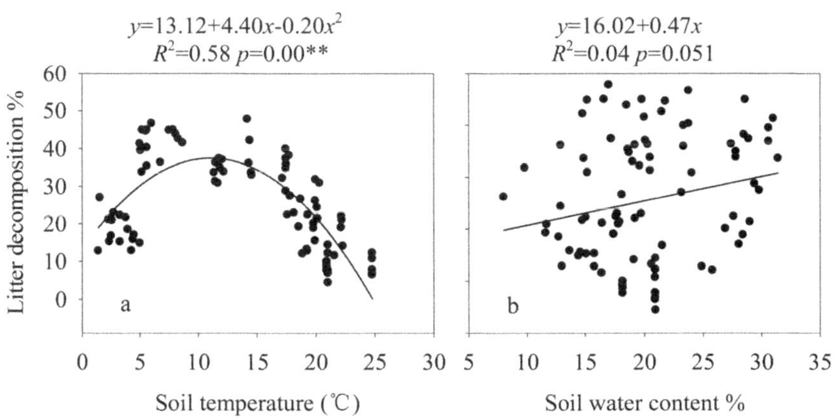

Figure 6. Relationships between litter decomposition (%) and soil temperature (a) or soil water content (b) in different-aged *Pinus massoniana* forests.

found that Q_{10} was 2.25 for the total soil respiration, 2.10 for the litter respiration, and 2.36 for the litter-free soil respiration.

Soil respiration is generally assumed to be strongly controlled by water availability [48]. In our study, we did not find any clear relationships between respirations and soil water contents (Fig. 5b, d and f). Previous studies indicated a wide range of relationships between respiration and moisture including linear [49], quadratic [50], exponential [28], logarithmic [51], and hyperbolic [52] relations, indicating that the physical (e.g.. diffusion), physiological (osmoregulation), and biochemical (enzyme dynamics) factors interact to affect the respiration-moisture relationships [48]. Recently, Moyano et al. [53] stated that further studies should concentrate on reducing uncertainties in the moisture-respiration relationships.

Soil temperature significantly interacted with soil water to affect the soil respiration (Table 4), it is probably because higher temperature may lead to lower soil water content and reverse [28]. Borken et al. [54] showed that the combined effects of temperature and water content on soil respiration were multiple linear regression. Lellei-Kovács et al. [55] showed that the interaction between temperature and moisture on soil respiration was straightforward for a linear model.

Davidson [28] found that soil temperature explained 80% of the variation of soil respiration, and there was no obvious relationship between soil water content and soil respiration. Peng et al. [56] pointed out that the sensitivity of soil respiration to temperature or moisture has not yet been adequately quantified, because most of the published results tend to be site-specific and no models have been widely accepted and commonly used.

Consistent with the results of Fiere et al. [20], we found that the effects of soil temperature on litter decomposition rates were quadratically well-described ($R^2 = 0.58$, $P<0.001$), which indicates that both lower or higher temperature lead to decreased decomposition rates (Fig. 6a). However, our results are inconsistent with the results of Aerts [57] who stated that warming resulted in increased decomposition rate.

Litter decomposition rates were found to positively respond to increased soil moisture [58]. In the present study, higher temperature resulted in lower soil moisture in the 30-year-old stands (Fig. 4). Higher temperature usually stimulates but lower soil moisture decreases the respiration and decomposition rates [55]. Cortze [21] and De Santo et al. [59] reported that soil moisture was most important during the early decomposition stage rather than the late stages. Under field experimental conditions, however, it is difficult to detect the net effects of moisture on respiration or decomposition because all factors interact to affect the litter decomposition and the effects of temperature might become stronger with increasing moisture [19].

Climate factors such as temperature and moisture alone and combined affect litter decomposition [26]. Butenschoen et al. [60] suggested that litter decomposition increased with increasing temperature in the high moisture treatment and decreased with increasing temperature in the low moisture treatment. Many studies failed to find positive responses of litter decomposition to warming when moisture is limited [57]. Cortez [21] reported that the relationships between litter decomposition rate and the ratio of soil humidity to temperature (H/T) showed a polynomial function ($y = ax^2+bx+c$), and soil temperature seemed to be the main determining factor in wet sites, while soil moisture was the most important factor during the early decomposition stages in dry period.

Relationships between soil respiration and litter decomposition

Litter provides the major C source for soil respiration [6]. Subke et al. [7] showed that an increase in litter input promotes rhizosphere respiration and rhizosphere activity, leading to increases in soil respiration. Reynolds and Hunter [6] showed that the soil respiration was significantly reduced by litter removal. Increased litter input stimulated soil microbial activity and soil C loss by microbial respiration [61]. In our study, the total soil respiration (Fig. 3a) was positively quadratically correlated with litter decomposition rate, which indicates that the soil respiration rates may be reliable indicators for long-term litter decomposability and litter carbon dynamics. Similarly, Aerts and de Caluwe [23] reported that the litter respiration rates were positively correlated with litter mass loss rates.

We found that both litter decomposition rate and soil respiration had similar temperature sensitivity and both peaked in summer. These may imply that both soil respiration and litter decomposition in the present study are controlled by similar environmental factors [26]. Moreover, we found that litter respiration peaked nearly one month earlier than the litter-free soil respiration and the total soil respiration (Fig. 1), which may be partly caused by the seasonality of the microbial community composition [46] associated with the environmental conditions, as reported by Berryman et al. [62].

The addition or exclusion of fresh litter was found to significantly increase or decrease the total soil CO_2 efflux, respectively [4,61]. The present study found that the mean contribution of the litter respiration to the total soil respiration ranged from 31.0% (30-year-old stands) to 45.9% (20-year-old stands) (see also [1,27]). Buchmann [16] estimated that soil respiration rates reduced by 10–20% when the litter and semi-decomposed litter layer were removed, and even reduced by up to 30-40% when the humus layer was also additionally removed.

The contribution of the litter respiration to the total soil respiration seems to vary with study area, tree species, and soil fertility [63]. Berger et al. [10] suggested that decomposing litter contributed 22–32% (base-rich sites) and 11–28% (base-poor sites) to the total soil respiration. In temperate coniferous forest ecosystems, the estimated contribution of leaf litter respiration to the total soil respiration exhibit a large seasonal variation from 2% (early spring) to 20% (mid summer) [4]. Cisneros-Dozal et al. [66] also found that the contribution of leaf litter decomposition to the total soil respiration increased from 5±2% (6±3 mg C m^{-2} hr^{-1}) during a transient drought to 37±8% (63±18 mg C m^{-2} hr^{-1}) immediately following water addition, indicating the effects of water availability on the contribution. In our study, the contribution of the leaf litter respiration to the total soil respiration was larger in winter than in summer. A possible explanation is that lower winter temperature limits the root respiration, and the variation of litter respiration is less important to the total soil respiration than the root respiration does [63]. Therefore, the soil respiration is obviously affected by many biotic and abiotic factors including root biomass and root activity across time [64].

In conclusion, the present paper revealed that soil moisture and temperature play complex roles in determining the respiration and decomposition across spatial and temporal scales in the *P. massoniana* forests studied. The apparent temperature sensitivity of soil respiration and litter decomposition is influenced not only by soil water but also by a wide range of factors including soil nutrients and litter quality [55,65]. The contribution of litter decomposition to the total soil respiration varies across spatial and temporal scales. These findings emphasize the need for separate consideration of soil and litter respiration when assessing climate impacts on forest carbon cycling [62].

Acknowledgments

We are grateful to the staff of the Key Laboratory of Forest Ecology and Environment for chemical analysis. We also thank anonymous reviewers for helpful suggestions on the manuscript. The funders had no role in study design, data collection and analysis, decision to publish, or preparation of the manuscript.

Author Contributions

Conceived and designed the experiments: WX XG. Performed the experiments: XG LZ ZH JL. Analyzed the data: XG ML. Contributed reagents/materials/analysis tools: XG WX ML. Wrote the paper: XG ML BZ.

References

1. Zimmermann M, Meir P, Bird M, Malhi Y, Ccahuana A (2009) Litter contribution to diurnal and annual soil respiration in a tropical montane cloud forest. Soil Biology and Biochemistry 41: 1–3.
2. Raich JW (1998) Aboveground productivity and soil respiration in three Hawaiian rain forests. Forest Ecology and Management 107: 309–318.
3. Raich JW, Schlesinger WH (1992) The global carbon dioxide flux in soil respiration and its relationship to vegetation and climate. Tellus 44B: 81–89.
4. Prévost-bouré NC, Soudani K, Damesin C, Berveiller D, Lata JC, et al. (2010) Increase in aboveground fresh litter quantity over-stimulates soil respiration in a temperate deciduous forest. Applied Soil Ecology 46: 26–34.
5. Sayer EJ, Powers JS, Tanner EVJ (2007) Increased litterfall in tropical forests boosts the transfer of soil CO_2 to the atmosphere. PLoS ONE 12: e1299.
6. Reynolds BC, Hunter MD (2001) Responses of soil respiration, soil nutrients, and litter decomposition to inputs from canopy herbivores. Soil Biology and Biochemistry 33: 1641–1652.
7. Subke JA, Hahn V, Battipaglia G, Linder S, Buchmann N, et al. (2004) Feedback interactions between needle litter decomposition and rhizosphere activity. Oecologia 139: 551–559.
8. Smolander A, Levula T, Kitunen V (2008) Response of litter decomposition and soil C and N transformations in a Norway spruce thinning stand to removal of logging residue. Forest Ecology and Management 256: 1080–1086.
9. Fioretto A, Musacchio A, Andolfi G, De Santo AV (1998) Decomposition dynamics of litters of various pine species in a Corsican pine forest. Soil Biology and Biochemistry 30: 721–727.
10. Berger TW, Inselsbacher E, Zechmeister-Boltenstern S (2010) Carbon dioxide emissions of soils under pure and mixed stands of beech and spruce, affected by decomposing foliage litter mixtures. Soil Biology and Biochemistry 42: 986–997.
11. Meier CL, Bowman WD (2008) Links between plant litter chemistry, species diversity, and below-ground ecosystem function. Proceedings of the National Academy of Sciences of the United States of America 105: 19780–19785.
12. Schaefer DA, Feng WT, Zou XM (2009) Plant carbon inputs and environmental factors strongly affect soil respiration in a substropical forest of southwestern China. Soil Biology and Biochemistry 41: 1000–1007.
13. Rubino M, Dungait JAJ, Evershed RP, Bertolini T, Angelis PD, et al. (2010) Carbon input belowground is the major C flux contributing to leaf litter mass loss: Evidences from a ^{13}C labelled-leaf litter experiment. Soil Biology and Biochemistry 42: 1009–1016.
14. Kammer A, Schmidt MWI, Hagedorn F (2012) Decomposition pathway of ^{13}C-depleted leaf litter in forest soils of the Swiss Jura. Biogeochemistry 108: 395–411.
15. Tu TTN, Egasse C, Zeller B, Bardoux G, Biron P, et al. (2011) Early degradation of plant alkanes in soils: a litterbag experiment using C-13-labelled leaves. Soil Biology and Biochemistry 11: 2222–2228.
16. Buchmann N (2000) Biotic and abiotic factors controlling soil respiration rates in *Picea abied* stands. Soil Biology and Biochemistry 32: 1625–1635.
17. Yuste JC, Janssens IA, Carrara A, Meiresonne L, Ceulemans R (2003) Interactive effects of temperature and precipitation on soil respiration in a temperate maritime pine forest. Tree Physiology 23: 1263–1270.
18. Bond-Lamberty B, Thomson A (2010) Temperature-associated increased in the global soil respiration record. Nature 464: 579–583.
19. Berbeco MR, Melillo JM, Orians CM (2012) Soil warming accelerates decomposition of fine woody debris. Plant and soil 344: 10–1007.
20. Fiere N, Craine JM, McLauchlan K, Schiml JP (2005) Litter quality and the temperature sensitivity of decomposition. Ecology 86: 320–326.
21. Cortez J (1998) Field decomposition of leaf litters: relationship between decomposition rates and soil moisture, soil temperature and earthworm activity. Soil Biology and Biochemistry 30: 783–793.
22. Xiao WF, Li JW, Yu CQ, Ma J, Cheng RM, et al. (2000) Terrestrial animal and plant ecology of the Three Gorges of Yangtze River. Chongqing: Southwest China Normal University Press. 320–321.
23. Aerts R, Caluwe H (1997) Initial litter respiration as indicator for long-term leaf litter decomposition of *Carex* species. Oikos 80: 353–361.
24. Gong ZT (2003) Chinese soil taxonomy (Revised Proposal). Beijing: China Science Press.
25. Yan LM, Chen SP, Huang JH, Lin GH (2010) Differential responses of auto- and heterotropic soil respiration to water and nitrogen addition in a semiard temperature steppe. Global Change Biology 16: 2345–2357.
26. Aerts R (1997) Climate, leaf litter chemistry and leaf litter decomposition in terrestrial ecosystems: a triangular relationship. Oikos 79: 439–449.
27. Rey A, Pegoraro E, Tedeschi V, De Parri I, Jarvis PG, et al. (2002) Annual variation in soil respiration and its components in a coppice oak forest in Central Italy. Global Change Biology 8: 851–866.
28. Davidson EA, Belk E, Boone RD (1998) Soil water content and temperature as independent or confounded factors controlling soil respiration in a temperate mixed hardwood forest. Global Change Biology 4: 217–227.
29. Kirschbaum MUF (1995) The temperature dependence of soil organic matter decomposition, and the effect of global warming on soil organic C storage. Soil Biology and Biochemistry 27: 753–760.
30. Olson JS (1963) Energy storage and the balance of producers and decomposers in ecological systems. Ecology 44: 322–331.
31. Ewel KC, Cropper WP, Glolz HL (1986) Soil CO_2 evolution in Florida slash pine plantains.I. Changes through time. Canadian Journal of Forest Research 17: 325–329.
32. Luan JW, Liu SR, Wang JX, Zhu XL, Shi ZM (2011) Rhizospheric and heterotrophic respiration of a warm-temperate oak. Soil Biology and Biochemistry 43: 503–512.
33. Wiseman PE, Seiler JR (2004) Soil CO_2 efflux across four age classed of plantation loblolly pine (*Pinus taeda* L.) on the Virginia Piedmont. Forest Ecology and Management 192: 297–311.
34. Tang JW, Bolstad PV, Martin JG (2009) soil carbon fluxes and stocks in a Great Lakes forest chronosequence. Global Change Biology 15: 145–155.
35. Bolstad PV, Davis KJ, Martin J, Cook BD, Wang W (2004) Component and whole-system respiration fluxes in northern deciduous forests. Tree Physiology 24: 493–504.
36. Saiz G, Byrne KA, Butterbach-Bahl K, Kiese R, Blujden V, et al. (2006) Forest age-related effects on soil respiration in a first rotation Sitka spruce chronosequence in central Ireland. Global Change Biology 12: 1007–1020.
37. Li FL, Liu M, Li ZP, Jiang CY, Han FX, et al. (2013) Changes in soil microbial biomass and functional diversity with a nitrogen gradient in soil columns. Applied Soil Ecology 64: 1–6.
38. Hättenschwiler S, Gasser P (2005) Soil animals alter plant litter diversity effects on decomposition. Proceedings of the National Academy of Sciences of the United States of America 102: 1519–1524.
39. Bini D, Dos Santos CA, Carmom KBD, Kishino N, Andrade G, et al. (2013) Effects of land use on soil organic carbon and microbial processes associated with soil health in southern Brazil. European Journal of Soil Biology 55: 117–123.
40. Fanin N, Hättenschwiler S, Barantal S, Schimann H, Fromin N (2011) Does variability in litter quality determine soil microbial respiration in an Amazonian rainforest? Soil Biology and Biochemistry 43: 1014–1022.
41. Xuluc-Tolosa FJ, Vester HFM, Ramírez-Marcial N, Castellanos-Albores J, Lawrence D (2003) Leaf litter decomposition of tree specied in three successional phases of Tropical dry secondary forest in Campeche, Mexico. Forest Ecology and Management 174: 401–412.
42. Vasconcelos HL, Laurance WF (2005) Influence of habitat, litter type, and soil invertebrates on leaf-litter decomposition in a fragmented Amazonian landscape. Oecologia 144: 456–462.
43. Barlow J, Gardner TA, Ferreira LV, Peres CA (2007) Litter fall and decomposition in primary, secondary and plantation forests in the Brazilian Amazon. Forest Ecology and Management 247: 91–97.
44. Pandey RR, Sharma G, Tripathi SK, Singh AK (2007) Litterfall, litter decomposition and nutrient dynamics in a subtropical natural oak forest and managed plantation in northeastern India. Forest Ecology and Management 240: 96–104.
45. Fang C, Moncrieff JB (2001) The dependence of soil CO_2 efflux on temperature. Soil Biology and Biochemistry 33: 155–165.
46. Shi WY, Zhang JG, Yan MJ, Yamanaka N, Du S (2012) Seasonal and diurnal dynamic s of soil respiration fluxes in two typical forests on the semiarid Loess Plateau of China: temperature sensitivities of autotrophs and heterotrophs and analyses of integrated driving factors. Soil Biology and Biochemistry 52: 99–107.
47. Luan JW, Liu SR, Wang JX, Zhu XL (2013) Factors affecting spatial variation of annual apparent Q_{10} of soil respiration in two warm temperature forests. PLoS ONE 8: e64167.
48. Moyano FE, Manzoni S, Chenu C (2013) Responses of soil heterotrophic respiration to moisture availability: an exploration of processes and models. Soil Biology and Biochemistry 59: 72–85.
49. Epron D, Farque L, Lucot É, Badot PM (1998) Soil CO_2 efflux in a beech forest: dependence on soil temperature and soil water content. Annual of Forest Science 56: 221–226.

50. Tang JW, Baldocchi DD (2005) Spatial-temporal variation in soil respiration in an oak-grass savanna ecosystem in California and its partitioning into autotrophic and heterotrophic component. Biogeochemistry 73: 183–207.
51. Raich JW, Potter CS, Bhagawati D (2002) Interannual variability in global soil respiration, 1980-94. Global Change Biology 8: 800–812.
52. Hanson PJ, Wullschleger SD, Bohlman SA, Todd DE (1993) Seasonal and topographic patterns of forest floor CO_2 efflux from an upland oak forest. Tree Physiology 13: 1–15.
53. Moyano FE, Vasilyeva N, Bouckaert L, Cook F, Craine J, et al. (2012) The moisture response of soil heterotrophic respiration: interaction with soil properties. Biogeochemistry 9: 1173–1182.
54. Borken W, Davidson EA, Savage K, Gaudinski J, Trumbore SE (2003) Drying and wetting effects on carbon dioxide release from organic Horizons. Soil Science Society of America Journal 67: 1888–1896.
55. Lellei-Kovács E, Kovács-Láng E, Botta-Dukát Z, Kalapos T, Emmett B, et al. (2011) Thresholds and interactive effects of soil moisture on the temperature response of soil respiration. European Journal of Soil Biology 47: 247–255.
56. Peng SS, Piao SL, Wang T, Sun JY, Shen ZH (2009) Temperature sensitivity of soil respiration in different ecosystems in China. Soil Biology and Biochemistry 41: 1008–1014.
57. Aerts R (2006) The freezer defrosting: global warming and litter decomposition rates in cold biomes. Journal of Ecology 94: 713–724.
58. Lee H, Fitzgerald J, Hewins DB, McCulley RL, Archer SR, et al. (2014) Soil moisture and soil-litter mixing effects on surface litter decomposition: A controlled environment assessment. Soil Biology and Biochemistry 72: 123–132.
59. De Santo AV, Berg B, Rutigliano FA, Alfani A, Floretto A (1993) Factors regulating early-stage decomposition of needle litters in five different coniferous forests. Soil Biology and Biochemistry 25: 1423–1433.
60. Butenschoen O, Scheu S, Eisenhauer N (2011) Interactive effects of warming, soil humidity and plant diversity on litter decomposition and microbial activity. Soil Biology and Biochemistry 43: 1902–1907.
61. Liu LL, King JS, Booker FL, Giardina CP, Allen HL, et al. (2009) Enhanced litter input rather than changes in litter chemistry drive soil carbon and nitrogen cycles under elevated CO_2: a microcosm study. Global Change Biology 15: 441–453.
62. Berryman EM, Marshall JD, Kavanagh K (2014) Decoupling litter respiration from whole-soil soil respiration along an elevation gradient in a Rochy Mountain mixed conifer forest. Canadian Journal of Forest Research 44. Doi.org/10.1139/cjfr-2013-0334.
63. Wang CK, Yang JY, Zhang QZ (2006) Soil respiration in six temperate forests in China. Global Change Biology 12: 2103–2114.
64. Bowden RD, Nadelhoffer KJ, Boone RD, Melillo JM, Garrison JB (1993) Contribution of aboveground litter, belowground litter, and root respiration to total soil respiration in a temperate mixed hardwood forest. Canadian Journal of Forest Research 23: 1402–1407.
65. Kirschbaum MUF (2004) Soil respiration under prolonged soil warming: are rate reductions caused by acclimation or substate loss? Global Change Biology 10: 1870–1877.
66. Cisneros-Dozal LM, Trumbore SE, Hanson PJ (2007) Effect of moisture on leaf litter decomposition and its contribution to soil respiration in a temperature forest. Journal of Geophysical Research 112: G1013.

Richness of Lichen Species, Especially of Threatened Ones, Is Promoted by Management Methods Furthering Stand Continuity

Steffen Boch[1,2]*, Daniel Prati[1,2], Dominik Hessenmöller[3], Ernst-Detlef Schulze[3], Markus Fischer[1,2]

1 Institute of Plant Sciences and Botanical Garden; University of Bern, Bern, Switzerland, **2** Oeschger Center, University of Bern, Bern, Switzerland, **3** Max-Planck-Institute for Biogeochemistry, Jena, Germany

Abstract

Lichens are a key component of forest biodiversity. However, a comprehensive study analyzing lichen species richness in relation to several management types, extending over different regions and forest stages and including information on site conditions is missing for temperate European forests. In three German regions (Schwäbische Alb, Hainich-Dün, Schorfheide-Chorin), the so-called Biodiversity Exploratories, we studied lichen species richness in 631 forest plots of 400 m^2 comprising different management types (unmanaged, selection cutting, deciduous and coniferous age-class forests resulting from clear cutting or shelterwood logging), various stand ages, and site conditions, typical for large parts of temperate Europe. We analyzed how lichen species richness responds to management and habitat variables (standing biomass, cover of deadwood, cover of rocks). We found strong regional differences with highest lichen species richness in the Schwäbische Alb, probably driven by regional differences in former air pollution, and in precipitation and habitat variables. Overall, unmanaged forests harbored 22% more threatened lichen species than managed age-class forests. In general, total, corticolous, and threatened lichen species richness did not differ among management types of deciduous forests. However, in the Schwäbische-Alb region, deciduous forests had 61% more lichen species than coniferous forests and they had 279% more threatened and 76% more corticolous lichen species. Old deciduous age classes were richer in corticolous lichen species than young ones, while old coniferous age-classes were poorer than young ones. Overall, our findings highlight the importance of stand continuity for conservation. To increase total and threatened lichen species richness we suggest (1) conserving unmanaged forests, (2) promoting silvicultural methods assuring stand continuity, (3) conserving old trees in managed forests, (4) promoting stands of native deciduous tree species instead of coniferous plantations, and (5) increasing the amount of deadwood in forests.

Editor: Francesco de Bello, Institute of Botany, Czech Academy of Sciences, Czech Republic

Funding: The DFG (German Research Foundation; http://www.dfg.de/en/index.jsp) funded the study in the framework of the Biodiversity Exploratories SSP 1374 "Infrastructure-Biodiversity-Exploratories" (FI 1246/6-1, FI 1246/9-1). The funders had no role in study design, data collection and analysis, decision to publish, or preparation of the manuscript.

Competing Interests: The authors have declared that no competing interests exist.

* E-mail: steffen.boch@ips.unibe.ch

Introduction

The main reasons for the ongoing decline of lichen species richness are habitat degradation caused by human interference such as disturbances by management activities, and air pollution. Together these factors have resulted in large numbers of threatened lichen species [1–3]. Because of their sensitivity to land-use and habitat changes lichens are very important environmental indicators [2]. In particular, rare and threatened species should be considered in conservation-oriented forest-management plans [4,5]. Between the 1950's and 1980's, air pollution was high in certain parts of Europe (e.g. [6]). In particular the extent of sulfur dioxide (SO_2) deposition, which reduced bark pH and regional lichen species pools [7–9], differed strongly among Central European regions. Hence, studying different regions is required to reach general conclusions about forest-management effects on lichen diversity.

Temperate European forests have long management histories, and forests without human influence are restricted to remote or inaccessible areas [10]. Hence, European forests are fragmented with large areas dominated by economically profitable age-class forests. These contain stands with even-aged tree structure, which had either resulted from clear cuts or from shelterwood logging. In the second case all trees of a stand are removed in a series of two cuts: first, approximately 60% of the tree cover is harvested, leaving scattered shelter trees for seedlings and secondly, after establishment of a closed stand cover of young trees, the remaining shelter trees are also removed. Such even-aged forest stands are often managed as plantations of fast growing conifers [11]. As both methods replace the whole stand by a new tree layer in a relatively short time period, stands are non-continuous in both cases. In recent years, silvicultural methods were promoted which attempt to mimic the natural-forest cycles typical for a particular vegetation zone. These methods focus on reproducing natural gap dynamics and regeneration and on promoting original vegetation, such as mixed forests dominated by deciduous tree species [10,12]. In comparison to conventional clear cuts and shelterwood logging, which result in age-class forests, silvicultural methods such as selection cuttings assure continuous tree cover by harvesting single trees or small groups of trees, thus promoting uneven-aged stands. Selection cuttings should maintain and enhance biodiversity in forests whilst producing timber in an economically efficient way [12].

In Europe the area of protected forest reserves without management has recently been increased, mainly to conserve vulnerable and rare forest ecosystems, and to establish a reserve network [10]. However, because nearly all of these forests were formerly more or less intensively managed. Central European unmanaged forests are not comparable with natural forests in North America, Siberia or some parts of Eastern and Northern Europe which have been largely untouched by man for centuries [12]. In temperate European forests the efficacy of forest protection and of different silvicultural systems for maintaining lichen diversity is still poorly investigated, calling for a comprehensive analysis [13].

Lichens, symbiotic associations between mycobiontic and photobiontic partners, occur on a wide range of substrates in most terrestrial ecosystems of the world, including the bark of trees (corticolous lichens), soil (terricolous lichens), rocks (saxicolous lichens) and deadwood (lignicolous lichens). Nevertheless, many particular lichen species are restricted to a narrow ecological niche with specific requirements concerning substrate (e.g. bark, deadwood, rocks, soil), pH value, and nutrient status. Thus, particular habitats, and even successional stages within habitats, harbor distinctive lichen communities with successional variation in their lichen composition [2].

Case studies in North America and Europe showed higher corticolous and threatened lichen species richness in unmanaged than in managed forests [13–15]. In addition, it was suggested that silvicultural systems assuring stand continuity of forests, such as selection cutting or prolonged rotation periods, might maintain and increase lichen species richness and should therefore be favored over conventional forestry methods including clear cuts [16,17].

Unfortunately, to date there have been no comprehensive, comparative studies from temperate European forests on the response of lichen species richness to management. No studies have included different regions, management types and developmental stages, along with detailed information on site conditions [13]. Furthermore, rather than addressing all lichen species in defined areas of differently managed forests, only corticolous species on individual trees were usually recorded [18,19]. In addition, studies on the effects of stand age or stand characteristics on lichen species richness are rare outside Fennoscandia, where case studies have been carried out (e.g. [20–23]).

We present a comparative study analyzing the response of the species richness of all lichens, of lichens separated by substrate (corticolous, lignicolous, saxicolous), and of threatened lichens to management and habitat variables (standing biomass, cover of deadwood, cover of rocks). This is the most extensive lichen dataset from Central Europe to date.

Our main questions are:

(1) How does lichen species richness respond to forest management?

(2) How does lichen species richness respond to habitat variables?

Materials and Methods

Study system

This study was conducted as part of the Biodiversity Exploratories project (www.biodiversity-exploratories.de) in three German regions: (1) the UNESCO Biosphere area Schwäbische Alb (Swabian Jura), situated in the low mountain ranges of South-western Germany (2) the National Park Hainich and its surrounding areas, situated in the hilly lands of Central Germany, and (3) the UNESCO Biosphere Reserve Schorfheide-Chorin, situated in the young glacial lowlands of North-eastern Germany. The three study regions differ in climate, geology, and topographical situations and harbor land uses as well as species pools typical for large parts of temperate Europe ([24]; Table 1). Past mean annual SO_2 depositions had been low in the Schwäbische Alb, high in Schorfheide-Chorin, and very high in Hainich-Dün ([25]; Table 1).

Plot selection

Each region, of at least 20 km by 30 km, contains more than 500 forest plots selected from the intersection points of a 100 m×100 m grid, after discarding plots fully or partially overlapping with settlements, grasslands, agricultural fields, water bodies and plots intersected by roads [24]. From these plots, we randomly selected 631 plots for this study, which cover all management types in each region: 152 in the Schwäbische Alb, 172 in Hainich-Dün, and 307 in Schorfheide-Chorin. Thus, we consider our plot sample as unbiased with regard to studying differences in forest management.

Management data

To assess the management system and stand characteristics of forests, a forest inventory had been conducted on a circular area of 500 m^2 (radius 12.62 m) in each plot. Unmanaged forests were mature, deciduous forests dominated mainly by European beech (*Fagus sylvatica)*. Age-class forests were dominated by European beech, Norway spruce *(Picea abies)* or Scots pine *(Pinus sylvestris)* and had different developmental stages of even-aged structure due to harvests at 80- to 120-year intervals. Selection forests were uneven-aged deciduous stands dominated by European beech, in which single or small groups of trees were harvested selectively. As stand characteristics, we counted the number of trees (>7 cm diameter at breast height; DBH), measured their DBH, and their height using an ultrasonic tree height meter (Vertex III Forester, Haglöf, Langsele, Sweden). We then calculated standing biomass (m^3/ha) using height and DBH of each occurring tree accounting for tree species specific trunk shapes (for details see [26]). Standing biomass can be used as a combined indicator for both tree densities (negative relation with standing biomass; [27]) and stand age (positive relation with standing biomass). We additionally recorded the percentage ground covered by rocks and deadwood respectively as indicators of substrate quantity. Furthermore, we recorded the occurrence of logging trails. Spacing of logging trails turned out to be about 20 m in coniferous and about 40 m in deciduous forests.

Vegetation data

During 2007 and 2008 the first author recorded lichens in all 631 plots. In each case lichens were recorded on 20 m×20 m (in the center of each plot and concentric with the forest inventory circle). All lichen species per plot were identified and lichens were also recorded separately for each of the four substrate categories: bark (corticolous species, up to 2.5 m height on tree trunks and branches of shrubs), rocks (saxicolous species), deadwood (lignicolous species), and soil (terricolous species), resulting in total and substrate-specific richness values. Further, we obtained the number of lichen species classified as critically endangered to vulnerable in the red list of threatened lichens of Germany [1]. As we recorded very few terricolous lichen species we considered

Table 1. Main geographic and habitat characteristics of the three Biodiversity Exploratories.

	Schwäbische Alb	Hainich-Dün	Schorfheide-Chorin
Location	**SW Germany**	**Central Germany**	**NE Germany**
Size	~422 km^2	~1300 km^2	~1300 km^2
Geology	Calcareous bedrock	Calcareous bedrock	Young glacial landscape
Altitude a.s.l.	460–860 m	285–550 m	3–140 m
Annual mean temperature	6.0–7.0 °C	6.5–8.0 °C	8.0–8.5 °C
Annual mean precipitation	700–1000 mm	500–800 mm	500–600 mm
SO_2 deposition			
until 1985	25–<50 μg/m^3	>150 μg/m^3	25–<50 μg/m^3
1985 to 1990	<25 μg/m^3	>150 μg/m^3	50–<75 μg/m^3
1990 to 1995	<25 μg/m^3	25–<50 μg/m^3	50–<75 μg/m^3
since 1995	<25 μg/m^3	<25 μg/m^3	<25 μg/m^3
Number of plots	152	172	307
Standing biomass [m^3/ha]			
mean (SD)	336.7 (187.8)	408.6 (188.9)	444.8 (187.0)
range	1.2–1017.4	3.9–881.5	18.3–1001.5
Cover deadwood [%]			
mean (SD)	3.6 (3.0)	3.1 (2.8)	3.8 (3.6)
range	0.5–20.0	0.5–15.0	0.5–25.0
Cover rocks [%]			
mean (SD)	1.0 (1.6)	0.2 (0.5)	0.3 (0.6)
range	0.0–11.0	0.0–4.0	0.0–5.5

them when analyzing total lichen species richness but, in contrast to the other groups, did not analyze them separately.

As an additional measurement of habitat diversity for corticolous lichen species, we recorded the number of tree species (>5 m height) per plot, estimated their percentage cover and summed these cover estimates (cumulative tree cover). Across age-class forests we used the proportion of coniferous tree cover, to separate coniferous (≥70%; N = 115) and deciduous age-class forests (including mixed and pure deciduous stands; N = 379).

Statistical analysis

We analysed the response of lichen species richness to management and habitat variables. Response variables were the species richness of all lichens and the species richness of the separate lichen groups (corticulous, saxiculous and ligniculous). Explanatory variables were management type, total cover of rocks, total cover of deadwood, standing biomass, age of the oldest tree per plot (max. DBH/plot), the number of tree species, and cumulative tree cover. As we were analyzing count data we used GLM models with quasi-Poisson errors to correct for over-dispersion. We used subsets of the data for three separate analyses to compare effects of different management types: (1) unmanaged vs. age-class forests (all regions), (2) unmanaged vs. deciduous age-class vs. selection forests (without Schorfheide-Chorin, where no selection forests were available in the dataset), and (3) coniferous age-class vs. deciduous age-class forests (all regions). Furthermore, we included interactions with region, management type and standing biomass. Sequential F-tests were used to test the significance of deviance changes associated factors added progressively to the model (the sequence is shown in Tables 2, 3, 4). Covariables (cover of rocks and cover of deadwood) were fitted before management, meaning that the management effect is corrected for these variables and effects of management on lichens are not due to differences in the amount of deadwood or rocks between management types. As the number of tree species per plot and the occurrence of logging trails had no effect on lichen species richness in our study, we removed these variables from the analyses. We also excluded maximum DBH and cumulative tree cover because these were correlated with standing biomass (for both variables: r = 0.623, p < 0.0001). Data were analyzed using *R*, Version 2.13.1 [28].

Results

Overall and regional lichen species richness

We recorded 202 lichen species, including 73 which are threatened in Germany. Of these 202 species, 124 were corticolous, 84 lignicolous, 59 saxicolous and 18 terricolous. Across all 631 plots the species richness of corticolous lichens was positively correlated with the species richness of lignicolous and saxicolous lichens (lignicolous versus corticolous, r = 0.0838, p = 0.0354; corticolous versus saxicolous, r = 0.3091, p < 0.0001).

In the Schwäbische Alb region we recorded 177 species, in Hainich-Dün 59, and in Schorfheide-Chorin 70. Thirty-three of the recorded species were found in all three regions, 52 species were shared by the Schwäbische Alb and Hainich-Dün regions, 47 by the Schwäbische Alb and Schorfheide-Chorin regions, and 36 by the Hainich-Dün and Schorfheide-Chorin regions. One-hundred and eleven species were recorded exclusively in the Schwäbische Alb region, 7 in Hainich-Dün, and 23 in Schorfheide-Chorin.

Species richness per plot of all lichens and of threatened lichens was significantly higher in the Schwäbische Alb than in both other

Table 2. GLM results for differences in lichen species richness between 86 unmanaged and 494 age-class forests.

		Species richness									
		All lichens		Threatened lichens		Corticolous lichens		Saxicolous lichens		Lignicolous lichens	
Source of variation	df	*F*	*p*	*F*	*p*	*F*	*p*	*F*	*p*	*F*	*p*
Region	2	435.42	**<0.001**	275.47	**<0.001**	301.85	**<0.001**	57.135	**<0.001**	41.538	**<0.001**
Rock cover	1	23.75	**<0.001**	16.60	**<0.001**	--	--	138.140	**<0.001**	--	--
Deadwood cover	1	1.68	0.196	16.87	**<0.001**	--	--	--	--	11.704	**0.001**
Management											
Management (unmanaged vs. age class)	1	0.32	0.572	20.61	**<0.001**	0.424	0.515	23.047	**<0.001**	5.460	**0.020**
Standing biomass	1	3.43	0.065	7.11	**0.008**	14.970	**<0.001**	0.005	0.944	25.587	**<0.001**
Management × standing biomass	1	0.06	0.812	0.25	0.615	0.160	0.690	0.697	0.404	2.265	0.133
Regional interactions											
Region × rock cover	2	0.18	0.839	3.76	**0.024**	--	--	13.711	**<0.001**	--	--
Region × deadwood cover	2	13.00	**<0.001**	2.85	0.059	--	--	--	--	0.036	0.965
Region × management	2	1.24	0.290	4.38	**0.013**	2.332	0.098	10.800	**<0.001**	0.412	0.663
Region × standing biomass	2	1.25	0.287	5.07	**0.007**	1.710	0.182	2.330	0.098	4.003	**0.019**
Residual Deviance	≥564	1072.40		814.86		1045.13		605.33		840.47	

Significant differences are indicated by bold *p* values.

regions (Figures 1, 2). Total species richness per plot was on average 18.6 (SD 8.6) in the Schwäbische Alb, 5.0 (2.7) in Hainich-Dün, and 6.0 (3.0) in Schorfheide-Chorin.

Management effects on lichen species richness

Total and corticolous lichen species richness did not differ between unmanaged and age-class forests. However, unmanaged forests harbored 21.9% more threatened lichen species than managed age-class forests (Tables 2, 5).

In general, total and corticolous lichen species richness did not differ among management types of deciduous stands. However, among those stands the species richness of threatened lichens was highest in unmanaged, intermediate in selection and lowest in age-class forests (Tables 3, 6). In general, threatened lichen species richness was higher in deciduous than in coniferous age-class

Table 3. GLM results for differences in lichen species richness among 516 deciduous forests (86 unmanaged vs. 379 age-class vs. 51 selection forests).

		Species richness									
		All lichens		Threatened lichens		Corticolous lichens		Saxicolous lichens		Lignicolous lichens	
Source of variation	df	*F*	*p*	*F*	*p*	*F*	*p*	*F*	*p*	*F*	*p*
Region	1	717.40	**<0.001**	479.84	**<0.001**	495.94	**<0.001**	133.81	**<0.001**	26.51	**<0.001**
Rock cover	1	4.85	**0.028**	5.70	**0.018**	--	--	83.47	**<0.001**	--	--
Deadwood cover	1	2.54	0.112	0.65	0.423	--	--	--	--	6.33	**0.012**
Management											
Management (Unmanaged vs. age class vs. selection)	2	0.31	0.731	7.15	**0.001**	0.34	0.714	2.14	0.119	0.23	0.794
Standing biomass	1	10.19	**0.002**	19.74	**<0.001**	22.12	**<0.001**	0.14	0.712	22.72	**<0.001**
Management × standing biomass	2	3.81	**0.023**	1.40	0.248	1.66	0.193	3.00	0.052	1.34	0.263
Regional interactions											
Region × rock cover	1	0.03	0.859	4.39	**0.037**	--	--	14.26	**<0.001**	--	--
Region × deadwood cover	1	10.20	**0.002**	1.33	0.250	--	--	--	--	0.41	0.521
Region × management	2	4.45	**0.013**	5.46	**0.005**	7.97	**<0.001**	4.68	**0.010**	0.10	0.908
Region × standing biomass	1	0.02	0.901	3.83	0.051	0.50	0.481	0.14	0.709	0.00	0.983
Residual Deviance	≥272	540.05		370.10		493.88		359.15		332.63	

Significant differences are indicated by bold *p* values.

Table 4. GLM results for differences in lichen species richness between 379 deciduous and 115 coniferous age-class forests.

		Species richness									
		All lichens		**Threatened lichens**		**Corticolous lichens**		**Saxicolous lichens**		**Lignicolous lichens**	
Source of variation	df	*F*	*p*	*F*	*p*	*F*	*p*	*F*	*p*	*F*	*p*
Region	2	392.86	**<0.001**	284.51	**<0.001**	277.88	**<0.001**	47.73	**<0.001**	35.86	**<0.001**
Rock cover	1	27.88	**<0.001**	19.58	**<0.001**	--	--	83.02	**<0.001**	--	--
Deadwood cover	1	1.44	0.231	14.39	**<0.001**	--	--	--	--	7.28	**0.007**
Management											
Management (deciduous vs. coniferous age class)	1	16.20	**<0.001**	63.94	**<0.001**	39.13	**<0.001**	5.28	**0.022**	26.00	**<0.001**
Standing biomass	1	4.34	**0.038**	14.15	**<0.001**	16.81	**<0.001**	0.29	0.592	25.82	**<0.001**
Management × standing biomass	1	11.62	**0.001**	20.67	**<0.001**	10.12	**0.002**	3.11	0.078	0.68	0.411
Regional interactions											
Region × rock cover	2	0.35	0.704	3.22	**0.041**	--	--	6.21	**0.002**	--	--
Region × deadwood cover	2	9.30	**<0.001**	2.26	0.105	--	--	--	--	0.20	0.817
Region × management	2	6.07	**0.002**	0.92	0.401	8.34	**<0.001**	3.04	**0.049**	1.10	0.334
Region × standing biomass	2	0.20	0.821	2.06	0.129	0.05	0.953	1.03	0.357	7.00	**0.001**
Residual Deviance	≥478	896.93		582.89		846.33		435.47		688.15	

Significant differences are indicated by bold *p* values.

forests. This was also the case for corticolous lichen species richness in the Schwäbische Alb and Schorfheide-Chorin, and for the total lichen species richness only in the Schwäbische Alb. In contrast, the species richness of lignicolous lichens was generally higher in coniferous than in deciduous age-class forests (Tables 4, 5; Figures 1, 2). Thus, promoting stands with site-typical composition of tree species appears important for promoting lichen species richness and threatened lichen species in forests.

In the Schwäbische Alb, species richness of saxicolous lichens was 428% higher in unmanaged than in deciduous age-class forests and also 73% higher than in selection forests (Tables 3, 6).

Standing biomass and lichen species richness

Overall, old forests with large quantities of standing biomass were slightly richer in corticolous lichen species than were young forests with low quantities of standing biomass: we found an increase of 1.2 species with an increase of 500m^3 standing biomass per ha (Table 2). Among deciduous stands this relationship was even more pronounced (+2.0 species/500 m^3). Furthermore, threatened lichen species richness increased with standing biomass, more strongly for the Schwäbische Alb (+2.6 species/ 500 m^3) than for the Hainich-Dün (+0.6 species/500 m^3), as

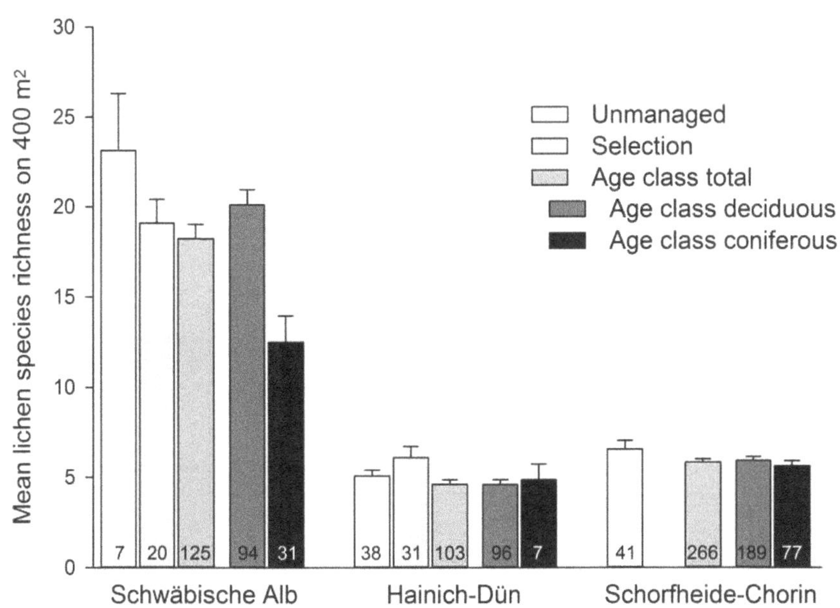

Figure 1. Mean lichen species richness (+SE) per plot for each of the forest management types in the three study regions.

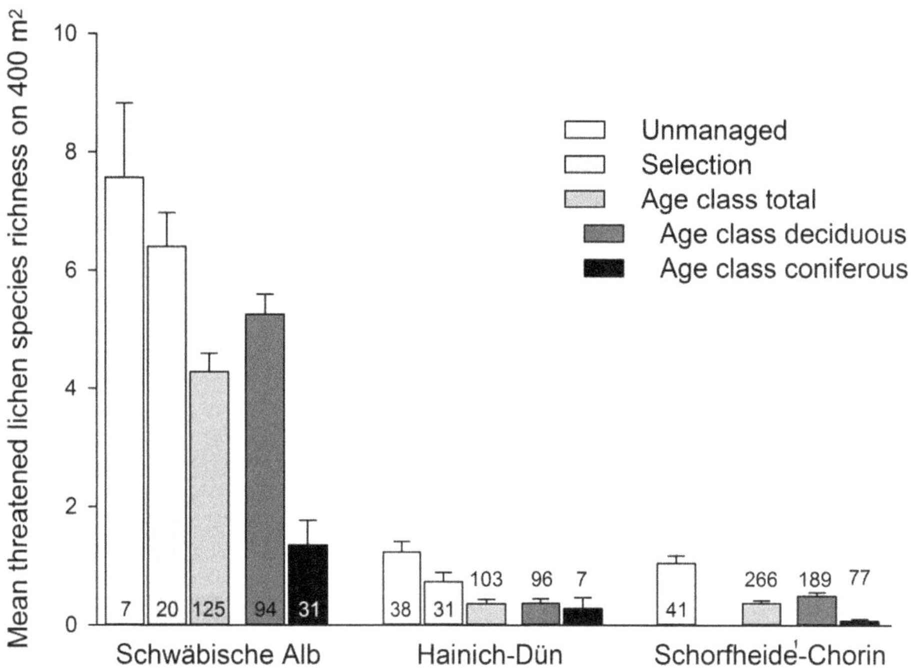

Figure 2. Mean species richness of threatened lichens (+SE) per plot for each of the forest management types in the three study regions.

indicated by the significant region - by - standing biomass interaction (Table 3). Overall, we found opposing effects of standing biomass on the richness of corticolous and threatened lichen species between deciduous and coniferous age-class forests (significant standing biomass - by - management interaction; Table 4). High standing biomass was associated with higher

Table 5. Mean lichen species richness per 400 m^2 (untransformed mean ±SE) in unmanaged and age-class forests, and in deciduous and coniferous age-class forests, in total and for the three study regions.

	N	All lichens		Threatened lichens		Corticolous lichens		Saxicolous lichens		Lignicolous lichens	
Total											
Unmanaged	86	7.3	(0.6)	1.7	(0.2)	5.4	(0.4)	1.3	(0.3)	0.6	(0.1)
Age class$_{total}$	494	8.7	(0.3)	1.4	(0.1)	6.9	(0.3)	0.5	(0.1)	1.2	(0.1)
Age class$_{deciduous}$	379	9.1	(0.4)	1.6	(0.1)	7.3	(0.3)	0.5	(0.1)	0.9	(0.1)
Age class$_{coniferous}$	115	7.4	(0.5)	0.4	(0.1)	5.4	(0.4)	0.2	(0.1)	2.0	(0.2)
Schwäbische Alb											
Unmanaged	7	23.1	(3.2)	7.6	(1.3)	14.0	(2.6)	6.9	(2.2)	0.7	(0.4)
Age class$_{total}$	125	18.2	(0.8)	4.3	(0.3)	13.8	(0.7)	1.2	(0.2)	1.2	(0.2)
Age class$_{deciduous}$	94	20.1	(0.9)	5.3	(0.3)	15.4	(0.7)	1.3	(0.2)	1.0	(0.2)
Age class$_{coniferous}$	31	12.5	(1.4)	1.4	(0.4)	8.7	(1.2)	0.7	(0.2)	1.8	(0.4)
Hainich-Dün											
Unmanaged	38	5.1	(0.3)	1.2	(0.2)	4.7	(0.3)	0.1	(0.1)	0.3	(0.1)
Age class$_{total}$	103	4.6	(0.3)	0.4	(0.1)	3.8	(0.2)	0.4	(0.1)	0.3	(0.1)
Age class$_{deciduous}$	96	4.6	(0.3)	0.4	(0.1)	3.8	(0.2)	0.4	(0.1)	0.3	(0.1)
Age class$_{coniferous}$	7	4.9	(0.9)	0.3	(0.2)	4.0	(1.0)	0.0	(0.0)	0.4	(0.4)
Schorfheide-Chorin											
Unmanaged	41	6.6	(0.5)	1.0	(0.1)	4.5	(0.3)	1.6	(0.2)	0.9	(0.2)
Age class$_{total}$	266	5.9	(0.2)	0.4	(0.0)	4.8	(0.2)	0.2	(0.0)	1.5	(0.1)
Age class$_{deciduous}$	189	6.0	(0.2)	0.5	(0.1)	5.1	(0.2)	0.2	(0.0)	1.2	(0.1)
Age class$_{coniferous}$	77	5.6	(0.3)	0.1	(0.0)	4.2	(0.2)	0.0	(0.0)	2.2	(0.2)

Table 6. Mean lichen species richness per 400 m² (untransformed mean ±SE) in unmanaged and differently managed deciduous forests in total, and for the Schwäbische Alb and Hainich-Dün regions.

	N	All lichens		Threatened lichens		Corticolous lichens		Saxicolous lichens		Lignicolous lichens	
Total											
Unmanaged	45	7.9	(1.1)	2.2	(0.4)	6.2	(0.7)	1.2	(0.5)	0.4	(0.1)
Age class$_{deciduous}$	190	12.3	(0.7)	2.8	(0.2)	9.5	(0.6)	0.8	(0.1)	0.7	(0.1)
Selection$_{deciduous}$	51	11.2	(1.1)	3.0	(0.5)	8.1	(0.8)	1.7	(0.3)	0.5	(0.1)
Schwäbische Alb											
Unmanaged	7	23.1	(3.2)	7.6	(1.3)	14.0	(2.6)	6.9	(2.2)	0.7	(0.4)
Age class$_{deciduous}$	94	20.1	(0.9)	5.3	(0.3)	15.4	(0.7)	1.3	(0.2)	1.0	(0.2)
Selection$_{deciduous}$	20	19.1	(1.3)	6.4	(0.6)	12.7	(1.1)	4.0	(0.5)	0.7	(0.2)
Hainich-Dün											
Unmanaged	38	5.1	(0.3)	1.2	(0.2)	4.7	(0.3)	0.1	(0.1)	0.3	(0.1)
Age class$_{deciduous}$	96	4.6	(0.3)	0.4	(0.1)	3.8	(0.2)	0.4	(0.1)	0.3	(0.1)
Selection$_{deciduous}$	31	6.1	(0.6)	0.7	(0.2)	5.1	(0.5)	0.2	(0.1)	0.4	(0.1)

richness of corticolous (+2.0 species/500 m^3) and of threatened lichen species (+0.7 species/500 m^3) in deciduous stands but with lower species richness of corticolous (-1.3 species/500 m^3) and of threatened lichen species (-1.2 species/500 m^3) in coniferous stands (Table 4). These findings indicate that the conservation of old forests dominated by native broadleaved tree species might enhance species richness of corticolous lichens and promote suitable habitats for threatened lichen species. Interestingly, the richness of lignicolous lichen species decreased (-0.7 species/500 m^3) with an increase in standing biomass (Tables 2, 3, 4). This was probably because of higher amounts of deadwood in younger stands due to recent timber harvesting.

Cover by deadwood and rocks and lichen species richness

Overall, deadwood cover decreased with standing biomass (N = 631, r = -0.1762; $p < 0.0001$). Lignicolous lichen species richness was generally positively related to the cover of deadwood, increasing by 0.7 species per 10% increased deadwood cover (Tables 2, 3, 4). These findings indicate that increasing the amount of deadwood in forests may lead to an increase of lichen species richness.

Overall, species richness of saxicolous lichens increased by 1.2 species per 1% increased rock cover (Tables 2, 3, 4). However, effects varied among regions and management types due to varying cover values (Table 1).

Discussion

Differences in lichen species richness among the regions

Our results showed strong differences in lichen species richness between the Schwäbische Alb and the two other regions, although all forests have similar management methods. These differences are not related to the protection status of the regions or to management activities but rather to the former intensity of atmospheric pollutants, especially SO_2, which was responsible for the decline of many lichen species in Germany [1]. Until the strong decrease of SO_2 pollution around 1990 in Germany, SO_2 deposition was low in the Schwäbische Alb, high in Schorfheide-Chorin and very high in Hainich-Dün ([25]; Table 1). Regionally, this resulted in a shift in lichen species composition to "lichen deserts" with only a few toxitolerant species [7,8]. Today, levels of SO_2 pollution have considerably decreased and many species are re-colonizing these areas [7,9]. However, the characteristic lichen communities have not yet recovered completely and therefore current lichen community composition in these areas might still be influenced by former air pollution. Furthermore, the higher mean rock cover and mean annual precipitation in the Schwäbische Alb than in the other regions (Table 1), might also contribute to the much higher lichen species richness in the Schwäbische Alb.

Management effects on lichen species richness

In our study, total and corticolous lichen species richness did not differ between unmanaged and age-class forest sites. Bergamini et al. [15] and Paillet et al. [13], who used much smaller datasets from other European countries, found higher total and corticolous lichen species richness in unmanaged compared with managed forests. Our results also contrast with those of Friedel et al. [18], who found higher species richness of corticolous lichens in 45 unmanaged than in 45 managed European beech forests in northeastern Germany, and with the findings of Rudolphi and Gustafsson [29], who compared 19 unmanaged and 19 young stands, originating from clear-cutting, in boreal forests in Sweden. The differences among the studies might be because the investigated unmanaged stands that we surveyed had not yet reached the degeneration phase and still showed signs of former management, such as a fairly even-aged structure and a dense canopy cover (as outlined in [30]). However, overall, unmanaged forests harbored more threatened lichen species than managed age-class forests. This confirms the particular importance of unmanaged forests for the conservation of lichen species with long generation times [3] or with specific habitat requirements that mean they are restricted to old growth forests [21]. Other management-related disturbances such as logging trails had no negative effects on lichen species richness, showing that management involving forestry equipment does not need to reduce the species richness of lichens if suitable habitats are spared.

We found no differences in total and corticolous lichen species richness among deciduous forests. However, we did find highest species richness of threatened lichens in unmanaged forests, with intermediate richness in selection forests, and the lowest richness in age-class forests. This result underlines the importance of silvicultural methods which ensure temporal forest continuity for many lichen species. This is line with several other studies pointing out the importance of stand continuity to preserve lichen species richness and communities with rare or threatened lichen species [13,22,31–34]. Alternatively, patches of old trees could be retained in forests to conserve lichen species richness in managed stands. Peterson and McCune [35] compared 51 forest stands in Oregon and found that retention patches or old trees were essential for the persistence of lichen communities that depend on old-growth forest conditions.

Our study provides further evidence for the idea that older forests harbor more lichen species and particularly more threatened species [5,36–39]. These positive effects might probably be explained by characteristics of old trees which make them better lichen habitats. These include, a larger bark surface which increases the probability of colonization [40], pronounced bark textures including crevices [22,40] and rot holes [41], which provide a range of microhabitats, and the fact that many lichen species associated with old-growth conditions can only establish on older trees [42]. The retention of numerous mature to over-mature trees in forests appears important for maintaining a high species richness of epiphytic lichens [43–45] and to conserve rare and threatened lichen species in managed forests [21,29,46,47]. These trees can act as population centers for the dispersal of lichen propagules and provide refuges for those species which depend on old trees [3,48–50].

Interestingly, species richness of lignicolous lichens decreased with higher standing biomass. Most likely, this finding was due to the higher amounts of deadwood, following recent timber harvesting, in our younger stands of low standing biomass than in our older stands of high standing biomass. Thus, we suggest that increasing the amount of deadwood in older forests may lead to an increase in lichen species richness, in agreement with the findings of Moning et al. [19] based on 113 plots within the National Park Bavarian Forest (Germany).

Clear cutting generally replaced site-characteristic forests with even-aged and homogeneous plantations often with different tree species and this had a very pronounced effect on lichen species richness and composition [3,44,51]. In general, we found more threatened lichen species in deciduous forests than in coniferous forests. In the Schwäbische Alb we also found more corticolous lichen species in deciduous forests. This is in line with the findings of Humphrey et al. [36] who observed higher lichen species richness in native deciduous stands than in conifer plantations in different forest sites in Britain. Furthermore, Neitlich and McCune [52] pointed out the importance of deciduous tree patches interspersed in young coniferous plantations for both lichen species richness and the species richness of specialized lichens with high conservation value. Thus, promoting stands or retaining patches with a site-typical composition of tree species appears important for promoting lichen species richness and threatened lichen species in forests.

Rock cover and lichen species richness

Overall we found positive relations between rock cover and the species richness of saxicolous lichens. Clearly, not just the presence but also the quantity of this substrate matters for lichen species richness. As managed stands occur on more easily accessible, flat sites where rocks are scarce, the higher species richness of saxicolous lichens in unmanaged than in age-class or selection forests might well be related to the lower rock quantity in unmanaged forests, rather than to the absence of management per se.

Deadwood cover and lichen species richness

In our study, the species richness of lignicolous lichens generally increased with increasing cover of deadwood. Similarly, in their deadwood-focused study Caruso et al. [53] reported a positive relationship between deadwood volume and the species richness of lignicolous lichens in 30 stands of planted boreo-nemoral Swedish forests. Thus, similar to Moning et al. [19] and Humphrey et al. [36] we recommend actively enhancing deadwood quantity, diversity of types and decay stages of standing deadwood and deadwood on the ground.

Lichen species richness in tree crowns

We probably have underestimated the overall lichen richness because we could not assess species restricted to tree crowns, as this would have required lengthy tree climbing of thousands of trees. According to this difficulty, we are not aware of any other study that attempted to assess lichen diversity in tree crowns of differently managed forests.

Conclusions

Our findings demonstrate the importance of management systems that ensure stand continuity for lichen conservation. Clearly, the conservation of old forests with high standing biomass is absolutely necessary to maintain a high species richness of lichens and to promote threatened lichen species. To increase total and threatened lichen species richness without overly compromising timber production, our results, combined with those of the other studies discussed above, lead us to the following recommendations for managed forests: (1) to promote silvicultural methods that assure stand continuity, e.g. by selection cutting rather than clear cutting and shelterwood logging, (2) to conserve retention patches with groups of old, mature to over-mature trees in managed forests, (3) to conserve at least single mature to over-mature trees which may serve as sources for colonization, (4) to select, as far as visibly obvious, such retention trees of high lichen abundance and species richness composition and population size rather than selecting them at random, (5) to promote stands of native deciduous tree species instead of coniferous plantations, and (6) to increase the amount of deadwood in forests.

Acknowledgments

We thank S. Gockel, A. Hemp, K. Wells, and S. Pfeiffer for maintaining the plot and project infrastructure, and the late E.K.V. Kalko, K.E. Linsenmair, J. Nieschulze, I. Schöning, F. Buscot, and W.W. Weisser for their role in setting up the Biodiversity Exploratories project. Furthermore, we thank V. Otte and L.B. Sparrius for verification of the identity of some critical species, and E. Allan for revising the English. Field work permits were issued by the responsible state environmental offices of Baden-Württemberg, Thüringen, and Brandenburg (according to § 72 BbgNatSchG).

Author Contributions

Provided forest inventory data: DH EDS. Conceived and designed the experiments: SB MF DH DP EDS. Performed the experiments: SB. Analyzed the data: SB DP MF. Contributed reagents/materials/analysis tools: SB. Wrote the paper: SB DP MF.

References

1. Wirth V, Schöller H, Scholz P, Feuerer T, Gnüchtel P, et al. (1996) Rote Liste der Flechten (*Lichenes*) der Bundesrepublik Deutschland. In: Bundesamt für Naturschutz, editor. Schriftenreihe Vegetationskunde 28, Rote Liste gefährdeter Pflanzen Deutschlands. Bundesamt für Naturschutz, pp. 307–368.
2. Seaward MRD (2008) Environmental role of lichens. In: Nash THIII, editor. Lichen Biology Cambridge: Cambridge Univ. Press, Ed. 2. pp. 274–298.
3. Scheidegger C, Werth S (2009) Conservation strategies for lichens: insights from population biology. Fungal Biology Reviews 23: 55ss66.
4. Thor G (1995) Red Lists–aspects of their compilation and use in lichen conservation. Mitt Eidgenöss Forsch anst Wald Schnee Landsch 70: 29–39.
5. Gustafsson L, Appelgren L, Jonsson F, Nordin U, Persson A, et al. (2004) High occurrence of red-listed bryophytes and lichens in mature managed forests in boreal Sweden. Basic Appl Ecol 5: 123–129.
6. Berge E, Jakobsen HA (1998) A regional scale multi-layer model for the calculation of long-term transport and deposition of air pollution in Europe. Tellus 50: 205–223.
7. Gilbert OL (1992) Lichen reinvasion with declining air pollution. In: Bates JW,Farmer A, editors. Bryophytes and Lichens in a Changing Environment.Oxford:Clarendon Press, pp. 159–177.
8. Kirschbaum U, Windisch U, Vorbeck A, Hanewald K (2006) Mapping lichen diversity in Wetzlar and Giessen as an indicator of air quality–Comparison between the surveys of 1970, 1985, 1995 and 2005. Gefahrenst Reinhalt Luft 66: 272–280.
9. Purvis OW, Tittley I, Chimonides PDJ, Bamber R, Hayes PA, et al. (2010) Long-term biomonitoring of lichen and bryophyte biodiversity at Burnham Beeches SAC and global environmental change. Syst Biodivers 8: 193–208.
10. Parviainen J, Bücking W, Vandekerkhove K, Schuck A, Päivinen R (2000) Strict forest reserves in Europe: efforts to enhance biodiversity and research on forests left for free development in Europe (EU-COST-Action E4). Forestry 73: 107–118.
11. Ellenberg H, Leuschner C (2010) Vegetation Mitteleuropas mit den Alpen.–6 th edition. Stuttgart:Ulmer.1357 p.
12. Parviainen J, Frank G (2003) Protected forests in Europe–approaches harmonising the definitions for international comparison and forest policy making. J Environ Manage 67: 27–36.
13. Paillet Y, Berges L, Hjalten J, Odor P, Avon C, et al. (2010) Biodiversity differences between managed and unmanaged forests: meta-analysis of species richness in Europe. Conserv Biol 24: 101–112.
14. Lesica P, McCune B, Cooper SV, Hong WS (1991) Differences in lichen and bryophyte communities between old-growth and managed second-growth forests in the Swan Valley, Montana. Can J Bot 69: 1745–1755.
15. Bergamini A, Scheidegger C, Stofer S, Carvalho P, Davey S, et al. (2005) Performance of macrolichens and lichen genera as indicators of lichen species richness and composition. Conserv Biol 19: 1051–1062.
16. Horstkotte T, Moen J, Lämås T, Helle T (2011) The Legacy of Logging–Estimating Arboreal Lichen Occurrence in a Boreal Multiple-Use Landscape on a Two Century Scale. PLoS ONE 6(12): e28779. doi:10.1371/journal.pone.0028779.
17. Rolstad J, Gjerde I, Storaunet KO, Rolstad E (2001) Epiphytic lichens in Norwegian coastal spruce forest historic logging and present forest structure. Ecol Appl 11: 421–436.
18. Friedel A, von Oheimb G, Dengler J, Haerdtle W (2006) Species diversity and species composition of epiphytic bryophytes and lichens–a comparison of managed and unmanaged beech forests in NE Germany. Feddes Repert 117: 172–185.
19. Moning C, Werth S, Dziock F, Bässler C, Bradtka J, et al. (2009) Lichen diversity in temperate montane forests is influenced by forest structure more than climate. Basic Appl Ecol 258: 745–751.
20. Lõhmus A, Lõhmus P, Vellak K (2007) Substratum diversity explains landscape-scale co-variation in the species-richness of bryophytes and lichens. Biol Conserv 135: 405–414.
21. Fritz Ö, Niklasson M, Churski M (2008) Tree age is a key factor for the conservation of epiphytic lichens and bryophytes in beech forests. Appl Veg Sci 12: 93–106.
22. Fritz Ö, Gustafsson L, Larsson K (2008) Does forest continuity matter in conservation?–A study of epiphytic lichens and bryophytes in beech forests of southern Sweden. Biol Conserv 141: 655–668.
23. Jüriado I, Liira J, Paal J, Suija A (2009) Tree and stand level variables influencing diversity of lichens on temperate broad-leaved trees in boreo-nemoral floodplain forests. Biodivers Conserv 18: 105–125.
24. Fischer M, Bossdorf O, Gockel S, Hänsel F, Hemp A, et al. (2010) Implementing large-scale and long-term functional biodiversity research: The Biodiversity Exploratories. Basic Appl Ecol 11: 473–485.
25. Umweltbundesamt (2005) Daten zur Umwelt–Der Zustand der Umwelt in Deutschland. Berlin: Erich Schmidt Verlag.352 p
26. Hessenmöller D, Nieschulze J, von Lüpke N, Schulze E-D (2011) Identification of forest management types from ground-based and remotely sensed variables and the effects of forest management on forest structure and composition. Forstarchiv 82: 171–183.
27. Luyssaert S, Schulze E-D, Börner A, Knohl A, Hessenmöller D, et al. (2008) Old-growth forests as global carbon sinks. Nature 455: 213–215.
28. R Development Core Team (2011) R: a language and environment for statistical computing. R Foundation for Statistical Computing, Vienna.
29. Rudolphi J, Gustafsson L (2011) Forests regenerating after clear-cutting function as habitat for bryophyte and lichen species of conservation concern. PLoS ONE 6(4): e18639. doi:10.1371/journal.pone.0018639.
30. Graae BJ, Heskjær VS (1997) A comparison of understorey vegetation between untouched and managed deciduous forest in Denmark. Forest Ecol Manag 96: 111–123.
31. Nascimbene J, Marini L, Nimis PL (2007) Influence of forest management on epiphytic lichens in a temperate beech forest of northern Italy. Forest Ecol Manag 247: 43–47.
32. Nascimbene J, Marini L, Nimis PL (2010) Epiphytic lichen diversity in old growth and managed *Picea abies* stands in Alpine spruce forests. Forest Ecol Manag 260: 603–609.
33. Jairus K, Lõhmus A, Lõhmus P (2009) Lichen acclimatization on retention trees: a conservation physiology lesson. J Appl Ecol 46: 930–936.
34. Marmor L, Tõrra T, Saag L, Randlane T (2011) Effects of forest continuity and tree age on epiphytic lichen biota in coniferous forests in Estonia. Ecol Ind 11: 1270–1276.
35. Peterson EB, McCune B (2001) Diversity and succession of epiphytic macrolichen communities in low-elevation managed conifer forests in Western Oregon. J Veg Sci 12: 511–524.
36. Humphrey JW, Davey S, Peace AJ, Ferris R, Harding K (2002) Lichens and bryophyte communities of planted and semi-natural forests in Britain: the influence of site type, stand structure and dead wood. Biol Conserv 107: 165–180.
37. Pykälä J, Heikkinen RK, Toivonen H, Jääskeläinen K (2006) Importance of Forest Act habitats for epiphytic lichens in Finnish managed forests. Basic Appl Ecol 223: 84–92.
38. Moning C, Müller J (2009) Critical forest age thresholds for the diversity of lichens, molluscs and birds in beech (*Fagus sylvatica* L.) dominated forests. Ecol Ind 9: 922–932.
39. Svoboda D, Peksa O, Veselá J (2010) Epiphytic lichen diversity in central European oak forests: Assessment of the effects of natural environmental factors and human influences. Environ Pollut 158: 812–819.
40. Ranius T, Johansson P, Berg N, Niklasson M (2008) The influence of forest age and microhabitat quality on the occurrence of crustose lichens associated with old oaks. J Veg Sci 19: 653–662.
41. Fritz Ö, Heilmann-Clausen J (2010) Rot holes create key microhabitats for epiphytic lichens and bryophytes on beech (*Fagus sylvatica*). Biol Conserv 143: 1008–1016.
42. Werth S, Wagner HH, Gugerli F, Holderegger R, Csencsics D, et al. (2006) Quantifying dispersal and establishment limitation in a population of an epiphytic lichen. Ecology 87: 2037–2046.
43. Lõhmus A, Lõhmus P (2010) Epiphyte communities on the trunks of retention trees stabilize in 5 years after timber harvesting, but remain threatened due to tree loss. Biol Conserv 143: 891–898.
44. Rose F (1992) Temperate forest management: its effects on bryophyte and lichen floras and habitats. In: Bates JW, Farmer A, editors. Bryophytes and Lichens in a Changing Environment.Oxford:Clarendon Press, pp. 211–233.
45. Uliczka H, Angelstam P (2000) Assessing conservation values of forest stands based on specialised lichens and birds. Biol Conserv 95: 343–351.
46. Edmann M, Eriksson A-M, Villard M-A (2008) Effects of selection cutting on the abundance and fertility of indicator lichens *Lobaria pulmonaria* and *Lobaria quercizans*. J Appl Ecol 45: 26–33.
47. Fedrowitz K, Kuusinen M, Snäll T (2012) Metapopulation dynamics and future persistence of epiphytic cyanolichens in a European boreal forest ecosystem. J Appl Ecol 49: 493–502.
48. Sillett SC, McCune B, Peck JE, Rambo TR, Ruchty A (2000) Dispersal limitations of epiphytic lichens result in species dependent on old-growth forests. Ecol Appl 10: 789–799.
49. Muir PS, Rambo TR, Kimmerer RW, Keon DB (2006) Influence of overstory removal on growth of epiphytic mosses and lichens in western Oregon. Ecol Appl 16: 1207–1221.
50. Perhans K, Appelgren L, Jonsson F, Nordin U, Söderström B, et al. (2009) Retention patches as potential refugia for bryophytes and lichens in managed forest landscapes. Biol Conserv 142: 1125–1133.
51. Wolseley PA, Stofer S, Mitchell R, Truscott AM, Vanbergen A, et al. (2006) Variation of lichen communities with landuse in Aberdeenshire, UK. Lichenologist 38: 307–322.
52. Neitlich PN, McCune B (1997) Hotspots of epiphytic lichen diversity in two young managed forests. Conserv Biol 11: 172–182.
53. Caruso A, Rudolphi J, Thor G (2008) Lichen species diversity and substrate amounts in young planted boreal forests: A comparison between slash and stumps of Picea abies. Biol Conserv 141: 47–55.

Dissecting Variation in Biomass Conversion Factors across China's Forests: Implications for Biomass and Carbon Accounting

Yunjian Luo[1*¶], Xiaoquan Zhang[2¶], Xiaoke Wang[3*], Yin Ren[1]

1 Key Laboratory of Urban Environment and Health, Institute of Urban Environment, Chinese Academy of Sciences, Xiamen, China, **2** The Nature Conservancy China Program, Beijing, China, **3** State Key Laboratory of Urban and Regional Ecology, Research Center for Eco-Environmental Sciences, Chinese Academy of Sciences, Beijing, China

Abstract

Biomass conversion factors (BCFs, defined as the ratios of tree components (i.e. stem, branch, foliage and root), as well as aboveground and whole biomass of trees to growing stock volume, Mg m^{-3}) are considered as important parameters in large-scale forest biomass carbon estimation. To date, knowledge of possible sources of the variation in BCFs is still limited at large scales. Using our compiled forest biomass dataset of China, we presented forest type-specific values of BCFs, and examined the variation in BCFs in relation to forest type, stand development and environmental factors (climate and soil fertility). BCFs exhibited remarkable variation across forest types, and also were significantly related to stand development (especially growing stock volume). BCFs (except Stem BCF) had significant relationships with mean annual temperature (MAT) and mean annual precipitation (MAP) ($P<0.001$). Climatic data (MAT and MAP) collectively explained 10.0–25.0% of the variation in BCFs (except Stem BCFs). Moreover, stronger climatic effects were found on BCFs for functional components (i.e. branch, foliage and root) than BCFs for combined components (i.e. aboveground section and whole trees). A general trend for BCFs was observed to decrease and then increase from low to high soil fertility. When qualitative soil fertility and climatic data (MAT and MAP) were combined, they explained 14.1–29.7% of the variation in in BCFs (except Stem BCFs), adding only 4.1–4.9% than climatic data used. Therefore, to reduce the uncertainty induced by BCFs in forest carbon estimates, we should apply values of BCFs for a specified forest type, and also consider climatic and edaphic effects, especially climatic effect, in developing predictive models of BCFs (except Stem BCF).

Editor: Keping Ma, Institute of Botany, Chinese Academy of Sciences, China

Funding: This study was funded by Forestry Public Welfare Programme of China (201304205), Strategic Priority Programme of Chinese Academy of Sciences (XDA05060102 and XDA05050602) and Fujian Provincial S&T Programme (2013Y0083 and 2011Y0052), and partly by National Natural Science Foundation of China (31200363 and 31270588) and International Partnership Programme for Creative Research Teams (KZCX2-YW-T08). The funders had no role in study design, data collection and analysis, decision to publish, or preparation of the manuscript.

Competing Interests: The authors have declared that no competing interests exist.

* E-mail: yunjianluo@gmail.com (YL); wangxk@rcees.ac.cn (XW)

¶ These authors contributed equally to this work.

Introduction

Reducing uncertainties in estimates of forest biomass carbon storage and their changes is a prerequisite not only for resolving a long-standing controversy about the role of forests in the global carbon cycle [1], but also for improving the accuracy of forest carbon accounting [2,3]. With a large number of statistically valid plots covering nearly all forest types with different site conditions, widely available forest inventory data (FID) are recognized as the best ground-based observations for identifying the magnitude and spatiotemporal patterns of forest biomass carbon storage and its changes at large scales (e.g. landscape, regional and global) [4,5]. Moreover, FID-based estimates are usually used to calibrate ones obtained by remote sensing, modeling or their combinations [6,7]. Unfortunately, traditional FID generally provides information on tree volume (e.g. merchantable volume or growing stock volume) but no available biomass data. Therefore, the requirement to accurately obtain FID-based biomass carbon storage and its changes draws attention in order to assess forest carbon stocks and their dynamics.

Generally, FID-based biomass carbon storages are estimated using a volume-to-biomass conversion method and carbon concentration (CC) [8,9]. The method has gained prominence in biomass estimation [2,8]. A convenient descriptor of the volume-to-biomass conversions is biomass conversion (or expansion) factors (BCFs) (i.e. the ratios of biomass to volume) [10–16]. BCFs are increasingly used in favor of volume-to-biomass conversion models (i.e. the biomass as a function of the volume) and the product of wood density times biomass expansion factors (i.e. ratios of some or all of tree biomass to stem biomass) [2]. BCFs have been identified as a large source of uncertainty in FID-based carbon estimates [8], because inventories of stem volumes tend to be precise [4,17], and CC is a relatively constant variable (e.g. 0.45 and 0.50) [18].

Knowledge of sources of the variation in BCFs is crucial to reduce uncertainty in BCFs, thus FID-based carbon estimates. Furthermore, this has important implications for how resources should be invested to determine BCFs. Indeed many early studies used static values of BCFs [19–21]. Recent studies have resulted in significant advances in reducing the uncertainty of BCFs. They focused on tree age- and size-related variation in BCFs for specific forest types [11–16,22,23]. Now local-level evidences indicate that BCFs vary with site conditions (e.g. temperature, water and nutrients) [24]. However, the effects and relative importance of environmental factors on BCFs remains unclear at large scales (e.g. landscape and national).

China has diverse range of forest types, ranging from tropical rainforest to boreal forest [25]. It has a good representation of global biome heterogeneity and environmental gradients, and thus provides an excellent opportunity to explore possible sources of variation in BCFs. During recent years, a comprehensive forest biomass dataset of China has been compiled based on a literature survey. Our objective was to use the dataset to determine values of various BCFs specific to forest types, and then to explore the variation in BCFs in relation to forest type, stand development and environmental factors (climate and soil fertility). Our results will contribute to accurately determining the values and predictive models of BCFs, thereby improving the reliability of the estimates of forest biomass and carbon stocks.

Materials and Methods

The dataset and data categorization

To date, we have compiled a comprehensive forest biomass dataset of China from available published studies and previous datasets [26–28]. For details, see File S1. For calculating BCFs and exploring their possible sources of the variation, only paired data with both growing stock volume and biomass measurements (for at least tree components, i.e. stem, branch and foliage) were selected from our dataset. As a result, a total of 1,099 paired data from 252 sites were available for subsequent analyses. These sites showed a sound spatial distribution across China (Fig. S1 in File S2), ranging from -5.1 to 23.5°C in mean annual temperature (MAT) and from 223 to 2300 mm in mean annual precipitation (MAP).

According to dominant tree species and ecogeographical zones given by a Chinese Vegetation Classification System [25], we categorized these stands into 15 forest types with a sample size of at least 10 (Table 1). For details, see Table S1 in File S2. We also categorized these stands by leaf form into coniferous forest and broadleaved forest, by leaf lifespan into deciduous forest and evergreen forest, and by stand origin into natural forest and planted forest.

Calculation of BCFs

Here we defined BCFs (in Mg m^{-3}) as stand-level ratios for estimating stand-level living biomass (the oven-dried mass per unit area):

$$BCF_i = B_i/V \quad (1)$$

where B_i is living biomass (in Mg ha^{-1}) of the *i*th tree component (i.e. stem, branch, foliage, and root), aboveground (the sum of all living aboveground tree components) and whole tree (the sum of all living tree components), and V is growing stock volume (in m^3 ha^{-1}). BCFs were developed for tree components (i.e. stem, branch, foliage, and root), as well as aboveground and whole tree biomass. Biomass of dead organic matter (i.e. dead wood and dead branch) and understory vegetation was not included in the calculation of BCFs.

Statistical analysis

Stands less than 20 years old were only used in analyses for the effects of stand development (expressed by stand age and tree size), but not for other effects (e.g. climatic and edaphic) to minimize the influences of stand age and other unidentifiable factors (e.g. previous management practices and local soil conditions at early growth stages). Differences in BCFs between groups were examined using *t*-test for two groups, or using one-way analysis of variance (ANOVA) followed by Duncan *post hoc* test for more than two groups. Correlation analyses and linear regression were performed to identify the relationships of BCFs with stand development and climatic variables (MAT and MAP). Nonlinear relationships of BCFs with climatic variables were also explored. Using data of older stands (≥20 years), furthermore, multiple regression analyses with backward stepwise procedure were performed to explore the influences of environmental variables (i.e. MAT, MAP and soil fertility) on BCFs. The explanatory variables were selected from MAT, MAP, soil fertility class and interactions between these variables, as well as the quadratic terms of MAT and MAP if they had nonlinear relationships with BCFs. The model selection was based on AIC (Akaike's information criterion) value. A substantial change is considered when a change in AIC of >2 in the performance of the final model over the alternatives [29]. Explanatory power of the model was assessed using the coefficient of determination R^2.

As ratio data are usually not normally distributed, data of BCFs were log_{10}-transformed to normalize the distribution prior to statistical analyses. Statistical significance was determined at $P \leq 0.05$. Statistical analyses were performed using software SYSTAT version 13.0 (Systat Software Inc., Chicago, Illinois).

Results

As expected, BCFs exhibited great variation in China's forests (Table 1). Values of BCFs were highest for whole tree and to a lesser extent, above-ground biomass, averaging 0.890 and 0.740 Mg m^{-3}, respectively (Table 1). They were much smaller for stems, roots, branches and foliage, averaging 0.505, 0.172, 0.137, and 0.096 Mg m^{-3}, respectively. The coefficients of variation (i.e. the ratio of the standard deviation to the mean, in percentage) in these estimates ranged between 49% (for Stem BCF) and 146% (for Foliage BCF).

Biotic influences on BCFs

Large variations in BCFs existed across forest types (Table 1). The largest variation in BCFs across forest types was in the branch and foliage components, with these varying by 2.7–3.2 times among forest types. The variation in BCFs for other components was only 1.5–1.8 times (Table 1). Also, BCFs varied greatly between functional groups, although not all BCFs differed significantly (Table 2). Coniferous forest had larger Foliage BCFs than broadleaved forest ($P<0.001$), while other BCFs were smaller ($P<0.001$). Compared with evergreen forest, deciduous forest had larger Stem BCFs, Aboveground BCFs, Root BCFs and Whole BCFs ($P \leq 0.002$), and smaller Foliage BCFs ($P<0.001$), but there were no significant differences in Branch BCFs ($P=0.249$). Natural forest generally had larger Root BCFs ($P=0.030$) than planted forest, and smaller Foliage BCFs ($P<0.001$), while other BCFs showed no significant differences ($P>0.05$). When major genera were analyzed separately, *Quercus*, *Castanopsis* and *Phoebe* forests had relatively larger BCFs (except Foliage BCF) than *Cunninghamia*, *Larix*, *Picea* and *Pinus* forests, while *Picea* and *Pinus* forests had larger Foliage BCFs than *Cunninghamia*, *Larix*,

Table 1. Statistics of biomass conversion factors (BCFs, Mg m^{-3}) for forest types (groups) in China *.

Forest type (group) †	*n*	Stem BCF	Branch BCF	Foliage BCF	Aboveground BCF	*n*	Root BCF	Whole BCF
All data	1099	0.505 (0.247)	0.137 (0.112)	0.096 (0.141)	0.740 (0.412)	796	0.172 (0.125)	0.890 (0.489)
Forest type (group)								
Abies and *Picea*	37	0.513 (0.226)	0.136 (0.110)	0.141 (0.277)	0.792 (0.573)	25	0.164 (0.108)	1.042 (0.775)
Cunninghamia lanceolata	239	0.394 (0.161)	0.086 (0.090)	0.136 (0.222)	0.617 (0.428)	194	0.152 (0.132)	0.741 (0.488)
Cupressus and *Fokienia*	22	0.451 (0.103)	0.111 (0.049)	0.111 (0.087)	0.674 (0.191)	18	0.141 (0.047)	0.799 (0.244)
Larix	85	0.512 (0.127)	0.119 (0.101)	0.044 (0.035)	0.674 (0.219)	38	0.186 (0.065)	0.915 (0.280)
Pinus koraiensis	35	0.427 (0.082)	0.140 (0.070)	0.102 (0.064)	0.668 (0.187)	33	0.156 (0.062)	0.830 (0.243)
P. massoniana	80	0.550 (0.177)	0.139 (0.125)	0.066 (0.078)	0.755 (0.321)	63	0.139 (0.067)	0.914 (0.391)
P. tabuliformis	105	0.442 (0.101)	0.165 (0.106)	0.121 (0.113)	0.733 (0.278)	74	0.149 (0.044)	0.861 (0.283)
Other temperate pines and conifers	47	0.567 (0.323)	0.230 (0.112)	0.131 (0.091)	0.933 (0.467)	39	0.225 (0.160)	1.110 (0.591)
Other subtropical pines and conifers	69	0.442 (0.123)	0.130 (0.077)	0.099 (0.084)	0.672 (0.228)	53	0.155 (0.090)	0.782 (0.260)
Betula and *Populus*	49	0.633 (0.410)	0.205 (0.168)	0.093 (0.126)	0.930 (0.645)	28	0.178 (0.155)	0.879 (0.651)
Other deciduous broadleafs	74	0.640 (0.187)	0.196 (0.108)	0.049 (0.046)	0.888 (0.278)	50	0.225 (0.107)	1.068 (0.342)
Acacia, *Casuarina* and *Eucalyptus*	95	0.618 (0.221)	0.102 (0.098)	0.067 (0.086)	0.790 (0.356)	59	0.175 (0.103)	0.940 (0.404)
Typical evergreen broadleafs	56	0.577 (0.456)	0.158 (0.107)	0.061 (0.065)	0.795 (0.584)	43	0.246 (0.274)	1.071 (0.925)
Other evergreen broadleafs	30	0.675 (0.581)	0.202 (0.156)	0.098 (0.162)	0.975 (0.858)	22	0.229 (0.151)	1.127 (0.689)
Mixed coniferous and broadleaved forest	76	0.481 (0.175)	0.137 (0.084)	0.088 (0.080)	0.707 (0.260)	57	0.172 (0.089)	0.911 (0.363)

* Data are presented as means (standard deviation), and sample sizes (*n*) are also given.
† Forest types (groups) are described in Table S1 in File S2.

Table 2. Comparison of biomass conversion factors (BCFs, Mg m^{-3}) between functional groups *.

Functional group	*n*	Stem BCF	Branch BCF	Foliage BCF	Aboveground BCF	*n*	Root BCF	Whole BCF
Leaf form								
Coniferous forest	365	0.447 (0.121) b	0.107 (0.078) b	0.063 (0.063) a	0.620 (0.216) b	258	0.131 (0.059) b	0.741 (0.251) b
Broadleaved forest	110	0.589 (0.145) a	0.160 (0.087) a	0.036 (0.023) b	0.786 (0.202) a	82	0.195 (0.077) a	0.956 (0.243) a
Leaf lifespan †								
Deciduous forest	117	0.556 (0.152) a	0.126 (0.087) a	0.032 (0.021) b	0.714 (0.215) a	62	0.197 (0.079) a	0.944 (0.254) a
Evergreen forest	376	0.460 (0.131) b	0.116 (0.081) a	0.064 (0.062) a	0.642 (0.224) b	296	0.136 (0.061) b	0.762 (0.256) b
Stand origin								
Natural forest	171	0.499 (0.137) a	0.120 (0.072) a	0.041 (0.029) b	0.661 (0.185) a	105	0.159 (0.067) a	0.820 (0.224) a
Planted forest	336	0.477 (0.147) a	0.119 (0.087) a	0.063 (0.064) a	0.662 (0.243) a	264	0.142 (0.070) b	0.789 (0.285) a
Species genera								
Cunninghamia	63	0.355 (0.062) e	0.040 (0.018) e	0.026 (0.014) b	0.422 (0.080) d	55	0.083 (0.020) d	0.511 (0.093) d
Larix	56	0.476 (0.102) d	0.068 (0.029) de	0.027 (0.013) b	0.572 (0.110) c	21	0.149 (0.052) c	0.758 (0.173) c
Picea	30	0.480 (0.147) cd	0.113 (0.054) bc	0.083 (0.072) a	0.679 (0.212) c	18	0.145 (0.065) c	0.868 (0.280) bc
Pinus	182	0.463 (0.121) d	0.133 (0.083) b	0.078 (0.067) a	0.678 (0.218) c	134	0.140 (0.58) c	0.795 (0.239) c
Castanopsis	12	0.569 (0.072) b	0.187 (0.076) a	0.038 (0.019) b	0.794 (0.138) b	9	0.186 (0.038) b	0.993 (0.165) b
Phoebe	13	0.550 (0.158) bc	0.090 (0.063) cd	0.039 (0.023) b	0.679 (0.216) c	13	0.189 (0.066) b	0.868 (0.261) bc
Quercus	22	0.709 (0.166) a	0.179 (0.065) a	0.033 (0.02) b	0.921 (0.191) a	17	0.242 (0.053) a	1.161 (0.140) a

* Data of older stands (≥20 years) were used in this table. Data are presented as means (standard deviation), and sample sizes (*n*) are also given. Different small letters indicate significant ($P<0.05$) differences between functional groups.

† Data for stands dominated jointly by deciduous and evergreen trees were not included in this category.

Castanopsis, *Phoebe* and *Quercus* forests. Among the seven genera, *Cunninghamia* forest had the smallest BCFs.

All BCFs had significant relationships with stand development (Table 3). They decreased significantly ($P<0.001$) with mean DBH, mean tree height and growing stock volume. Moreover, most BCFs decreased with stand age ($P<0.01$). However, increasing trends with stand density were only for Foliage BCFs, Root BCFs and Whole BCFs ($P<0.001$). Among these biometric stand variables, growing stock volume was generally the most powerful variable, which can explain 17.6–54.3% of total variances of BCFs.

Environmental influences on BCFs

BCFs except Stem BCF had significant relationships with MAT and MAP ($P<0.001$) (Fig. 1 and 2). Branch BCFs, Foliage BCFs and Aboveground BCFs had humped-shaped relationships with MAT ($P<0.001$) (i.e. they increased with increasing MAT and then declined beyond *ca.* 7.5–8.2°C) (Fig. 1B, 1C and 1D), and they had negatively linear relationship with MAP ($P<0.001$) (Fig. 2B, 2C and 2D). Root BCFs and Whole BCFs were negatively correlated with MAT ($P<0.001$) (Fig. 1E and 1F), but they showed different changes with MAP (Fig. 2E and 2F). Root BCFs showed a U-shaped relationships with MAP ($P<0.001$) (i.e. it decreased and then increased above *ca.* 1400 mm) (Fig. 2E), whereas Whole BCFs did a negatively linear relationship with MAP ($P<0.001$) (Fig. 2F). MAT and MAP explained 5.6–22.9% and 5.5–18.6% of variation in BCFs (except Stem BCFs), respectively (Table 4). When climatic data (MAT and MAP) were combined, they explained 10.0–25.0% of the variation in BCFs (except Stem BCFs) (Table 4), and only MAP had significant effects on both Root BCFs and Whole BCFs (Table S2 in File S2). Furthermore, stronger climatic effects were found on BCFs for functional components (i.e. branch, foliage and root) than BCFs for combined components (i.e. aboveground section and whole trees) (Table 4).

A general trend for BCFs was observed to decrease and then increase from low to high soil fertility, though Stem BCFs did not differ significantly among fertility classes ($P=0.253$) (Fig. 3). BCFs except Stem BCF were greater ($P<0.001$) for lower fertility (Class I and II) than for intermediate fertility(Class III or IV), while generally smaller ($P<0.01$) for intermediate fertility than higher fertility (Class IV and V). Moreover, soil fertility class explained only 3.5–12.1% of variation in BCFs (except Stem BCF) (Table 4). When qualitative soil fertility and climatic data (MAT and MAP) were combined, they explained 14.1–29.7% of the variation in in BCFs (except Stem BCF), adding only 4.1–4.9% than climatic data used (Table 4).

Discussion

Factors influencing BCFs

Using the statistics (mean, standard deviation and sample size) in Table 1, coefficients of variation (expressed as a percentage) for BCFs were calculated and were in an increasing order: Stem BCF (49%), Whole BCF (55%), Aboveground BCF (56%), Root BCF (73%), Branch BCF (81%) and Foliage BCF (146%). This suggests that BCFs, especially Foliage BCF, Branch BCF and Root BCF, could be sensitive to biotic and environmental variables, and also might have inherent variation in growth rhythm and biomass allocation strategy.

(1) Influence of stand development on BCFs. Our study found that BCFs generally decreased with stand age and size (e.g. DBH, tree height and growing stock volume) (Table 3). Similar results for one or more BCFs are observed in specific forest types, e.g. *Betula pubescens* forest [12], *Larix* forest [23], *Picea abies* forest [12,14], *Pinus densiflora* forest [13], *Pinus sylvestris* forest [12,15], major forest types of China [11] and temperate broadleaved forest of USA [16].

As trees become larger, stem volume and stem biomass proportion in whole tree biomass increases, meanwhile biomass proportions of other tree compartments decrease proportionally (or remain keep more-or-less unchanged) [30]. In addition, wood density slowly decreases as the trees grow older, although it increases eventually again in older stands as the annual rate of growth abates [14]. Therefore, the trend for BCFs to decrease with stand development appears to be a joint result of changes in the growth rhythm and biomass allocation as stands grow, as well as the definition (i.e. the ratios of biomass to volume) of BCFs.

(2) Influence of forest traits on BCFs. Our study found that all BCFs exhibited large variation across forest types (Table 1), while not all of them varied significantly with functional groups (leaf form, leaf lifespan, stand origin, and species genera) (Table 2). Some factors are considered to regulate growth rhythm and biomass allocation strategy of trees in order to accommodate the changes in environments [30–33]: photosynthetic rate (PR), relative growth rate (RGR), leaf traits (e.g. leaf form and lifespan), inherent growth rhythm and soil characteristics (e.g. soil water and nutrient availability).

Generally, plants with higher PRs and RGRs show larger stem and root proportions and smaller foliage proportions to whole tree biomass in order to acquire more water and nutrients for achieving similar growth than plants with lower RGRs and PRs [31,32,34]. Studies have found that broadleaved trees and deciduous trees generally have higher PRs and RGRs than coniferous and evergreen counterparts [32,34]. These can explain why broad-

Table 3. Pearson correlations between biomass conversion factors (BCFs) and stand variables.

BCFs (Mg m^{-3})	Stand age (years)	Mean DBH (cm)	Mean tree height (m)	Stand density (trees ha^{-1})	Growing stock volume (m^3 ha^{-1})
Stem BCF	−0.043 *ns*	−0.284 *	−0.186 *	−0.058 *ns*	−0.420 *
Branch BCF	−0.109 *	−0.411 *	−0.550 *	−0.007 *ns*	−0.629 *
Foliage BCF	−0.404 *	−0.656 *	−0.789 *	0.311 *	−0.734 *
Aboveground BCF	−0.181 *	−0.498 *	−0.512 *	0.032 *ns*	−0.682 *
Root BCF	−0.238 *	−0.566 *	−0.588 *	0.140 *	−0.697 *
Whole BCF	−0.213 *	−0.599 *	−0.585 *	0.139 *	−0.737 *

All data were $\log_{10}$-transformed to linearize relationships between variables and also to reduce the influence of outlying data. *ns*, not significant ($P>0.05$); *, $P<0.001$.

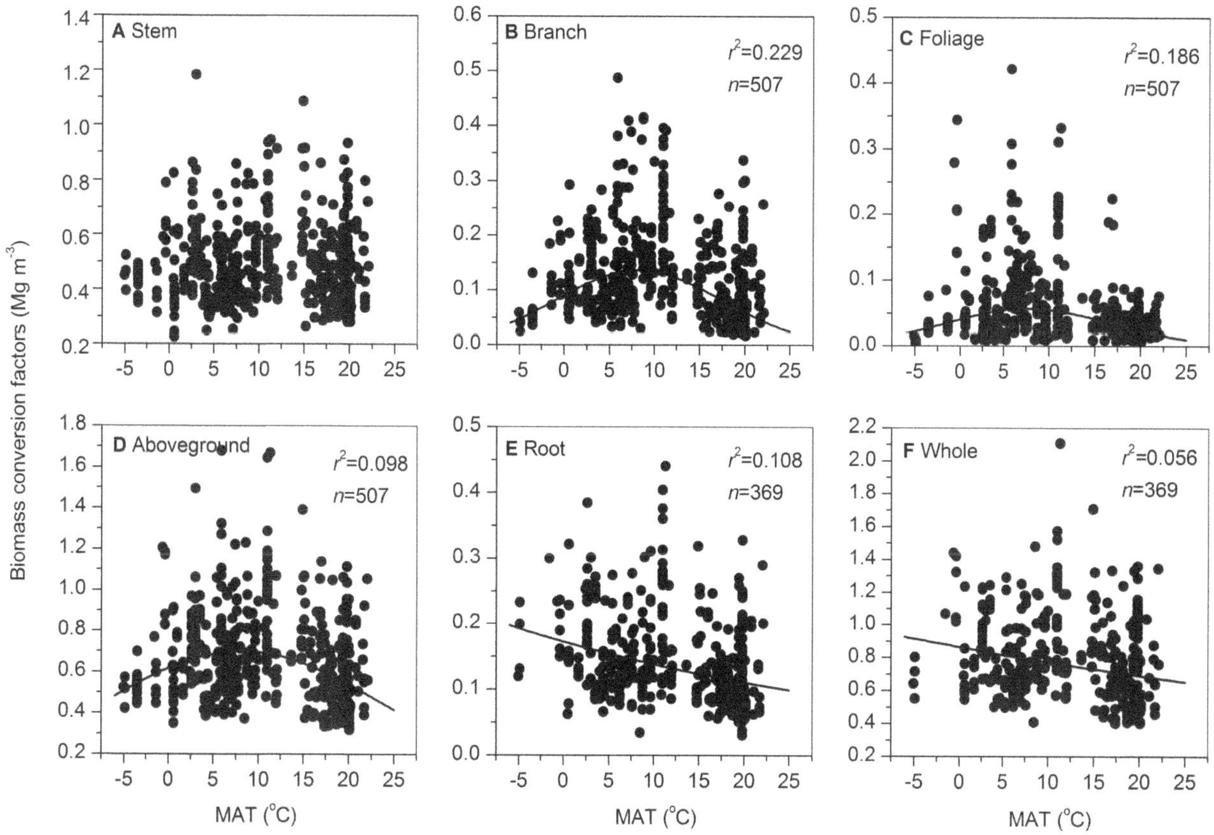

Figure 1. Changes in biomass conversion factors with mean annual temperature (MAT). The regression equations are presented when there are significant relationships between variables ($P<0.05$). Regression equations are given in Table S2 in File S2.

leaved forest and deciduous forest have larger Stem BCFs and Root BCFs and smaller Foliage BCFs than coniferous forest and evergreen forest, and also why *Quercus*, *Castanopsis* and *Phoebe* forests had relatively larger BCFs (except Foliage BCF).

In addition, plantations are usually established in more productive environments (e.g. more fertile and moister soil) than natural forests, indicating that they can achieve optimal growth with larger foliage proportions and lower root proportions [33,35]. This could explain why planted forest has larger Foliage BCFs and smaller Root BCFs than natural forest.

(3) Influence of environmental factors on BCFs. At broad scales, temperature and precipitation are considered as very important factors in shaping distribution, structure and ecological process of terrestrial biomes, and also influence soil microbial activity and nutrient availability [36]. Luo et al. [37] found that higher temperature and precipitation favor higher tree growth and more photosynthate allocated to tree stem, suggesting that the size and biomass proportion of stems have positive relationships with MAT and MAP. This promotes the defensive capacity of trees against environmental perturbation, and might contribute to ensure their better soil-to-leaf hydraulic conductance [38,39].

Our study found that relationships of BCFs with MAT and MAP varied with tree components (e.g. branch, foliage, and root) (Figs. 1 and 2). Branch BCFs, Foliage BCFs and Aboveground BCFs decreased with MAP (Fig. 2), which can also be explained by higher tree growth and light use efficiency, and hence decreasing biomass proportions of photosynthetic compartments (i.e. foliage and branch) with increasing MAP. However, a humped-shaped change of BCFs with increasing MAT was observed for three BCFs (i.e. Branch BCFs, Foliage BCFs and Aboveground BCFs) (Fig. 1), which can be explained by the following causes. At lower temperatures, increasing temperature favors tree growth by better growth conditions (e.g. higher soil microbial activity and nutrient availability), and thus results in more branches and leaves for more water and photosynthate at a tree size; at higher temperatures, larger transpiration and respiration lead to a reduction in biomass allocation to branches and leaves to minimize maintenance costs at a tree size. The decreasing trend of Root BCFs with MAT (Fig. 1) can be attributed to higher tree growth and lower root biomass proportion with increasing MAT [37], and the effect of MAT might be indirect because the effect disappeared when both MAT and MAP were considered (Table S2 in File S2). Root BCFs had a U-shaped pattern with increasing MAP across China's forests (Fig. 2). With increasing precipitation, water supply and soil fertility becomes abundant (Fig. S2 in File S2), and thus tree growth will be accelerated and trees can allocate less biomass to roots to absorb soil water and nutrients. However, when precipitation exceeds the demand of trees, tree growth will be inhibited by soil nutrient availability rather than water (Fig. S2 in File S2), resulting in more biomass would be allocated into roots to absorb more nutrients from relatively infertile soil with high precipitation [37].

Terrestrial plants take up most of their nutrients and water directly from soils. Soil chemical attributes (e.g. pH and mineral nutrient availability) are critical to plant growth and thus affect biomass allocation patterns [33,40,41]. Moreover, climate influ-

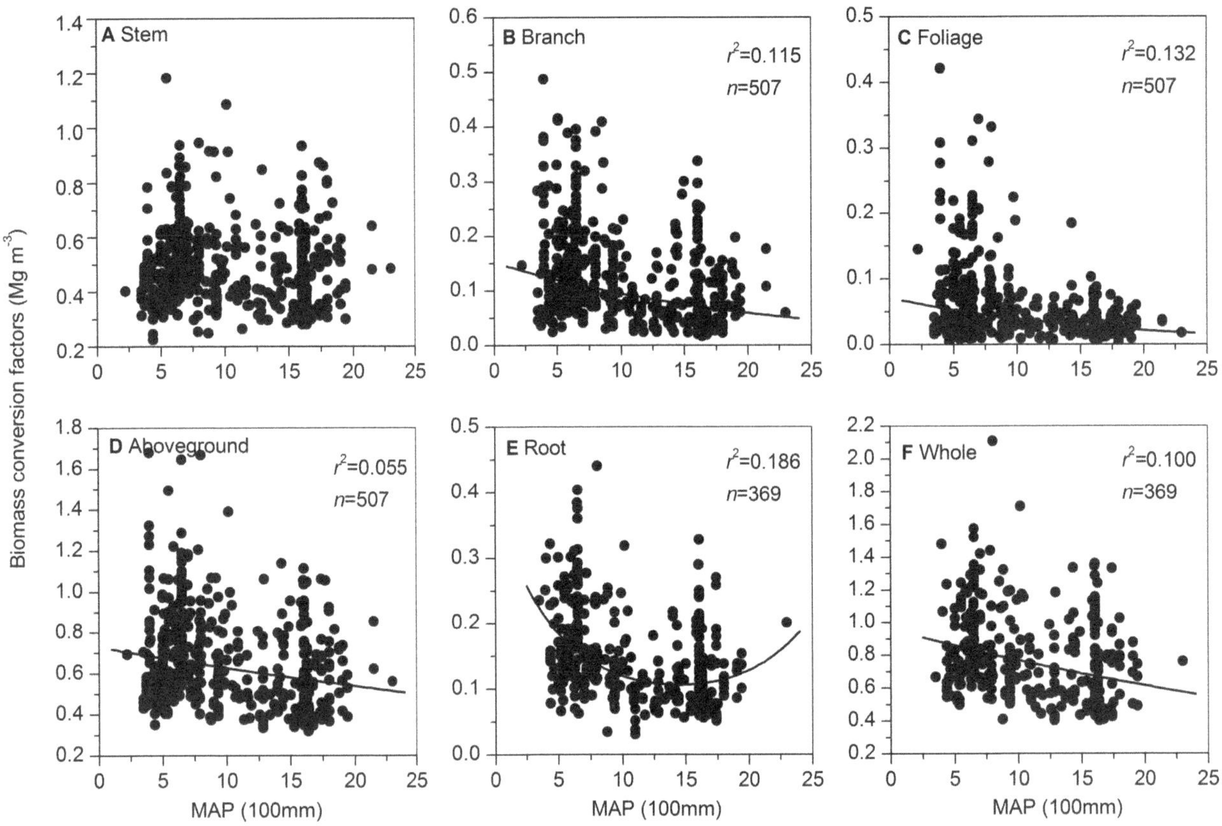

Figure 2. Changes in biomass conversion factors with mean annual precipitation (MAP). The regression equations are presented when there are significant relationships between variables ($P<0.05$). Regression equations are given in Table S2 in File S2.

ences soil chemical attributes and thus shapes the vegetation biogeography [36]. Across China's forests, MAT and MAP showed a humped-shaped change with increasing soil fertility (Fig. S2 in File S2). As soil fertility increases, soil nutrients become abundant and thus trees show rapider growth in the size and broader tree rings. That might lead to smaller wood density and less foliage and root proportions in tree biomass, which can explain a decreasing trend in BCFs with increasing soil fertility at lower soil fertility. However, when soil nutrients exceed the demand of trees, tree growth will be inhibited by temperature and precipitation, and trees show slower growth in the size and narrower tree rings. That might lead to larger wood density and more foliage and root proportions in tree biomass, which can explain an increasing trend in BCFs with increasing soil fertility at higher fertility. In addition, soil fertility showed only a weak effect on BCFs, which add only 4.1–4.9% in explaining the variation of BCFs meanwhile climatic data (MAT and MAP) were considered (Table 4). And the effect of soil fertility was reflected by the interactions between soil fertility and climate (MAT and MAP) (Table S3 in File S2).

Table 4. Explanatory powers (R^2 values) of final models for the effects of climate and soil fertility on biomass conversion factors (BCFs, Mg m^{-3}).

Model	Stem BCF	Branch BCF	Foliage BCF	Aboveground BCF	Root BCF	Whole BCF
MAT	0.000 ns	0.229 **	0.186 **	0.098 **	0.108 **	0.056 **
MAP	0.002 ns	0.115 **	0.132 **	0.055 **	0.186 **	0.100 **
Climate	0.009 ns	0.250 **	0.210 **	0.112 **	0.186 **	0.100 **
Soil fertility class	0.011 ns	0.104 **	0.121 **	0.077 **	0.037 *	0.043 *
Climate+Soil	0.028 ns	0.297 **	0.259 **	0.159 **	0.233 **	0.141 **

Models 'Climate' denoted that the explanatory variables were selected by a backward stepwise procedure from mean annual temperature (MAT, °C), mean annual precipitation (MAP, 100 mm) and interactions between these variables, as well as the quadratic terms of MAT and MAP, which had nonlinear relationships with several BCFs (see Fig. 1 and 2). Similarly, models 'Climate+Soil' denoted that the explanatory variables were selected by a backward stepwise procedure from MAT, MAP, soil fertility class, interactions between these variables and quadratic terms of continuous variables (MAT and MAP). Details of final models 'Climate' and 'Climate+Soil' were given in Table S2 and S3 in File S2, respectively. Significance of a model: *ns*, not significant ($P>0.05$);

*, $P<0.01$;

**, $P<0.001$.

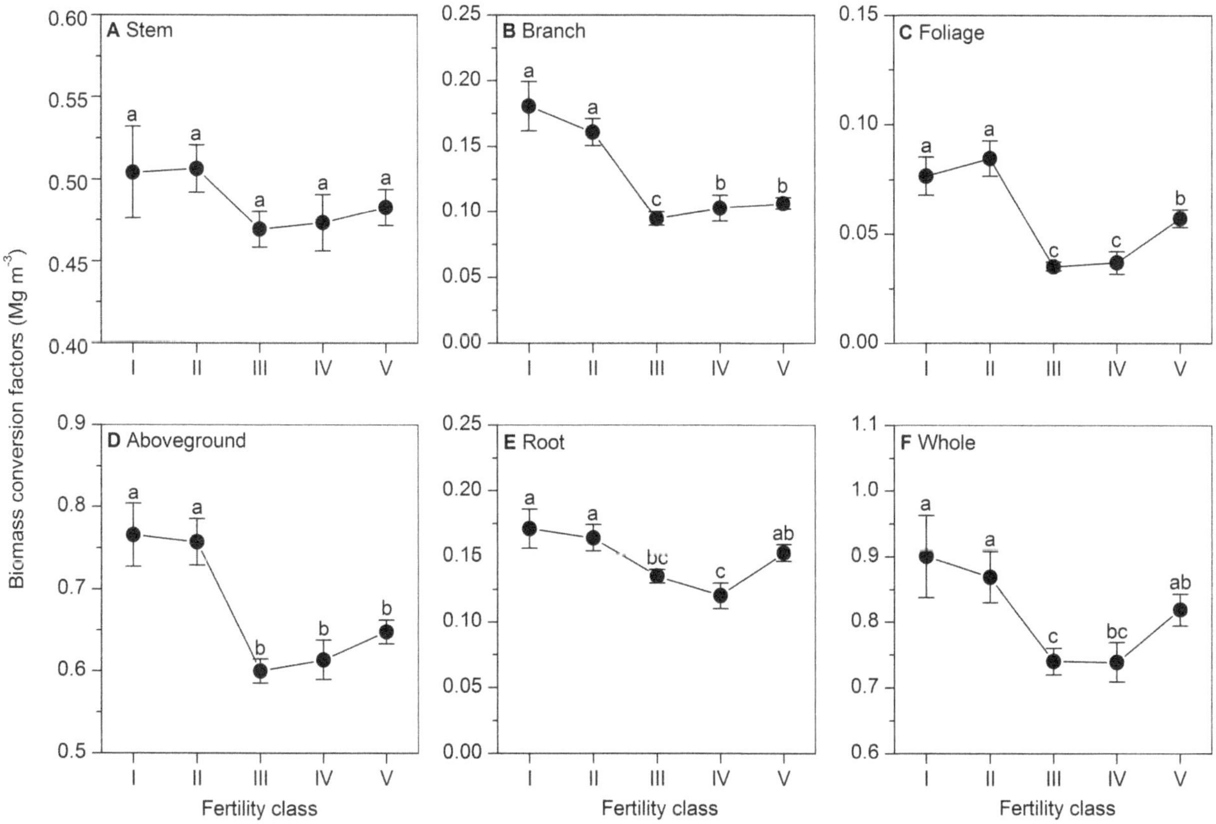

Figure 3. Changes in biomass conversion factors with soil fertility. According to the background values of soil organic matter content, soil fertility is divided into five classes: (I) ≤1.0 g (100 g)$^{-1}$, (II) 1.0–2.0 g (100 g)$^{-1}$, (III) 2.0–3.0 g (100 g)$^{-1}$, (IV) 3.0–4.0 g (100 g)$^{-1}$, and (V) ≥4.0 g (100 g)$^{-1}$. Mean and standard error are shown for each class. Different small letters indicate significant ($P<0.05$) differences between fertility classes.

Apart from the above mentioned influencing factors, management practices (e.g. thinning, pruning and fertilization) might lead to changes in growth rhythm and biomass allocation of trees, and thus affect BCFs. In this study, however, we did not analyze effects of management practices, largely due to the lack of management descriptions in original data sources.

Implications for biomass and carbon accounting

Our study presented forest type-specific values of various BCFs in Table 1. When comparing our values for BCFs with others available [2], means of Aboveground BCFs for China's forests and for most forest types (Table 1) were smaller than IPCC default values (mean = 1.405 Mg m^{-3}; range = 0.692–2.913 Mg m^{-3}) [2], indicating that a large overestimation would result if IPCC default values was used in the estimation of China's forest carbon storage. In addition, China conducted seven consecutive national forest inventories between 1973 and 2008, which can provide statistical data of forest area and growing stock volume by age class and forest type [17]. To be compatible with these FIDs, forests were categorized by tree species, growing region and stand origin into five age classes (young, middle-aged, premature, mature and overmature) (Table S4 in File S2), and then forest type-specific BCFs by age class were given in Table S5–S10 in File S2. However, the use of the values of BCFs for premature, mature and overmature forests might induce somewhat uncertainty in forest biomass and carbon estimates, due to limited data for the three growth stages.

Considering large coefficients of variation for BCFs, the use of constant values of BCFs, especially Foliage BCF, Branch BCF and Root BCF, would lead to high uncertainties in the estimates of forest biomass and carbon. Our results showed that high variation in BCFs were related to forest types (Table 1), functional groups (leaf form, leaf lifespan, stand origin and species genera) (Table 2), stand development (Table 3), climate (MAT and MAP) (Fig. 1 and 2) and soil fertility (Fig. 3). In order to reduce uncertainties induced by BCFs, we recommend that: (i) different values of BCFs should be selected for forest types and functional groups, and (ii) climatic and edaphic factors, especially climatic factors, should be considered in developing predictive models of BCFs (except Stem BCF). To further reduce uncertainty induced by BCFs in forest carbon estimates, future emphases should be placed on obtaining BCF estimates for region and forest types where there is currently little data available, and also on developing the predictive models of BCFs (except Stem BCF) for a specified forest type integrating environmental factors, especially climatic factors (MAT and MAP). Certainly, if the predictive model of a certain BCF integrating stand (e.g. growing stock volume) and environmental factors (e.g. MAT, MAP, and soil fertility) has poor performance (e.g. low R^2), it is unsuitable for the model to estimate values of the BCF which are used to calculate forest biomass and carbon stock.

Conclusions

BCFs are considered as important parameters in quantifying forest biomass carbon stock and dynamics. Knowledge of the sources of variation in BCFs is prerequisite to accurately determining the values of BCFs, thereby reducing the uncertainties in forest biomass carbon estimates. However, prior to this study there has been little information on possible causes of the variation in BCFs, especially environmental effects.

Our study was the first to give China's forest type-specific values of BCFs, and to examine the variation in BCFs in relation to forest type, stand development, climate and soil fertility. Results showed that BCFs generally varied significantly with forest types, stand development (stand age and size), climate (MAT and MAP) and soil fertility (fertility class). These results indicate that: (i) different values of BCFs should be selected for forest types and functional groups, and (ii) climatic and edaphic factors, especially climatic factors, should be considered in developing predictive models of BCFs (except Stem BCF).

Supporting Information

File S1 Description of forest biomass dataset of China.

File S2 It included 2 figures and 10 tables. (DOC). **Figure S1** Spatial distribution of study sites used in this study; **Figure S2** Relationships between soil fertility and climate: (A) mean annual temperature (MAT) and (B) mean annual precipitation (MAP); **Table S1** Categorization of forest type by dominant species and ecogeographical zone; **Table S2** Summary of multiple regression equations for the effects of climate on biomass conversion factors (BCFs, Mg m^{-3}); **Table S3** Summary of final models for the effects of climate and soil fertility on biomass conversion factors (BCFs, Mg m^{-3}); **Table S4** Categorization of age class by tree species, growing region and stand origin; **Table S5** Statistics of stem biomass conversion factors (Mg m^{-3}) by forest type and age class; **Table S6** Statistics of branch biomass conversion factors (Mg m^{-3}) by forest type and age class; **Table S7** Statistics of foliage biomass conversion factors (Mg m^{-3}) by forest type and age class; **Table S8** Statistics of aboveground biomass conversion factors (Mg m^{-3}) by forest type and age class; **Table S9** Statistics of root biomass conversion factors (Mg m^{-3}) by forest type and age class; **Table S10** Statistics of whole biomass conversion factors (Mg m^{-3}) by forest type and age class.

Acknowledgments

We thank anonymous reviewers for their helpful comments on the earlier manuscript.

Author Contributions

Conceived and designed the experiments: YL XZ XW. Analyzed the data: YL XZ YR. Wrote the paper: YL XZ.

References

1. Houghton RA, Hall F, Goetz SJ (2009) Importance of biomass in the global carbon cycle. Journal of Geophysical Research 114: G00E03.
2. IPCC (Intergovernmental Panel on Climate Change) (2006) 2006 IPCC guidelines for national greenhouse gas inventories (vol. 4): Agriculture, forestry and other land use. Available: http://www.ipcc-nggip.iges.or.jp/public/2006gl/vol4.html. Accessed 2013 Oct 10.
3. Zhang XQ, Wu SH (2010) Forest carbon mitigation programme: Principles and practices. Beijing: China Forestry Publishing House. 375pp.
4. Brown S (2002) Measuring carbon in forests: Current status and future challenges. Environmental Pollution 116: 363–372.
5. Guo ZD, Fang JY, Pan YD, Birdsey R (2010) Inventory-based estimates of forest biomass carbon stocks in China: A comparison of three methods. Forest Ecology and Management 259: 1225–1231.
6. Ren Y, Yan J, Wei XH, Wang YJ, Yang YS, et al. (2012) Effects of rapid urban sprawl on urban forest carbon stocks: Integrating remotely sensed, GIS and forest inventory data. Journal of Environmental Management 113: 447–455.
7. Stinson G, Kurz WA, Smyth CE, Neilson ET, Dymond CC, et al. (2011) An inventory-based analysis of Canada's managed forest carbon dynamics, 1990 to 2008. Global Change Biology 17: 2227–2244.
8. Somogyi Z, Cienciala E, Mäkipää R, Muukkonen P, Lehtonen A, et al. (2007) Indirect methods of large-scale forest biomass estimation. European Journal of Forest Research 126: 197–207.
9. Luo YJ, Zhang XQ, Wang XK, Zhu JH, Hou ZH, et al. (2009) Forest biomass estimation methods and their prospects. Scientia Silva Sinicae 45(8): 129–134.
10. Brown S, Lugo AE (1984) Biomass of tropical forests: A new estimate based on forest volumes. Science 223: 1290–1293.
11. Fang JY, Chen AP, Peng CH, Zhao SQ, Ci LJ (2001) Changes in forest biomass carbon storage in China between 1949 and 1998. Science 292: 2320–2322.
12. Lehtonen A, Mäkipää R, Heikkinen J, Sievänen R, Liski J (2004) Biomass expansion factors (BEFs) for Scots pine, Norway spruce and birch according to stand age for boreal forests. Forest Ecology and Management 188: 211–224.
13. Li XD, Yi MJ, Son Y, Park PS, Lee KH, et al. (2010) Biomass expansion factors of natural Japanese red pine (*Pinus densiflora*) forests in Korea. Journal of Plant Biology 53: 381–386.
14. Pajtík J, Konôpka B, Lukac M (2008) Biomass functions and expansion factors in young Norway spruce (*Picea abies* [L.] Karst) trees. Forest Ecology and Management 256: 1096–1103.
15. Pajtík J, Konôpka B, Lukac M (2011) Individual biomass factors for beech, oak and pine in Slovakia: A comparative study in young naturally regenerated stands. Trees 25: 277–278.
16. Schroeder P, Brown S, Mo J, Birdsey R, Cieszewski C (1997) Biomass estimation for temperate broadleaf forests of the United States using inventory data. Forest Science 43(32): 424–434.
17. Xiao XW (2005) Forest resource inventory of China. Beijing: China Forestry Publishing House. 312pp.
18. Lamlom SH, Savidge RA (2003) A reassessment of carbon content in wood: Variation within and between 41 North American species. Biomass and Bioenergy 25: 381–388.
19. Alexeyev V, Birdsey R, Stakanov V, Korotkov I (1995) Carbon in vegetation of Russian forests: Methods to estimate storage and geographical distribution. Water, Air and Soil Pollution 82: 271–282.
20. Fang JY, Wang GG, Liu GH, Xu SL (1998) Forest biomass of China: An estimate based on the biomass-volume relationship. Ecological Applications 8(4): 1084–1091.
21. Turner DP, Koerper GJ, Harmon ME, Lee JJ (1995) A carbon budget for forests of the conterminous United States. Ecological Applications 5(2): 421–436.
22. Lindner M, Karjalainen T (2007) Carbon inventory methods and carbon mitigation potentials of forests in Europe: A short review of recent progress. European Journal of Forest Research 126: 149–156.
23. Luo YJ, Zhang XQ, Hou ZH, Yu PT, Zhu JH (2007) Biomass carbon accounting factors of *Larix* forests in China based on literature data. Chinese Journal of Plant Ecology 31(6): 1111–1118.
24. Albaugh TJ, Bergh J, Lundmark T, Nilsson U, Stape JL, et al. (2009) Do biological expansion factors adequately estimate stand-scale aboveground component biomass for Norway spruce? Forest Ecology and Management 258: 2628–2637.
25. Zhang XS (2007) Vegetation map of China and its geographic patterns. Beijing: Geological Publishing House, pp. 91–124.
26. Feng ZW, Wang XK, Wu G (1999) Biomass and productivity of forest ecosystems in China. Beijing: Science Press. 241pp.
27. Luo TX (1996) Patterns of net primary productivity for Chinese major forest types and their mathematical models (Ph.D. dissertation). Beijing: Commission for Integrated Survey of Natural Resources, Chinese Academy of Sciences, 211pp.
28. Wang XP, Fang JY, Zhu B (2008) Forest biomass and root-shoot allocation in northeast China. Forest Ecology and Management 255: 4007–4020.
29. Chatterjee S, Hadi AS (2006) Regression analysis by example, 4th Edition. New Jersey: John Wiley & Sons. 375pp.
30. Landsberg J, Sands P (2011) Physiological ecology of forest production: Principles, processes and models. London: Academic Press, 331pp.
31. Antúnez I, Retamosa EC, Villar R (2001) Relative growth rate in phylogenetically related deciduous and evergreen woody species. Oecologia 128: 172–180.
32. Cornelissen JHC, Castro-Díez P, Carnelli AL (1998) Variation in relative growth rate among woody species. In: Lambers H, Poorter H, van Vuuren MMI, editors. Inherent variation in plant growth: Physiological mechanisms and ecological consequences. Leiden: Backhuys Publishers. pp 363–392.
33. Poorter H, Niklas KJ, Reich PB, Oleksyn J, Poot P, et al. (2012) Biomass allocation to leaves, stems, and roots: Meta-analysis of interspecific variation and environmental control. New Phytologist 193: 30–50.
34. Reich PB (1998) Variation among plant species in leaf turnover rates and associated traits: Implications for growth at all life stages. In: Lambers H, Poorter H, van Vuuren MMI, editors. Inherent variation in plant growth: Physiological mechanisms and ecological consequences. Leiden: Backhuys Publishers. pp. 467–487.

35. Luo YJ, Wang XK, Zhang XQ, Booth TH, Lu F (2012) Root:shoot ratios across China's forests: Forest type and climatic effects. Forest Ecology and Management 269: 19–25.
36. Chapin FS, Matson PA, Mooney HA (2002) Principles of terrestrial ecosystem ecology, 2nd Edition. New York: Springer-Verlag. 529pp.
37. Luo YJ, Wang XK, Zhang XQ, Ren Y, Poorter H (2013) Variation in biomass expansion factors for China's forests in relation to forest type, climate and stand development. Annals of Forest Science 70(6): 589–599.
38. Mencuccini M, Martinez-Vilalta J, Vanderklein D, Hamid HA, Korakaki E, et al. (2005) Size-mediated aging reduces vigor in trees. Ecology Letters 8: 1183–1190.
39. Ryan MG, Phillips N, Bond BJ (2006) The hydraulic limitation hypothesis revisited. Plant, Cell and Environment 29: 367–381.
40. Han WX, Fang JY, Reich PB, Woodward FI, Wang ZH (2011) Biogeography and variability of eleven mineral elements in plant leaves across gradients of climate, soil and plant functional type in China. Ecology Letters 14: 788–796.
41. Ren Y, Wei XH, Wang DR, Luo YJ, Song XD, et al. (2013) Linking landscape patterns with ecological functions: A case study examining the interaction between landscape heterogeneity and carbon stock of urban forests in Xiamen, China. Forest Ecology and Management 293: 122–131.

19

Scaling Relationships between Leaf Mass and Total Plant Mass across Chinese Forests

Shanshan Xu, Yan Li, Genxuan Wang*

The state key laboratory of plant physiology and biochemistry, Institute of ecology, College of Life Sciences, Zhejiang University, Hangzhou, China

Abstract

Biomass partitioning is important for illustrating terrestrial ecosystem carbon flux. West, Brown and Enquist (WBE) model predicts that an optimal 3/4 allometric scaling of leaf mass and total biomass of individual plants will be applied in diverse communities. However, amount of scientific evidence suggests an involvement of some biological and environmental factors in interpreting the variation of scaling exponent observed in empirical studies. In this paper, biomass information of 1175 forested communities in China was collected and categorized into groups in terms of leaf form and function, as well as their locations to test whether the allocation pattern was conserved or variable with internal and/or environmental variations. Model Type II regression protocol was adopted to perform all the regressions. The results empirically showed that the slopes varied significantly across diverse forested biomes, between conifer and broadleaved forests, and between evergreen and deciduous forests. Based on the results, leaf form and function and their relations to environments play a significant role in the modification of the WBE model to explore more accurate laws in nature.

Editor: Dafeng Hui, Tennessee State University, United States of America

Funding: This work was supported by the National High-Tech R & D Program (863 Program) for the 12th Five-Year Plan (2011AA100503), Natural Science Fund of China (31330010), and Zhejiang Provincial Natural Science Foundation of China (LZ13C030002). The funders had no role in study design, data collection and analysis, decision to publish, or preparation of the manuscript.

Competing Interests: The authors have declared that no competing interests exist.

* E-mail: fzstsys2@zju.edu.cn

Introduction

Metabolism, which dictates material and energy fluxes through organisms, is a fundamental process in plant life [1]. As the basis for many life-history strategies and ecological theories interested in ecological dynamics and global change, the scaling exponent for the partitioning of total body biomass among metabolic organs is of major interest in basic and applied ecology [2–6]. Foliar leaves are the principle or sole light-harvesting organs for metabolic reactions in photosynthesis, which is the primary metabolic activity involved in carbon fluxes. Therefore, a quantitative understanding of biomass patterns of leaf biomass to total mass is of fundamental importance to terrestrial carbon cycle. WBE model (West, Brown and Enquist, hence WBE) has outlined a set of allometric derivations interrelating individual plant growth as well as vegetative organ biomass partitioning, covering the broadest allometric phenomena at multiply scales of ecological organizations [1,7]. Based on the constraints of maximizing photosynthetic harvesting capacity and resource transport while minimizing hydrodynamic resistance and transport times, the allometric partitioning model predicts that allometric relationships are governed by 1/4 power rules and leaf mass should scale to the $3/4^{th}$ power of total mass [1,4,7–12]. Extensive data for conifers, monocots and dicots spanning 12 orders of magnitude fit the predicted relationships remarkably well, with two empirical exponents of 0.739 (95% CIs: 0.646–0.831) for angiosperms and 0.756 (95% CIs: 0.664–0.846) for gymnosperms, both of which were very close to 3/4 [13,14]. Although the 3/4 scaling exponent has been sustained in many theoretical and empirical studies, an extension of WBE model based on the "fractal continuum" theory predicts that the intraspecific allometry of plant biomass partitioning should fall along a continuum of variation rather than being a single optimal exponent [15]. Nevertheless, deviated from the predictions, a lot of variations in plant allometry have been observed in empirical studies [16–28] and discussed in reviews [29–31]. These variations are assumed to result from altering multiple and/or different physiological and morphological factors in response to ontogeny [17,24,27,28], environmental variability [21,25,26] and species specificity [16,22,23,30]. The universal applications of the WBE are challenged [33–35].

In contrast to the central assumption of WBE model that leaf properties keep consistent with whole-plant size across orders of magnitude of plant size [7,10,36], leaves do vary in form and function with species variety and environmental heterogeneity. As both cause and consequence for plant adaptive strategy, leaves always have different chlorophyll concentrations, light compensation points and light use efficiencies in response to environmental variation or phyletic affiliation [20]. These variations in leaf properties are commonly present in nature and influence plant behavior at a variety of levels of complexity. In addition, leaves are more variable when compared with other component biomass, ascribed to the high sensitivity of leaf mass to light, water, nutrient and soil conditions [37,38]. Therefore, variability in allometry of plant biomass partitioning can be understood by considering specific leaf growth forms or functional groups. Leaf habit and leaf form are vital leaf properties, the substantial difference of which could exert profound impact on populations and communities with interactions of heterogeneous environmental availability. For example, plants in harsh environments prefer relatively high ratio of photosynthetic mass to body mass as a consequence of thicker leaves and relatively smaller body size [21]; in tropical dry forests, broadleaved deciduous trees tend to have greater nutrient resorption efficiencies, specific leaf area and photosynthetic rate compared with broadleaved evergreen trees [39]; boreal coniferous forests are prone to show lower scaling law of M_L with M_T than deciduous forests ascribed to lower light transmittance of the denser canopy in conifers, in which asymmetric competition for light is dominant [40].

Despite the progress made so far, our quantitative understanding of the allocation patterns of M_L and M_T in relation to leaf form and function in trees still remains limited. By analyzing long-term data collected from 1175 tree-dominant large plots located across the Chinese, we address several important questions to the trees growing in Chinese forests: 1) whether there is a general pattern of biomass partitioning for M_L and M_T across diverse forestry biomes in terms of location, leaf habit and leaf form; 2) whether there are different allometric scaling exponents between conifer and broadleaved forest; 3) whether the allocation patterns of leaf biomass and total mass vary across evergreen and deciduous forest. Although our contribution here to the ongoing debate about allometry of biomass partitioning is more empirical evidence rather than theories, the answers to the questions will help understand the biomass partitioning pattern in tree-dominant biomes, which is important to future inquiries to mechanisms underlying these scaling exponents appearing to govern important ecological and evolutionary phenomena.

Materials and Methods

Data resource and collection

In this research, we used a Chinese forest biomass dataset complied from continuous forest-inventory plots of the Forestry Ministry of China in different time-periods during 1970–1994, published forest reports in over 60 Chinese journals and additional investigations by the author himself [41]. The dataset provides standing biomass and production estimates for all plant components (including stem, branch, leaf, root and total plant mass) associated with climatic characters and age information for the selected 1,266 forest plots covering 17 forest types representative of the entire forest vegetation across China. The database has been applied to abundant researches ascribed to its great contribution to biomass estimation in the past decade. More detailed information describing the dataset can be traced in Luo [41], Ni et al. [42] and Zhang et al. [43]. Here, we described only information relevant to the current analysis.

In the field investigations, "allometric equation method" was widely used. Plant biomass of these forest communities were determined by harvesting quadrates [44]. In the forest plots (0.04–0.5 ha), tree height (H, m) and diameter at breast height (D, cm) were measured for all trees with D greater than 2.5 cm. Six to eight trees of different diameters in a plot were cut down as samples (authorized and approved by Changbai Mountain Forest Ecosystem Research Station, Chinese Academy of Sciences and Huitong Forest Ecosystem Research Station in Changbai Mountain and Guangping, respectively; no endangered or protected species included); the separated stem, branch, leaf and root organs, the combination of whose biomass equaled to plant total mass (M_T), were given a direct measurement. The biomass of the rest trees were estimated by species-specific allometric equations derived from these harvested trees based on H and D [44,45]. The equations are always expressed as $M = aD^b$ or $M = a(D^2H)^b$ [41,43]. Leaf biomass (M_L) estimation was conducted in the growing seasons with fully expanded leaves in early July or August across evergreen and deciduous species [44]. Trees with diameter at breast height (D, cm) smaller than 2.5 cm were excluded in the dataset. This method was widely immersed in the database. Moreover, "mean tree method" (chose "mean trees" in plots as samples) and "clear-cutting method" (all trees were cut and measured) were used for biomass estimation as well. Further detailed information about data collection was described in Luo *et al.* [45] and Zhang *et al.* [43].

Data grouping

Group 1. Leaf morphology and physiology are closely associated with environmental variation and species specification. To test whether the M_L versus M_T scaling exponents were conserved or varied with these factors, we divided the dataset into five biomes. The biomes were defined according to temperature- and precipitation-based biome classifications in addition to leaf habit and leaf form. Not all observations in the original dataset of Luo [41] had sufficient information to be placed into a biome category. Sufficient data were available only to evaluate relationships for the temperate (including boreal) and subtropical biomes. Therefore, 1175 selected plots of forests in the database were sorted into the following five biomes based on the combination of latitude (temperate/subtropical), leaf habit (evergreen/deciduous) and leaf form (conifer/broadleaf): Biome 1, Temperature Deciduous Coniferous Forest (TDCF), consisting of 48 plots of Temperature *Larix* Forest; Biome 2, Temperature Evergreen Coniferous Forest (TECF), consisting of 168 plots of Boreal/alpine *Picea-Abies* Forest, 10 plots of Boreal *Pinus sylvestris* var. *mongolica* Forest, 154 plots of Temperate *Pinus tabulaeformis* Forest; Biome 3, Temperature Deciduous Broadleaved Forest (TDBF), consisting of 165 plots of Temperate Typical Deciduous Broadleaved Forest, 127 plots of Temperate/subtropical Montane *Populus-Betula* Deciduous Forest; Biome 4, Subtropical Evergreen Broadleaved Forest (SEBF), consisting of 238 plots of Subtropical Evergreen Broadleaved Forest (without detailed species information); Biome 5, Subtropical Evergreen Coniferous Forest (SECF), consisting of 66 plots of Subtropical *Pinus massoniana* Forest, 98 plots of Subtropical *Cunninghamia lanceolata* Forest, 46 plots of Subtropical Montane *Pinus yunnanensis* and *P. khasya* Forest, 55 plots of Subtropical montane *Pinus armandii*, *P. taiwanensis* and *P. densada* Forest. Detail information about the biomes, including vegetation, location, climate characters and ages could be traced in Table 1.

Group 2. To illustrate the effect of leaf form on biomass partitioning pattern of M_L and M_T, we divided the dataset (the same as in Group 1) into two categories: conifer and broadleaved forests. Conifer Forest (CF) consisted of 48 plots of Temperature *Larix* Forest, 168 plots of Boreal/alpine *Picea-Abies* Forest, 10 plots of Boreal *Pinus sylvestris* var. *mongolica* Forest, 154 plots of Temperate *Pinus tabulaeformis* Forest, 66 plots of Subtropical *Pinus massoniana* Forest, 98 plots of Subtropical *Cunninghamia lanceolata* Forest, 46 plots of Subtropical Montane *Pinus yunnanensis* and *P. khasya* Forest, 55 plots of Subtropical montane *Pinus armandii*, *P. taiwanensis* and *P. densada* Forest. Broadleaved Forest (BF) consisted of 165 plots of Temperate Typical Deciduous Broadleaved Forest, 127 plots of Temperate/subtropical Montane *Populus-Betula* Deciduous Forest, 238 plots of Subtropical Evergreen Broadleaved Forest (without detailed species information).

Group 3. To test the influence of leaf habit on the biomass partitioning of M_L and M_T, we divided the dataset (the same as in Group 1) into two categories: evergreen and deciduous forest. Evergreen Forest (EF) consisted of 168 plots of Boreal/alpine *Picea-Abies* Forest, 10 plots of Boreal *Pinus sylvestris* var. *mongolica* Forest, 154 plots of Temperate *Pinus tabulaeformis* Forest, 238 plots of Subtropical Evergreen Broadleaved Forest (without detailed

Table 1. Vegetation, climatic characteristics and ages of biomes in Luo (1996).

Biome	Altitude (m)	Longitude (°)	Latitude (°)	MAT (°C)	MAP (mm)	PET (mm)	Age (years)
TDCF	441–4,240	86.4–131.8	28.6–52.6	−6.2–4.2	370.9–1274.1	340.3–522.9	30–195
TECF	240–4,200	81.1–131.8	26.14–53.0	−6.6–13.9	369.6–1937.3	328.7–820.5	15–350
TDBF	150–3,640	85.2–134.0	25.75–52.5	−5.5–16.0	241.0–1282.8	358.3–939.9	20–222
SEBF	80–4,160	85.2–120.17	20.7–30.25	3.5–24.2	636.6–2323.1	386.4–1131.7	3–200
SECF	10–3,558	85.2–121.57	18.7–36.4	5.7–24.0	369.7–2989.1	503.8–1130.3	15–160
Total	10–4,240	81.1–134.0	18.7–53.0	−6.6–24.2	241.0–2989.1	328.7–1131.7	3–350

TDCF Temperate Deciduous Coniferous Forest, *TECF* Temperate Evergreen Coniferous Forest, *TDBF* Temperate Deciduous Broadleaved Forest, *SEBF* Subtropical Evergreen Broadleaved Forest, *SECF* Subtropical Evergreen Cniferous Forest. MAT (°C) stands for Mean Annual Temperature (°C). MAP (mm) stands for Mean Annual Precipitation (mm). PET (mm) stands for Potential Evapotransporation (mm).

species information), 66 plots of Subtropical *Pinus massoniana* Forest, 98 plots of Subtropical *Cunninghamia lanceolata* Forest, 46 plots of Subtropical Montane *Pinus yunnanensis* and *P. khasya* Forest, 55 plots of Subtropical montane *Pinus armandii*, *P. taiwanensis* and *P. densada* Forest. Deciduous Forest (DF) consisted of 48 plots of Temperature *Larix* Forest, 165 plots of Temperate Typical Deciduous Broadleaved Forest, 127 plots of Temperate/subtropical Montane *Populus-Betula* Deciduous Forest.

Statistical protocols

As was common in cross-species analysis, data for analyzing the scaling relations of mean individual M_L and M_T (Tons/individual), which were obtained as stand mass/density in plots, were $\log_{10}$-transformed to normalize the distributions and minimize patterns in residuals [46]. Because functional rather than predictive relationships were sought [47], Model Type II (RMA) regression was used to determine the slope (scaling exponent, β) and y-intercept (allometric constant, K) of log-log linear functions. A standardized major axis (RMA) fit was the line minimizing sums of squares in X and Y dimensions simultaneously; and these routines were run using the 'Standardized Major Axis Tests and Routines' (SMTR) computer package [48]. The software package SMTR was also used to determine whether the numerical values of β for $\log_{10} M_L$ vs. M_T differed between contrasted data subsets. Heterogeneity between SMA slopes within and among biomes was tested via a permutation test in SMATR [48].

Results

For all the forest plots pooled (n = 1175), M_L increased with M_T ($P<0.001$, $R^2 = 0.806$; Table 2 and Fig. 1) with the scaling exponent of 0.873 (95% CIs: 0.851–0.895), which was statistically higher than a predicted 2/3 or 3/4 and lower than 1.This exponents varied significantly in diverse biomes, with a range of 0.655 to 1.016. TDBF and TDCF had the highest scaling exponents, which were very close to 1. SEBF and SECF had the medium exponents, the values of which were significantly higher than 3/4. TECF shared a value of 0.655, which was close to 2/3 (Table 2; Fig. 1). The average exponents of the five biomes were 0.885, which deviated from 3/4.

Regressions performed in Group 2 showed that the biomass allocation pattern of M_L and M_T varied from conifer to broadleaved forests. The exponent of CF (0.795^b) was lower than BF (0.964^a). (Table 3; Fig. 2)These results were mostly consistent with the findings in Group 1 that in both temperate and subtropical forests, the scaling exponents of broadleaved biomes (TDBF, 1.016^a; SEBF, 0.941^b) were commonly higher than the conifer ones (SECF, 0.847^c, TECF, 0.655^d; except TDCF, 0.968^{ab}) (Table 2; Fig. 1). Regressions assessed in Group 3 suggested that the scaling exponents of leaf mass and total mass of deciduous forests (1.012^b) were higher than evergreen forests (0.805^a) (Table 4; Fig. 3). These findings agreed with the observation that the scaling exponents in deciduous forests (TDBF, 1.016; TDCF, 0.968) were higher than in evergreen forests (SEBF, 0.941; SECF, 0.847; TECF, 0.655) (Table 2; Fig. 1).

Discussion

Scaling has been treated as a particularly powerful tool in physical sciences for revealing "emergent" phenomena in complex systems [49]. WBE theory aims to provide the broadest coverage of allometric phenomena with 1/4 rules. West *et al.* [9,10] and Enquist and Niklas [12] and Niklas [4] predicted that the leaf mass scaled as the 3/4 power of body mass based on an internalized fractal delivery networks for energy and mass transfer. Although their predictions were validated with a global database [11,14], it was not the same case in our results. Our results showed that the biomass partitioning of leaf organs and total plants were much variable across diverse forested biomes. The discrepancy may partly result from the different scale-levels of investigations. The data cited by Enquist [14] spans over 20 orders of magnitude of body size from unicellular to multi-cellular plants. The large-scale process tends to show relative independence of phylogenetic affiliation, growth habit and abiotic environmental features [50]. In contrast, the tight clustering of our data covers only three orders of magnitude in standing tree mass [41], which necessarily accentuate the biological or environmental characters in local sample [51]. Despite the potential statistical weakness in local-sample regression [11], the limitation is of minor relevance to our result, as all the regressions in our report have high r^2 values (Table 2; Fig. 1) [31,32,34,52]. However, the distribution of the exponents in our result (from 0.655 to 1.016) was similar to the combining 95% CIs of scaling exponents regressed for angiosperms and gymnosperms from the global database (from 0.646 to 0.946) [14]. These ranges were consistent with the intra-specific scaling relationships predicted by Kozlowski *et al.* [53], which was from 2/3 to 1, addressing the impact of species specification on the biomass partitioning pattern. In fact, WBE has been modified with the consideration of ontogeny and taxonoxy [54,55]. Price et al.[15,55,56] suggested a signigicant variability between speceis and predicted that the the scaling exponent should fall along a continuum of variation based on more extension models of WBE, such as leaf venation networks. All these attempts have greatly

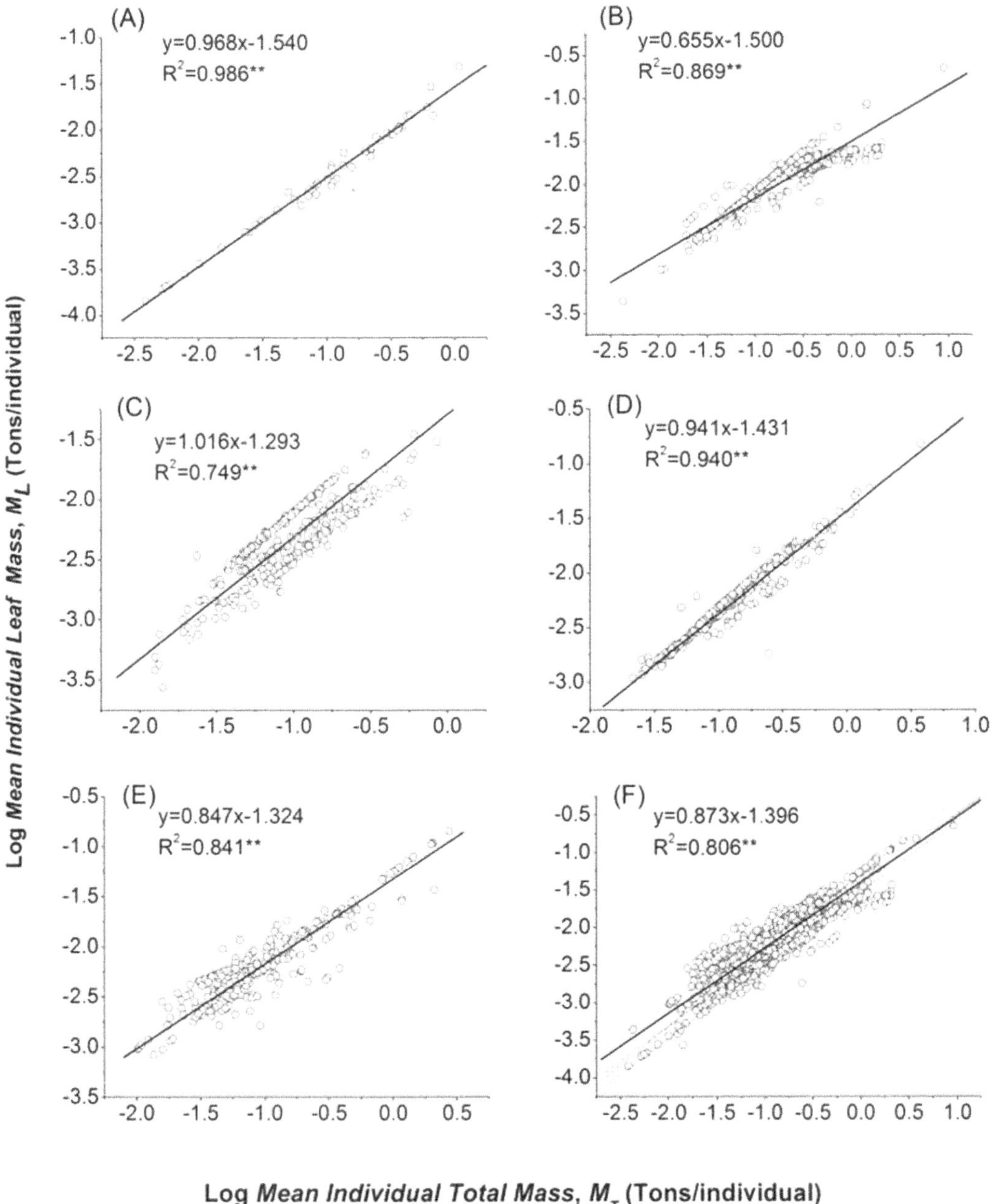

Figure 1. Scaling relations of leaf mass (*M_L*, Ton/individual) and total biomass (*M_T*, Ton/individual) in differing forested biomes. M_T includes leaves, branches, stems and roots. The lines are linear RMA fits to the log-transformed data. For each biome, data include dominant trees from stands of variable age: (A) TDCF, (B) TECF, (C) TDBF, (D) SEBF, (E) SECF. All data are pooled in (F).

promoted WBE theories to embrace more varibable emperical findings.

As sessile organisms, plants have to be evolved with highly variable leaf form and function as both cause and consequence for adaptive strategy in response to species variety and environmental heterogeneity. In our research, the differing leaf mass and total mass partitioning relationships within groups indicated that leaf form and leaf habit played essential roles in adjusting the biomass allocation patterns (Table 2, 3, 4; Fig. 1, 2, 3). The conifer forests always produced needle-like leaves with lower special leaf area (SLA). SLA is positively correlated with potential photosynthetic rate, decomposition rate, leaf toughness and relative leaf growth rate [57]. Broadleaves with higher SLA can provide high metabolism (gas change rates) and energy capture (light harvesting and CO_2 fixation) per unit leaf mass [39,16]. The discrepancy of SLA is consistent with the different resource consumption strategies adapted by CF and BF, which are resource-conserving and resource-demanding, respectively [58]. That may explain why the scaling exponent of leaf mass and total mass was lower in CF than in BF (Table 3; Fig. 2). In addition, the conifer trees commonly achieve their growing space by more lateral and less vertical oriented expansion, leading to a broom-like crown. In contrast, the broadleaved trees always arrange their leaf area in an umbrella-like shape, which demands more space compared to broom-like crowns [59]. The different crown structure challenges the assumption by WBE that leaf area ∝ leaf mass, which may as

Table 2. Allometric scaling relationships of leaf mass and total mass across biomes as estimated by standardized Major Axis Estimation and Testing Routines (SMATR).

Biome	β	SE	95%CIs	K	SE	95%CIs	R^2	P-value	No.
TDCF	0.968^{ab}	0.017	0.935, 1.003	−1.540	0.020	−1.581, −1.500	0.986	0.000	48
TECF	0.655^{d}	0.013	0.630, 0.681	−1.500	0.012	−1.524, −1.476	0.869	0.000	332
TDBF	1.016^{a}	0.030	0.959, 1.077	−1.293	0.033	−1.357, −1.229	0.749	0.000	292
SEBF	0.941^{b}	0.015	0.912, 0.972	−1.431	0.015	−1.461, −1.402	0.940	0.000	238
SECF	0.847^{c}	0.021	0.807, 0.889	−1.324	0.023	−1.369, −1.278	0.841	0.000	265
Total	0.873	0.011	0.851, 0.895	−1.396	0.012	−1.419, −1.374	0.806	0.000	1175

β is the exponent as a consequence of individual M_L (leaf mass, tons/individual) scales with M_T (total mass, tons/individual). 95%CIs and SE are confidence intervals and standard error for β and K, respectively. No. is the number of plots. *TDCF, TECF, TDBF, SEBF, SECF* are defined as in Table 1.

well cause a failure of universal biomass partitioning. We also observed that the M_L versus M_T scaling exponents were commonly lower in evergreen forest biomes than those in deciduous ones (Table 2, 4; Fig. 1, 3). These phenomena are consistent with the adaptive strategies of evergreen and deciduous functional types in warm temperate to tropical regions. Evergreen trees always produce leaves that have a longer lifespan, lower SLA and higher tensile strength than deciduous trees. A longer lifespan of leaves requires investment in structural integrity and/or defense against disturbances, especially with resource constraint, such as sun light [60]. Thereby the evergreen plant promotes greater allocation of biomass to structural rather than metabolic components [61], resulting in lower scaling exponents of M_L and M_T as plant grows. These phenomena explicate the patterns of leaf mass and total mass partitioning occurring in deciduous and evergreen forests. Noticeably, the scaling exponents of leaf mass and total mass are higher deciduous forests than in evergreen forests, in spite of conifer or broadleaved species (Table 2; fig. 1), suggesting the importance of leaf lifespan in determining biomass allocation patterns. Although the forests we selected commonly distribute in temperature or subtropical area, the variation of allocation patterns keeps consistent with forests in tropical dry forests [39]. Therefore, the influence of leaf form and functions could not be neglected in metabolism and the construction of biomass allocation patterns.

Forest is the most important carbon pool in terrestrial ecosystem. Given the common perception of leaves as engines of metabolic activity, scaling relation of M_L with M_T performs as a consequence of the synergy between physiological, biophysical and evolutionary constraints on leaf phenotypes of all species in terrestrial ecosystems. In the present research, the majority of regression curves failed the test for slope homogeneity, suggesting there is hardly a uniform pattern of biomass partitioning for the differing forests in terms of leaf form and function. Although the numerical values do not conform to a WBE prediction, a simple right or wrong is not our intention. Our analysis in this research

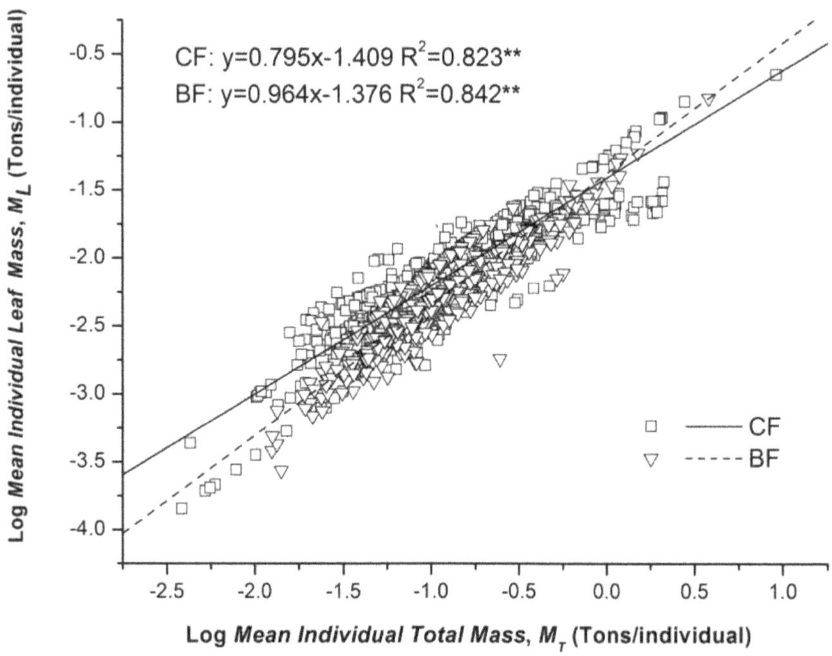

Figure 2. Scaling relations of leaf mass (M_L, Ton/individual) and total biomass (M_T, Ton/individual) in conifer forest (CF) and broadleaved forest (BF). M_T includes leaves, branches, stems and roots. The lines are linear RMA fits to the log-transformed data.

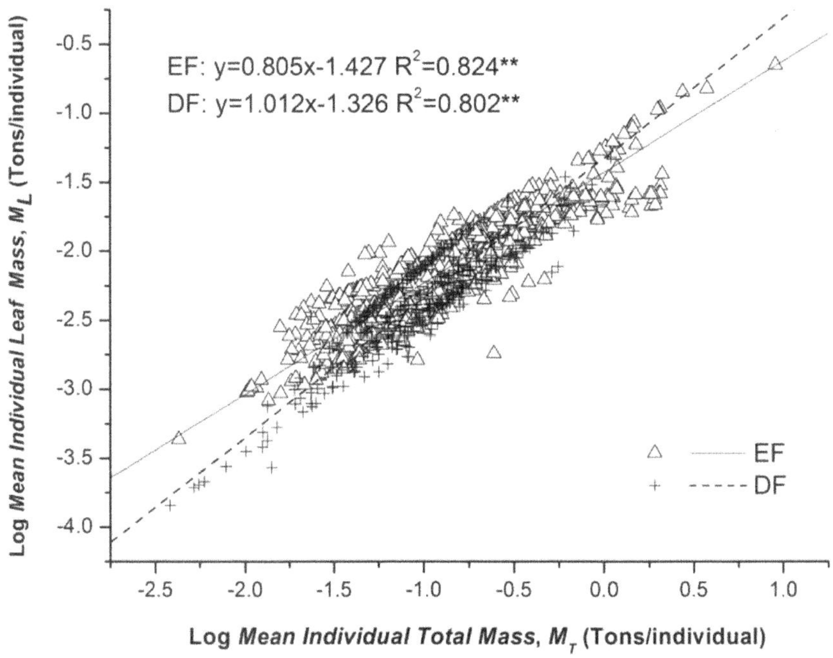

Figure 3. Scaling relations of leaf mass (M_L, Ton/individual) and total biomass (M_T, Ton/individual) in evergreen forest (EF) and deciduous forest (DF). M_T includes leaves, branches, stems and roots. The lines are linear RMA fits to the log-transformed data.

mainly focused on the variations within the sorted forests, while in-depth understanding of mechanism underlying the allometric biomass patterns, especially the relations with leaf morphological, physiological and anatomical traits, will be explored in future work.

Conclusions

Biomass partitioning is considered a strong driver of the capacity of plants to take up carbon, water and nutrients for future use. A visual and statistical examination of our collective data from a Chinese forest biomass dataset empirically reveals a striking heterogeneity in scaling laws of M_L with M_T across diverse tree-dominated biomes, between conifer and broadleaved forests, and between evergreen and deciduous forests. The results of the present study highlight the effects of leaf form and function on the behavior of individuals and consequently on populations and communities, questioning the application of unique scaling law of biomass partitioning to all terrestrial plants. Many details of plant morphology, physiology as well as ecology characters (just like

Table 3. Allometric scaling relationships of leaf mass and total mass between conifer and broadleaved forests as estimated by standardized Major Axis Estimation and Testing Routines (SMATR).

Category	β	SE	95%CIs	*K*	SE	95%CIs	R^2	*P*-value	No.
CF	0.795[b]	0.018	0.770, 0.821	−1.409	0.014	−1.436, −1.383	0.823	0.000	645
BF	0.964[a]	0.015	0.932, 0.998	−1.376	0.017	−1.410, −1.341	0.842	0.000	530

β is the exponent as a consequence of individual M_L (leaf mass, tons/individual) scales with M_T (total mass, tons/individual). 95%CIs and SE are confidence intervals and standard error for β and *K*, respectively. No. is the number of plots. *CF* and *BF* represent Conifer Forest and Broadleaved Forest, respectively.

Table 4. Allometric scaling relationships of leaf mass and total mass between evergreen and deciduous forests as estimated by standardized Major Axis Estimation and Testing Routines (SMATR).

Category	β	SE	95%CIs	*K*	SE	95%CIs	R^2	*P*-value	No.
EF	0.805[b]	0.012	0.782, 0.828	−1.427	0.012	−1.450, −1.404	0.824	0.000	835
DF	1.012[a]	0.023	0.965, 1.061	−1.326	0.027	−1.379, −1.272	0.802	0.000	340

β is the exponent as a consequence of individual M_L (leaf mass, tons/individual) scales with M_T (total mass, tons/individual). 95%CIs and SE are confidence intervals and standard error for β and *K*, respectively. No. is the number of plots. *EF* and *DF* represent Evergreen Forest and Deciduous Forest, respectively.

light, water and nutrient availability, plant density) can be incorporated into the allometric coefficients to link pattern and process across multiple scales and biological organization. This work may invoke further research about the morphological, physiological, biochemical and phylogenic mechanism underlying scaling.

Acknowledgments

We thank Dr. Tianxiang Luo (Institute of Geographic Sciences and Natural Resources Research, the Chinese Academy of Sciences, Beijing, China) for providing valuable information for this study. We thank Dr. Xiaowei Wang, Dr. Yanyuan Bai and Dr. Weiping Zhang for comments and suggestions in organizing this manuscript. Finally, we are grateful to the anonymous reviewers for their helpful comments.

Author Contributions

Conceived and designed the experiments: SX YL GW. Performed the experiments: SX YL GW. Analyzed the data: SX. Contributed reagents/materials/analysis tools: SX YL GW. Wrote the paper: SX YL GW.

References

1. Brown JH, Gillooly JF, Allen AP, Savage VM, West GB (2004) Toward a metabolic theory of ecology. Ecology 85: 1771–1789.
2. Enquist BJ, Economo EP, Huxman TE, Allen AP, Ignace DD, et al. (2003) Scaling metabolism from organisms to ecosystems. Nature 423: 639–642.
3. Denny M, Benedetti-Cecchi L (2012) Scaling up in ecology: mechanistic approaches. Annu Rev Ecol Evol Syst 43: 1–22.
4. Niklas KJ (2004) Plant allometry: is there a grand unifying theory? Biol Rev 79: 871–889.
5. Hui D, Jackson RB (2006) Geographical and interannual variability in biomass partitioning in grassland ecosystems: a synthesis of field data. New Phytol 169: 85–93.
6. Poorter H, Niklas KJ, Reich PB, Oleksyn J, Poot P, et al. (2012) Biomass allocation to leaves, stems and roots: meta-analyses of interspecific variation and environmental control. New Phytol 193: 30–50.
7. West GB, Brown JH, Enquist BJ (1999) A general model for the structure, function, and allometry of plant vascular systems. Nature 400: 664–667.
8. West GB, Brown JH (2004) Life's universal scaling laws. Phys Today 57: 36–43.
9. West GB, Brown JH, Enquist BJ (1997) A general model for the origin of allometric scaling laws in biology. Science 276: 122–126.
10. West GB, Brown JH, Enquist BJ (1999) The fourth dimension of life: fractal geometry and allometric scaling of organisms. Science 284: 1677.
11. Niklas KJ, Enquist BJ (2001) Invariant scaling relationships for interspecific plant biomass production rates and body size. Proc Natl Acad Sci 98: 2922–2927.
12. Enquist BJ, Niklas KJ (2002) Global allocation rules for patterns of biomass partitioning in seed plants. Science 295: 1517–1520.
13. Cannell MG (1982) World forest biomass and primary production data: Academic Press London.
14. Enquist BJ (2003) Scaling the macroecological and evolutionary implications of size and metabolism within and across plant taxa. In: Blackburn TM, Gaston KJ, editors. Macroecology: Concepts and Consequences. London: Cambridge University Press. pp. 321–341.
15. Price CA, Enquist BJ (2007) Scaling mass and morphology in leaves: an extension of the WBE model. Ecology 88: 1132–1141.
16. Li HT, Han XG, Wu JG (2005) Lack of evidence for 3/4 scaling of metabolism in terrestrial plants. J Integr Plant Biol 47: 1173–1183.
17. Reich PB, Tjoelker MG, Machado J-L, Oleksyn J (2006) Universal scaling of respiratory metabolism, size and nitrogen in plants. Nature 439: 457–461.
18. Finkel ZV, Irwin AJ, Schofield O (2004) Resource limitation alters the 3/4 size scaling of metabolic rates in phytoplankton. Mar Ecol Prog Ser 273: 269–279.
19. Reich PB, Tjoelker MG, Machado JL, Oleksyn J (2006) Universal scaling of respiratory metabolism, size and nitrogen in plants. Nature 439: 457–461.
20. Chen X, Li BL (2003) Testing the allometric scaling relationships with seedlings of two tree species. Acta Oecol 24: 125–129.
21. Deng JM, Li T, Wang GX, Liu J, Yu ZL, et al. (2008) Trade-offs between the metabolic rate and population density of plants. PloS ONE 3: e1799.
22. Pretzsch H (2006) Species-specific allometric scaling under self-thinning: evidence from long-term plots in forest stands. Oecologia 146: 572–583.
23. Pretzsch H, Mette T (2008) Linking stand-level self-thinning allometry to the tree-level leaf biomass allometry. Trees 22:611–622.
24. Cheng DL, Li T, Zhong QL, Wang GX (2010) Scaling relationship between tree respiration rates and biomass. Biol Lett 6: 715–717.
25. Dai XF, Jia X, Zhang WP, Bai YY, Zhang JJ, et al. (2009) Plant height-crown radius and canopy coverage-density relationships determine above-ground biomass-density relationship in stressful environments. Biol Lett 5: 571–573.
26. Deng JM, Wang GX, Morris EC, Wei XP, Li DX, et al. (2006) Plant mass-density relationship along a moisture gradient in north-west China. J Ecol 94: 953–958.
27. Peng Y, Niklas KJ, Reich PB, Sun S (2010) Ontogenetic shift in the scaling of dark respiration with whole-plant mass in seven shrub species. Funct Ecol 24: 502–512.
28. Mori S, Yamaji K, Ishida A, Prokushkin SG, Masyagina OV, et al. (2010) Mixed-power scaling of whole-plant respiration from seedlings to giant trees. Proc Natl Acad Sci 107: 1447–1451.
29. Coomes DA (2006) Challenges to the generality of WBE theory. Trends Ecol Evol 21: 593–596.
30. Isaac NJ, Carbone C (2010) Why are metabolic scaling exponents so controversial? Quantifying variance and testing hypotheses. Ecol Lett 13: 728–735.
31. Price CA, Weitz JS, Savage VM, Stegen J, Clarke A, et al. (2012) Testing the metabolic theory of ecology. Ecol Lett 15: 1465–1474.
32. Dodds PS, Rothman DH, Weitz JS (2001) Re-examination of the "3/4-law" of metabolism. J Theor Biol 209: 9–27.
33. Kozłowski J, Konarzewski M (2004) Is West, Brown and Enquist's model of allometric scaling mathematically correct and biologically relevant? Funct Ecol 18: 283–289.
34. Kozłowski J, Konarzewski M (2005) West, Brown and Enquist's model of allometric scaling again: the same questions remain. Funct Ecol 19: 739–743.
35. Glazier DS (2006) The 3/4-power law is not universal: evolution of isometric, ontogenetic metabolic scaling in pelagic animals. Bioscience 56: 325–332.
36. Price CA, Enquist BJ (2006) Scaling of mass and morphology in plants with minimal branching: an extension of the WBE model. Funct Ecol 20: 11–20.
37. Wang C (2006) Biomass allometric equations for 10 co-occurring tree species in Chinese temperate forests. For Ecol Manage 222: 9–16.
38. Bond-Lamberty B, Wang C, Gower S, Norman J (2002) Leaf area dynamics of a boreal black spruce fire chronosequence. Tree physiol 22: 993–1001.
39. Powers JS, Tiffin P (2010) Plant functional type classifications in tropical dry forests in Costa Rica: leaf habit versus taxonomic approaches. Funct Ecol 24: 927–936.
40. Lin Y, Berger U, Grimm V, Huth F, Weiner J (2013) Plant interactions alter the predictions of metabolic scaling theory. PloS ONE 8: e57612.
41. Luo TX (1996) Patterns of biological production and its mathematical models for main forest types of China (in Chinese with English abstract). PhD Dissertation. Committee of SynthesisInvestigation of Natural Resources. Chinese Academy of Sciences, Beijing.
42. Ni J, Zhang XS, Scurlock JM (2001) Synthesis and analysis of biomass and net primary productivity in Chinese forests. Ann For Sci 58: 351–384.
43. Zhang WP, Jia X, Bai YY, Wang GX (2011) The difference between above-and below-ground self-thinning lines in forest communities. Ecol Res 26: 819–825.
44. Luo TX, Zhang L, Zhu HZ, Daly C, Li MC, et al. (2009) Correlations between net primary productivity and foliar carbon isotope ratio across a Tibetan ecosystem transect. Ecography 32: 526–538.
45. Luo TX, Li WH, Zhu HZ (2002) Estimated biomass and productivity of natural vegetation on the Tibetan Plateau. Ecol Appl 12: 980–997.
46. Reich PB, Oleksyn J, Wright IJ, Niklas KJ, Hedin L, et al. (2010) Evidence of a general 2/3-power law of scaling leaf nitrogen to phosphorus among major plant groups and biomes. Proc R Soc B: Biol Sci 277: 877–883.
47. Niklas KJ (1994) Plant allometry: the scaling of form and process. Chicago: University of Chicago Press.
48. Falster D, Warton D, Wright I (2006) User's guide to SMATR: standardised major axis tests and routines version 2.0, copyright 2006. http://www.bio.mq.edu.au/ecology/SMATR/.
49. Enquist BJ, Tiffney BH, Niklas KJ (2007) Metabolic scaling and the evolutionary dynamics of plant size, form, and diversity: toward a synthesis of ecology, evolution, and paleontology. Int J Plant Sci 168: 729–749.
50. Deng JM, Zuo WY, Wang ZQ, Fan ZX, Ji MF, et al. (2012) Insights into plant size-density relationships from models and agricultural crops. Proc Natl Acad Sci 109: 8600–8605.
51. Cheng DL, Niklas KJ (2007) Above-and below-ground biomass relationships across 1534 forested communities. Ann Bot 99: 95–102.
52. Muller-Landau HC, Condit RS, Chave J, Thomas SC, Bohlman SA, et al. (2006) Testing metabolic ecology theory for allometric scaling of tree size, growth and mortality in tropical forests. Ecol Lett 9: 575–588.
53. Kozlowski J, Konarzewski M, Gawelczyk AT (2003) Intraspecific body size optimization produces interspecific allometries. In: Blackburn TM, Gaston KJ, editors. Macroecology: Concepts and Consequences. London: Cambridge University Press. pp. 299–320.
54. Enquist BJ, West GB, Brown JH (2009) Extensions and evaluations of a general quantitative theory of forest structure and dynamics. Proc Natl Acad Sci 106: 7046–7051.
55. Price CA, Enquist BJ (2007) Scaling mass and morphology in leaves: an extension of the WBE model. Ecology 88: 1132–1141.

56. Price CA, Wing S, Weitz JS (2012) Scaling and structure of dicotyledonous leaf venation networks. Ecol Lett 15:87–95.
57. Castro-Díez P, Puyravaud J, Cornelissen J (2000) Leaf structure and anatomy as related to leaf mass per area variation in seedlings of a wide range of woody plant species and types. Oecologia 124: 476–486.
58. Rahman MM, Tsukamoto J (2013) Leaf traits, litter decomposability and forest floor dynamics in an evergreen- and a deciduous-broadleaved forest in warm temperate Japan. Forestry 86: 441–451.
59. Pretzsch H, Dieler J (2012) Evidence of variant intra-and interspecific scaling of tree crown structure and relevance for allometric theory. Oecologia 169: 637–649.
60. Pan S, Liu C, Zhang WP, Xu SS, Wang N, et al. (2013) The Scaling Relationships between Leaf Mass and Leaf Area of Vascular Plant Species Change with Altitude. PloS ONE 8: e76872.
61. Reich PB, Ellsworth DS, Walters MB, Vose JM, Gresham C, et al. (1999) Generality of leaf trait relationships: a test across six biomes. Ecology 80: 1955–1969.

PERMISSIONS

All chapters in this book were first published in PLOS ONE, by The Public Library of Science; hereby published with permission under the Creative Commons Attribution License or equivalent. Every chapter published in this book has been scrutinized by our experts. Their significance has been extensively debated. The topics covered herein carry significant findings which will fuel the growth of the discipline. They may even be implemented as practical applications or may be referred to as a beginning point for another development.

The contributors of this book come from diverse backgrounds, making this book a truly international effort. This book will bring forth new frontiers with its revolutionizing research information and detailed analysis of the nascent developments around the world.

We would like to thank all the contributing authors for lending their expertise to make the book truly unique. They have played a crucial role in the development of this book. Without their invaluable contributions this book wouldn't have been possible. They have made vital efforts to compile up to date information on the varied aspects of this subject to make this book a valuable addition to the collection of many professionals and students.

This book was conceptualized with the vision of imparting up-to-date information and advanced data in this field. To ensure the same, a matchless editorial board was set up. Every individual on the board went through rigorous rounds of assessment to prove their worth. After which they invested a large part of their time researching and compiling the most relevant data for our readers.

The editorial board has been involved in producing this book since its inception. They have spent rigorous hours researching and exploring the diverse topics which have resulted in the successful publishing of this book. They have passed on their knowledge of decades through this book. To expedite this challenging task, the publisher supported the team at every step. A small team of assistant editors was also appointed to further simplify the editing procedure and attain best results for the readers.

Apart from the editorial board, the designing team has also invested a significant amount of their time in understanding the subject and creating the most relevant covers. They scrutinized every image to scout for the most suitable representation of the subject and create an appropriate cover for the book.

The publishing team has been an ardent support to the editorial, designing and production team. Their endless efforts to recruit the best for this project, has resulted in the accomplishment of this book. They are a veteran in the field of academics and their pool of knowledge is as vast as their experience in printing. Their expertise and guidance has proved useful at every step. Their uncompromising quality standards have made this book an exceptional effort. Their encouragement from time to time has been an inspiration for everyone.

The publisher and the editorial board hope that this book will prove to be a valuable piece of knowledge for researchers, students, practitioners and scholars across the globe.

LIST OF CONTRIBUTORS

Hao Chen
Key Laboratory of Vegetation Restoration and Management of Degraded Ecosystems, South China Botanical Garden, Chinese Academy of Sciences, Guangzhou, China

Geshere A. Gurmesa
Department of Geosciences and Natural Resource Management, University of Copenhagen, Copenhagen, Denmark
University of Chinese Academy of Sciences, Beijing, China
Sino-Danish Center for Education and Research, Aarhus, Denmark

Lei Liu
Key Laboratory of Vegetation Restoration and Management of Degraded Ecosystems, South China Botanical Garden, Chinese Academy of Sciences, Guangzhou, China
State Key Laboratory of Urban and Regional Ecology, Research Center for Eco-Environmental Sciences, Chinese Academy of Sciences, Beijing, China

Tao Zhang
Key Laboratory of Vegetation Restoration and Management of Degraded Ecosystems, South China Botanical Garden, Chinese Academy of Sciences, Guangzhou, China
Institute of Tropical Pratacultural Science, Zhanjiang Normal University, Zhanjiang, China

Shenglei Fu, Zhanfeng Liu, Shaofeng Dong, Chuan Ma and Jiangming Mo
Key Laboratory of Vegetation Restoration and Management of Degraded Ecosystems, South China Botanical Garden, Chinese Academy of Sciences, Guangzhou, China

Ralf Lauterbach and Elisabeth K. V. Kalko
Institute of Experimental Ecology, University of Ulm, Ulm, Germany

Konstans Wells
Institute of Experimental Ecology, University of Ulm, Ulm, Germany
Biodiversity and Climate Research Centre, Frankfurt, Germany

Swen C. Renner
Institute of Experimental Ecology, University of Ulm, Ulm, Germany
Smithsonian Conservation Biology Institute, National Zoological Park, Front Royal, United States of America

Robert B. O'Hara
Biodiversity and Climate Research Centre, Frankfurt, Germany

Hisatomo Taki and Kimiko Okabe
Department of Forest Entomology, Forestry and Forest Products Research Institute, Tsukuba, Ibaraki, Japan

Isamu Okochi and Shun'ichi Makino
Forestry and Forest Products Research Institute, Tsukuba, Ibaraki, Japan

Takenari Inoue
Tama Forest Science Garden, Forestry and Forest Products Research Institute, Hachioji, Tokyo, Japan

Hideaki Goto
Kyushu Research Centre, Forestry and Forest Products Research Institute, Kumamoto, Kumamoto, Japan

Takeshi Matsumura
Nasushiobara, Tochigi, Japan

Tihomir Rugani, Jurij Diaci and David Hladnik
Biotechnical Faculty, Department of Forestry and Renewable Forest Resources, University of Ljubljana, Ljubljana, Slovenia

Guohua Liang, Xingzhao Liu, Xiaomei Chen and Qingyan Qiu
South China Botanical Garden, Chinese Academy of Sciences, Guangzhou, China
University of Chinese Academy of Sciences, Beijing, China

Deqiang Zhang, Guowei Chu, Juxiu Liu, Shizhong Liu and Guoyi Zhou
South China Botanical Garden, Chinese Academy of Sciences, Guangzhou, China

Billie T. Lazenby
School of Biological Sciences, University of Sydney, Sydney, New South Wales, Australia
Department of Primary Industries Parks Water and Environment, Hobart, Tasmania, Australia

Christopher R. Dickman
School of Biological Sciences, University of Sydney, Sydney, New South Wales, Australia

Denise C. McGregor, Sarah E. Kerr and Andrew K. Krockenberger
Centre for Tropical Environmental and Sustainability Science, and School of Marine and Tropical Biology, James Cook University, Cairns, Queensland, Australia

Deanna H. Olson, George Weaver and Kathryn L. Ronnenberg
Pacific Northwest Research Station, U.S. Forest Service, Corvallis, Oregon, United States of America

David M. Aanensen, Christopher I. Powell, Susan F. Walker and Matthew C. Fisher
Department of Infectious Disease Epidemiology, Imperial College, London, United Kingdom

Jon Bielby
Institute of Zoology, Zoological Society of London, London, United Kingdom

Trenton W. J. Garner
Department of Statistics, Oregon State University, Corvallis, Oregon, United States of America

Jiabing Wu, Dexin Guan, Fenhui Yuan, Anzhi Wang and Changjie Jin
State Key Laboratory of Forest and Soil Ecology, Institute of Applied Ecology, Chinese Academy of Sciences, Shenyang, P.R. China

Lei Liu
State Key Laboratory of Urban and Regional Ecology, Research Center for Eco-Environmental Sciences, Chinese Academy of Sciences, Beijing, China

Tao Zhang, Wei Zhang, Hao Chen and Jiangming Mo
Key Laboratory of Vegetation Restoration and Management of Degraded Ecosystems, South China Botanical Garden, Chinese Academy of Sciences, Guangzhou, China

Frank S. Gilliam
Department of Biological Science, Marshall University, Huntington, West Virginia, United States of America

Per Gundersen
Danish Centre for Forest, Landscape and Planning, University of Copenhagen, Copenhagen, Denmark

Zhongliang Huang, Wanhui Ye and Linfang Wu
Key Laboratory of Vegetation Restoration and Management of Degraded Ecosystems, South China Botanical Garden, Chinese Academy of Sciences, Guangzhou, People's Republic of China

Guojun Lin and Guiquan Gong
Key Laboratory of Vegetation Restoration and Management of Degraded Ecosystems, South China Botanical Garden, Chinese Academy of Sciences, Guangzhou, People's Republic of China
University of Chinese Academy of Sciences, Beijing, People's Republic of China

Diana Stralberg
Department of Biological Sciences, University of Alberta, Edmonton, Alberta, Canada

Limei Wang, Longhui Li, Xi Chen and Geping Luo
State Key Laboratory of Desert and Oasis Ecology, Xinjiang Institute of Ecology and Geography, Chinese Academy of Sciences, Urumqi, China

Xin Tian
Institute of Forest Information Resource Techniques, Chinese Academy of Forestry, Beijing, China

Xiaoke Wang
Beijing Urban Ecosystem Research Station, State Key Lab of Urban and Regional Ecology Research Center for Eco-Environmental Sciences, Chinese Academy of Sciences, Beijing, China

Chihiro Takahata
Interdisciplinary Graduate School of Science and Technology, Shinshu University, Nagano, Japan

Scott Eric Nielsen
Department of Renewable Resources, University of Alberta, Edmonton, Alberta, Canada

Akiko Takii and Shigeyuki Izumiyama
Education and Research Center of Alpine Field Science, Faculty of Agriculture, Shinshu University, Nagano, Japan

Henry Loescher and Hongyan Luo
Science Team, National Ecological Observatory Network (NEON), Boulder, Colorado, United States of America
Institute of Arctic and Alpine Research (INSTAAR), University of Colorado, Boulder, Colorado, United States of America

Edward Ayres
Science Team, National Ecological Observatory Network (NEON), Boulder, Colorado, United States of America
Institute of Arctic and Alpine Research (INSTAAR), University of Colorado, Boulder, Colorado, United States of America

Natural Resource Ecology Laboratory, Colorado State University, Fort Collins, Colorado, United States of America

Paul Duffy
Neptune and Company Inc., Lakewood, Colorado, United States of America

Max Brunke
Science Team, National Ecological Observatory Network (NEON), Boulder, Colorado, United States of America

Yakun Tang
Key Laboratory of Ecosystem Network Observation and Modeling, Institute of Geographic Sciences and Natural Resources Research, Chinese Academy of Sciences, Beijing, China
University of Chinese Academy of Sciences, Beijing, China

Xuefa Wen, Xiaomin Sun and Huimin Wang
Key Laboratory of Ecosystem Network Observation and Modeling, Institute of Geographic Sciences and Natural Resources Research, Chinese Academy of Sciences, Beijing, China

Wenfa Xiao, Lixiong Zeng and Zhilin Huang
State Forestry Administration Key Laboratory of Forest Ecology and Environment, Research Institute of Forest Ecology, Environment and Protection, Chinese Academy of Forestry, Beijing, PR China
State Forestry Administration Key Laboratory of Forest Silviculture, Research Institute of Forestry, Chinese Academy of Forestry, Beijing, PR China

Xiaogai Ge
State Forestry Administration Key Laboratory of Forest Ecology and Environment, Research Institute of Forest Ecology, Environment and Protection, Chinese Academy of Forestry, Beijing, PR China
Research Institute of Subtropical Forestry, Chinese Academy of Forestry, Fuyang, Zhejiang, PR China

Jingpin Lei
State Forestry Administration Key Laboratory of Forest Silviculture, Research Institute of Forestry, Chinese Academy of Forestry, Beijing, PR China

Benzhi Zhou
Research Institute of Subtropical Forestry, Chinese Academy of Forestry, Fuyang, Zhejiang, PR China

Maihe Li
Swiss Federal Research Institute WSL, Birmensdorf, Switzerland
State Key Laboratory of Forest and Soil Ecology, Institute of Applied Ecology, Chinese Academy of Sciences, Shenyang, PR China

Steffen Boch, Daniel Prati and Markus Fischer
Institute of Plant Sciences and Botanical Garden; University of Bern, Bern, Switzerland
Oeschger Center, University of Bern, Bern, Switzerland

Dominik Hessenmöller and Ernst-Detlef Schulze
Max-Planck-Institute for Biogeochemistry, Jena, Germany

Yunjian Luo and Yin Ren
Key Laboratory of Urban Environment and Health, Institute of Urban Environment, Chinese Academy of Sciences, Xiamen, China

Xiaoquan Zhang
The Nature Conservancy China Program, Beijing, China

Xiaoke Wang
State Key Laboratory of Urban and Regional Ecology, Research Center for Eco-Environmental Sciences, Chinese Academy of Sciences, Beijing, China

Shanshan Xu, Yan Li and Genxuan Wang
The state key laboratory of plant physiology and biochemistry, Institute of ecology, College of Life Sciences, Zhejiang University, Hangzhou, China

Index

A

Aerial Film Photographs, 29
Allometric Relationship, 113, 118
American Ecosystems, 135
Amphibian Chytrid Fungus, 73, 75, 78, 85
Amphibian Pathogen, 73, 84-85
Animal Communities, 48
Anthropogenic Landscapes, 122
Arboreal Folivores, 64
Asiatic Black Bears, 122-133

B

Batrachochytrium Dendrobatidis, 73, 75-77, 79-81, 83-85
Beech Forest Remnants, 25-26
Belowground Biomass (bgb), 113, 118
Biogeochemical Cycling, 93
Biomass Allocation Patterns, 113, 201
Biomass Conversion Factors, 196, 198-204
Biomass Partitioning, 113, 120-121, 206-208, 210-211
Bowen Ratio, 166-174
Broadleaved Forest (bf), 38, 41-42, 44-45, 207, 210

C

Canopy Gaps, 25-26, 28, 33, 36, 140, 159
Carbon Accounting, 196, 203
Carbon Assimilation, 86, 90-91
Charismatic Megafauna, 64
Cohabiting Predators, 48
Color Infrared (cir), 29
Coniferous Field Stations, 86
Coniferous Plantation, 125, 166-167, 174

D

Database Utility, 73
Dinghushan Biosphere Reserve (dhsbr), 2, 93
Dispersal Limitation, 103-104, 106-108, 110-111
Dissecting Variation, 196

E

Edaphic Factors, 106, 108-110, 203-204
Energy Consumption, 166
Environmental Heterogeneity, 103, 107, 110, 206

F

Fecal Standing Crop Method (fscm), 65
Forest Biodiversity, 15, 104, 187
Forest Inventory Data (fid), 196
Forest Stand Attributes, 10

G

Gap Dynamics, 25, 28-29, 36, 187
Geographic Range, 64, 71
Global Emergence, 73, 84
Gross Photosynthesis, 38

H

Habitat Degradation, 187
Habitat Selection, 63, 122-127, 129-130, 132-133
Human Land Domination, 122
Human-wildlife Boundaries, 122-123, 132

I

Interactive Effects, 93-94, 99, 183, 185-186
Interannual Variation, 166-171, 173-174
Island Population, 64-65
Ixodid Ticks, 10, 15-16

K

Koalas (phascolarctos Cinereus), 64, 72
Korean Pine, 86-92

L

Large Carnivore, 122, 132
Leaf Mass, 206, 208-211, 213
Lichen Species, 187-195
Light Response Curves, 87-89
Litter Decomposition, 1-5, 7-9, 42, 47, 94, 100-101, 112, 176-178, 180-181, 183-186
Litter Manipulation, 1-8

M

Management Methods, 187, 193
Masson Pine Forest (mpf), 1
Multiple-host Infectious Diseases, 73

N

Nitrogen (n) Deposition, 93

O

Occupancy Modelling, 48
Occurrence Patterns, 10, 74, 111
Old-growth Forests, 25, 33, 35, 100, 195

P
Photosynthetic Mass, 207
Pinus Massoniana Forests, 176, 178-183
Plant Litter Fluxes, 176
Planted Forests, 17, 20-22, 114, 118
Precipitation Frequency, 166-168, 172-173
Principal Coordinate Analysis (pcoa), 105

Q
Questing Activity, 10, 12, 15

R
Root To Shoot Ratio (rs), 113, 115-117, 119-120

S
Sampling Designs, 135, 138, 140, 161-163
Sar Treatments, 38-39, 41-42, 44-45
Scaling Relationships, 206, 208, 210-212
Simulated Acid Rain (sar), 38
Soil Microbial Communities, 93, 97, 101
Soil Properties, 3, 39, 46, 104-105, 108, 110, 118, 120, 135, 137-138, 140, 159-160, 163-165
Soil Respiration, 38-42, 44, 46-47, 94-98, 101, 164-165, 174, 176-186
Soil Sampling, 3, 94, 140, 162-163
Soil Temperature, 3, 5-7, 39-42, 44, 86-92, 135, 138, 140, 150, 155-156, 163-165, 176-178, 180-185
Spatial Factors, 103-104, 108, 110
Species Richness, 9, 17-18, 20-22, 73-74, 76, 78-83, 110, 187-194
Species' Distributions, 48, 103, 108
Spring Photosynthesis, 86-87, 90, 92
Stand Continuity, 187-188, 194
Successional Gradients, 1-2, 7
Successional Stages, 2, 8, 10-11, 13, 17, 20-22, 47, 188

T
Temperate Forest, 9, 17, 20, 23, 38, 42, 46-47, 54, 87, 97, 100-101, 103-104, 108, 111, 174, 195
Temperate Zones, 86, 91
Terrestrial Biomes, 113-114, 120-121, 201
Terrestrial Forest Ecosystems, 17, 22
Three Gorges Reservoir, 176-177
Time-stamping Photographs, 48
Tree Species Distribution, 103-104
Tropical Forests, 1-2, 5, 8-9, 64, 93-94, 99-102, 121, 185, 204, 212

V
Variable Strength, 10

W
Wasteful Over-sampling, 135
Weather Conditions, 10-15
Wild Bees, 17-18, 20, 22-23